Comparative Social Evolution

Darwin famously described special difficulties in explaining social evolution in insects. More than a century later, the evolution of sociality – defined broadly as cooperative group living – remains one of the most intriguing problems in biology. Providing a unique perspective on the study of social evolution, this volume synthesizes the features of animal social life across the principle taxonomic groups in which sociality has evolved. The chapters explore sociality in a range of species, from ants to primates, highlighting key natural and life history data and providing a comparative view across animal societies. In establishing a single framework for a common, trait-based approach towards social synthesis, this volume will enable students and investigators new to the field to systematically compare taxonomic groups and reinvigorate comparative approaches to studying animal social evolution.

Dustin R. Rubenstein is an Associate Professor of Ecology, Evolution and Environmental Biology at Columbia University, New York and Director of the Center for Integrative Animal Behavior. His research focuses on the causes and consequences of sociality and how animals adapt to environmental change. He has been recognized by the US National Academy of Sciences for his research accomplishments, and for his innovation in STEM teaching.

Patrick Abbot is an Associate Professor and Vice-Chair of Biological Sciences at Vanderbilt University, Tennessee. His research includes the evolution and ecology of social insects. He has served as an Associate Editor for the journal *Evolution* and is a recipient of the Jeffrey Nordhaus Award for Excellence in Undergraduate Teaching.

Comparative Social Evolution

Edited by

DUSTIN R. RUBENSTEIN
Columbia University

PATRICK ABBOT
Vanderbilt University

CAMBRIDGE
UNIVERSITY PRESS

University Printing House, Cambridge CB2 8BS, United Kingdom

One Liberty Plaza, 20th Floor, New York, NY 10006, USA

477 Williamstown Road, Port Melbourne, VIC 3207, Australia

4843/24, 2nd Floor, Ansari Road, Daryaganj, Delhi - 110002, India

79 Anson Road, #06-04/06, Singapore 079906

Cambridge University Press is part of the University of Cambridge.

It furthers the University's mission by disseminating knowledge in the pursuit of education, learning and research at the highest international levels of excellence.

www.cambridge.org
Information on this title: www.cambridge.org/9781107043398

First published 2017

A catalogue record for this publication is available from the British Library

Library of Congress Cataloging in Publication data
Names: Rubenstein, Dustin R., editor. | Abbot, Patrick, editor.
Title: Comparative social evolution / edited by Dustin R. Rubenstein, Associate Professor, Department of Ecology, Evolution, and Environmental Biology, Columbia University, and Patrick Abbot, Associate Professor, Department of Biological Sciences, Vanderbilt University.
Description: New York : Cambridge University Press, 2017. | Includes bibliographical references and index.
Identifiers: LCCN 2016047637 | ISBN 9781107043398 (hardback : alk. paper) | ISBN 9781107647923 (pbk. : alk. paper)
Subjects: LCSH: Social evolution in animals.
Classification: LCC QL775 .C645 2017 | DDC 591–dc23 LC record available at https://lccn.loc.gov/2016047637

ISBN 978-1-107-04339-8 Hardback
ISBN 978-1-107-64792-3 Paperback

Contents

Contributors

Patrick Abbot
Department of Biological Sciences
Vanderbilt University
Nashville, TN, USA

Leticia Avilés
Department of Zoology & Biodiversity Research Centre
University of British Columbia
Vancouver, BC, Canada

Tom Chapman
Department of Biology
Memorial University of Newfoundland
St. John's, NL, Canada

Andrew Cockburn
Division of Evolution, Ecology and Genetics, Research School of Biology
Australian National University
Canberra, Australia

J. Emmett Duffy
Tennenbaum Marine Observatories Network
Smithsonian Institution
Washington, DC, USA

Jennifer H. Fewell
School of Life Sciences & Social Insect Research Group
Arizona State University
Tempe, AZ, USA

Jennifer Guevara
Department of Zoology & Biodiversity Research Centre
University of British Columbia
Vancouver, BC, Canada

Ben J. Hatchwell
Department of Animal and Plant Sciences
University of Sheffield
Sheffield, UK

Loren D. Hayes
Department of Biology, Geology, and Environmental Sciences
University of Tennessee at Chattanooga
Chattanooga, TN, USA

Jürgen Heinze
Zoology / Evolutionary Biology
University of Regensburg
Regensburg, Germany

Kristin Hultgren
Department of Biology
Seattle University
Seattle, WA, USA

James H. Hunt
Department of Biological Sciences and W. M. Keck Center For Behavioral Biology
North Carolina State University
Raleigh, NC, USA

Peter M. Kappeler
Behavioral Ecology & Sociobiology Unit
German Primate Center
Göttingen, Germany

Katrin Kellner
Department of Biology
University of Texas Tyler
Tyler, TX, USA

Walter D. Koenig
Lab of Ornithology & Department of Neurobiology and Behavior
Cornell University
Ithaca, NY, USA

Judith Korb
Faculty of Biology, Evolutionary Biology & Ecology
University of Freiburg
Freiburg, Germany

Eileen A. Lacey
Museum of Vertebrate Zoology & Department of Integrative Biology
University of California, Berkeley
Berkeley, CA, USA

Dustin R. Rubenstein
Department of Ecology, Evolution and Environmental Biology
Columbia University
New York, USA

Jon Seal
Department of Biology
University of Texas Tyler
Tyler, TX, USA

Joan B. Silk
School of Human Evolution and Social Change
Arizona State University
Tempe, AZ, USA

Jennifer E. Smith
Biology Department
Mills College
Oakland, CA, USA

Michael Taborsky
Institute of Ecology and Evolution,
University of Bern
Bern, Switzerland

Barbara Thorne
University of Maryland
Department of Entomology
College Park, MA, USA

Amy L. Toth
Department of Ecology, Evolution, and Organismal Biology & Department of Entomology
Iowa State University
Ames, IA, USA

William Wcislo
Smithsonian Tropical Research Institute
Panama City, Panama

Geoffrey M. While
School of Biological Sciences
University of Tasmania
Hobart, Australia

Martin J. Whiting
Department of Biological Sciences
Macquarie University
Sydney, Australia

Marian Wong
School of Biological Sciences
University of Wollongong
Wollongong, Australia

Foreword

W. D. Hamilton deeply admired Ronald Fisher's fundamental theorem, and developed one of his own, usually crystallized as $r\mathrm{B} - \mathrm{c} > 0$, that now serves as a core conceptual framework for understanding the evolution of social behavior. Unlike Fisher, Hamilton was a keen natural historian as well as a maths aficionado, not just admiring natural diversity but also using it as raw material for developing his theories of eusociality, sex ratios, asexuality, senescence, and beyond. Hamilton's theories were, and still are, general and inspirational. But are they fundamental enough to be predictive?

I was academically bred as a behavioral ecologist, orbiting both Hamilton and Richard Alexander at the University of Michigan in the 1980s. The behavioral ecological paradigm, at least back then, was the prediction of behavioral from ecological variation, for mating systems, social systems, and any other systems one cared to study. On one hand, the paradigm was hugely successful, as the ideas of Hamilton, Alexander, and others such as Robert Trivers motivated test after successful empirical test. On the other hand, or rather dropped from its grasp, the paradigm may be failing. More than three decades later, we still cannot truly predict social systems from ecology, can we, with any substantial degree of confidence? If we try, what specific set of predictors are, or should we, be using? Despite much work on social evolution since then, we still do not really have an answer to this question.

For the study of social evolution, we have as data sets far-flung convergences, between for example, social thrips and aphids or naked mole rats and sweat bees, and data on the correlates of social divergences between closely related species. This book brings convergences, divergences, and social diversity to the fore, and will certainly serve as a leaping-off point, but to where? I am hoping to a more predictive social behavioral ecology, where we at least try to determine what collection of life historical, demographic, morphological, physiological, functional genetic, and genetic structural data can best, and quantitatively, predict among species patterns of social diversity, from humans to slime molds. Experts in each social taxon, who are well-represented in this book, have the components of such prediction in their brains and their publications, but reaching across taxa becomes more and more difficult as our accumulated knowledge of social biodiversity deepens and broadens – enough to drown anyone but a next Hamilton, Alexander, or Trivers.

Hamilton started with genes "for" social behavior, torn from the frayed social academic fabric of post-Nazi eugenics. Researchers are now finding such genes, for example, in honeybees, *Polistes* wasps, and even *Homo sapiens*. Can these genes, or genome-wide patterns of expression and variation, be leveraged to help make social behavioral ecology a predictive science? I fear, and reason, that for some decades genomes will remain far too complex and too far from social systems for any tight connections to be drawn. But perhaps we can begin to focus more upon specific genes and their pathways, social hormones, and social neurobiology across disparate taxa to fill and connect the gaps.

Indeed, neural and reproductive hormones may connect nicely with genes from below, and components of life history from above. And perhaps such interdisciplinary work that combines proximate with ultimate approaches, adding the spirit of Tinbergen to our pantheon, can link with bottom-up approaches guided by the taxa in this book. At this point, I can only exhort younger readers to become experts in specific social taxa, cognizant of other social taxa much more generally, and integrative in their methods of data collection. Scientifically, I have loved nothing more than learning about the social thrips of Australia, hearing about remarkable new discoveries of social cooperation in shrimps, aphids, or parasitic wasps, and collecting new forms of data on such creatures. I think this is all because we, as so-social humans, see echoes of ourselves among other social animals, each with their own amazing stories to tell.

Bernard Crespi
Simon Fraser University

1 The Evolution of Social Evolution

Dustin R. Rubenstein and Patrick Abbot

Overview

Why do animals live in cooperative groups? How do these societies function? These are the types of questions that motivated both of us in graduate school to study the evolutionary causes and consequences of sociality. We became part of a large and diverse group of scientists studying animal social behavior, a group that today spans the biological sciences, even extending into mathematics, engineering, and more. There are national and international societies, specialized journals, graduate programs, and institutes – each devoted in one way or another to studying animal social behavior. There is a vast body of knowledge about sociality in a diversity of animal species. As we come to know more about social diversity, synthesis has become more challenging. Ambitious, comprehensive narratives in the vein of *Sociobiology* (Wilson, 1975) are all but absent today (Sapp, 1994). But that was not always the case. Early naturalists once composed sweeping treatments of cooperation in nature (Cronin, 1991; Dugatkin, 2006; Dixon, 2008). If you have ever wondered how a wasp is like a bird, with notable exceptions (e.g. Brockmann, 1997; Korb & Heinze, 2008; Székely, *et al.*, 2010), you would have to dig deep into the literature to even find them discussed on the same pages. Modern animal behaviorists have become more specialized when it comes to studies of animal societies.

In editing this volume, the two of us were motivated to return to our original graduate school questions, to look beyond the organisms we are familiar with (birds, shrimps, aphids, ants), and begin to synthesize the features of social life that unite disparate animal taxa. In doing so, we take an admittedly optimistic view of animal sociality, arguing that there are convergent and common themes that span vertebrate and invertebrate societies. There is room for debate, and our goal with this book is to re-energize the conversation between scholars who think comparatively about the major animal lineages containing species that form societies.

In this introductory chapter, we begin with a short description of what it means to study sociality and social evolution in animals, and then provide a brief retrospective of studies of animal social behavior from Darwin to the present. We are not historians, and

Dustin Rubenstein was supported by the US National Science Foundation (IOS-1121435, IOS-1257530, IOS-1439985). Patrick Abbot was supported by the US National Science Foundation (IOS-1147033). We thank Jim Costa, Paul Sherman, and Joan Silk for helpful comments on previous versions of this chapter.

our retrospective is incomplete and reflects our own biases. We recommend to the interested reader works such as Crook (1970), Brown (1994), Cronin (1991), Dugatkin (1997, 2006), Costa (2006), Dixon (2008), Gibson, *et al.* (2013) as a start. Instead of a complete history of this field, we emphasize some of the key empirical and theoretical insights that led to advances in the study of social evolution, as well as some of the scientists responsible for these discoveries. We further discuss those researchers whose theories, empirical studies, or published volumes have attempted to bridge the divide between social vertebrates and invertebrates. We then highlight previous attempts to synthesize animal sociality, and discuss the structure of the chapters in this book as a way to begin analyzing these ideas in a new way. We lay out a "bottom-up" approach to dissecting animal societies, first by summarizing the distinct terminologies that researchers studying sociality use to describe the different forms of animal societies, then by discussing the various reasons that groups form and the numerous factors that influence their formation. Finally, we discuss the types of life history traits that are important for characterizing both social species and the groups that they form. Ultimately, this chapter is but a starting point for the book, which is itself a call for a renewed focus upon an empirical and theoretical unification of animal social life.

1.1 Sociality and the Definition of Animal Societies

What is sociality? This is not an obvious question. After all, nearly all animals are social at some point during their lives: individuals often exhibit affiliative or aggressive social interactions with members of their own species, and individuals in nearly all species must come together to mate (Trivers, 1985; Kokko, 2007). Many species also form groups, either ephemerally or permanently, including colonies of nesting seabirds, herds of migrating ungulates, schools of swimming fish, aggregations of feeding insects, or assemblages of mating amphibians (Alexander, 1974; Krause & Ruxton, 2002). The most advanced groups – eusocial insect societies – are described by three criteria first introduced by Batra (1966) and later expanded by Michener (1969) and Wilson (1971): (1) overlapping generations; (2) cooperative care of young; and (3) reproductive division of labor (i.e. many individuals in a group are temporarily or permanently sterile). Although originally created to define eusocial societies in insects, these criteria have also been used to describe cooperatively breeding societies in vertebrates (Sherman, *et al.*, 1995), as well as the societies of other invertebrates like aphids and thrips (Chapter 6), spiders (Chapter 7), and shrimps (Chapter 8). Most of these societies consist of kin (i.e. they are family groups), though this is also not a prerequisite for cooperative societies (Riehl, 2013). For many researchers, social groups consist of individuals who cooperate, but for others, cooperation is not a defining characteristic. What then are the characteristics of animal societies, and how should sociality be defined?

It turns out that researchers have a difficult time defining the term sociality, as it means different things in different fields. We define sociality for the purposes of this book simply as cooperative group living. This intentionally broad definition encompasses species that have at least some form of reproductive division of labor, and thus is similar to the broad approach taken by others in their treatments of species that grade from simple to complex social living (e.g. Korb & Heinze, 2008; Bourke, 2011). However, we also asked our authors to consider species that form non-ephemeral groups lacking reproductive division of labor. Moreover, we asked that they consider species that exhibit kin structure (i.e. form family groups), but also those that do not live with kin. Our definition also describes species where some group members cooperatively care for young that are not their own (i.e. alloparental care), but others where there is no communal care of offspring. Thus, our definition of sociality by necessity encompasses a range of forms of social organization. While we may have erred on the side of being too broad, the key element these species share is that some form stable groups within which various cooperative behaviors are typically expressed.

While this definition will not entirely satisfy all of the readers – or even all of the authors – it is a starting point from which we can begin to explore the similarities and differences among animals that are often described as being social. Indeed, one of the primary goals of this book is to detail the diversity of social lives that animals exhibit. It is therefore not surprising that the scientists who study social animals define sociality a bit differently. Sociality as we define it here occurs in less than 2 percent of insects (eusociality), and in only about 5 percent of mammals and 9 percent of birds (cooperative breeding) (Wilson, 1971; Cockburn, 2006; Lukas & Clutton-Brock, 2012). Sociality is even rarer among fishes, shrimps, lizards, spiders, and most of other taxonomic groups covered in this book. But while sociality is rare, what is obvious is that there are common features to social life in animals. The same patterns show up repeatedly in disparate animal lineages. The unique features of different animal groups help to explain why empirical studies of social vertebrates and invertebrates have largely taken divergent, though often parallel, paths over the last century.

1.2 The Importance of Studying Sociality

Why should you care about sociality? Whether or not you think ants or meerkats are fascinating, how convincing you find any argument for studying animal social behavior may depend upon your perspective. Even if you like to avoid bees or wasps, you may be persuaded by that fact that social insects probably account for about half of all of the biomass of the planet's biological diversity, and that ants and termites dominate the terrestrial habitats in which they occur (Wilson, 1990; Hölldobler & Wilson, 1990; Wilson, 2012) or that an unmistakable pattern in the evolution of all life on earth is that transitions in levels of organization have repeatedly occurred, in which formerly independent units (e.g. genes, chromosomes, cells, individuals, and so on) bind their fates together in a social enterprise, overcoming freeloaders (i.e. individuals who reap the benefit but pay no cost) in the process (Maynard Smith & Szathmáry, 1995; Queller & Strassman, 2009; Bourke, 2011). Or maybe you would be convinced by arguments closer to home: how, for example, studying sociality informs our understanding of ourselves (Pinker, 2010) and the development of our own societies (Fukuyama, 2011). Ultimately, the goal of those studying sociality and social evolution is to account for the special nature of social organisms (Strassmann & Queller, 2007), and to the extent that all organisms are, one way or another, social, the study of sociality is not only integral to the study of biology, but to all life on earth.

1.3 A History of Taxonomic Divergence

William D. Hamilton's (1936–2000) publications in 1964 on kin selection (Hamilton, 1964a,b), followed by the publication of Edward O. Wilson's *Sociobiology* in 1975, mark the start of what many would recognize as the beginning of the modern study of social behavior (Dugatkin, 1997; Costa, 2006; Clutton-Brock, *et al.* 2009). However, comparative approaches to social behavior have their origins almost a century earlier in the years following *On the Origin of Species* (Darwin, 1859), in the emerging fields of sociology and ecology (Crook, 1970). As many scholars have noted, Charles Darwin (1809–1882) was not a neo-Darwinian (Cronin, 1991; Browne, 2002; Dixon, 2008), particularly when it came to social behavior. Altruism was a word that neither he nor most of his contemporaries used (Dixon, 2008). Yet, in the decades that followed the publication of *On the Origin of Species* (Darwin, 1859), studies of animal social behavior proliferated, as moral philosophers, natural historians, political economists, theologians, and nascent sociologists turned to nature to discover the biological roots of moral philosophies, or to gauge the merit of positivist theories of human social development and progress. Their interests were largely philosophical and political, and only later did they become zoological. Was social behavior red in tooth and claw, as the prevailing interpretation of Darwin held in the late nineteenth century? Or was social life a regular, even progressive, outcome of natural laws?

In both Europe and America, the two decades that immediately followed the publication of *On the Origin of Species* were characterized by economic and social upheaval. Rapid changes in industrialization were accompanied by efforts to realign political, economic, and social orders. Darwin's book was a bestseller, but it was not uncommon for his theories on natural selection to be misunderstood, ignored, or even ridiculed; a "law of higgledy-piggledy," as Herschel famously put it (Browne, 2006; Hull, 2011). Even before the publication of *On the Origin of Species,* philosophers and political theorists turned to the natural world for both definition, validation, and critiques of Victorian society (Clark, 2009). A growing middle class, industrialization, urbanization, secularization, and a social reform movement were among the drivers of a growing fascination with the lessons and curiosities of animal life. Herbert Spencer (1820–1903) coined the phrase "survival of the fittest," and advocated for the natural progression of societies from simple to more complex via a mixture of Darwinian and Lamarkian reasoning. Spencer is largely forgotten today, but he was enormously influential, because he inspired a generation of devotees who melded sociological inquiry with avid study of animal social life (Francis & Taylor, 2015). One such devotee was Alfred Espinas (1844–1922), a French doctoral student who took a great interest in animal social behavior and in 1877, published *On Animal Societies* (Espinas, 1877), which summarized the existing knowledge of animal sociality at the time. His goal was to justify the Spencerian vision of an organic progression of society based upon natural laws, rather than moral imperatives (Brooks, 1998). Espinas' book, like those of other nascent sociologists, influenced a number of early biologists working on social behavior in the early and mid-twentieth century. Similarly, the Russian, Petr Kropotkin, published *Mutual Aid: A Factor in Evolution* in 1902 (Kropotkin, 1902), which posited a universally cooperative principle organizing the natural world, in opposition to the struggle for existence that was the prevailing, though incorrect, synopsis of Darwin's thesis. Other important texts from this era included *Animal Life: A First-Book of Zoology,* first published in 1900 by David Jordan (1851–1931) and Vernon Kellogg (1867–1937), which provided a broad synopsis of animal ecology, including mutualistic and social interactions in animals, and Benjamin Kidd's (1858–1916) bestseller *Social Evolution*, an ultimately discredited volume published in 1894 that popularized evolutionary ethics and became associated with social Darwinist movements (Crook, 1980).

Enthusiasm waxed and waned in the early decades of the twentieth century for the positivist ideals and naturalistic moralisms of the French philosopher Auguste Comte (1798–1857) and the late Victorians, preempted by the emergence of reductionism and the "fissioning" of the life sciences (Sapp, 1994). The result was, as Crook (1970) notes, a considerable gap in the social behavior literature in the period before World War II. Yet, holism gathered momentum both in Europe and America, at places like the University of Chicago and other campuses and field stations (Gibson, 2012). William Morton Wheeler (1865–1937) was an American entomologist who, having already published extensively on ants and other insects for nearly two decades, formulated views of ant and termite societies as more than the sum of their parts, as superorganisms (Wheeler, 1911; 1928). Since the Greeks, social insects had been used as mirrors of human societies (Costa, 2002). Victorian naturalists and political economists found in ants and bees validations of social order and the merits of division of labor (Clark, 2009). Wheeler, influenced by Weismann's cell theories, made remarkable descriptions of the organizations of ant societies that fueled the first elaborations of what we now recognize as the levels of selection (Gibson, 2012; Gibson, *et al.* 2013).

Meanwhile, Julian Huxley (1887–1975) was laying the foundations for the new field of ethology with his studies of courting behavior in grebes and other birds (Huxley, 1914; Brooke, 2014). The ethological tradition, fostered by Huxley and fellow ornithologists Niko Tinbergen (1907–1988), Konrad Lorenz (1903–1989), and David Lack (1910–1973), posited a comparative and experimental approach to animal behavior built upon both mechanistic and evolutionary principles (Brown, 1994; Burkhardt, 2014), and most importantly, close observation. This was the milieu out of which Alexander Skutch (1904–2004) made the first detailed observations of the family lives of neotropical birds (Stiles, 2005). Skutch, a polymath, naturalist, and leading ornithologist, published *Helpers at the Nest* in 1935, which included the first systematic observations in birds of cooperative care of young by non-breeding auxiliaries who aid in raising others' offspring (i.e. helpers). Skutch's descriptions of cooperative breeding behavior in three species of Central American birds (Skutch, 1935), followed by his much broader accounts a few decades later (Skutch, 1961), inspired later generations of scientists to study vertebrate social behavior, and although largely ignored for decades, marked the origin of thinking about cooperative breeding in vertebrates as an evolutionary and ecological problem (Brown, 1978).

Wheeler and Skutch were both expert naturalists whose traditions of fieldwork helped the study of social behavior re-emerge after World War II. Wheeler died in 1937, and though he mentored many students, many moved on to other fields, such as Alfred Kinsey of "human sexology" fame. Thus, despite the work of ethologists and entomologists like Wheeler on ants and Karl von Frisch (1886–1982) on bees, the roots of a taxonomic schism emerged in the study of social evolution. Animal behavior was founded as a discipline by ornithologists, while social insect biologists devoted their attention to taxonomy, morphology, and especially physiology. The focus on mechanisms accompanied the rejection of holism and "emergent evolution" by scientists newly minted in atomic age reductionism (Gibson, *et al.*, 2013). In essence, there were not many entomologists at the table at the time when studies of social behavior and comparative evolutionary biology were beginning to blossom. Warder Allee (1885–1955) stands out during this period as the inheritor of a tradition for comparative studies of social evolution, for which he received only modest recognition for most of his career (Mitman, 1992; Dugatkin, 2006). His focus, however, was centered upon developing Kropotkin-like universals about the overarching role of cooperation in nature, and many of his works, couched in the language of demography and the nascent field of ecology, had little influence on his contemporaries until late in his life.

This taxonomic schism during the postwar years was exemplified in the difficulties that a young Hamilton had in convincing his graduate mentors in the early 1960s that there were indeed interesting evolutionary problems posed by altruism in ants and bees (Segerstrale, 2013). However, Hamilton eventually not only succeeded in convincing his mentors that individually costly, but helpful, behaviors were worth thinking about, but he changed forever how we study social behavior. As Brown (1994) pointed out, Hamilton's theoretical advances on the genetics and evolution of altruism, with insects in mind, came at a time when Lack and other ornithologists were developing an empirically informed ecological framework for comparing different social systems in birds. At the same time, Crook and Gartlan (1966) were establishing ecological comparative studies of primates and other mammals. Thus, by 1970, the ingredients were in place for the emergence of a socioecological framework for comparative social behavior and evolution. And yet, empirical studies of social insects and cooperatively breeding birds and mammals proceeded largely independently, despite the emergence of this new unifying theory of inclusive fitness. Empirical insights from testing Hamilton's theory would arrive piecemeal from an early fission into vertebrate and invertebrate camps that Wilson (1971), building upon a tradition inherited from the likes of Crook and Allee (Allee, *et al.*, 1949), would bridge.

1.4 Attempts at Social Synthesis

Inclusive fitness theory is an explicit framework that governs the evolution of social traits, irrespective of taxonomy (Hamilton, 1964a,b; Bourke, 2011). The broad utility of

Hamilton's rule – and inclusive fitness theory more generally – is born out in its simplicity. According to Hamilton's Rule, a social action will be favored when its positive effect on indirect fitness is greater than its direct fitness cost. In simple math, rb more than c, or the product of the relatedness (r) between two individuals and the fitness benefit (b) an individual receives from the action valued against the fitness cost (c) to the individual expressing the action. Hamilton's concept – termed kin selection by Maynard Smith (1964) – was the foundation for Wilson's *Sociobiology*, as well as many of the later theoretical contributions to the field. Most of these subsequent theoretical contributions were synthetic in their nature, generating hypotheses that were later tested in organisms as diverse as birds, wasps, and microbes (e.g. Eberhard, 1972; Emlen & Wrege, 1988; Griffin, *et al.* 2004). The theoretical contributions of Trivers (1971, 1974) and others on cooperation and conflict also apply equally well across all taxonomic groups, and ultimately shaped the thinking of many empiricists. The development of reproductive skew theory (Vehrencamp, 1977), an extension of Hamilton's rule, has also been tested empirically in birds, mammals, and numerous species of insects (Keller & Reeve, 1994; Hager & Jones, 2009). The generality of these theories, the alluring examples of convergence across disparate taxa, and a Darwinian/Spencerian tradition of unification and comparative approaches spurred a number of researchers to generate broadly synthetic summaries of social life. For example, Alexander (1974) discussed forms of social behavior (kin-selected or otherwise) in very general terms, largely avoiding specific taxonomic language in much of his review. Trivers (1985) and Bourke (2011) discussed aspects of social evolution common to all animal societies. Andersson (1984) and later Brockmann (1997) recognized the similarities between cooperatively breeding birds and eusocial insects. Sherman, *et al.* (1995) argued that species from different taxonomic groups could be arrayed along a continuum of reproductive sharing, or reproductive skew, though this synthetic idea was criticized by those who saw eusociality as something unique to insects and were reluctant to recognize potential parallels between insects and vertebrates (e.g. Crespi & Yanega, 1995; Costa & Fitzgerald, 1996a,b; Wcislo, 1997a,b; Costa & Fitzgerald, 2005; Crespi, 2005; Wcislo, 2005). Other researchers have suggested alternative views on how disparate social taxa could be linked (Aviles & Harwood, 2012), and some have extended what we know about cooperation in animals to human societies (e.g. Crespi, 2013). Edited volumes by Rubenstein and Wrangham (1986) and later Choe and Crespi (1997) explored social evolution in vertebrates (birds and mammals) and invertebrates (insects and arachnids), respectively, as did the book *The Other Insect Societies*, by Costa (2006). In addition to these treatments comparing social behavior within animal lineages, edited volumes by Slobodchikoff (1988) and Korb and Heinze (2008) began to explore the similarities and differences in social vertebrates and invertebrates, largely from an ecological perspective.

In summary, the very earliest studies of social behavior were comparative, but they were conducted by moral philosophers, natural historians, political economists, theologians, and nascent sociologists, not by biologists. It was the early decades of the twentieth century, when academic departments were constructed around taxonomy and natural history, that entomological and vertebrate research agendas initially diverged. It was not until the theoretical advances of the mid-century that the first attempts at a modern social synthesis began. These mathematical models applied equally well to all social organisms – invertebrates, vertebrates, or even microbes. However, to a large extent, the breadth of social evolutionary theory has not been matched by equally broad empirical research programs. Given the wealth of new comparative methods for analyzing large datasets of life history traits and biogeographic data, as well as the explosion and growing affordability of comparative genomic tools, the time is right to reconsider a social synthesis, from both a theoretical and empirical perspective. This book is an attempt to do just that: to comparatively survey the diversity of vertebrate and invertebrate societies, and lay the groundwork for a new generation of theoretical, empirical, and comparative studies of animal social evolution.

1.5 Comparative Social Evolution: Social Diversity, Traits and Synthesis

The goal of this book is to synthesize the features of animal social life across the principle taxonomic groups in which sociality has evolved, and do so in a cohesive and comparative manner by centralizing the review within a single volume and with a unified format. Our book differs from previous treatments in that it takes a "bottom-up" rather than a "top-down" approach to explore social evolution. That is, instead of emphasizing the theoretical advances that seemingly link disparate taxa (e.g. kin selection) or a consideration of shared evolutionary histories (e.g. the Hymenoptera), the bulk of each chapter on a given taxonomic group instead discusses the traits and characteristics of social (and non-social) species and the groups that they form. Our intent is to highlight much of the interesting natural and life history data that attracted us and other scientists to study social organisms in the first place. Identifying a suite of traits shared among social species and groups may ultimately allow researchers to better define the social phenotype, and then study the proximate and ultimate factors that shape its evolution. Ultimately, each chapter is comparable in both structure and content, so scholars of one type of organism can readily compare and contrast with those from other taxonomic groups.

To achieve our bottom-up approach to exploring social evolution, each chapter in the book is structured similarly. Using a uniform chapter structure was challenging, since some sections apply to some taxa better than they do to others. Yet, exploring the same life history traits, ecological factors, and other characteristics of social species and groups across all taxonomic groups provides opportunities to observe previously unrecognized patterns. The chapters themselves are grouped into three parts. Part I briefly explores the social diversity within a taxonomic group, highlighting the frequency, forms, and reasons that social groups develop, as well as the evolutionary and ecological factors that shape social living. Part II describes the social phenotype in greater detail, highlighting both the traits of social species and of the social groups that they form. Finally, Part III begins to synthesize social diversity, both within a taxonomic group, as well as among groups and across lineages. Next, we describe briefly why chapters are structured the way that they are, and how doing so can begin to shed light upon the comparative social evolution of such disparate organisms.

1.5.1 Social Diversity

Both the frequency of occurrence and diversity of form of animal societies vary greatly within and among taxonomic groups. Sociality is ubiquitous in groups like the ants where all of the more than 15,000 species are eusocial (Chapter 2), but it is extremely rare, for example, in spiders (Chapter 7), shrimps (Chapter 8), and freshwater fishes (Chapter 12). Animal societies also vary widely in form, and we know that the most well-studied species often do not exhibit the most common form of society. For example, it can be a common assumption that most bees are like honeybees and live in eusocial societies with a single queen and hundreds of workers, but in fact, most bee species are solitary or live in small groups consisting of several adult females (Chapter 3). Additionally, within cooperatively breeding birds, most species live in family groups with helpers, though some family-living species lack helpers and some species form social groups of unrelated individuals (Chapter 11).

A synthetic discussion about the variation in form and structure of animal societies is often difficult because the criteria and terminology that researchers use to describe animal societies varies among taxonomic groups. In addition to the jargon that uniquely describes specific taxonomic groups (see the glossary), at even the broadest scale, the terms that describe basic reproductive structures in vertebrates and invertebrates differ greatly (Rubenstein, *et al.*, 2016). For example, vertebrate societies are often divided into those with a single breeding pair (singular breeding) and those with more than one breeding pair (plural breeding) (Brown, 1987; Solomon & French, 1997). Although the same categorization is often used to describe eusocial insect societies, the terminology used by entomologists is very different. Invertebrate societies with a single queen are typically referred to as being monogynous, whereas those with multiple queens are referred to as polygynous. Sometimes, the multi-queen/multi-breeder societies are called communal in both invertebrate and vertebrates, but even in cooperatively breeding birds and mammals, communal breeding is used to describe different behaviors (Chapter 14). In other words, the lineage-specific terminology used to categorize the most basic social structures, as well as the taxon-specific jargon that describes the forms of animal societies, often hinders our ability to compare and contrast disparate social animal groups. Taking a bottom-up approach to describe the demographic and breeding structures of groups may ultimately help alleviate some of this terminological confusion by quantifying societies continuously rather than categorically.

Despite the terminological differences in how animal societies are described and defined, there are many similarities across taxonomic groups in why societies form. At the most basic level, animal societies often form because the benefits (either direct or indirect) of grouping outweigh the costs of breeding independently (Hamilton, 1964a,b; Bourke, 2011). The relative importance of direct versus indirect benefits has been debated in the literature (e.g. Clutton-Brock, 2002). While there are a variety of potential direct benefits individuals receive by living in groups (Krause & Ruxton,

2002), individuals most often group to gain access to resources and avoid predators (Alexander, 1974), and in some species, to maintain homeostasis, to gain access to mates, or to provide offspring care. Although not all organisms achieve all of these benefits, it is often surprising to learn which species reap which rewards. For example, the benefits of homeostasis obviously apply to many eusocial insects that maintain a constant temperature inside the nest, but homeostatic benefits are also important to a number of avian species (Chapter 11), cavity-nesting primates (Chapter 9), denning mammals (Chapter 10), and even some lizards (Chapter 13).

Ecological factors have long been thought to play a key role in shaping social behavior and mating systems in vertebrates and insects (Alexander, 1974; Jarman, 1974; Emlen & Oring, 1977; Bourke & Heinze, 1994). Ecology can constrain the formation of groups by limiting dispersal and independent breeding (Emlen, 1982), as well as influence the biogeographic distribution and niches of social groups and species (Jetz & Rubenstein, 2011). Climate – considered to be part of ecology – can also influence the broad-scale distribution of social species in vertebrates (Rubenstein & Lovette, 2007; Jetz & Rubenstein, 2011) and invertebrates (Kocher, *et al.*, 2014; Purcell, *et al.*, 2015; Sheehan, *et al.*, 2015). Indeed, climate, ecology, and biogeography often interact to influence the evolution and distribution of social species. For example, the high incidence of cooperative breeding in Australian birds is the product of both the continent's variable and semi-arid environment (Jetz & Rubenstein, 2011), as well the biogeographic and evolutionary history of its avifauna (Cockburn, 1996; Cockburn 2013; Chapter 11).

Finally, evolutionary history also plays an important role in explaining the distribution of social species within a taxonomic group. For example, within the Hymenoptera (ants, bees, and wasps), the evolutionary histories of social clades vary among orders. Eusociality has evolved independently and repeatedly in numerous clades of bees (Chapter 3) and wasps (Chapter 4), but only once in ants (Chapter 2). However, in groups like the birds (Chapter 11), cooperative breeding evolved early in the radiation, but has since been lost and regained many times across the avian tree of life (Ligon & Burt, 2004; Chapter 11). Moreover, life history constraints resulting from shared evolutionary history often combine with ecology to influence the evolution of sociality in birds (Arnold & Owens, 1998; Hatchwell & Komdeur, 2000).

1.5.2 Social Traits

A discussion of the potential direct benefits, costs, and evolutionary constraints on social grouping is arguably a top-down approach that summarizes what we already know – or think we know – about social diversity within a given group of animals. In contrast, taking a bottom-up approach that searches for the similarities and differences in animal societies across disparate animal lineages requires a systematic summary of the traits that characterize both social species and the groups that they form. However, choosing which traits to summarize is not an easy task. The most thoroughly studied life history traits of social species for comparative studies represent different components of the breeding life histories of social animals – lifespan and longevity, fecundity, age at first reproduction, dispersal, and so on. We also consider cognition and communication, traits that define many social species (e.g. brain size) and that allow societies to function efficiently and effectively. After all, most social organisms require mechanisms to recognize group mates, identify individuals or kin, and then coordinate their actions (Hauber & Sherman, 2001). Finally, when considering the life histories of social species, it is important to do so within the broader context of non-social species, particularly within taxonomic groups where sociality is rare. Thus, wherever possible, we asked the authors to frame the life histories of social species in the context of closely related non-social species.

Although many of the life history traits of social species have been quantified in a variety of taxonomic groups, it turns out that they are not always comparable. That is, specific life history traits often mean different things in different lineages and are sometimes quantified in very different ways. For example, longevity in vertebrates is typically measured in terms of individual lifespan (median or maximum), but many social insects are measured in terms of colony or queen lifespan (maximum).

Defining the traits of social groups is also not easy. Doing so is complicated by the fact that the terminology researchers use to define animal societies varies immensely across taxonomic groups. For example, primate and other mammalian societies are often defined by three components: (1) social organization; (2) mating system; and (3) social structure (Kappeler & van Schaik, 2002; see also Chapters 9 and 10). Within this framework, social organization refers to who lives with whom, mating system refers to who mates with whom, and social structure refers to the social relationships among group members. However, these same terms are often defined very differently in other taxonomic groups, even in the following chapters. In our bottom-up approach to editing this book, we chose not to force a single terminology onto the authors of each chapter. Instead, we begin to highlight some of these terminological differences here, and then return to this point at the book's conclusion (Chapter 14).

We believe that separating social and mating interactions and bonds – as Kappeler & van Schaik (2002) did – needs to be one of the primary goals of any definition of animal sociality. Most animal behaviorists recognize the clear distinction between the social system (e.g. cooperative breeding, eusociality, etc.) and the mating system (e.g. monogamy, polygyny, polyandry, polygynandry). After all, within a given type of social system, a variety of mating systems can occur (e.g. different cooperatively breeding bird species can be monogamous, polygynous, polyandrous, or polygynandrous [Cockburn, 2004]; or within and among ant species, queens can be either monogamous or polyandrous [Keller, 1993]). Additionally, researchers often emphasize the role of genetic relatedness in the study of social evolution, a tradition that derives from Hamilton's (1964a, b) seminal work. Genetic structure, often referred to as kin structure, within a group is greatly influenced by the mating decisions of female breeders (Boomsma 2007, 2009, 2013). Lifetime monogamy by females results in high relatedness among offspring, which is thought to influence the evolution of eusociality in insects (Hughes, *et al.*, 2008) and cooperative breeding in birds (Cornwallis, *et al.*, 2010) and mammals (Lukas & Clutton-Brock, 2012). However, kin structure within a social group is also influenced by the number of females that breed in a group (Rubenstein, 2012; Boomsma, *et al.*, 2014). Indeed, in some insects, there is a trade-off between the number of queens in a colony and the number of mates each queen has (Kronauer & Boomsma, 2007). Thus, it is important to keep the concepts of social structure and mating system separate when defining animal societies because the genetic structure of social groups can be influenced by both the number of mates a breeding female has (the mating system), as well as the number of breeding females in the group (the social structure).

To keep these two ideas – social structure and mating system – apart, there is a section in each chapter devoted to each concept. We first explore the variation in genetic structures within societies of each taxonomic group. Although some taxonomic groups are much better studied than others, we explore not only the genetic structure of societies, but also their mating patterns and systems. Following this, our exploration of the social structure focuses upon the demographic structures of groups (e.g. group structure, breeding structure, and the sex ratio). This includes discussion of the variation in the number of breeders/reproductive versus helpers/workers in animal societies, as well as the different sex ratios within and among these social categories. Ultimately, categorizing the variation in demographic and breeding structure is, easier for some groups than others: shrimps (Chapter 8) are easier than ants (Chapter 2), for example, because the former have an order of magnitude fewer social species than do the latter. Nonetheless, it remains important to consider the variation in demographic and breeding structure of animal societies for any comparative approach.

1.5.3 Social Synthesis

By exploring both the traits that define the phenotype of social organisms as well as those that describe the form and structure of the groups that they form, we can begin to synthesize the diversity of social life within and across taxonomic groups. The final section of each chapter is devoted to doing exactly this. Authors were given more leeway here to explore patterns of social life within their taxonomic group, as well as to begin to make links to other groups within this book. Ultimately, the goal of this section is for readers to start to synthesize the material in each chapter and begin to make further connections between disparate animal groups. We then continue and expand upon these syntheses at the end of the book in the concluding chapter (Chapter 14).

1.6 The Prospect of Social Convergence

Theory has always been a critical component of the field of animal social behavior, often working hand-in-hand with empirical studies. As we discussed earlier, Hamilton's theory of kin selection (Hamilton, 1964a,b) not only set the stage for a generation of empirical studies, but those same empirical studies helped refine our theoretical predictions. For example, the discovery of diploid eusocial naked mole-rats (Jarvis, 1991), coupled with the appreciation of diploidy in eusocial termites, helped to reframe kin selection as something more than a synonym for haplodiploidy. Reproductive skew theory (Vehrencamp, 1977) extended Hamilton's rule and led empiricists to focus upon two of its key components, the roles that genetic relatedness and ecological constraints play in shaping the formation of groups and altruistic behaviors. Arguably, insect biologists have emphasized the genetic relatedness in their studies of social evolution, whereas vertebrate biologists have emphasized ecological constraints (Elgar, 2015). Yet more recently, entomologists have begun to consider the role that ecology plays in shaping eusocial insect societies (e.g. Kocher, *et al.*, 2014; Purcell, *et al.*, 2015; Sheehan, *et al.*, 2015), and vertebrate biologists have begun to study the role of genetic relatedness in driving the evolution of cooperative breeding behavior independently from any ecological factors (e.g. Cornwallis, *et al.*, 2010; Lukas & Clutton-Brock, 2012).

Although cross-fertilization is occurring between biologists who study invertebrate or vertebrate sociality, the two sub-fields remain largely distinct. Can we ultimately achieve the social synthesis that kin selection and reproductive skew theory promised decades ago? We remain optimistic that not only can a unified theory of sociality be developed, but that empirical studies of social animals can become more integrated across taxonomic boundaries. This book is a first attempt to do this by beginning to compile relevant concepts and summarize key social traits across all animals. We return to this goal at the end of the book (Chapter 14) and discuss what can be learned from thinking about sociality in different taxonomic groups in the same way.

References

Alexander, R. D. (1974) The evolution of social behavior. *Annual Review of Ecology and Systematics*, **5**, 325–383.

Allee, W. C., Emerson, A. E., Park, O., Park, T., & Schmidt, K. (1949) *Principles of Animal Ecology*. Philadelphia: Saunders Co.

Andersson, M. (1984) Evolution of eusociality. *Annual Review of Ecology and Systematics*, **15**, 165–189.

Arnold, K. E. & Owens, I. P. F. (1998) Cooperative breeding in birds: A comparative test of the life history hypothesis. *Proceedings of the Royal Society of London B*, **265**, 739–745.

Aviles, L. & Harwood, G. (2012) A quantitative index of sociality and its application to group-living spiders and other social organisms. *Ethology*, **118**, 1219–1229.

Batra, S. W. T. (1966) Nests and social behavior of halictine bees of India (Hymenoptera: Halictidae). *Indian Journal of Entomology*, **28**, 375–393.

Boomsma, J. J. (2007) Kin selection vs. sexual selection: Why the ends do not meet. *Current Biology*, **17**, R673–R683.

(2009) Lifetime monogamy and the evolution of eusociality. *Philosophical Transactions of the Royal Society B*, **364**, 3191–3207.

(2013) Beyond promiscuity: Mate-choice commitments in social breeding. *Philosophical Transactions of the Royal Society B*, **368**, 20120050.

Boomsma, J. J., Huszar, D. B., & Pedersen, J. S. (2014) The evolution of multiqueen breeding in eusocial lineages with permanent physically differentiated castes. *Animal Behaviour*, **92**, 241–252.

Bourke, A. F. G. (2011) *Principles of Social Evolution*. Oxford: Oxford University Press.

Bourke, A. F. G. & Heinze, J. (1994) The ecology of communal breeding: The case of multiple-queen leptothoracine ants. *Philosophical Transactions of the Royal Society B*, **345**, 359–372.
Brockmann, H. J. (1997) Cooperative breeding in wasps and vertebrates: The role of ecological constraints. *In*: Choe, J. C. & Crespi, B. J. (eds.) *Evolution of Social Behavior in Insects and Arachnids*. Cambridge: Cambridge University Press, pp. 347–371.
Brooks, J. I. (1998) *The Eclectic Legacy: Academic Philosophy and the Human Sciences in Nineteenth-Century France*. Delaware: University of Delaware Press.
Brooke, M. (2014) In retrospect: The courtship habits of the Great Crested Grebe. *Nature*, **513**, 484–485.
Brown, J. L. (1978) Avian communal breeding systems. *Annual Review of Ecology and Systematics*, **9**, 123–155.
(1987) *Helping and Communal Breeding in Birds: Ecology and Evolution*. Princeton: Princeton University Press.
(1994) Historical patterns in the study of avian social behavior. *Condor*, **96**, 232–243.
Browne, J. (2002) *Charles Darwin: vol. 2 The Power of Place*. London: Jonathan Cape.
(2006) *Darwin's Origin of Species: A Biography*. London: Atlantic Books.
Burkhardt, R. W. (2014) Tribute to Tinbergen: Putting Niko Tinbergen's 'Four Questions' in historical context. *Ethology*, **120**, 215–223.
Choe, J. C. & Crespi, B. J. (eds.) (1997) *Social Behavior in Insects and Arachnids*, Cambridge: Cambridge University Press.
Clark, J. F. M. (2009) *Bugs and the Victorians*. New Haven: Yale University Press.
Clutton-Brock, T. (2002) Breeding together: Kin selection and mutualism in cooperative vertebrates. *Science*, **296**, 69–72.
Clutton-Brock, T., West, S., Ratnieks, F., & Foley, R. (2009) The evolution of society. *Philosophical Transactions of the Royal Society of London B*, **364**, 3127–3133.
Cockburn, A. (1996) Why do so many Australian birds cooperate: Social evolution in the Corvida? *In:* Floyd, R. B., Sheppard, A. W., & De Barro, P. J. (eds.) *Frontiers of Population Ecology*. East Melbourne: CSIRO, pp. 451–472.
(2004) Mating systems and sexual conflict. *In*: Koenig, W. D. & Dickinson, J. L. (eds.) *Ecology and Evolution of Cooperative Breeding in Birds*. Cambridge: Cambridge University Press, pp. 81–101.
(2006) Prevalence of different modes of parental care in birds. *Proceedings of the Royal Society of London B*, **273**, 1375–1383.
(2013) Cooperative breeding in birds: Towards a richer conceptual framework. *In*: Sterelny, K., Joyce, R., Calcott, B., & Fraser, B. (eds.) *Cooperation and its Evolution*. Cambridge, MA: MIT Press, pp. 223–245.
Cornwallis, C. K., West, S. A., & Davis, K. E. (2010) Promiscuity and the evolutionary transition to complex societies. *Nature*, **466**, 969–972.
Costa, J. T. (2002) Scale models? What insect societies teach us about ourselves. *Proceedings of the American Philosophical Society*, **146**,170–180.
(2006) *The Other Social Insect Societies*. Cambridge, MA: Harvard University Press.
Costa, J. T. & Fitzgerald, T. D. (1996a) Developments in social terminology: Semantic battles in a conceptual war. *Trends in Ecology & Evolution*, **11**, 285–289.
(1996b) Social terminology revisited–reply. *Trends in Ecology & Evolution*, **11**, 472–473.
(2005) Social terminology revisited: Where are we ten years later? *Annales Zoologici Fennici*, **42**, 559–564.
Crespi, B. (2013) The insectan apes. *Human Nature*, **25**, 6–27.

Crespi, B. J. (2005) Social sophistry: Logos and mythos in the forms of cooperation. *Annales Zoologici Fennici*, **42**, 569–571.

Crespi, B. J. & Yanega, D. (1995) The definition of eusociality. *Behavioral Ecology*, **6**, 109–115.

Cronin, H. (1991). *The Ant and The Peacock: Altruism and Sexual Selection from Darwin to Today*. Cambridge: Cambridge University Press.

Crook, D. P. (1980) Bio-politics in the 1890s: Benjamin Kidd and social evolution. *Australian Journal of Politics & History* **26**, 212–227.

Crook, J. H. (Ed.) (1970) *Social Behaviour in Birds and Mammals: Essays on the Social Ethology of Animal and Man*. London: Academic Press.

Crook, J. H. & Gartlan J. S. (1966) Evolution of primate societies. *Nature*, **210**, 1200–1203.

Darwin, C. (1859) *On the Origin of Species by Natural Selection*. London: John Murray.

Dixon, T. (2008). *The Invention of Altruism: Making Moral Meanings in Victorian Britain*. Oxford: Oxford University Press.

Dugatkin, L. A. (1997) *Cooperation Among Animals*. Oxford: Oxford University Press.

(2006) *The Altruism Equation: Seven Scientists Search for the Origins of Goodness*. Princeton: Princeton University Press.

Eberhard, W. G. (1972) Altruistic behavior in a sphecid wasp: Support for kin-selection theory. *Science*, **175**, 1390–1391.

Elgar, M. A. (2015) Integrating insights across diverse taxa: Challenges for understanding social evolution. *Frontiers in Ecology and Evolution*, **3**, 124.

Emlen, S. T. (1982) The evolution of helping. I. An ecological constraints model. *The American Naturalist*, **119**, 29–39.

Emlen, S. T. & Oring L. (1977) Ecology, sexual selection, and the evolution of mating systems. *Science*, **197**, 215–223.

Emlen, S. T. & Wrege, P. H. (1988). The role of kinship in helping decisions among white-fronted bee-eaters. *Behavioural Ecology and Sociobiology*, **23**, 305–315.

Esplnas, A. (1878) *Des Sociétés Animales*. Paris: Ballliere.

Francis, M. & Taylor, M. W. (2015). *Herbert Spencer: Legacies*. London: Routledge Press.

Fukuyama, F. (2011) *The Origins of Political Order: From Prehuman Times to the French Revolution*. New York: Farrar, Straus and Giroux.

Gibson, A. H. (2012) Edward O. Wilson and the organicist tradition. *Journal of Historical Biology*, **46**, 599–630.

Gibson, A. H., Kwapich, C. L., & Lang, M. (2013) The roots of multilevel selection: Concepts of biological individuality in the early twentieth century. *History and Philosophy of the Life Sciences*, **35**, 505–532.

Griffin, A. S., West, S. A., & Buckling, A. (2004) Cooperation and competition in pathogenic bacteria. *Nature*, **430**, 1024–1027.

Hager, R. & Jones, C. B. (eds.) (2009) *Reproductive Skew in Vertebrates: Proximate and Ultimate Causes*. Cambridge: Cambridge University Press.

Hamilton, W. D. (1964a) The genetical evolution of social behaviour. *I. Journal of Theoretical Biology*, **7**, 1–16.

(1964b) The genetical evolution of social behaviour. *II. Journal of Theoretical Biology*, **7**, 17–52.

Hatchwell, B. J. & Komdeur, J. (2000) Ecological constraints, life history traits and the evolution of cooperative breeding. *Animal Behavior*, **59**, 1079–1086.

Hauber, M. E. & Sherman, P. W. (2001) Self-referent phenotype matching: Theoretical considerations and empirical evidence. *Trends in Neuroscience*, **24**, 609–616.

Hofmann, H., Couzin, I. D., Earley, R. L., *et al.*, (2014) An evolutionary framework for studying mechanisms of social behavior. *Trends in Ecology & Evolution*, **29**, 581–589.

Hölldobler, B. & Wilson, E. O. (1990) *The Ants*. Berlin: Springer.

Hughes, W. O. H., Oldroyd, B. P., Beekman, M., & Ratnieks, F. L. W. (2008) Ancestral monogamy shows kin selection is key to the evolution of eusociality. *Science*, **320**, 1213–1216.

Hull, D. L. (2011) Darwin's science and Victorian philosophy of science. *In:* Hodge, J. & Radick, G. (eds.) *The Cambridge Companion to Darwin*. Cambridge: Cambridge University Press.

Huxley, J. S. (1914) The courtship habits of the Great Crested Grebe *(Podiceps cristatus)*: With an addition to the theory of sexual selection. *Proceedings of the Zoological Society of London*, **84**, 491–562

Jarman, P. J. (1974) The social organisation of antelope in relation to their ecology. *Behaviour*, **48**, 215–267.

Jarvis, J. U. M. (1991) Reproduction in naked mole-rats. *In*: Sherman, P. W., Jarvis, J. U. M., & Alexander, R. D. (eds.) *The Biology of the Naked Mole-rat*. Princeton: Princeton University Press, pp. 384–425.

Jetz, W. & Rubenstein, D. R. (2011) Environmental uncertainty and the global biogeography of cooperative breeding in birds. *Current Biology*, **21**, 72–78.

Kappeler, P. M. & van Schaik, C. P. (2002) Evolution of primate social systems. *International Journal of Primatolgy*, **23**, 707–740.

Keller, L. (Ed). (1993) *Queen Number and Sociality in Insects*. Oxford: Oxford University Press.

Keller, L. & Reeve, H. K. (1994) Partitioning of reproduction in animal societies. *Trends in Ecology & Evolution*, **9**, 98–102.

Kocher, S. D., Pellissier, L., Veller, C., *et al.*, (2014) Transitions in social complexity along elevational gradients reveal a combined impact of season length and development time on social evolution. *Proceedings of the Royal Society of London B*, **281**, 20140627.

Kokko, H. (2007) Cooperative behaviour and cooperative breeding: What constitutes an explanation? *Behavioural Processes*, **76**, 81–85.

Korb, J. & Heinze, J. (eds.) (2008) *Ecology of Social Evolution*. Berlin: Springer-Verlag.

Krause, J. & Ruxton, G. D. (2002) *Living in Groups*. Oxford: Oxford University Press.

Kronauer, D. J. C. & Boomsma, J. J. (2007) Multiple queens means fewer mates. *Current Biology*, **17**, R753–R755.

Kropotkin, P. (1902) *Mutual Aid*. London: William Heinemann.

Ligon, J. D. & Burt, D. B. (2004) Evolutionary origins. *In*: Koenig, W. D. & Dickinson, J. L. (eds.) *Ecology and Evolution of Cooperative Breeding in Birds*. Cambridge: Cambridge University Press, pp. 5–34.

Lukas, D. & Clutton-Brock, T. H. (2012) Cooperative breeding and monogamy in mammalian societies. *Proceedings of the Royal Society of London B*, **279**, 2151–2156.

Maynard Smith, J. (1964) Group selection and kin selection. *Nature*, **201**, 1145–1147.

Maynard Smith, J. & Szathmáry, E. (1995) *The Major Transitions in Evolution*. Oxford: Oxford University Press.

Michener, C. D. (1969) The evolution of social behavior of bees. *Annual Review of Entomology*, **14**, 299–342.

Mitman, G. (1992) *The State of Nature: Ecology, Community, and American Social Thought, 1900–1950*, Chicago: University of Chicago Press.

Pinker, S. (2010) The cognitive niche: Coevolution of intelligence, sociality, and language. *Proceedings of the National Academy of Sciences USA*, **107**, 8893–8999.

Purcell, J., Pellissier, L., & Chapuisat, M. (2015) Social structure varies with elevation in an Alpine ant. *Molecular Ecology*, **24**, 498–507.

Queller, D. C. & Strassman, J. E. (2009) Beyond society: The evolution of organismality. *Philosophical Transactions of the Royal Society B*, **364**, 3143–3155.

Riehl, C. (2013) Evolutionary routes to non-kin cooperative breeding in birds. *Proceedings of the Royal Society of London B*, **280**, 20132245.

Rubenstein, D. R. (2012) Family feuds: Social competition and sexual conflict in complex societies. *Philosophical Transactions of the Royal Society B*, **367**, 2304–2313.

Rubenstein, D. R. & Lovette, I. J. (2007) Temporal environmental variability drives the evolution of cooperative breeding in birds. *Current Biology*, **17**, 1414–1419.

Rubenstein, D. I. & Wrangham, R. W. (1986) *Ecological Aspects of Social Evolution*. Princeton: Princeton University Press.

Rubenstein D. R., Botero, C. A., & Lacey, E. A. (2016) Discrete but variable structure of animal societies leads to the false perception of a social continuum. *Royal Society Open Science*, **3**, 160147.

Sapp, J. (1994) *Evolution by Association: A History of Symbiosis*. Oxford: Oxford University Press.

Segerstrale, U. (2013) *Nature's Oracle. The Life and Work of W.D. Hamilton*. Oxford: Oxford University Press.

Sheehan, M. J., Botero, C. A., Hendry, T., *et al.*, (2015) Different axes of environmental variation explain the presence vs. extent of cooperative nest founding associations in *Polistes* paper wasps. *Ecology Letters*, **18**, 1057–1067.

Sherman, P. W., Lacey, E. A., Reeve, H. K., & Keller, L. (1995) The eusociality continuum. *Behavioral Ecology*, **6**, 102–108.

Skutch, A. F. (1935) Helpers at the nest. *Auk*, **52**, 257–273.

(1961) Helpers among birds. *Condor*, **63**, 198–226.

Slobodchikoff, C. N. (ed.) 1988. *The Ecology of Social Behavior*. New York: Academic Press.

Solomon, N. G. & French, J. A. (1997) *Cooperative Breeding in Mammals*. Cambridge, MA: Cambridge University Press.

Stiles, F. G. (2005) In Memoriam: Alexander F. Skutch, 1904–2004. *The Auk*, **122**, 708–710.

Strassmann, J. & Queller, D. (2007) Insect societies as divided organisms: The complexities of purpose and cross-purpose. *Proceedings of the National Academy of Sciences USA*, **104**, 8619–8626.

Székely, T., Moore, A. J., & Komdeur, J. (eds.). (2010) *Social Behaviour: Genes, Ecology and Evolution*. Cambridge: Cambridge University Press

Trivers, R. L. (1971) The evolution of reciprocal altruism. *Quarterly Review of Biology*, **46**, 35–57.

(1974) Parent-offspring conflict. *American Zoologist*, **14**, 249–264.

(1985) *Social Evolution*. Menlo Park, NJ: Benjamin/Cummings.

Vehrencamp, S. L. (1977) A model for the evolution of despotic versus egalitarian societies. *Animal Behaviour*, **31**, 667–682.

Wcislo, W. T. (1997a) Are behavioral classifications blinders to natural variation? *In*: Choe, J. C. & Crespi, B (eds.) *The Evolution of Social Behavior in Insects and Arachnids*. Cambridge: Cambridge University Press, pp. 8–13.

(1997b) Behavioral environments of sweat bees (Hymenoptera: Halictinae) in relation to variability in social organization. *In*: Crespi, B. & Choe, J. C. (eds.). *The Evolution of Social Behaviour in Insects and Arachnids*. Cambridge: Cambridge University Press, pp. 316–322.

(2005) Social labels: We should emphasize biology over terminology and not vice versa. *Annales Zoologici Fennici*, **42**, 565–568.

Wheeler, W. M. (1911) The ant-colony as an organism. *Journal of Morphology*, **22**, 307–325.

(1928) *The Social Insects: Their Origin and Evolution*. New York: Harcourt, Brace.

Wilson, E. O. (1971) *The Insect Societies*. Cambridge: The Belknap Press of Harvard University Press.

(1975) *Sociobiology: The New Synthesis*. Cambridge: Belknap Press of Harvard University Press.

(1990) *Success and Dominance in Ecosystems: The Case of the Social Insects*. Oldendorf/Luhe, Germany: Ecology Institute.

(2012) *The Social Conquest of Earth*. New York: Liverlight Publishing Corporation.

Part I

Invertebrates

2 Sociality in Ants

Jürgen Heinze, Katrin Kellner, and Jon Seal

Overview

Ants (Formicidae) are one of the greatest successes in social evolution. Ants live in almost all terrestrial biomes, from rainforests to deserts and from cold to hot environments. Wherever they occur they abound and play key roles in ecosystems as predators of small invertebrates, in nutrient turnover, soil modification, seed dispersal, and more (e.g. Lach, *et al.*, 2009). All presently described species of ants – close to 14,000 by the end of 2015 (Agosti & Johnson 2005) – live in colonies, but their life histories vary enormously. It is futile to try condensing all that is worth knowing about the fascinating world of ants into a single chapter. Instead, our aim here is to highlight their diversity and how it evolved. Readers interested in a more complete review are referred to such informative volumes as *The Ants* (Hölldobler & Wilson, 1990), *Les Fourmis* (Passera & Aron, 2005), or *Ant Ecology* (Lach, *et al.*, 2009). Adapting our chapter to the structure of this book provided some difficulties because no solitary ants exist and many aspects of their life histories fit into several sections. We begin Part I with a general description of social life in ants, with emphasis on variation in queen number and its causes and consequences, and proceed to discuss the ecological and genetic causes of sociality. In Part II, we examine traits of ants that evolved with their sociality, such as longevity and communication, and then return to the stunning diversity of ant life histories.

I SOCIAL DIVERSITY

2.1 How Common is Sociality in Ants?

All ants are eusocial, i.e. they have overlapping generations, cooperative brood care, and division of reproductive labor (Wilson, 1971). Social life and division of labor among individuals highly specialized for different tasks underlie the enormous evolutionary and ecological success of ants. In contrast to solitary animals, individual ants are not forced to make compromises among foraging, avoiding predators, reproduction, and brood care.

We thank the editors and two referees for helpful comments on earlier drafts of this paper and DFG and the US National Science Foundation (IOS 0920138, DEB 1354629) for funding.

As in other social Hymenoptera (bees and wasps), ant societies are composed of one or several female reproductives ("queens") and their female helpers ("workers"), which both are diploid and develop from fertilized eggs. Ant males are haploid and arise from unfertilized eggs (haplodiploid sex determination, e.g. Cook, 1993; van Wilgenburg, *et al.*, 2006); they do not actively engage in social life, are short-lived, and die after inseminating one or a few mating partners (reviewed in Boomsma, *et al.*, 2005). Female reproductives store the sperm obtained during mating and use it for the fertilization of all female-destined eggs throughout their lives without ever re-mating.

2.2 Forms of Sociality in Ants

Whereas in bees and wasps, queens and workers often are similar in morphology, in ants the two castes may differ strikingly. Queens of many species are morphologically specialized for dispersal, mating, founding a new society, and egg laying, thus retaining the traits of their solitary ancestors: they have bulky wing muscles, four wings that they shed after mating, ocelli for navigation, and ovaries with a spermatheca for sperm storage and numerous ovarioles (Figure 2.1). Workers, in contrast, are more or less cheaply built all-purpose tools for non-reproductive tasks, wingless, with a slender thorax and less complex ovaries, if at all. With a few exceptions, caste is determined by the environment: depending upon external and social conditions, each fertilized egg can develop into a worker or a queen (Schwander, *et al.*, 2010). Both queens and workers may show additional within-caste variation, such as wing or size polyphenism in queens, and morphologically and functionally discrete types of workers (e.g. tiny "minor" workers, large-headed soldiers, etc., reviewed in Hölldobler & Wilson, 1990).

The ancestral ant colony presumably consisted of a simple family with a single, singly mated queen (i.e. monogyny and monandry, Hughes, *et al.*, 2008; Boomsma,

Figure 2.1. Founding queen of the ant *Pachycondyla inversa* carrying its eggs. Photo by D. Nash.

Figure 2.2. Workers and brood of *Platythyrea punctata.* The individual on top of the cocoon is a reproductive, which produces female offspring from unfertilized eggs through thelytokous parthenogenesis. All nestmates are clonally identical. Photo by B. Barth.

2013) and its non-reproductive worker daughters. Among extant ants, however, there is great variation in social structure and reproductive biology (Heinze, 2008). About 50 percent of ant species may have multiple queens per colony (i.e. polygyny) and in many species, queens can mate with multiple males (i.e. polyandry). In some species, all individuals – both queens and workers – can mate, store sperm, and lay fertilized eggs. Reproductives and non-reproductives may then be morphologically indistinguishable (Peeters, 1997) and who reproduces may be determined by physical aggression or dominance (Peeters & Higashi, 1989; Monnin & Peeters, 1999; Heinze, 2004). While the production of males from unfertilized eggs is the rule in species with haplodiploid sex determination, several ant species have additionally evolved thelytokous parthenogenesis and produce female offspring from unfertilized eggs (reviewed by Rabeling & Kronauer, 2013; Figure 2.2). Thelytoky may be combined with normal sexual reproduction, such that queens produce female sexuals that are clonally identical to themselves. In contrast, workers, which are exposed to environmental variability and hence benefit from genetic diversity, develop from fertilized eggs. In addition, males of a few of these species arise from fertilized eggs from which the maternal nucleus has been expelled. This leads to a complete separation of female and male lineages and to sib mating without the detrimental effects of inbreeding (Fournier, *et al.*, 2005; Ohkawara, *et al.*, 2006; Pearcy, *et al.*, 2011).

The number of queens per colony is tightly interrelated with many other life history traits of individual ants and whole colonies, such as queen morphology, dispersal, lifespan and fecundity, and colony structure (Hölldobler & Wilson, 1977). Much of this variation results from differences in mating behavior and the method of colony founding.

2.2.1 Solitary Colony Founding and Monogyny

In most monogynous species, solitary queens found new colonies independently without the help of workers. Young queens disperse on the wing and mate with one or several males during nuptial flights (Hölldobler & Bartz, 1985; Boomsma, *et al.*, 2005). After mating, they shed their wings, hide away in excavated or preformed cavities, and produce their first batch of workers. Founding queens of some species forage for food for their first offspring (semi-claustral founding, Brown & Bonhoeffer, 2003; Seal, 2009). The risks of colony failure through foraging mortality can be reduced when queens use their own fat and storage proteins (e.g. wing muscles) to nourish their first workers without leaving their nests again. This strategy of claustral founding is typically associated with an impressive size difference between queens and workers (e.g. Hölldobler & Wilson, 1990; Seal, 2009).

2.2.2 Pleometrosis, Dependent Founding and Polygyny

Unrelated queens may also team up for cooperative colony founding (i.e. pleometrosis). Foundress associations produce workers more quickly than solitary queens and, in semi-claustral species, the risk of foraging is spread among queens. Before or shortly after first workers have enclosed, foundresses may fight and kill one another until only one queen per nest survives (Bernasconi & Strassmann, 1999). Workers appear to support the most fecund queen (e.g. Holman, *et al.*, 2010a), thus maximizing the chance that their own mother survives (Forsyth, 1980). Pleometrosis can be stable in evolution when the likelihood of a cofounding queen becoming the sole survivor in an established colony is larger than the success rate of a solitarily founding queen.

Multiple foundresses of a few species persist when the colony has become mature and produce sexual offspring (i.e. primary polygyny, Mintzer, 1987; Trunzer, *et al.*, 1998). More frequently, the presence of multiple queens in mature nests results from queen adoption (i.e. secondary polygyny) where young queens do not disperse far for mating but instead attract males with sexual pheromones in the immediate vicinity of their natal nests. After mating, they return to their natal nest and begin to lay eggs attended by their nestmates (Hölldobler & Bartz, 1985; Boomsma, *et al.*, 2005). In many polygynous, but also several monogynous species, a few queens and workers may leave the nest and establish a new nest nearby (colony fission or budding, Peeters & Ito, 2001). This may lead to a highly viscous population structure with temporarily or permanently interconnected nests (polydomy, Debout, *et al.*, 2007).

While queen dominance and fighting appears to be the rule during cooperative founding, overt queen to queen conflict in mature polygynous colonies is restricted to a handful of species in which habitat patchiness greatly increases the value of the established nest (e.g. Bourke & Heinze, 1994).

2.2.3 Social Parasitism

An even further derived type of dependent colony founding is social parasitism, i.e. young queens usurp colonies of their own or another species (e.g. Buschinger, 1986,

2009; Tschinkel, 1996; Foitzik & Heinze, 2000; Savolainen & Vepsäläinen, 2003). Parasitic founding may be associated with slave raiding. Rather than engaging in the everyday duties of the societies, the parasitic workers invade neighboring colonies of their host species and pillage brood, which after eclosion increases the stock of host workers in their own nests (Buschinger, 1986, 2009). Social parasites and their hosts may engage in a complex arms race leading to adaptations and counter-adaptations, such as sophisticated mechanisms to kill, drive away, or confuse workers that defend the host colony and slave rebellion against the slave-makers (Lenoir, *et al.*, 2001; Brandt, *et al.*, 2005; Achenbach & Foitzik, 2009).

2.3 Why Ants Form Social Groups

As all extant ants are eusocial, it remains speculative as to why and how they evolved sociality more than 100 million years ago (LaPolla, *et al.*, 2013). The first ants probably lived in warm climates and individually hunted small arthropod prey in leaf litter (e.g. Wilson & Hölldobler, 2005; Moreau & Bell, 2013). Several exaptations may have facilitated the evolution of social life in the ancestors of ants. For example, ancestral females presumably would have mated with a single male early during their adult lives without re-mating. As in extant solitary bees and wasps, mothers may have provisioned their offspring in stable nests and defended them with their sting. The production of several broods per season would have allowed first-generation daughters to stay at the nest, take over foraging and care for the second-generation offspring, the dispersing sexuals, while their mother concentrated on brood production (e.g. Hunt & Amdam, 2005).

In an analogy to Haeckel's "ontogeny recapitulates phylogeny," the presumed ancestral way in which extant ants initiate new societies, semi-claustral founding, recapitulates this hypothetical scenario for the origin of ant sociality. Hence, all evidence points to an origin of eusociality from simple family groups of mothers and daughters (subsocial route) and emphasizes the importance of kinship in the origin of insect societies (Linksvayer, 2010). Inclusive fitness theory (Hamilton, 1964) provides the solid mathematical foundation for this scenario (Chapter 1). It states that altruistic helping can be stable in evolution when the fitness costs C to the helper are lower than the benefit B for the recipient of the help weighted by the relatedness r between helper and recipient (C less than rB). Helpers gain fitness indirectly from boosting their mother's output of male and female sexuals that carry copies of the workers' genes identical by descent. Several features of Hymenoptera may shift the cost-benefit ratio towards helping: the presence of a sting, a stable nest, and progressive provisioning all increase the benefits of helping. In addition, ancestral monogyny and the "lifelong pair bond" between a female and its mate stabilize genetic relationships among their offspring and guarantee the close relatedness crucial for the evolution of eusociality (Boomsma, *et al.*, 2005; Boomsma, 2007, 2013; Hughes, *et al.*, 2008).

It has been argued that the high relatedness among sisters resulting from haplodiploid sex determination ($r = 0.75$) might have facilitated the multiple evolution of eusociality in the Hymenoptera. However, high relatedness between workers and their sexual sisters is offset by a lower relatedness between workers and their brothers ($r = 0.25$).

Hence, the relatedness asymmetries created by haplodiploidy bear little significance on the multiple evolution of eusociality in the Hymenoptera (e.g. Bourke & Franks, 1995; Bourke, 2011; Gardner, *et al.*, 2012). However, haplodiploidy might have prepared the way for eusociality through other mechanisms. For example, haplodiploidy allows mothers to manipulate the cost-benefit ratio of helping for their daughters simply through not producing any males early in the season (Hunt, 2007; Gardner & Ross, 2013). Rather than searching in vain for mates, daughters could then increase their fitness indirectly by helping their mothers to raise additional sisters. Parental manipulation of offspring to help (e.g. Alexander, 1974; Michener & Brothers, 1974) also resurfaces in the recent notion of helper altruism being at least in part enforced. Through queen aggression or worker attacks against nestmates (policing), workers are prevented from selfishly producing their own sons from unfertilized eggs (Wenseleers & Ratnieks, 2006a).

Worker selfishness, such as dominance interactions or "selfish policing" of workers that lay eggs in the presence of the queen (Bourke, 1988; Heinze, 2004; Stroeymeyt, *et al.*, 2007), indicates that even workers of advanced eusocial insects may occasionally gain considerable direct fitness benefits. In small ant colonies, the likelihood of a worker producing sons may not be much smaller than that of its female sexual sister to produce male and female sexuals. To fully understand the evolution and maintenance of ant sociality, it might therefore be helpful to simultaneously consider the direct and the indirect component of inclusive fitness. Where worker reproduction is extremely unlikely, e.g. when colony size is very large or queen number is high, workers may lose their ovaries in evolution and completely rely on indirect fitness (e.g. Bourke, 1999).

2.3.1 Resource Acquisition and Use

A stable nest, be it in the soil or in preformed cavities in plants, is a valuable resource – a fortress in which the brood can be defended against predators (e.g. Andersson, 1984; Korb & Heinze, 2008). At the same time, it limits the range in which individuals can collect food. Like many of their solitary relatives, all ants are central place foragers (i.e. foragers bring back food to the nest). In species with small colony size, workers may forage solitarily. In species with large worker numbers, patchily distributed profitable food sources are quickly located and efficiently exploited and defended through chemical communication among workers. Groups of workers may also overwhelm prey, which is too large for an individual ant, and cooperatively carry it back to the nest (e.g. Anderson & Franks, 2001; Lach, *et al.*, 2009).

2.3.2 Predator Avoidance

Reproductive division of labor between the mother queen and its worker daughters minimizes external mortality for the reproductive: the queen is sheltered off from external mortality risks in the stable nest and can focus on reproduction, while workers take over the more dangerous tasks of colony defense and foraging. Though ancestral ants presumably could defend themselves with a potent sting, mortality risks through predation may have selected for very short mating flights and claustral or dependent colony founding (e.g. Brown & Bonhoeffer, 2003).

2.3.3 Homeostasis

In larger ant colonies, cooperation and division of labor helps to maintain an internal equilibrium. In the short-term, individuals may adjust their behavior to changed environmental or social conditions. For example, the removal of a particular size class of workers from a colony of leaf-cutter ants was balanced immediately by increased activity of the remaining individuals of this class (Wilson, 1983). Over longer time spans, feedback loops may help maintaining a stable ratio among different types of workers or preventing immature colonies from producing costly female sexuals. For example, pheromones of adult *Pheidole* soldiers inhibit the development of new soldiers from the brood, and queen pheromones of the fire ant *Solenopsis invicta* prevent the production of new female sexuals (reviewed in Hartfelder & Emlen, 2012). The increased investment in a needed group may also change caste ratios: *Pheidole* colonies produced more soldiers in the presence of alien workers (Passera, *et al.*, 1996). Finally, colonies of some species, such as *Formica* wood ants (Figure 2.3), may keep the temperature of their nests independent of ambient conditions, e.g. through active heating and optimizing nest architecture for insolation, ventilation, and insulation (Kadochová & Frouz, 2012).

2.3.4 Mating

Queens of ants, like those of many other Hymenoptera, mate only early during their adult lives without re-mating (Boomsma, *et al.*, 2005; Boomsma, 2007, 2013; Hughes,

Figure 2.3. Nest of a *Formica* wood ant. Throughout much of the year, wood ants are capable of maintaining the temperature in their nest rather stable by increasing their metabolism, basking in the sun, and re-arranging the shape of the mound to optimize insolation. Photo by J. N. Seal.

Figure 2.4. Nuptial flight of *Acromyrmex* in Southern Arizona, during which hundreds of winged males and female sexuals from dozens of different colonies meet for a short period of mating activity. Photo by J. Heinze.

et al., 2008). In extant ants with large colony sizes, mating may occur during highly synchronized nuptial flights, in which sexuals from hundreds of nests gather in large clouds (Figure 2.4). In species with smaller colony sizes, female sexuals may move away from their natal nests one by one and attract males with sexual pheromones. Both tactics minimize the risk of inbreeding, which in most social Hymenoptera leads to the production of sterile or inviable diploid males because of their typical haplodiploid mechanism of sex determination (Cook, 1993; van Wilgenburg, *et al.*, 2006).

In a minority of species, queens mate in the confined space of their natal nests. While ant males do not overtly fight for access to female sexuals during nuptial flights, the "seraglio situation" arising from mating in the nest (Hamilton, 1978) allows males to monopolize a harem of queens (Boomsma, *et al.*, 2005). Mating in the nest has indeed

Figure 2.5. Two wingless males and a winged queen of *Cardiocondyla elegans.* While wingless males engage in lethal fighting in other species of this genus, those of *C. elegans* tolerate each other's presence. Photo by J. C. Lenoir.

occasionally selected for males with bizarre morphology, behavior, and physiology, including mate guarding, lethal combat, and, in *Cardiocondyla*, prolonged longevity, life-long spermatogenesis and a changed reproductive pattern of mother queens (Kinomura & Yamauchi, 1987; Heinze & Hölldobler, 1993; Yamauchi, *et al.*, 2001, 2006; Figure 2.5). Mating in the nest may involve regular inbreeding, and several species apparently have evolved alternative sex determining mechanisms that do not lead to the production of costly diploid males (Schrempf, *et al.*, 2006).

Sperm from a single male often suffices to fertilize all the eggs a queen can lay, and the prolonged exposure to predation during multiple mating favors single mating (e.g. Whitcomb, *et al.*, 1973). Nevertheless, multiple mating may be beneficial in species with very large colony sizes, such as army ants and leaf cutter ants, as it provides queens with the enormous number of sperm needed to fertilize the eggs during their long lives (Cole, 1983). In addition, polyandry, like polygyny, may be beneficial under conditions that select for increased genetic diversity. Pathogens may more easily devastate large, genetically homogenous than genetically diverse colonies (Sherman, *et al.*, 1988; Keller & Reeve, 1994; Bourke & Franks, 1995; Crozier & Pamilo, 1996; Brown & Schmid-Hempel, 2003; Hughes & Boomsma, 2004; Reber, *et al.*, 2008). Furthermore, patrilines or matrilines may be differently prone to develop into particular phenotypes or be differently adept in specific tasks (Hughes, *et al.*, 2003; Fournier, *et al.*, 2008; Helanterä *et al.*, 2013). Hence, genetically diverse colonies may be more efficient. Indirect evidence for this comes from a positive association between colony growth rate and queen mating frequency in *Pogonomyrmex occidentalis* (Cole & Wiernasz, 1999). Similarly, the higher mating frequency of queens of *Lasius niger* in southern populations was explained by more complex tasks in more diverse habitats (Corley & Fjerdingstad, 2011).

2.3.5 Offspring Care

Parental care is one of the three traits defining eusociality. With the exception of the workers of social parasites, ant workers take care of their immature relatives, they feed and groom them and assist during eclosion. Though often considered helpless and inactive, larvae may play important roles in the society in providing services in food processing, the production of silk, and perhaps the secretion of substances that help to maintain an optimal brood-worker ratio (Mamsch 1967; Wilson & Hölldobler, 1980; Cassill, *et al.*, 2005). Furthermore, ant larvae may selfishly cannibalize eggs and thus actively engage in social conflict (Rüger, *et al.*, 2008; Schultner, *et al.*, 2014).

2.4 The Role of Ecology in Shaping Sociality in Ants

Ecology has long been recognized as an important driver of social evolution in that ecological constraints determine the cost-benefit ratio of helping in Hamilton's rule (e.g. Hamilton, 1964; Bourke & Franks, 1995; Korb & Heinze, 2008; Bourke, 2011). It remains unknown what originally sparked the evolution of eusociality in the ancestors of ants in the Mesozoic era, but advances have been made to link variation in extant colony phenotypes with ecology. Environmental constraints on dispersal, mating and colony founding, and the demands of specific ecological niches are strongly interwoven with the social and genetic colony structure. Furthermore, the ecological dominance of many extant ants appears to be due to their complex symbioses with bacteria, fungi, or sap-sucking insects. Below we explore both how ecology has influenced the evolution of sociality in ants, but also how sociality has influenced the ability of ants to colonize the globe.

2.4.1 Habitat and Environment

Several fundamental traits of ant sociality, in particular queen number, mating frequency, and colony size, evolve rapidly with environmental conditions (e.g. Ross & Carpenter, 1991). Harsh environmental conditions (e.g. cold or dry climate, nest site limitation) appear to reduce the success of dispersal and solitary colony founding and instead select for local mating and dependent founding (Herbers, 1993; Bourke & Franks, 1995). For example, in five of the six ant genera found near the tree-line in alpine and boreal habitats, queens found their new nests in a dependent way (Heinze, 1993). Dispersal flights and solitary founding have also been lost in many genera in hot and dry habitats (Tinaut & Heinze, 1992). Different levels of nest site availability and habitat patchiness might explain variation in the partitioning of reproduction among nestmate queens in multi-queen societies of the genus *Leptothorax.*

One particularly variable feature of social organization is colony size, which ranges from less than five workers in the specialized millipede hunter *Thaumatomyrmex* to the impressive army ants with hundreds of thousands of workers per colony (reviewed by Dornhaus, *et al.*, 2012). Ecologically subordinate species with small colonies and few workers that forage solitarily for small, scattered food items can thus coexist with dominant species with large numbers of workers that exploit and defend reliable resource patches (e.g. Savolainen, *et al.*, 1989). Group size again is strongly linked with life history traits of both individual queens and workers and the whole colony. For example, queen-worker polyphenism and the specialization of workers for particular tasks tend to increase with group size (e.g. Bourke, 1999).

It remains unclear which ecological features have promoted the convergent evolution of several other traits of ant sociality, such as the complete loss of worker reproduction, mating in the nest, or thelytokous parthenogenesis. Remarkably, however, a considerable percentage of species combining these traits are invasive species that have spread worldwide. Favored by the absence of natural enemies and pathogens some of them have become dominant in their new habitats. As invasive species often differ from native species in prey spectrum, interactions with plant-sap feeding insects, and seed dispersal, they may have a devastating effect on ecosystems. Some of them, such as the Argentine ants, *Linepithema humile,* red imported fire ants, *Solenopsis invicta*, and little fire ants, *Wasmannia auropunctata*, are listed among the 100 worst invasive species, on a par with tiger mosquitoes, black rats, or Japanese knotweed (reviewed by Holway, *et al.*, 2002; Tsutsui & Suarez, 2003; see also King & Tschinkel, 2008).

2.4.2 Biogeography

Ants have conquered almost all biogeographic regions of the world, with exception of high altitudes, the Antarctic, and the Arctic. With a little help from human commerce, ants have recently colonized even isolated islands, such as Hawaii, St. Helena, or Iceland, from which they had been absent before the arrival of humans (e.g. Ólafsson & Richter, 1985; Wetterer, *et al.*, 2007; Morrison, 2014). While ants are abundant wherever they occur, their diversity decreases tremendously with latitude (Kusnezov, 1957; Kaspari, *et al.*, 2000). This supports the hypothesis that ants evolved in moist, tropical rainforests rather than in steppe or taiga habitats. The paucity of ant fossils in Cretaceous amber suggests that the earliest ants were ecologically inconspicuous predators in tropical forest litter. They may have reached the level of dominance they have today only in the Eocene with the advent of higher plants and the evolution of mutualistic interactions with aphids and other plant-sap feeders (Wilson & Hölldobler, 2005; Moreau, *et al.*, 2006; but see Pie & Tschà, 2009).

Several features of the life history of ants appear to have evolved convergently in different biogeographic areas (e.g. the storing of nectar in the guts of specialized workers in Australian *Melophorus* and American *Myrmecocystus* or the replacement of queens by mated workers in some Ponerinae and *Myrmecia pyriformis*, Dietemann, *et al.*, 2004). Other life history traits (e.g. interspecific slave raiding or the cultivation of fungus on chewed leaves) appear to be restricted to particular biogeographic regions.

2.4.3 Niches

Ants inhabit a multitude of ecological niches from generalized predation to highly specialized diets or interactions with other arthropods, plants, fungi, or bacteria. They perform nearly every conceivable ecological interaction and play keystone roles as seed predators and dispersers, primary consumers, predators, ecosystem engineers, and biological indicators (e.g. Lach, *et al.*, 2009; King, *et al.*, 2013). Through soil excavation, semi-permanent trails and stable above- or below-ground nests, ants may impact plant growth or soil characteristics. In the following section, we discuss two of the more striking life styles of ants in order to highlight the multiple facets of their ecological importance: seed dispersal and their symbioses with other organisms (e.g. fungi and bacteria).

Seed dispersal. Of the large number of ants that consume seeds and have evolved interactions with plants, harvester ants are probably the most ecologically significant. Seed gathering has evolved convergently in several genera in deserts and other arid or semi-arid habitats. Species of *Messor* harvester ants occurring in the Mediterranean region presumably inspired both Solomon's proverb 6,6 ("go to the ant, thou sluggard...") as well as Aesop's fable about the ant and the grasshopper (Aesop, 2011). In America, *Pogonomyrmex* ants collect and store vast amounts of seeds; for example, a single colony of *P. badius* may store up to 500 g, or 300,000 individual, seeds in their subterranean nests (Tschinkel, 1999). As a result, their activities are of immense importance by altering plant communities and soil structure (MacMahon, *et al.*, 2000). Experiments have demonstrated that seed harvesting ants compete with conspecific colonies, colonies of other seed harvesting ant species and even granivorous rodents (Davidson, 1985).

Symbioses. Many ants have evolved symbioses with microorganisms, which greatly expand the possible niches these ants could otherwise occupy. For example, ants that directly or indirectly, via aphids and other sap-sucking insects, feed on carbohydrate-rich phloem sap, exudates, or nectar have evolved obligate symbioses with bacteria that may upgrade the otherwise poor diet by synthesizing essential amino acids or fixing atmospheric nitrogen (Feldhaar, *et al.*, 2007; Stoll, *et al.*, 2007). It is probably safe to say that all ant species likely harbor rich communities of bacteria that serve important dietary or defensive functions (Anderson, *et al.*, 2012; Kellner, *et al.*, 2015). Other examples include ants that interact to various degrees with fungi. Some ants forage and feed directly on fungi (Witte & Maschwitz, 2008) and others live inside a nest constructed from fungi (e.g. Schlick-Steiner, *et al.*, 2008). Fungus-gardening ants (tribe Attini) show the most derived symbiosis with fungi in that they grow their own food by farming specific strains of fungi in monoculture (Mueller, *et al.*, 2005). They forage exclusively on plant matter, which, depending upon the species, may include dead or fresh leaves and flowers or insect excrement.

The coadaptation between fungus-gardening ants and fungi is remarkable because the ants have lost the vast majority of their own digestive machinery and rely instead on the metabolic activities of their symbiotic fungus. The ants in return have become essentially an enzyme delivery system for the fungi, where the ants ingest fungus-derived enzymes, which pass through the ant gut unmodified and are then deposited by defecation on newly acquired fungal substrate (Erthal, *et al.*, 2009; de Fine Licht & Boomsma, 2010; Bacci, *et al.*, 2013). The ancestral lineages of ants grow fungi that have retained much of the physiological machinery found in other saprotrophic fungi and can readily metabolize cellulose and other plant fibers (de Fine Licht, *et al.*, 2010; Aylward, *et al.*, 2013). At the other extreme, the most derived ant lineages (e.g. the leaf-cutting ants) cultivate fungi which appear to have higher activity toward starches and pectins–carbohydrates that are common and easily accessible in their main food source (i.e. fresh vegetation, Seal, *et al.*, 2014). An emergent property of this specialization in leaf-cutting ants is that they produce vast amount of refuse depots in the form of undigested structural carbohydrates (Bacci, *et al.*, 1995; Abril & Bucher, 2002; de Fine Licht, *et al.*, 2010). Frequently, the refuse depots of leaf-cutting ants are rich in rare essential minerals (e.g. nitrogen, phosphorus, potassium, and sulfur) and are host to distinct microbial communities (Scott, *et al.*, 2010). As a result, these ants have notable effects on the local plant community and influences on the regeneration of tropical rainforests (Hölldobler & Wilson, 2011; Meyer, *et al.*, 2013). Species inhabiting seasonal environments are also known to move massive amounts of soil by digging deeper nests under extremely hot or cold conditions (Mueller, *et al.*, 2011; Seal & Tschinkel, 2006).

2.5 The Role of Evolutionary History in Shaping Sociality in Ants

Eusociality in ants has been remarkably stable during their long evolutionary history. Not a single ant species has reverted to a solitary life, even though a few have secondarily lost some of the defining traits of eusociality. For example, in several socially parasitic ant queens no longer produce workers and host workers from another species take over all non-reproductive tasks. Similarly, in a few thelytokous ants, such as *Pristomyrmex punctatus*, most workers produce female offspring from unfertilized eggs when they are young and become sterile foragers later in life (Tsuji, 1988). In contrast, many of the details of social life, such as queen number of mating frequency, appear to be quite flexible and to evolve quickly with changing environmental conditions, as will be shown below.

II SOCIAL TRAITS

As all ants are eusocial, it is difficult to identify traits that evolved specifically because of their social life. We here investigate traits that vary between social and solitary species of other taxa and show how they have been refined in ants.

2.6 Traits of Social Species

2.6.1 Cognition and Communication

The cohesiveness of groups relies on the resolution of internal conflict, the closure of their societies against parasites, and efficient communication (e.g. Maynard Smith & Szathmáry, 1995). Compared to non-social insects, ants have an extremely sophisticated system of communication, which goes much beyond luring conspecifics to mate or chasing away rivals (Hölldobler, 1995, 1999). Ant communication is predominantly based on pheromones, which they produce in dozens of glands in almost all body parts (Billen & Morgan, 1998; d'Ettorre & Moore, 2008). Occasionally modulated by vibrational and tactile stimuli, pheromones convey information about the location of food sources, mark a colony's nest and territory, and signal an individual's colony of origin, caste, age, and reproductive status (Le Conte & Hefetz, 2008).

Many ants can direct nestmates to food sources by laying a scent trail, which often not only leads the way but also encodes food quality (Hölldobler & Wilson, 1990). The dynamics of trail laying and recruitment allow ants to find shortcuts, optimize path length and travel time (Goss, *et al.*, 1989; Dussutour, *et al.*, 2004; Oettler, *et al.*, 2013), or to select the better of two alternative nest sites (Pratt, *et al.*, 2002). Through "swarm intelligence," i.e. positive and negative feedback loops driven by communication among large numbers of individuals with only local information, ants may be able to construct bridges, rafts, and air-conditioned, waterproof nests with networks of brood chambers and garbage dumps, connected to profitable food sites by an elaborate trail system on which traffic flows almost without any congestion (Bonabeau, *et al.*, 1999; Dussutour, *et al.*, 2004; Garnier, *et al.*, 2007). Algorithms based on the dynamics of ant pheromone trails are currently employed to solve optimization problems in human information technology and logistics (Dorigo & Stützle, 2004).

Queen pheromones communicate the presence of highly fertile queens and are thus of vital importance for conflict resolution and the regulation of reproduction in the society (Liebig, *et al.*, 2000; Monnin, 2006; Heinze & d'Ettorre, 2009; Kocher & Grozinger, 2011). There is now growing evidence that ant workers may respond to particular hydrocarbons in the chemical bouquet of queens by refraining from egg laying (Holman, *et al.*, 2010b). Applying a cuticular compound specific of fertile individuals on non-reproductives induces worker aggression against the besmeared individual (Smith, *et al.*, 2009).

Chemical communication also serves to regulate non-reproductive tasks. Individual workers may differ in their propensity to take over specific tasks, either through age (e.g. Beshers & Fewell, 2001), experience (e.g. Ravary, *et al.*, 2007), or genetic predisposition (e.g. Hughes, *et al.*, 2003; Fournier, *et al.*, 2008; Helanterä, *et al.*, 2013). However, division of labor is most strongly affected by communication among individuals (e.g. Gordon, 1996; Pratt, *et al.*, 2002; Greene & Gordon, 2003; Gordon, *et al.*, 2008; Pinter-Wollman, *et al.*, 2013).

2.6.2 Longevity and Lifespan

Evolutionary theories of aging predict that intrinsic mortality evolves in response to the risks of dying from external hazards, predators, or pathogens. In many social animals, helpers shield the reproductives from external risks. Hence, sociality is expected to delay senescence and aging in reproductives (Keller & Genoud, 1997; Bourke, 2007; Heinze & Schrempf, 2008). Queens of ants and other perennial social animals, including termites (Chapter 5) and the naked mole-rat (Chapter 10), are extraordinarily long-lived compared to related solitary species and non-reproductive conspecifics. Ant queens live for up to 30 years, while the life expectancy of ant workers, depending upon species, ranges from a few weeks to a few years (Keller & Genoud, 1997; Heinze & Schrempf, 2008). Reproductives by far outlive their non-reproductive nestmates even when they do not differ in ontogeny or morphology (Tsuji, *et al.*, 1996; Hartmann & Heinze, 2003). The fecundity/longevity trade-off commonly observed in solitary insects therefore does not exist in ants, at least not on the level of the individual, as the costs of reproduction are born by the whole society.

In solitary insects, mating may have a negative impact on female life span. For example, the seminal fluids of *Drosophila* males contain substances, which prevent females from re-mating and force them to invest all available resources into the offspring resulting from the actual mating (sexual conflict, Chapman, *et al.*, 2003). Because ant queens do not re-mate later in life and first produce large numbers of workers before switching to the production of sexual offspring, ant males would not benefit from harming their mates. Instead, they might have evolved mechanisms that improve the physical condition of their mating partners ("sexual cooperation," Schrempf, *et al.*, 2005). Indeed, mating appears to increase the longevity of queens: queens of *C. obscurior* that mated with normal or sterilized males lived considerably longer than virgin queens (Schrempf, *et al.*, 2005).

A monogynous colony quickly declines when its queen dies. In contrast, polygynous colonies continuously recruit young queens and are potentially immortal. Queens from polygynous species are usually less long-lived and less fecund than queens from monogynous relatives (Keller & Genoud, 1997). Between species, this might reflect different extrinsic mortality levels, since many polygynous species inhabit ephemeral nest sites and frequently move. However, assigning young queens of *C. obscurior* to either single- or multi-queen colonies led to the same result: queens in eight-queen associations had a shorter life span and laid less eggs than queens in one-queen colonies (Schrempf, *et al.*, 2011). This suggests a social influence on longevity independent of external mortality.

The proximate mechanisms underlying delayed senescence in queens are not yet understood. At present, there is little support for the involvement of standard mechanisms of prevention of damage with age (e.g. antioxidants, Parker, 2010). However, like in honeybees (Münch, *et al.*, 2008), the insulin-like signaling pathway and vitellogenin might be involved in the regulation of both fecundity and longevity. In *Cardiocondyla* queens, gene expression patterns change with age in the opposite direction to age-related changes in solitary *Drosophila* (von Wyschetzki, *et al.*, 2015).

2.6.3 Fecundity

Considering only monogynous taxa, queen fecundity and colony size are obviously tightly interrelated, and as colony size varies, so does the queen's egg-laying capacity. In polygynous colonies, individual fecundity is lower than in monogynous colonies but this is compensated by higher queen number. Nestmate queens usually contribute more or less equally to the colony offspring and only in a few cases form social and reproductive hierarchies. Maximal differences in reproduction (high reproductive skew) are found in a few species that are "functionally monogynous," i.e. only one of several potential egg layers lays eggs. In accordance with theoretical models, how queens share reproduction is associated with the success of dispersal and solitary founding. Queens of *Leptothorax* spp. in extended coniferous forests with a homogeneous distribution of suitable nest sites tolerate each other's presence and reproductive skew is low. In contrast, in habitats, where suitable nest sites are patchily distributed and less abundant, queens establish rank orders and only the top-ranking individual lays eggs (Bourke & Heinze, 1994; Heinze, 2004).

2.6.4 Age at First Reproduction

Age at first reproduction – here referring only to the production of sexuals – varies among species, depending upon the size of mature colonies and queen number. While solitarily founding queens have to build up a large workforce before switching to the production of sexuals, queens that lay eggs in their natal nest or are otherwise assisted by workers very quickly begin with the production of sexuals. While this is obvious in the case of workerless social parasites, data on the reproductive trajectory of individual queens in multi-queen societies are currently lacking.

2.6.5 Dispersal

Queen mating and dispersal tactics are intricately linked with numerous features of the phenotypes of individual queens and the complete colony. With dependent founding and the loss of long-range dispersal, queens no longer need large flight muscles and rich body reserves for rearing their first offspring and can very quickly begin with the production of sexual offspring. Consequently, their morphology may lack specific adaptations for dispersal; in some species, they are short-winged or even wingless and "worker-like" in appearance (Peeters, 1991; Heinze, 1998). Ant species in which colony organization and founding tactics vary with environment or genetic composition may help to elucidate the ultimate and proximate factors underlying the various traits associated with queen dispersal (Heinze & Keller, 2000; Peeters & Ito, 2001).

2.7 Traits of Social Groups

2.7.1 Genetic Structure

As formalized by inclusive fitness theory (Hamilton, 1964), the evolutionary stability of worker altruism in eusocial animals requires that it benefits relatives. Most ant societies indeed consist of more or less complex families. Nevertheless, because of the tremendous variation in queen number and mating frequency described earlier, relatedness varies from just above zero in highly polygynous species to one in clonal ants with thelytokous parthenogenesis. While lower relatedness may lead to improved colony-level resistance against pathogens or an enhanced division of labor, it affects the indirect fitness gains of workers and therefore changes their interests concerning the maternity of male sexuals, investment ratio in colony maintenance versus reproduction, and sex ratio (Bourke & Franks, 1995; Crozier & Pamilo, 1996).

Workers, who have retained ovaries and are capable of producing males, would be most closely related to their own sons ($r = 0.5$) and, in a monogynous, monandrous society, also more closely related to the sons of other workers ($r = 0.375$) than to their own brothers ($r = 0.25$). All else being equal, natural selection should thus promote worker reproduction. However, with polygyny and polyandry, the average relatedness of workers to worker-produced males may drop below that of workers to their brothers. In such cases, workers are predicted to oppose worker reproduction (Ratnieks, 1988). Surprisingly, whether workers lay eggs or not appears to be relatively insensitive to the genetic composition of the colony: they typically refrain from reproduction in the presence of the queen(s) (Hammond & Keller, 2004; Heinze, 2004; but see Wenseleers & Ratnieks, 2006b). Workers that are experimentally allowed to lay eggs are "policed," i.e. attacked by other workers, and their eggs are destroyed, probably because egg laying by selfish workers introduces costs at the colony level (Ratnieks, 1988; Ratnieks & Reeve, 1992). Reproductive workers do not take part in costly non-reproductive tasks and aggressively compete with other workers for egg laying rights. Worker reproduction thus might decrease the average inclusive fitness of workers. This selects for the abandonment of selfish egg laying and mutual policing, hence, "enforced altruism" (Wenseleers & Ratnieks, 2006a).

Both multiple queening and multiple mating open the possibility of workers favoring female larvae of their own genetic lineage during brood care and thus manipulation of the contribution of individual mothers or fathers to the sexual offspring. Surprisingly, only a few studies could substantiate nepotism in the context of brood care (e.g. Hannonen & Sundström, 2003), probably because the constant exchange of odor cues among nestmates prevents the exact discrimination between more or less closely related kin (van Zweden, *et al.*, 2010). Here again, through incomplete information, individuals might suffer reduced fitness for the benefit of the whole colony (Queller & Strassmann, 2013).

2.7.2 Group Structure, Breeding Structure and Sex Ratio

One of the best-documented examples for an effect of variation in relatedness on breeding structure and within-group conflict comes from research on sex allocation. As haplodiploid sex determination creates relatedness asymmetries of workers to female and male sexuals, queens and workers may have different preferences concerning the production of female and male sexuals (Trivers & Hare, 1976). While the queen is equally related to its sons and daughters and prefers a balanced sex investment ratio of 1:1, workers in a monogynous, monandrous colony are three times more closely related to their sisters than their brothers. Workers may achieve their favorite investment ratio of 3:1 by cannibalizing male larvae or biasing the development of female larvae by overfeeding them (Sundström, *et al.*, 1996; Hammond, *et al.*, 2002). Polygyny and polyandry reduce the relatedness asymmetry and thus attenuate the queen-worker conflict about sex allocation (Bourke & Franks, 1995; Crozier & Pamilo, 1996). Intraspecific variation in queen number or mating frequency may lead to split sex ratios (Boomsma, 1996). Though numerous studies have demonstrated the importance of variation in relatedness asymmetries for sex allocation, other factors, such as resource limitation, local resource competition, or local mate competition may be similarly or even more relevant than exact relatedness (e.g. Cremer & Heinze, 2002; Pearcy & Aron, 2006).

Over the last centuries, human activities have opened what might be a new chapter in ant evolution. Concealed in potted plants, fruit boxes, or packaging material, colonies of numerous ant species have been introduced accidentally into new ecosystems. Many invasive species have lost colony boundaries and form supercolonies with hundreds of thousands of nests scattered over dozens or even hundreds of kilometers, among which individuals can be exchanged freely (e.g. Helanterä, *et al.*, 2009). Many supercolonies originate from one or a few imported ancestor colonies, and the depletion of genetic diversity associated with these founder effects may underlie the loss of nestmate discrimination (Tsutsui, *et al.*, 2000; but see Giraud, *et al.*, 2002). Uniformity of colony odor allows invasive species to reach enormous densities and greatly promotes their successful spread at the cost of the native ant community.

The evolution of supercolonies resembles the transitions from unicellular to multicellular organisms and from solitary to social animals. Relatedness is relative to a reference population in which competition occurs, and though many supercolonies are family-based, nestmate relatedness may be essentially zero. At low relatedness, altruism is no longer favored by selection, and it remains to be seen whether supercolonies are an evolutionary cul-de-sac (Queller, 2000; Helanterä, *et al.*, 2009) or a next major transition in social evolution. Meanwhile, however, the ecological damage caused by invasive ants and the global homogenization of ant faunas convincingly illustrates the global importance of ants as key stone species.

III SOCIAL SYNTHESIS

2.8 A Summary of Ant Sociality

Our review documents that ants are one of the most diverse taxa of social animals and that their social evolution has hundreds of different aspects. There is almost no trait in ants – other than sociality itself – that does not show some variation, be it breeding structure, life span, sex ratio, or communication. From an ancestral life style of solitary foragers in rainforest litter ants have spread to such diverse niches as granivory, fungivory, and specialized predation on collembolans or ant brood. Ants have invented complex, air-conditioned nests, agriculture, animal husbandry, and slave-raids millions of years before *Homo sapiens* appeared on the scene.

This enormous diversity of life histories makes ants highly suitable model organisms for the study of numerous fundamental problems in biology beyond social evolution.

Even though ants, like bees and wasps, may appear somewhat exotic because of their haplodiploid sex determination and eusociality, they feature prominently in studies of ecological competition, the channels of communication, the spread of diseases in networks, the pace of senescence, and genome evolution.

2.9 Comparative Perspectives on Ant Sociality

In comparison to other taxa that exhibit the whole range of solitary to social species, studying social evolution in ants is restricted to explaining the variation of group structures, the occurrence and resolution of kin conflict. Despite recurrent misunderstandings and criticisms (e.g. Wilson, 2012), all available data support Hamilton's theory of inclusive fitness as solid foundation of social evolution and highlight the importance of relatedness (e.g. Dawkins, 1979; Foster, *et al.*, 2006; Abbot, *et al.*, 2011; Rousset & Lion, 2011; Bourke, 2011). With rare exceptions, such as foundress associations, ant societies consist of related individuals, and though measuring relatedness alone often fails to explain colony-level traits, taking all three parameters of Hamilton's equation into account provides a robust basis for explaining ant sociality.

The societies of ant species, in which reproductives and non-reproductives are identical in ontogeny and morphology, may be remarkably similar in structure to those of social vertebrates. Division of labor may rely on rank order, and policing maintains the cohesiveness of the group. Like in vertebrate societies (Singh & Boomsma, 2015), police workers often rank high in the colony's hierarchy and may obtain the breeding position when the old alpha is removed (e.g. Brunner, *et al.*, 2009; but see Monnin, *et al.*, 2002).

The average individual's probability of obtaining direct fitness decreases with the evolution of pronounced caste polyphenism and the loss of a spermatheca in the workers. The structure of ant colonies, in which workers can only produce sons from unfertilized eggs, is often similar to that of honeybees and higher termites. Like in the honeybees (Ratnieks, 1988), worker selfishness may still be visible in such ant societies, e.g. through young individuals refraining from costly tasks (Bourke, 1988) or occasionally sneaking unfertilized, male-destined eggs onto the egg pile (e.g. Wenseleers, *et al.*, 2006b). This documents that options for direct fitness affect the structure of most societies throughout the animal kingdom (Korb & Heinze, 2016).

The complete loss of ovaries in a few ant genera removes the conflict about reproduction. Even though other conflicts may still exist, the interests of all individuals widely converge and the societies achieve a "superorganismic" integration that most closely resembles that of cells in a multicellular organism or of polyps in a Portuguese man o' war (e.g. Hölldobler & Wilson, 2008; Bourke, 2011; Hunt, 2012; Korb & Heinze, 2016).

2.10 Concluding Remarks

Although some of the mechanisms that link ecological, social, and genetic traits in ants have been analyzed in detail (e.g. Keller & Passera, 1989; Stille 1996; Keller &

Genoud, 1997; Peeters & Ito, 2001), many other aspects remain unclear. It has long been recognized that a general life history theory of social insects is missing (e.g. Bourke & Franks, 1995; Starr, 2006; see also Chapters 1 and 13). Similarly, though important advances have been made to unravel the endocrine, genomic and epigenetic pathways underlying caste polyphenism, our understanding of "socio-evo-devo" (e.g. Toth & Robinson, 2007; Khila & Abouheif, 2010) is still far from complete. Several ant genomes have recently become available (Gadau, *et al.*, 2012) and comparative analyses might help to better understand what makes a worker a worker or a queen a queen (e.g. Schrader, *et al.*, 2015). Even in traditional fields of ant research, which after decades of research were thought to hold few secrets, new and fascinating observations can be made. This is documented by the recent discoveries of no-entry signals in ant trails (e.g. Robinson, *et al.*, 2005), the precocious production of males in species with local mate competition (e.g. Yamauchi, *et al.*, 2006), colonies building rafts to survive flooding (Mlot, *et al.*, 2011), or the controlled gliding behavior of ants knocked off tree branches (Yanoviak, *et al.*, 2005). Thus, after thousands of years of ant studies, from the Greek philosophers to present-day myrmecologists, the secrets of ants still have not been fully unveiled.

References

Abbot, P., Abe, J., Alcock, J., *et al.* (2011) Inclusive fitness theory and eusociality. *Nature*, **471**, E1–E4.

Abril, A. B. & Bucher, E. H. (2002) Evidence that the fungus cultured by leaf-cutting ants does not metabolize cellulose. *Biology Letters*, **5**, 325–328.

Achenbach, A. & Foitzik, S. (2009) First evidence for slave rebellion: Enslaved ant workers systematically kill the brood of their social parasite *Protomognathus americanus*. *Evolution*, **63**, 1068–1075.

Aesop (2011) *Aesop's Fables*. London: Penguin Classics.

Agosti, D., & Johnson, N. F. (editors) (2005) *Antbase*. World Wide Web electronic publication. antbase.org, version *(05/2005)*.

Alexander, R. D. (1974) The evolution of social behavior. *Annual Review of Ecology and Systematics*, **5**, 325–383.

Anderson, C. & Franks, N. R. (2001) Teams in animal societies. *Behavioral Ecology*, **12**, 535–540.

Anderson, K. E., Russell, J. A., Moreau, C. S., *et al.* (2012) Highly similar microbial communities are shared among related and trophically similar ant species. *Molecular Ecology*, **21**, 2282–2296.

Andersson, M. (1984) The evolution of eusociality. *Annual Review of Ecology and Systematics*, **15**, 165–189.

Aylward, F. O., Burnum-Johnson, K. E., Tringe, S. G., *et al.* (2013) *Leucoagaricus gongylophorus* produces diverse enzymes for the degradation of recalcitrant plant polymers in leaf-cutter ant fungus gardens. *Applied and Environmental Microbiology*, **79**, 3770–3778.

Bacci, M., Anversa, M. M., & Pagnocca, F. C. (1995) Cellulose degradation by *Leucocoprinus gongylophorus*, the fungus cultured by the leaf-cutting ant *Atta sexdens rubropilosa*. *Antonie van Leeuwenhoek Journal of Microbiology*, **67**, 385–386.

Bacci, M. J., Bueno, O. C., Rodrigues, A., *et al.* (2013) A metabolic pathway assembled by enzyme selection may support herbivory of leaf-cutter ants on plant starch. *Journal of Insect Physiology*, **59**, 525–531.

Bernasconi, G. & Strassmann, J. E. (1999) Cooperation among unrelated individuals: The ant foundress case. *Trends in Ecology and Evolution*, **14**, 477–482.

Beshers, S. N. & Fewell, J. W. (2001) Models of division of labor in social insects. *Annual Review of Entomology*, **46**, 413–440.

Billen, J. & Morgan, E. D. (1998) Pheromone communication in social insects: Sources and secretions. *In:* Vander Meer, R. K., Breed, M. D., Espelie, K. E., & Winston, M. L. (eds). *Pheromone Communication in Social Insects*. Boulder, CO: Westview Press, pp. 3–33.

Bonabeau, E., Dorigo, M., & Theraulaz, G. (1999) *Swarm Intelligence: From Natural to Artificial Systems*. Oxford: Oxford University Press.

Boomsma, J. J. (1996) Split sex ratios and queen–male conflict over sperm allocation. *Proceedings of the Royal Society of London B*, **263**, 697–704.

(2007) Kin selection versus sexual selection: Why the ends do not meet. *Current Biology*, **17**, R673–R683.

(2013) Beyond promiscuity: Mate-choice commitments in social breeding. *Philosophical Transactions of the Royal Society of London B*, **368**, 20120050.

Boomsma, J. J., Baer, B. & Heinze, J. (2005) The evolution of male traits in social insects. *Annual Review of Entomology*, **50**, 395–420.

Bourke, A. F. G. (1988) Dominance orders, worker reproduction, and queen–worker conflict in the slave-making ant *Harpagoxenus sublaevis*. *Behavioral Ecology and Sociobiology*, **23**, 323–333.

(1999) Colony size, social complexity and reproductive conflict in social insects. *Journal of Evolutionary Biology*, **12**, 245–257.

(2011) *Principles of Social Evolution*. Oxford: Oxford University Press.

(2007) Kin selection and the evolutionary theory of aging. *Annual Review of Ecology, Evolution and Systematics*, **38**, 103–128.

Bourke, A. F. G. & Franks, N. R. (1995) *Social Evolution in Ants*. Princeton, N.J.: Princeton University Press.

Bourke, A. F. G. & Heinze, J. (1994) The ecology of communal breeding: The case of multiply-queened leptothoracine ants. *Philosophical Transactions of the Royal Society B*, **345**, 359–372.

Brandt, M., Foitzik, S., Fischer-Blass, B. & Heinze, J. (2005) The coevolutionary dynamics of obligate ant social parasite systems — between prudence and antagonism. *Biological Reviews*, **80**, 251–267.

Brunner, E., Kellner, K., & Heinze, J. (2009) Policing and dominance behaviour in the parthenogenetic ant *Platythyrea punctata*. *Animal Behaviour*, **78**, 1427–1431.

Brown, M. J. F. & Bonhoeffer, S. (2003) On the evolution of claustral colony founding in ants. *Evolutionary Ecology Research*, **5**, 305–313.

Brown, M. J. F. & Schmid-Hempel, P. (2003) The evolution of female multiple mating in social Hymenoptera. *Evolution*, **57**, 2067–2081.

Buschinger, A. (1986) Evolution of social parasitism in ants. *Trends in Ecology and Evolution*, **1**, 155–160.

(2009) Social parasitism among ants: A review (Hymenoptera: Formicidae). *Myrmecological News*, **12**, 219–235.

Cassill, D. L., Butler, J., Vinson, S. B., & Wheeler, D. E. (2005) Cooperation during prey digestion between workers and larvae in the ant, *Pheidole spadonia*. *Insectes Sociaux*, **52**, 339–343.

Chapman, T., Arnqvist, G., Bangham, J., & Rowe, L. (2003) Sexual conflict. *Trends in Ecology and Evolution*, **18**, 41–47.

Cole, B. J. (1983) Multiple mating and the evolution of social behavior in the Hymenoptera. *Behavioral Ecology and Sociobiology*, **12**, 191–201.

& Wiernasz, D. C. (1999) The selective advantage of low relatedness. *Science*, **285**, 891–893.

Cook, J. M. (1993) Sex determination in the Hymenoptera: A review of models and evidence. *Heredity*, **71**, 421–435.

Corley, M. & Fjerdingstad, E. J. (2011) Mating strategies of queens in *Lasius niger* ant: Is environment type important? *Behavioral Ecology and Sociobiology*, **65**, 889–897.

Cremer, S. & Heinze, J. (2002) Adaptive production of fighter males: Queens of the ant *Cardiocondyla* adjust the sex ratio under local mate competition. *Proceedings of the Royal Society of London B*, **269**, 417–422.

Crozier, R. H. & Pamilo, P. (1996) *Evolution of Social Insect Colonies*. Oxford: Oxford University Press.

Davidson, D. W. (1985) An experimental study of diffuse competition in harvester ants. *The American Naturalist*, **125**, 500–505.

Dawkins, R. (1979) Twelve misunderstandings of kin selection. *Zeitschrift für Tierpsychologie*, **51**, 184–200.

Debout, G., Schatz, B., Elias, M., & McKey, D. (2007) Polydomy in ants: What we know, what we think we know, and what remains to be done. *Biological Journal of the Linnean Society*, **90**, 319–348.

de Fine Licht, H. H. & Boomsma, J. J. (2010) Forage collection, substrate preparation, and diet composition in fungus-growing ants. *Ecological Entomology*, **35**, 259–269.

de Fine Licht, H. H., Schiøtt, M., Mueller, U. G., & Boomsma, J. J. (2010) Evolutionary transitions in enzyme activity of ant fungus gardens. *Evolution*, **64**, 2055–2069.

d'Ettorre, P. & Moore, A. J. (2008) Chemical communication and the coordination of social interactions in insects. *In:* d'Ettorre, P & D. P. Hughes, D. P. (eds). *Sociobiology of Communication: An Interdisciplinary Perspective*. Oxford: Oxford University Press, pp. 81–96.

Dietemann, V., Peeters, C. & Hölldobler, B. (2004) Gamergates in the Australian ant subfamily Myrmeciinae. *Naturwissenschaften*, **91**, 432–435.

Dorigo, M. & Stützle, T. (2004) *Ant Colony Optimization*. Boston, MA: MIT Press.

Dornhaus, A., Powell, S. & Bengston, S. (2012) Group size and its effects on collective organization. *Annual Review of Entomology*, **57**, 123–141.

Dussutour, A., Fourcassié, V., Helbing, D., & Deneubourg, J. L. (2004) Optimal traffic organization in ants under crowded conditions. *Nature*, **428**, 70–73.

Erthal, M., Silva, C., Cooper, R., & Samuels, R. (2009) Hydrolytic enzymes of leaf-cutting ant fungi. *Comparative Biochemistry and Physiology B*, **152**, 54–59.

Feldhaar, H., Straka, J., Krischke, M., *et al.* (2007) Nutritional upgrading for omnivorous carpenter ants by the endosymbiont *Blochmannia*. *BMC Biology*, **5**, 48.

Foitzik, S. & Heinze, J. (2000) Intraspecific parasitism and split sex ratios in a monogynous and monandrous ant (*Leptothorax nylanderi*). *Behavioral Ecology and Sociobiology*, **47**, 424–431.

Forsyth, A. (1980) Worker control of queen density in hymenopteran societies. *The American Naturalist*, **116**, 895–898.

Foster, K. R., Wenseleers, T., Ratnieks, F. L. W., & Queller, D. C. (2006) There is nothing wrong with inclusive fitness. *Trends in Ecology and Evolution*, **21**, 599–600.

Fournier, D., Estoup, A., Orivel, J., *et al.* (2005) Clonal reproduction by males and females in the little fire ant. *Nature*, **435**, 1230–1234.

Fournier, D., Battaille, G., Timmermans, I. & Aron, S. (2008) Genetic diversity, worker size polymorphism and division of labour in the polyandrous ant *Cataglyphis cursor*. *Animal Behaviour*, **75**, 151–158.

Gadau, J., Helmkampf, M., Nygaard, S., *et al.* (2012) The genomic impact of 100 million years of social evolution in seven ant species. *Trends in Genetics*, **28**, 14–21.

Gardner, A. & Ross, L. (2013) Haplodiploidy, sex-ratio adjustment, and eusociality. *The American Naturalist*, **181**, E60–E67.

Gardner, A., Alpedrinha, J. & West, S. A. (2012) Haplodiploidy and the evolution of eusociality: Split sex ratios. *The American Naturalist*, **179**, 240–256.

Garnier, S., Gautrais, J. & Theraulaz, G. (2007) The biological principles of swarm intelligence. *Swarm Intelligence*, **1**, 3–31.

Giraud, T., Pedersen, J. S., & Keller, L. (2002) Evolution of supercolonies: The Argentine ants of southern Europe. *Proceedings of the National Academy of Sciences USA*, **99**, 6075–6079.

Gordon, D. M. (1996) The organization of work in social insect colonies. *Nature*, **380**, 121–124.

Gordon, D. M., Holmes, S. & Nacu, S. (2008) The short-term regulation of foraging in harvester ants. *Behavioral Ecology*, **19**, 217–222.

Goss, S., Aron, S., Deneubourg, J. L., & Pasteels, J. M. (1989) Self-organized shortcuts in the Argentine Ant. *Naturwissenschaften*, **76**, 579–581.

Greene, M. J. & Gordon, D. M. (2003) Social insects: Cuticular hydrocarbons inform task decisions. *Nature*, **432**, 32.

Hamilton, W. D. (1964) The genetical evolution of social behaviour. *I & II. Journal of Theoretical Biology*, **7**, 1–52.

(1978) Evolution and diversity under bark. *In:* Mound, L. A. & Waloff, N. (eds.) *Diversity of Insect Faunas*. Oxford: Blackwell Scientific, pp. 154–175.

Hammond, R. L. & Keller, L. (2004) Conflict over male parentage in social insects. *PLoS Biology*, **2**, e248.

Hammond, R. L., Bruford, M. W. & Bourke, A. F. G. (2002) Ant workers selfishly bias sex ratios by manipulating female development. *Proceedings of the Royal Society of London B*, **269**, 173–178.

Hannonen, M. & Sundström, L. (2003) Sociobiology: Worker nepotism among polygynous ants. *Nature*, **421**, 910.

Hartfelder, K. & Emlen, D. J. (2012) Endocrine control of insect polyphenism. *In:* Gilbert, L. I. (ed.) *Insect Endocrinology*. London: Academic Press, pp. 651–703.

Hartmann, A. & Heinze, J. (2003) Lay eggs, live longer: Division of labor and life span in a clonal ant species. *Evolution*, **57**, 2424–2429.

Heinze, J. (1993) Life history strategies of subarctic ants. *Arctic*, **46**, 354–358.

(1998) Intercastes, intermorphs, and ergatoid queens: Who is who in ant reproduction? *Insectes Sociaux*, **45**, 113–124.

(2004) Reproductive conflict in insect societies. *Advances in the Study of Behavior*, **34**, 1–57.

(2008) The demise of the standard ant. *Myrmecological News*, **11**, 9–20.

Heinze, J. & d'Ettorre, P. (2009) Honest and dishonest communication in social Hymenoptera. *Journal of Experimental Biology*, **212**, 1775–1779.

Heinze, J. & Hölldobler, B. (1993) Fighting for a harem of queens: Physiology of reproduction in *Cardiocondyla* male ants. *Proceedings of the National Academy of Sciences USA*, **90**, 8412–8414.

Heinze, J. & Keller, L. (2000) Alternative reproductive strategies: A queen perspective in ants. *Trends in Ecology and Evolution*, **15**, 508–512.

Heinze, J. & Schrempf, A. (2008) Aging and reproduction in social insects. *Gerontology*, **54**, 160–167.

Helanterä, H., Strassmann, J. E., & Queller, D. C. (2009) Unicolonial ants: Where do they come from, what are they and where are they going? *Trends in Ecology and Evolution*, **24**, 341–349.

Helanterä, H., Aehle, O., Roux, M., Heinze, J., & d'Ettorre, P. (2013) Family-based guilds in the ant *Pachycondyla inversa*. *Biology Letters*, **9**, 20130125.

Herbers, J. M. (1993) Ecological determinants of queen number in ants. *In:* Keller, L. (ed.) *Queen Number and Sociality in Insects*. Oxford: Oxford University Press, pp. 262–293.

Hölldobler, B. (1995) The chemistry of social regulation: Multicomponent signals in ant societies. *Proceedings of the National Academy of Sciences USA*, **92**, 19–22.

(1999) Multimodal signals in ant communication. *Journal of Comparative Physiology A*, **184**, 129–141.

Hölldobler, B. & Bartz, S. H. (1985) Sociobiology of reproduction in ants. In Hölldobler, B. & Lindauer, M. (eds.) *Experimental Behavioral Ecology and Sociobiology*. Stuttgart: Gustav Fischer, pp. 237–257.

Hölldobler, B. & Wilson, E. O. (1977) The number of queens: An important trait in ant evolution. *Naturwissenschaften*, **64**, 8–15.

(1990) *The Ants*. Cambridge, MA: Harvard University Press.

(2008) *The Superorganism: The Beauty, Elegance, and Strangeness of Insect Societies*. New York, N.Y.: W.W. Norton & Co.

(2011) *The Leafcutter Ants*. New York, NY: W.W. Norton.

Holman, L., Dreier, S. & d'Ettorre, P. (2010a) Selfish strategies and honest signalling: Reproductive conflicts in ant queen associations. *Proceedings of the Royal Society of London B*, **277**, 2007–2015.

Holman, L., Jørgensen, C. G., Nielsen, J., & d'Ettorre, P. (2010b) Identification of an ant queen pheromone regulating worker sterility. *Proceedings of the Royal Society of London B*, **277**, 3793–3800.

Holway, D. A., Lach, L., Suarez A. V., Tsutsui, N. D., & Case, T. D. (2002) The causes and consequences of ant invasions. *Annual Review of Ecology and Systematics*, **33**, 181–233.

Hughes, W. O. H & Boomsma, J. J. (2004) Genetic diversity and disease resistance in leaf-cutting ant societies. *Evolution*, **58**, 1251–1260.

Hughes, W. O. H., Sumner, S., Van Borm, S., & Boomsma, J. J. (2003) Worker caste polymorphism has a genetic basis in *Acromyrmex* leaf-cutting ants. *Proceedings of the National Academy of Sciences USA*, **100**, 9394–9397.

Hughes, W. O. H., Oldroyd, B. P., Beekman, M., & Ratnieks, F. L. W. (2008) Ancestral monogamy shows kin selection is key to the evolution of eusociality. *Science*, **320**, 1213–1216.

Hunt, J. H. (2007) *The Evolution of Social Wasps*. Oxford: Oxford University Press.

(2012) A conceptual model for the origin of worker behaviour and adaptation of eusociality. *Journal of Evolutionary Biology*, **25**, 1–19.

Hunt, J. H. & Amdam, G. C. (2005) Bivoltinism as an antecedent to eusociality in the paper wasp genus *Polistes*. *Science*, **308**, 264–267.

Kadochova, Š. & Frouz, J. (2012) Thermoregulation strategies in ants in comparison to other social insects, with a focus on *Formica rufa*. *F1000Research*, **2**, 280.

Kaspari, M., O'Donnell, S., & Kercher, J. R. (2000) Energy, density, and constraints to species richness: Ant assemblages along a productivity gradient. *The American Naturalist*, **155**, 280–293.

Keller, L. & Genoud, M. (1997) Extraordinary lifespans in ants: A test of evolutionary theories of ageing. *Nature*, **389**, 958–960.

Keller, L. & Passera, L. (1989) Size and fat content of gynes in relation to the mode of colony founding in ants (Hymenoptera; Formicidae). *Oecologia*, **80**, 236–240.

Keller, L. & Reeve, H. K. (1994) Genetic variability, queen number, and polyandry in social Hymenoptera. *Evolution*, **48**, 694–704.

Kellner, K., Ishak, H. D., Linksvayer, T. A., & Mueller, U. G. (2015) Bacterial community composition and diversity in an ancestral ant fungus symbiosis. *FEMS Microbiologyl Ecology*, **91**, fiv073.

Khila, A. & Abouheif, E. (2010) Evaluating the role of reproductive constraints in ant social evolution. *Philosophical Transactions of the Royal Society B*, **365**, 617–630.

King, J. R. & Tschinkel, W. R. (2008) Experimental evidence that human impacts drive fire ant invasions and ecological change. *Proceedings of the National Academy of Science USA*, **105**, 20339–20343.

King, J. R., Warren, R. J., & Bradford, M. A. (2013) Social insects dominate eastern US temperate hardwood forest macroinvertebrate communities in warmer regions. *PLoS ONE*, **8**, e75843.

Kinomura, K. & Yamauchi, K. (1987) Fighting and mating behaviors of dimorphic males in the ant *Cardiocondyla wroughtoni*. *Journal of Ethology*, **5**, 75–81.

Kocher, S. D. & Grozinger, C. D. (2011) Cooperation, conflict and the evolution of queen pheromones. *Journal of Chemical Ecology*, **37**, 1263–1275.

Korb, J. & Heinze, J. (2008) The ecology of social life: A synthesis. *In:* Korb, J. & Heinze, J. (eds.). *Ecology of Social Evolution*. Berlin: Springer, pp. 245–259.

(2016) Major hurdles for the evolution of sociality. *Annual Review of Entomology*, **61**. 297–316.

Kusnezov, N. (1957) Numbers of species of ants in faunae of different latitudes. *Evolution*, **11**, 298–299.

Lach, L., Parr, C. L., & Abbott, K. L. (2009) *Ant Ecology*. Oxford: Oxford University Press.

LaPolla, J. S., Dlussky, G. M. & Perrichot, V. (2013) Ants and the fossil record. *Annual Review of Entomology*, **58**, 609–630.

Le Conte, Y. & Hefetz, A. (2008) Primer pheromones in social Hymenoptera. *Annual Review of Entomology*, **53**, 523–542.

Lenoir, A., d'Ettorre, P., Errard, C., & Hefetz, A. (2001) Chemical ecology and social parasitism in ants. *Annual Review of Entomology*, **46**, 573–599.

Liebig, J., Peeters, C., Oldham, N. J., Markstädter, C., & Hölldobler, B. (2000) Are variations in cuticular hydrocarbons of queens and workers a reliable signal of fertility in the ant *Harpegnathos saltator*? *Proceedings of the National Academy of Sciences USA*, **97**, 4124–4131.

Linksvayer T. A. (2010) Subsociality and the evolution of eusociality. *In:* Breed, M. D. & Moore, J. (eds.) *Encyclopedia of Animal Behavior*, vol. 3. Oxford: Academic Press, pp. 358–362.

MacMahon, J. A., Mull, J. F., & Crist, T. O. (2000) Harvester ants (*Pogonomyrmex* spp.): Their community and ecosystem influences. *Annual Review of Ecology and Systematics*, **31**, 265–291.

Mamsch, E. (1967) Quantitative Untersuchungen zur Regulation der Fertilität im Ameisenstaat durch Arbeiterinnen, Larven und Königin. *Zeitschrift für vergleichende Physiologie*, **55**, 1–25.

Maynard Smith, J. & Szathmáry, E. (1995) *The Major Transitions in Evolution*. Oxford: Oxford University Press.

Meyer, S. T., Neubauer, M., Sayer, E., *et al.* (2013) Leaf-cutting ants as ecosystem engineers: Topsoil and litter perturbations around *Atta cephalotes* nests reduce nutrient availability. *Ecological Entomology*, **38**, 497–504.

Michener, C. D. & Brothers, D. J. (1974) Were workers of eusocial Hymenoptera initially altruistic or oppressed? *Proceedings of the National Academy of Sciences USA*, **71**, 671–674.

Mintzer, A. C. (1987) Primary polygyny in the ant *Atta texana*: Number and weight of females and colony foundation success in the laboratory. *Insectes Sociaux*, **34**, 108–117.

Mlot, N. J., Tovey, C. A., & Hu, D. L. (2011) Fire ants self-assemble into waterproof rafts to survive floods. *Proceedings of the National Academy of Sciences USA*, **108**, 7669–7673.

Monnin, T. (2006) Chemical recognition of reproductive status in social insects. *Annales Zoologici Fennici*, **43**, 515–530.

Monnin, T. & Peeters, C. (1999) Dominance hierarchy and reproductive conflicts among subordinates in a monogynous queenless ant. *Behavioral Ecology*, **10**, 323–332.

Monnin, T., Ratnieks, F. L. W., Jones, G.R. & Beard, R. (2002) Pretender punishment induced by chemical signalling in a queenless ant. *Nature*, **419**, 61–65.

Moreau, C. S. & Bell, C. D. (2013) Testing the museum versus cradle tropical biological diversity hypothesis: Phylogeny, diversification, and ancestral biogeographic range evolution of the ants. *Evolution*, **67**, 2240–2257.

Moreau, C. S., Bell, C. D., Vila, R., Archibald, S. B., & Pierce, N. E. (2006) Phylogeny of the ants: Diversification in the age of angiosperms. *Science*, **312**, 101–104.

Morrison, L. W. (2014) The ants of remote Polynesia revisited. *Insectes Sociaux*, **61**, 217–228.

Mueller, U., Mikheyev, A., Hong, E., *et al.* (2011) Evolution of cold-tolerant fungal symbionts permits winter fungiculture by leafcutter ants at the northern frontier of a tropical ant-fungus symbiosis. *Proceedings of the National Academy of Sciences USA*, **108**, 4053–4056.

Mueller, U. G., Gerardo, N. M., Aanen, D. K., Six, D. L., & Schultz, T. R. (2005) The evolution of agriculture in insects. *Annual Review of Ecology and Systematics*, **36**, 563–595.

Münch, D., Amdam, G. V., & Wolschin, F. (2008) Ageing in a eusocial insect: Molecular and physiological characteristics of life span plasticity in the honey bee. *Functional Ecology*, **22**, 407–421.

Oettler, J., Schmid, V. S., Zankl, N., *et al.* (2013) Fermat's principle of least time predicts refraction of ant trails at substrate borders. *PLoS ONE*, **8**, e59739.

Ohkawara, K., Nakayama, M., Satoh, A., Trindl, A., & Heinze, J. (2006) Clonal reproduction and genetic caste differences in a queen-polymorphic ant, Vollenhovia emeryi. *Biology Letters*, **2**, 359–363.

Ólafsson, E. & Richter, S. H. (1985) Húsamaurinn (*Hypoponera punctatissima*). *Náttúrufræðingurinn*, **55**: 139–146.

Parker, J. D. (2010) What are social insects telling us about aging? *Myrmecological News*, **13**, 103–110.

Passera, L. & Aron, S. (2005) *Les Fourmis: Comportement, organisation sociale et évolution.* Ottawa: CNRC·NRC.

Passera, L., Roncin, E., Kaufmann, B., & Keller, L. (1996) Increased soldier production in ant colonies exposed to intraspecific competition. *Nature*, **379**, 630–631.

Pearcy, M. & Aron, S. (2006) Local resource competition and sex ratio in the ant *Cataglyphis cursor*. *Behavioral Ecology*, **17**, 569–574.

Pearcy, M., Goodisman, M. A. D., & Keller, L. (2011) Sib-mating without inbreeding in the Crazy ant. *Proceedings of the Royal Society of London B*, **278**, 2677–2681.

Peeters, C. (1991) Ergatoid queens and intercastes in ants: Two distinct adult forms which look morphologically intermediate between workers and winged queens. *Insectes Sociaux*, **38**, 1–15.

(1997) Morphologically 'primitive' ants: Comparative review of social characters, and the importance of queen-worker dimorphism. *In:* Choe, J. C. & Crespi, B. J. (eds.) *The Evolution of Social Behaviour in Insects and Arachnids*. Cambridge: Cambridge University Press, pp. 372–391.

Peeters, C. & Higashi, S. (1989) Reproductive dominance controlled by mutilation in the queenless ant *Diacamma australe*. *Naturwissenschaften*, **76**, 177–180.

Peeters, C. & Ito, F. (2001) Colony dispersal and the evolution of queen morphology in social Hymenoptera. *Annual Review of Entomology*, **46**, 601–630.

Pie, M. R. & Tschà M. K. (2009) The macroevolutionary dynamics of ant diversification. *Evolution*, **63**, 3023–3030.

Pinter-Wollman, N., Bala, A., Merrell, A., *et al.* (2013) Harvester ants use interactions to regulate forager activation and availability. *Animal Behaviour*, **86**, 197–207.

Pratt, S. C., Mallon, E. B., Sumpter, D. J., & Franks, N. R. (2002) Quorum sensing, recruitment, and collective decision-making during colony emigration by the ant *Leptothorax albipennis*. *Behavioral Ecology and Sociobiology*, **52**, 117–127.

Queller, D. C. (2000) Pax argentinica. *Nature*, **405**, 519–520.

Queller, D. C. & Strassmann, J.E. (2013) The veil of ignorance can favour biological cooperation. *Biology Letters*, **9**, 20130365.

Rabeling, C. & Kronauer, D. J. C. (2013) Thelytokous parthenogenesis in eusocial Hymenoptera. *Annual Review of Entomology*, **58**, 273–292.

Ratnieks, F. L. W. (1988) Reproductive harmony via mutual policing by workers in eusocial Hymenoptera. *The American Naturalist*, **132**, 217–236.

Ratnieks, F. L. W. & Reeve, H. K. (1992) Conflict in single-queen Hymenopteran societies: The structure of conflict and processes that reduce conflict in advanced eusocial species. *Journal of Theoretical Biology*, **158**, 33–65.

Ravary, F., Lecoutey, E., Kaminski, G., Châline, N., & Jaisson, P. (2007) Individual experience alone can generate lasting division of labor in ants. *Current Biology*, **17**, 1308–1312.

Reber, A., Castella, G., Christe, P., & Chapuisat, M. (2008) Experimentally increased group diversity improves disease resistance in an ant species. *Biology Letters*, **11**, 682–689.

Robinson, E. J. H., Jackson, D. E., Holcombe, M., & Ratnieks, F. L. W. (2005) Insect communication: 'No entry' signal in ant foraging. *Nature*, **438**, 442.

Ross, K. G. & Carpenter, J. M. (1991) Phylogenetic analysis and the evolution of queen number in eusocial Hymenoptera. *Journal of Evolutionary Biology*, **4**, 117–130.

Rousset, F. & Lion, S. (2011) Much ado about nothing: Nowak, *et al.*'s charge against inclusive fitness theory. *Journal of Evolutionary Biology*, **24**, 1386–1392.

Rüger, M.H., Fröba, J. & Foitzik, S. (2008) Larval cannibalism and worker-induced separation of larvae in *Hypoponera* ants: A case of conflict over caste determination? *Insectes Sociaux*, **55**, 12–21.

Savolainen, R. & Vepsäläinen, K. (2003) Sympatric speciation through intraspecific social parasitism. *Proceedings of the National Academy of Sciences USA*, **100**, 7169–7174.

Savolainen, R., Vepsäläinen, K. & Wuorenrinne, H. (1989) Ant assemblages in the taiga biome: Testing the role of territorial wood ants. *Oecologia*, **81**, 481–486.

Schlick-Steiner, B. C., Steiner, F. M., Konrad, H., *et al.* (2008) Specificity and transmission mosaic of ant nest-wall fungi. *Proceedings of the National Academy of Sciences USA*, **105**, 940–943.

Schrader, L., Simola, D. F., Heinze, J., & Oettler, J. (2015) Sphingolipids, transcription factors, and conserved toolkit genes: Developmental plasticity in the ant *Cardiocondyla obscurior*. *Molecular Biology and Evolution*, **32**, 1474–1486.

Schrempf, A., Heinze, J., & Cremer, S. (2005) Sexual cooperation: Mating increases longevity in ant queens. *Current Biology*, **15**, 267–270.

Schrempf, A., Aron, S. & Heinze, J. (2006) Sex determination and inbreeding depression in an ant with regular sib-mating. *Heredity*, **97**, 75–80.

Schrempf, A., Cremer, S., & Heinze, J. (2011) Social influence on age and reproduction: Reduced lifespan and fecundity in multi-queen ant colonies. *Journal of Evolutionary Biology*, **24**, 1455–1461.

Schultner, E., Gardner, A., Karhunen, M., & Helanterä, H. (2014) Ant larvae as players in social conflict: Relatedness and individual identity mediate cannibalism intensity. *The American Naturalist*, **184**, E161–E174.

Schwander, T., Lo, N., Beekman, M., Oldroyd B. P., & Keller, L. (2010) Nature versus nurture in social insect caste differentiation. *Trends in Ecology and Evolution*, **25**, 275–282.

Scott, J., Budsberg, K., Suen, G., *et al.* (2010) Microbial community structure of leaf-cutter ant fungus gardens and refuse dumps. *PLoS ONE*, **5**, e9922.

Seal, J. N. (2009) Scaling of body weight and fat content in fungus-gardening ant queens: Does this explain why leaf-cutting ants found claustrally? *Insectes Sociaux*, **56**, 135–141.

Seal, J. N. & Tschinkel, W. R. (2006) Colony productivity of the fungus-gardening ant, *Trachymyrmex septentrionalis* McCook, in a Florida pine forest (Hymenoptera: Formicidae). *Annals of the Entomological Society of America*, **99**, 673–682.

Seal, J. N., Schiøtt, M. & Mueller, U. G. (2014) Ant-fungal species combinations engineer physiological activity of fungus gardens. *Journal of Experimental Biology* **217**, 2540–2547.

Sherman, P. W., Seeley, T. D., & Reeve, H. K. (1988) Parasites, pathogens, and polyandry in social Hymenoptera. *The American Naturalist*, **131**, 602–610.

Singh, M. & Boomsma, J. J. (2015) Policing and punishment across the domains of social evolution. *Oikos*, **124**, 971–982.

Smith, A. A., Hölldobler, B., & Liebig, J. (2009) Cuticular hydrocarbons reliably identify cheaters and allow enforcement of altruism in a social insect. *Current Biology*, **19**, 78–81.

Starr, C. (2006) Steps toward a general theory of the colony cycle in social insects. *In:* Kipyatkov, V. (ed.) *Life Cycles in Social Insects: Behaviour, Ecology and Evolution*. St. Petersburg: St. Petersburg University Press, pp. 1–20.

Stille, M. (1996) Queen/worker thorax volume ratios and nest-founding strategies in ants. *Oecologia*, **105**, 87–92.

Stoll, S., Gadau, J., Gross, R., & Feldhaar, H. (2007) Bacterial microbiota associated with ants of the genus *Tetraponera*. *Biological Journal of the Linnean Society*, **90**, 399–412.

Stroeymeyt, N., Brunner, E., & Heinze, J. (2007) "Selfish worker policing" controls reproduction in a *Temnothorax* ant. *Behavioral Ecology and Sociobiology*, **61**, 1449–1457.

Sundström, L., Chapuisat, M., & Keller, L. (1996) Conditional manipulation of sex ratios by ant workers: A test of kin selection theory. *Science*, **274**, 993–995.

Tinaut, A. & Heinze, J. (1992) Wing reduction in ant queens from arid habitats *Naturwissenschaften*, **79**, 84–85.

Toth, A. L. & Robinson, G. E. (2007) Evo-devo and the evolution of social behaviour. *Trends in Genetics*, **23**, 334–341.

Trivers, R. L. & Hare, H. (1976) Haplodiploidy and the evolution of social insects. *Science*, **191**, 249–263.

Trunzer, B., Heinze, J., & Hölldobler, B. (1998) Cooperative colony founding and experimental primary polygyny in the ponerine ant *Pachycondyla villosa*. *Insectes Sociaux*, **45**, 267–276.

Tschinkel, W. R. (1996) A newly-discovered mode of colony founding among fire ants. *Insectes Sociaux*, **43**, 267–276.

(1999) Sociometry and sociogenesis of colony-level attributes of the Florida harvester ant (Hymenoptera: Formicidae). *Annals of the Entomological Society of America*, **92**, 80–89.

Tsuji, K. (1988) Obligate parthenogenesis and reproductive division of labor in the Japanese queenless ant *Pristomyrmex pungens*. *Behavioral Ecology and Sociobiology*, **23**, 247–255.

Tsuji, K., Nakata, K. & Heinze, J. (1996) Lifespan and reproduction in a queenless ant. *Naturwissenschaften*, **83**, 577–578.

Tsutsui, N. D. & Suarez, A. V. (2003) The colony structure and population biology of invasive ants. *Conservation Biology*, **17**, 48–58.

Tsutsui, N. D., Suarez, A. V., Holway, D. A., & Case, T. J. (2000) Reduced genetic variation and the success of an invasive species. *Proceedings of the National Academy of Sciences USA*, **97**, 5948–5953.

Van Wilgenburg, E., Driessen, G., & Beukeboom, L. W. (2006) Single locus complementary sex determination in Hymenoptera: An "unintelligent" design. *Frontiers in Zoology*, **3**, 1.

Van Zweden, J. S., Brask, J. B., Christensen, J. H., *et al.* (2010) Blending of heritable recognition cues among ant nestmates creates distinct colony gestalt odours but prevents within-colony nepotism. *Journal of Evolutionary Biology*, **23**, 1489–1508.

von Wyschetzki, K., Rueppell, O., Oettler, J., & Heinze, J. (2015) Transcriptomic signatures mirror the lack of the fecundity/longevity trade-off in ant queens. *Molecular Biology and Evolution*, **32**, 3173–3185.

Wenseleers, T. & Ratnieks, F. L. W. (2006a) Enforced altruism in insect societies. *Nature*, **444**, 50.

(2006b) Comparative analysis of worker reproduction and policing in eusocial Hymenoptera supports relatedness theory. *The American Naturalist*, **168**, E163–E179.

Wetterer, J. K., Espadaler, X., & Ashmole, P. (2007) Ants (Hymenoptera: Formicidae) of the South Atlantic islands of Ascension Island, St Helena, and Tristan da Cunha. *Myrmecological News*, **10**, 29–37.

Whitcomb, W. H., Bhatkar, A., & Nickerson, J. C. (1973) Predators of *Solenopsis invicta* queens prior to successful colony establishment. *Environmental Entomology*, **2**, 1101–1103.

Wilson, E. O. (1971) *The Insect Societies*. Cambridge, MA: Harvard University Press.

(1983) Caste and division of labor in leaf-cutter ants (Hymenoptera: Formicidae: *Atta* III. Ergonomic resiliency in foraging by *A. cephalotes*. *Behavioral Ecology and Sociobiology*, **14**, 47–54.

(2012) *The Social Conquest of Earth*. New York, NY: Liveright Publishing Co.

Wilson, E. O. & Hölldobler, B. (1980) Sex differences in cooperative silk-spinning by weaver ant larvae. *Proceedings of the National Academy of Sciences USA*, **77**, 2343–2347.

(2005) The rise of the ants: A phylogenetic and ecological explanation. *Proceedings of the National Academy of Sciences USA*, **102**, 7411–7414.

Witte, V. & Maschwitz, U. (2008) Mushroom harvesting ants in the tropical rain forest. *Naturwissenschaften*, **95**, 1049–1054.

Yamauchi, K., Oguchi, S., Nakamura, Y., *et al.* (2001) Mating behavior of dimorphic reproductives of the ponerine ant, *Hypoponera nubatama*. *Insectes Sociaux*, **48**, 83–87.

Yamauchi, K., Ishida, Y., Hashim, R. & Heinze, J. (2006) Queen–queen competition by precocious male production in multiqueen ant colonies. *Current Biology*, **16**, 2424–2427.

Yanoviak, S. P., Dudley, R., & Kaspari, M. (2005) Directed aerial descent in canopy ants. *Nature*, **433**, 624–626.

3 Sociality in Bees

William Wcislo and Jennifer H. Fewell

Overview

The most common image that comes to mind with the term "bee" is the honey bee, *Apis mellifera*, with its highly derived and well-studied eusocial organization. The diversity of social organization among bees and near relatives (Apoidea), however, is much richer and extensive. The "bees" (Anthophila, within the superfamily Apoidea) include approximately 20,000 species forming a monophyletic clade with a likely single ancestral origin (Danforth, *et al.*, 2006). Ecologically, all bees rely on floral resources to survive, and share a relatively common body plan, related to their reliance on flight as their primary form of locomotion. These commonalities belie their incredible social diversity. The basal (plesiomorphic) and most common life history form in this taxon is actually solitary living. Across the bee phylogeny, however, there are multiple and diverse taxa in which adult females form social alliances without reproductive castes (primarily communal sociality), and diverse taxa have independently evolved reproductive caste-based eusocial societies. Beyond these general categories, bee species vary considerably in how they form social groups and in the specific behaviors they perform cooperatively. Sociality has appeared and been lost multiple times in different lineages, and the expression of sociality can vary within a single species or even within populations. This diversity simultaneously makes the bees one of the most complex and most rewarding of taxa with which to explore questions of social evolution.

I SOCIAL DIVERSITY

3.1 How Common is Sociality in Bees?

In his nearly forgotten paper, "The application of Darwin's theory to the flowers and flower-visiting insects," Müller (1872) laid out the arguments that bees are essentially vegetarian digger wasps, most of which are solitary ground-nesters, with some species

The authors thank their fellow research colleagues at the Smithsonian Tropical Research Station (STRI) and in Arizona State University's Social Insect Research Group for the many valuable discussions. This chapter, and the field of bee sociality more generally, was fundamentally shaped by the lifetime of research provided by Charles Michener on the social bees. Thanks also go to the reviewers and to Patrick Abbot and Dustin Rubenstein for their work in constructing the volume.

showing tolerance for social interactions and nest cohabitation (reviewed in Wcislo & Tierney, 2009). The majority of bee species are solitary, or live in communal social groups consisting of adult females that independently engage in nest construction, defense, and offspring care, but without reproductive division of labor (O'Neil, 2001; Wcislo & Tierney, 2009). In a much more limited number of taxa, family groups (i.e. a mother and her female offspring) have evolved into eusocial societies, in which morphologically or functionally sterile female offspring remain at the nest to help their parent queen rear offspring. The very different phylogenetic and ecological contexts in which eusocial versus communal taxa have evolved make it likely that they represent two different paths to sociality. In other words, communal bees, and their similar social types, likely do not represent a transitional stage to eusociality, but are instead a different and ecologically relevant social strategy.

3.2 Forms of Sociality in Bees

Their wealth of diversity in social behavior, ecology, and phylogeny make the bees especially informative in testing questions about the origins and elaboration of social behavior, including the conditions leading to the formation of communally cooperative groups, as well as those with a reproductive division of labor. As such, they serve as an important group for comparative studies with both vertebrate and invertebrate taxa for understanding the evolution of cooperative sociality. Questions of how sociality evolved within the bees have been complicated by somewhat misplaced assumptions about the relationships between the different forms of sociality and social evolutionary pathways. These assumptions have also shaped the categorization of social behavior in Hymenoptera (and in most invertebrates). The categorization of bees' social behavior derives primarily from Michener's (1969, 1974) social taxonomy, which included: (1) *solitary* species, in which individual females construct nests and provision offspring alone; (2) *communal* and *quasisocial* groups, in which several adult females form groups in which individuals reproduce independently (quasisocial being differentiated from communal by the presence of alloparental care); and (3) *eusociality*, conventionally defined by an overlap of generations, reproductive division of labor, and cooperative rearing of young.

These categories are usefully descriptive, and based upon observed social behavior. Historically, however, they have been combined with a theoretical "levels of sociality" perspective (Wilson, 1971), which implicitly hypothesizes that non-eusocial societies represent a series of intermediate steps on an evolutionary path from solitary to eusociality. This hypothesis became a core assumption about the historical development (phylogeny) of eusociality, reinforced by a series of terms, such as "presocial" (parental care), "subsocial" (non-eusocial family groups with adult offspring), and "parasocial" (including communal and quasisocial groups). Unfortunately, these terms imply that non-eusocial groups are also non-social. The "levels of sociality" assumption is not generally supported by empirical data (Michener, 1985; Danforth, 2002; Danforth, *et al.*, 2006; Schwarz, *et al.*, 2011; Gibbs, *et al.*, 2012). Lineages that have given rise to communal forms rarely give rise to eusocial/semisocial forms, and *vice versa* (Michener, 1985; Wcislo & Tierney, 2009).

The established lexicon has complicated attempts to generate a more synthetic understanding of the ecological and evolutionary forces shaping social evolution. However, its revision faces the difficulties of re-defining deeply entrenched terms (see Costa & Fitzgerald, 2005; Wcislo, 2005; Kocher & Paxton, 2014; Dew, *et al.*, 2015). Although these typologies can be problematic for guiding readers through the evolution of social behavior in bees, they are also key to making sense of past literature. We will use *group living* as an umbrella term to cover all forms of sociality, either when it is unnecessary to make finer distinctions for particular questions, or impossible to differentiate. Given the limited empirical data available for many species (see Wcislo, 1997), these issues are often present.

Collapsing the above social taxonomies reveals a general pattern of variation in social behavior that roughly fits Michener's original ecological categorization, and divides into: (1) parasocial, in which often unrelated adult females share a nest, reproduce, and may (quasisocial) or may not (communal) engage in cooperative brood care; (2) semisocial groups, in which related females of the same generation share a nest (perhaps after hibernation, or perhaps a natal nest), with a reproductive division of labor, so reproduction is skewed within colonies; and (3) parent–offspring based societies, including subsocial and eusocial groups, with eusocial societies defined by showing a clear reproductive division of labor. This is somewhat different from Wilson's (1971) taxonomy of sociality, in which semisocial is presented as an extension of communal and quasisocial social evolution. A summary of the "current state of knowledge" for the group living distribution of social forms across the bee taxa is given in Table 3.1. It is worth looking at this table before reading through the rest of the chapter, because it provides an overview of the different kinds of bees, their taxonomic and common names, and their "typical" patterns of sociality.

3.2.1 Social Evolutionary Transitions

Bee social evolution entails three major evolutionary transitions. The first is from solitary to some sort of group living. This is likely initially to be facultative. Interestingly, even in species in which females are predominantly solitary, sometimes a small percentage of females live in social groups (e.g. Wcislo, *et al.*, 1993; Yagi & Hasegawa, 2012). This first transition results in one of two social states. In some taxa, groups are comprised of adult females that are not socially or morphologically differentiated and all are reproductively active (communal behavior). Facultatively group living bees can be exemplified by communal sweat bees (Halictidae), or some mining bees (Andrenidae) in which groups of mated females cooperatively excavate a nest together (Kukuk & Schwarz, 1987; Danforth, 1991a). Although they share a nest, these communally nesting females individually construct and provision their own brood cells. They cooperatively share the tasks of excavation and guarding the nest entrance (Kukuk & Schwarz, 1987; Jeanson, *et al.*, 2005). Unlike in solitary species that avoid other individuals, these bees show mutual tolerance, often both to nestmates and non-nestmates alike, suggesting a

Table 3.1 Occurrence of solitary, social and parasitic behavior within the major taxa of bees. 0 = absent; + = behavior is expressed in some species; ++ = essentially all species exhibit the behavior. The biology of many species in many taxa is not known, so absence of behavior should be considered provisional; ? = suggestive but not conclusive data for the occurrence of the behavior. Taxonomy follows Michener (2007).

Taxon	Common name	Biogeography	Solitary	Communal	Semisocial-eusocial	Socially parasitic
Apidae: Apinae						
Meliponini	stingless bees	Central and South America Tropical	0	0	++	0
Bombini	bumble bees	Europe, North America Temperate	0	0	++ (except parasitic species)	+
Apini	honey bees	Africa, Europe Tropical-temperate	0	0	++	0
Euglossini	orchid bees	Central and South America Tropical	+	+	+ (but rare)	+
Allodapini	twig-nesting bees	Africa, Asia, Australia	0	0	++ (except parasitic species)	+
Ceratinini	small carpenter bees	Cosmopolitan Temperate, Xeric	+	+	+	0
Xylocopini	large carpenter bees	Cosmopolitan Temperate, Xeric	+	+	0	0
other apine tribes	digger bees		+	+	0	+
Nomadinae	cuckoo bees	Cosmopolitan	0	0	0	++
Megachilidae	mason and leafcutting bees	Cosmopolitan	+	+	0	+
Colletidae	polyester bees	Primarily Australia and S. America, Few species in Europe, N. America	+	+	0	0
Stenotritidae		Australia Xeric	+	0	0	0
Halictidae	sweat bees	Cosmopolitan Temperate, Tropical Xeric				
Halictinae			+	+	+	+
Nomiinae			+	+	+?	0
Nomiodinae			+	+	0	0
Rophitinae			+	0	0	0
Andrenidae	mining bees	Cosmopolitan, Temperate, Xeric	+	+	0	0
Melittidae		Africa Temperate	+	+	0	0
Apoid wasps (outgroup)	digger wasps	Cosmopolitan	+	+	+ (but rare)	+

general acceptance of others (Paxton, *et al.*, 1999). Some communal *Lasioglossum* have even been observed mutually exchanging food via trophallaxis (i.e. sharing crop contents) (Kukuk & Schwarz, 1987).

In other cases, females become behaviorally differentiated into reproductive (i.e. "queen") and non-reproductive (i.e. "worker") forms. There may be statistical differences in body size between the forms, but otherwise these females generally have similar external morphologies. Generally in semisocial and eusocial groups, the females within the nests are present because they emerged in that nest and did not disperse (reviewed in Wcislo & Tierney, 2009).

Social transitions within the bees are surprisingly evolutionarily labile, making an assessment of the proportional number of solitary versus social species difficult to assess. Again, although exact numbers cannot be assigned, the vast majority of bee species actually exhibit solitary behavior (Linsley, 1958). Of the social species, many live in non-differentiated communal societies (Wcislo & Tierney, 2009), and much more rarely, a limited number of taxa exhibit eusocial behavior (Michener, 1974, 2007). It is notable that, across all social bee types or taxa, the transition to sociality stems either from cooperative females associating to help in maternal behavior, including offspring defense, or from adult female offspring remaining to help with brood care. No matter what type of group living they may take, bee social systems are almost entirely based around females; males generally have no social roles apart from mating.

3.2.2 Transitioning from Solitary to Group Living

The lability of transitions between solitary and group living is one of the most fascinating aspects of bee social evolution. Multiple species, whether eusocial or communal, display both solitary and group living strategies, and sometimes both strategies are found in the same population. This variation implies that early social transitions from solitary to group living may not require the evolution of novel behaviors. Instead, they may rely upon selective tempering of behaviors already in place, particularly the reduction of aggression and associated increase in tolerance. Selection also acts on social behaviors that emerge from resulting within-group dynamics (Jeanson, *et al.*, 2005).

Some species show facultative transitions between solitary and communal nesting, in which adult females of the same generation share a nest. This may occur at nest initiation or when a female invades the nest secondarily and is accepted by the resident (Dunn & Richards, 2003). Habitats that permit multiple broods per year or season, coupled with strong competition for nest sites and raiding among nests, likely facilitate this transition to communal sociality (Gerling, *et al.*, 1989).

In the case of the transition to semisocial living, young female bees generally stay at the maternal nest and direct brood care towards their siblings (Field, *et al.*, 2010), similarly to observations in wasps (Hunt & Amdam, 2005; Hunt, 2007; Chapter 4). An example is found in the large carpenter bees (Xylocopini), a widespread taxon of large-bodied bees with both temperate and tropical distributions, that nest by boring holes in wood. This group is generally considered solitary, with individual females provisioning multiple eggs within a single nest across the growing season (reviewed in Gerling, *et al.*, 1989).

However, family groups are also occasionally observed; in these, female offspring participate in brood rearing and nest defense (Hogendoorn & Velthius, 1999).

The mechanisms and conditions underlying the transition to semisociality can be examined perhaps most clearly in the halictids (sweat bees), a taxon with wide geographical range and corresponding diversity in social strategy. The sweat bee *Halictus rubicundus* displays solitary versus eusocial living (with female offspring helping to rear brood), depending on location. Bees can be induced to change between solitary and group living if translocated across geographic sites (Field, *et al.*, 2010). Thus, ecological context plays a central role in determining social type. In this case, eusociality is expressed where the season is long enough for females to rear multiple successive brood cycles, because it allows the first round of offspring to mature and help rear the second round within a single season.

It is important to note, however, that the appearance of sociality is not simply a plastic behavioral response to season length. It is mediated by underlying genetic and neurophysiological mechanisms. Soro, *et al.* (2010) demonstrated genetic differentiation between solitary and semisocial populations of *H. rubicundus*. Similarly, Plateaux-Quénu, *et al.* (2000) reared another facultatively semisocial sweat bee, *Lasioglossum albipes,* in common garden conditions under different regimes of temperature and photoperiod. Under these experimental conditions, bees retained their population-typical behavior. Thus, the transition to group living involves specific combinations of ecological conditions, such as competition and/or seasonality, in combination with variation in genetically based behavioral attributes, such as tolerance and/or maternal care.

3.2.3 Transitioning to Eusociality

The behavioral antecedents of eusociality appear widely across diverse taxa of bees, including in many clades that have not given rise to obligately eusocial forms. There is growing evidence that eusociality in the bees most likely evolved via a route in which female offspring remain at the nest facultatively and help to rear their full or half sisters and/or their sisters' offspring (Rehan, *et al.,* 2014; Kapheim, *et al.,* 2015). In some cases, these individuals are capable of reproducing, but are competitively suppressed, as in some facultatively eusocial sweat bees (e.g. *Megalopta genalis,* Smith, *et al.,* 2009). Thus, a facultative reproductive division of labor appears. Despite the contribution of relatedness to facilitating this transition, the evolution of true worker castes (indicating eusociality, rather than semisociality) is extremely rare (Schwarz, *et al.,* 2011).

Once eusociality is in place, a second key transition entails its evolutionary elaboration from a facultatively expressed trait to one in which eusociality is functionally a fixed trait, but all adults intrinsically remain capable of assuming any social role (primitive eusociality). A final transition is from primitive eusociality among totipotent individuals displaying functional sterility, to obligate eusociality in which all females are not reproductively totipotent. In these societies, members of the worker caste may be able to produce males via unfertilized eggs, but are no longer capable of becoming breeders and nest foundresses (Wilson, 1971; Alexander, 1974; Michener, 1974).

Obligate eusociality is rare, and phyletically restricted to two tribes of sweat bees (Halictidae: Augochlorini and Halictini) and two monophyletic lineages in Apidae: the Allodapini and the Corbiculate Apids (bees with concave "pollen baskets" on their hindlegs) which include the most well-known eusocial taxa, the bumble bees and honey bees. Although the sweat bees (halictids) are globally widespread, and show a wide diversity of social types, very few species are actually eusocial; those that are remain primitively eusocial (Michener, 1974). The allodapine bees are somewhat unusual in that they rear brood together in open tunnels (rather more like ants) rather than in individual cells. This, in combination with their provisioning behavior, contributes to variation in larval nutrition and to consequent morphological diversity. All known allodapine species are semisocial or eusocial, excepting parasitic species (reviewed by Michener, 2007; Tierney, *et al.*, 2008). With 250 described species (Michener, 2007; Schwarz, *et al.*, 2007), they vary from family groups in which daughters delay dispersal, to highly eusocial species (Schwarz, *et al.*, 2007). Allodapines are closely related to carpenter bees, but as noted earlier, carpenter bees are predominantly solitary. The distinct difference in social structure between the two groups exemplifies that the diversity of bee social systems extends beyond purely phylogenetic explanations.

The corbiculates include what we often think of as the typical eusocial species, including the honey bees (Apini), stingless bees (Meliponini), and bumble bees (Bombini); all of these taxa collect pollen and store honey. The group also includes the often beautifully iridescent orchid bees (Euglossini). Males of this taxon collect scent oils in their corbiculae, which they use to attract females for mating. Unlike the other corbiculate taxa, orchid bee species are generally solitary (Roubik, 1989). Thus, the different corbiculate lineages range from solitary to communal societies to diverse forms of eusociality, including functional but not necessarily morphological worker sterility (as illustrated by the bumblebees). Finally, the highly derived eusocial corbiculates lineages include those with morphologically distinct worker and reproductive castes (seen both in the stingless bees and the honey bees).

3.3 Why Bees Form Social Groups

As for all social taxa, sociality in any given bee species derives from specific combinations of historical, ecological, life-history, and genetic factors. The factor that has received perhaps the most attention for Hymenoptera is the high relatedness among sisters that is generated by their haplodiploid sex determination mechanism (Hamilton 1964). Relatedness, however, cannot be the single driving answer to the question of "why do bees form social groups?" because one most also take into account appropriate benefits and costs of behavioral alternatives. Females in many communal species form cooperative associations with non-relatives (Moore & Kukuk, 2002). Further, *all* hymenopteran are haplodiploid, but very *few* are social. Thus, although relatedness likely facilitates eusocial evolution, we must look to ecology, life history and behavior for drivers underlying the broad diversity of social types within the Apoidea.

3.3.1 Resource Acquisition and Use

Depending upon the ecological context, need for food can either enhance or constrain social evolution. Bees almost universally rely upon flowering plants to feed their young, usually collecting pollen, nectar, and in some cases floral oils (e.g. Linsley, 1958; Wcislo & Cane, 1996; Roubik, 2012). As the rare exception, a very small number of necrophagous stingless bees feed on carrion and brood of other species (Noll, *et al.*, 1996). Flowers provide abundant resources that can be stored, a likely facilitator of sociality; however, they are also spatially and temporally patchy. Floral preference may influence the types and level of bee sociality based on a combination of breadth of floral choice, and the relative patchiness of preferred flowers.

This is particularly the case for pollen foraging. Bees roughly fall into two groups: oligolectic and polylectic species (e.g. Linsley, 1958; Wcislo & Cane, 1996). Oligolectic bees are resource specialists that collect pollen (or oils) from a taxonomically restricted group of flowering plants, whereas polylectic bees generalize, collecting food from any of a number of locally available and taxonomically diverse flowering plants that are suitable for bees. The flight season of floral specialists must synchronize with the blooming times of their flowering plants, and they are thus restricted to a relatively small portion of the total active season for bees and flowers in a given locality. In contrast, the flight season of generalist species covers the entire period when conditions are favorable for foraging, regardless of the flowering phenology of any specific bee-associated plant species (Minckley, *et al.*, 1994). Perhaps unsurprisingly then all eusocial species are pollen generalists, with few exceptions.

The capacity for food storage also plays a role in shaping social evolution, particularly in eusocial taxa, by providing a buffer against food scarcity and thus facilitating the development of larger and more permanent colonies. The most derived eusocial bee species, including the Meliponini, Bombini, and Apini, all have pollen and nectar storage capacities that extend beyond the simple larval provisioning seen in communal or solitary species. This allows food sharing among adults, and in many cases progressive (i.e. ongoing) provisioning of offspring.

As colonies become larger, group size can also facilitate foraging success, which in turn enhances colony growth. The evolution of mass recruitment by some eusocial species allows those colonies to dominate floral resources. This capacity has been most intensively studied for honey bees, in which colonies distribute their massive foraging force dynamically as resources change in abundance and quality (Seeley, 1994, 1997), or as colony storage levels and brood requirements change (Fewell & Winston, 1992; Pankiw, *et al.*, 1998).

The evolution of the corbiculates bees (particularly the honey bees, stingless bees, and bumblebees) involves two major adaptations relevant to their success as social foragers. The first is the pollen basket itself, located on the rear legs and allowing transport of larger pollen loads (Winston & Michener, 1977). The second comprises a series of signaling adaptations for foraging recruitment, with increasingly complex information transmitted about floral odor cues, food quality, distance, and direction (von Frisch, 1967; Dornhaus & Chittka, 1999; Dyer, 2002). Again, these are most developed in the honey bees, which can symbolically communicate distance and direction, in addition to providing odor cue information on floral type (reviewed in von Frisch, 1967; Dyer, 2002). However, some stingless bees (Meliponini) also have been shown to communicate information concerning the location of floral resources, although the mechanism differs from that used by honey bees (Nieh, 1999). Bumblebees, in contrast, can stimulate increased foraging via recruitment, and transmit odor information on flower source. This information exchange can help regulate foraging activity level, but does not seem to communicate location (Dornhaus & Chittka, 1999). No capabilities for directional information are known for other social bee taxa either (e.g. halictine and xylocopine bees), but most taxa have not been studied.

Food and sociality also serve as a context for the complex relationships between bees and their microbiota. Sociality brings increased risk in the spread of microbial pathogens (Schmid-Hempel, 1998; Ayasse & Paxton, 2002), and in bees is associated with a need for increased microbial defenses (Stow, *et al.,* 2007). Studies suggest that disease risk in highly eusocial species, including bees, may have been a driver in the evolution of mechanisms to increase worker genetic diversity, including queen multiple mating or polyandry (Tarpy, 2003). On the other hand, beneficial microbes function as positive symbionts, especially for nutrient digestion and gut function (Engel & Moran, 2013). The question of how bee sociality (and sociality more generally) shapes and is shaped by relationships with microbiomes is an emerging and rapidly expanding area of research (Anderson, *et al.*, 2011; McFrederick, *et al.*, 2014; Moran, 2015).

3.3.2 Predator Avoidance

Predator avoidance (by adults) probably plays a limited role in the origins and maintenance of group living in most bees, for the simple reason that most bees possess a weapon that can be used against such predators – their modified ovipositor or stinger (Malyshev, 1968). Predators that attack foraging bees usually have relatively little impact on sociality, because both solitary and socially nesting bees generally collect at the resource site individually. Although social bees have evolved complex recruitment strategies to provide information on resource location and quality, a given predator usually encounters one or a few foragers at a time.

In contrast, the developing brood, being relatively immobile and individually defenseless, can face a significant threat from predators and parasites. The ability to store resources within the nest also carries with it a risk for eusocial bees because it attracts predators and parasites. These include predatory insects such as flies, beetles and wasps (Wcislo & Cane, 1996; Ayasse & Paxton, 2002), as well as social parasites that mimic social signals to live in colonies and co-opt resources (Wcislo, 1987; Lenoir, *et al.*, 2001). The largest eusocial colonies, especially *Apis*, are also regularly attacked by vertebrate predators, including humans, attracted to brood and food stores (Bromley, 1948; Winston, 1991; Breed, *et al.*, 2004).

Certainly, nest predation and parasitism represent a high enough cost for bees that social defense mechanisms have evolved. All apoids have some means of concealing or defending their nests against enemies, ranging from nesting in isolated sites and hiding nests (Wcislo & Cane, 1996), to nesting in aggregations under higher densities (Rosenheim, 1990), to coordinated social defense strategies (Michener, 1974). Most social bee species exhibit some form of nest-guarding behavior, in which a bee or group of bees sits at the nest entrance and challenges intruders. (Michener, 1974; Abrams & Eickwort, 1981; Breed, *et al.*, 2004). Nest guarding theoretically presents an early advantage in the transition from solitary behavior to group living; however, current data testing the effective relationship between sociality and nest defense are mixed. As examples, in field observations, cleptoparasitic bees (*Nomada*, Apidae) never entered multifemale nests of *Agapostemon virescens* (Halictidae), but successfully entered a singleton nest (Abrams & Eickwort, 1981). In contrast, however, field studies of facultatively communal versus solitary carpenter bees, *Xylocopa* (Prager, 2014), found no specific defense advantage of having multiple females in the nest. Similarly, field comparisons of solitary versus communal nests in *Lasioglossum (Dialictus) figueresi* (Wcislo, *et al.*, 1993) reported no difference in brood mortality rates in single- versus multi-queen nests, even though the former are unguarded when a foundress is foraging.

A coordinated nest defense is common in eusocial taxa, whose larger and more visible nests can face diverse invertebrate and vertebrate predators. The fitness costs of coordinated defense can be illustrated by a comparison of solitary versus primitively eusocial *Lasioglossum balecium*. Under some conditions the (indirect) fitness benefits of females that stayed in eusocial colonies as workers were higher than the comparative direct benefits if they had formed nests as solitary queens. The condition that contributed most strongly to this difference was the presence of predatory ants and their impact on brood survival (Yagi & Hasegawa, 2012). Thus, under some contexts predation, and the associated value of group nest defense, may contribute to the transition to eusociality.

Eusocial nest defense is often connected with self-sacrificing behaviors, sometimes spectacularly so (Shorter & Rueppell, 2012). Honey bees are well known to aggressively attack predators in massive defensive responses. This behavior reaches its apex in *Apis mellifera scutellata*, the African honey bee (reviewed in Winston, 1992; Breed, *et al.*, 2004), which faces intense predation pressure from vertebrates, including humans, and which can instantly mobilize thousands of workers to defend the colony. Some honey bees also use thermoregulation to defend against predators and parasites. In the Asian bee, *Apis cerana*, workers ball up around invading wasps (*Provespa*) to repel attacks. In these balls, the bees heat themselves to temperatures above the critical limits of other *Apis* species (Ken, *et al.*, 2005). In doing so, they effectively "cook" the invading wasp. Although lacking stingers, the eusocial Meliponini (the stingless bees) employ multiple defense strategies, from constructing nest entrance collars that minimize the size of the opening, to aggressive worker defense, including, in some species, the application of acids when biting the flesh of a predator (reviewed in Roubik, 2006). As one of the more charismatic examples, workers of the stingless "angel bee," *Tetragonsica angustula*, maintain a continuous hovering front of defenders outside the nest entrance to dissuade potential invaders (Grüter, *et al.*, 2011). Individual *Trigona* workers use their exceptionally sharp mandibles to aggressively bite invading predators and resource competitors (Neih, *et al.*, 2005); defenders latch on to attackers with a mandibular locking mechanism that has been described as a "death grip" (Buchwald & Breed, 2005).

3.3.3 Homeostasis

An integrated social group, within a sheltered nest, can buffer its members against multiple environmental onslaughts, from variation in food availability, to changes in thermal or weather conditions. For bees, as with many other invertebrates, temperature can severely constrain life history and physiology. As for most ectotherms, brood development rates and success are affected by the thermal environment and by moisture levels (reviewed in Harrison, *et al.*, 2012). The vast majority of bees build nests to help buffer against weather conditions. Depending upon the species, these are placed in soil, within sticks or other hollow cavities, and some build free-standing nests of mud or resin (reviewed in Michener, 1964, 1974; Wcislo & Engel, 1996). In general, these efforts would contribute little to maintaining homeostasis, until societies are large enough that workers can moderate local environmental conditions.

The corbiculate bees (honey bees, bumblebees and stingless bees) construct wax or resin-based nests, containing individual cells used for brood rearing, and nectar (honey) and pollen storage (except in orchid bees) (Michener, 1964). With exceptions, these are the "fortress nests" set up by many social insects (Queller & Strassmann, 1998). Most bumblebees are ground nesters, but stingless and honey bees utilize cavities in trees or other structures, within which they construct cells. Thermal temperatures in bumblebee and honey bee nests are carefully regulated to setpoints conducive to optimize larval development (Heinrich, 1985, 2004). Multiple behaviors contribute to this homeostasis. Individual workers can produce metabolic heat to maintain higher than ambient nest temperatures, and use their wings to fan when temperature or CO_2 levels rise above optimal (Heinrich, 1985; Weidenmüller, *et al.,* 2002). Workers pay particular attention to brood temperatures, and honey bee workers will act as heat shields between brood and sections of the nest wall that are elevated in temperature, by placing their bodies against that section of nest wall, and allowing their bodies to act as thermal buffers (Starks & Gilley, 1999; Starks, *et al.*, 2005).

The ability to regulate temperature to a specific setpoint despite environmental challenges is just one context in which group living bees, and especially eusocial colonies, regulate their social organization and local environments. Complex bee societies also regulate their internal status to successfully exploit and buffer the dynamic environments in which they live. A honey bee colony, for example, regulates pollen intake homeostatically around immediate needs for brood production and around pollen stores already in the colony (Fewell & Winston, 1992; Pankiw, *et al.*, 1998). Brood production itself is in turn regulated around current brood levels, food availability, and seasonality (reviewed in Schmikl & Crailsheim, 2004). The colony can do this because it functions as a distributed network, with strong lines of communication across bees involved in these interconnected tasks (reviewed in Fewell, 2003). These homeostatically regulated activities occur as a result of a network of feedback loops from local information; individual workers encountering this information are stimulated to perform tasks based on indicators of task need or availability (reviewed in Beshers & Fewell, 2001; Fewell, 2003; Seeley, 2009).

3.3.4 Mating

Although they form societies primarily made up of females, the social behavior of bee species is strongly connected to mating strategy, both in terms of whether or not a female mates to begin with, as well as to the degree of polyandry (i.e. multiple mating) by those who do mate. With some exceptions (e.g. Danforth, 1991b), most bees mate outside the nest. Males generally compete for access to sexually receptive females (Alcock, 1980; Wcislo, 1987, 1992; reviewed in Ayasse, *et al.*, 2001), such that the probability of multiple mating by most males of social species is probably low.

The question of whether a given female mates (and thus is capable of producing female offspring) is central to eusociality, as it is defined by the presence of functionally sterile individuals (Wilson, 1971). In Hymenoptera, males are produced via unfertilized eggs, and so haplodiploidy generates the unusual condition that unmated females (including sterile workers) can produce male offspring. This means that workers with functional ovaries retain some capacity for reproduction even if unmated, opening the door for social conflict over male production and sex ratio. The mechanisms underlying reproductive division of labor are primarily mediated by nutritional status (reviewed in Wheeler, 1986), and are connected ecologically to seasonal constraints on sociality (Wcislo, 1997). In temperate (and highly seasonal tropical or xeric) climates, primitively eusocial bee colonies generally start from previously mated foundresses that emerge from diapause at the beginning of the season. These new queens are often constrained in the amount of food that they can collect for offspring production, and so produce smaller workers earlier in the season (reviewed in Michener, 1990, 2007). As workers emerge and the colony begins to generate a worker force, it can increase food provisions for later brood, which are nutritionally capable of mating and establishing new colonies.

Nutrition and seasonality do not capture the entire explanation for eusocial variation, however. The percentage of "workers" that mate varies considerably across species. Evidence for temperate halictine bees (Yanega, 1992, 1997) suggests that once eusociality is established, mating limitations may shape a female's social role. In many halictines, workers will actively reject the attempts of males (Wcislo, 1987), even though they may mate later, in the event that a queen dies. As Michener (1990) noted, it is as a puzzle why reproductively capable workers do not mate, since it often takes more time to repel males than to mate, which suggests that reproductively dominant females are actively inhibiting the sexual behavior of subordinates. Empirical evidence supports this, especially at smaller colony sizes (Richards, 2000; Strohm & Bordon-Hauser, 2003). In advanced eusocial colonies, such as the honey bees, workers (rather than queens) police cells to eliminate male-destined eggs laid by other workers. They also select which eggs or young larvae will be reared in queen cells, to replace the current queen upon her reproductive deterioration of the current queen or before swarming (Ratnieks & Visscher, 1989; Tarpy, *et al.*, 2004). The shift to worker policing is indicative of a transition to colony-level selection, in which individual behaviors are directed towards successful colony-level queen replacement, rather than to nepotism (Page, *et al.*, 1989; Carlin & Frumhoff, 1990; Châline, *et al.*, 2005).

3.3.5 Offspring Care

With the exception of brood parasitic species (Wcislo, 1987), all bees provide parental care; this occurs minimally in the form of a nest and an adequate supply of food (Linseley, 1958; Michener, 1974). Among bees, there are two distinct ways of provisioning food (Michener, 1974). In mass-provisioning species, such as halictid bees and indeed most solitary bees, a mixture of nectar and pollen is cached in a brood cell prior to oviposition. Progressive provisioning species, in contrast, periodically provide a quantity of food directly to larvae. These resources may be collected directly prior to feeding, as in some allodapine bees (Schwarz, *et al.*, 2005), or can be collected and stored, allowing them also to be transferred between adult members of the colony, as in the stingless bees, bumblebees, and honey bees.

The connection between parental care and eusociality has long been discussed. Wheeler (1928) originally argued that eusociality originated in family groups containing one or both parents that care for offspring (rather than sister–sister or non-relative associations). Michener (1958, 1985) and Lin & Michener (1972) contrasted this route to eusociality with the alternative of a cohort of same-generation individuals. The distinction between them focused attention upon specifically *who* was providing the care for the young, and hence the relative importance of genetic relatedness among nestmates (Hamilton, 1972; Gadagkar, 1991). Alexander, *et al.* (1991), however, argued that this distinction takes attention away from the more fundamental fact that *somebody* is providing brood care, and argued that parental care ("subsociality" in a broad sense) is a near-universal precursor to eusociality. Although alloparental care is not universally a component of eusociality in other taxa (see Chapters 5 and 6), its presence is certainly near-universal in the eusocial bees.

3.4 The Role of Ecology in Shaping Sociality in Bees

The ecological diversity among the bees matches their variation in social behavior. However, it is useful to summarize the fundamental commonalities that shape their life-histories and social pathways. Because behavior and ecology form a tightly interwoven whole, a large part of the discussion on behavior presented earlier is also relevant to the issues of ecological context discussed here.

3.4.1 Habitat and Environment

As ectothermic insects, most bees are relatively short-lived and have primarily annual or seasonal lifespans. Seasonality also constrains the availability of floral food sources, restricting sociality in all but the perennial eusocial species. The importance of seasonality is evident in several ways. Egg to adult development times vary widely in bees, from several weeks to months (Michener, 1974, 2007; Yanega, 1997; Roubik, 2012). For annual species to develop a eusocial society with overlap of generations, there must be a sufficiently long flowering season to produce at least one generation of workers between the foundress generation and the production of new reproductive gynes and males (i.e. bivoltinism). For perennial species to evolve, they must also solve the challenges of collecting and storing enough food to survive periods of resource dearth, and for temperate species especially, maintaining thermal homeostasis when external temperatures fall below the tolerances of individual bees. These constraints are reflected in the social biogeography of bees. The most diverse clade of highly eusocial bees (Meliponini) occur in the tropics, where floral diversity is greatest and food is available nearly year-round in many localities (Roubik, 2006). Even so, tropical bees are limited by seasonal shifts in resource availability, unless they can access multiple food sources and/or store food (Roubik, 2012).

Seasonality can shape intraspecific variation in social structure, particularly the transition from solitary to group living (Wcislo, 1997; Schürch, *et al.*, 2016). In many species displaying primitively eusociality, the production of a worker generation often drops out at high latitudes (and altitudes in these regions) because the reduced active season does not allow for the addition of a worker generation. Those populations become effectively solitary. Some temperate halictine species (e.g. in *Halictus* and *Lasioglossum)* are fully social at low elevations and solitary at higher ones (e.g. Eickwort, *et al.*, 1996; Soucy & Danforth, 2002).

Seasonal limitations also must be overcome for the development of perennial societies (generally eusocial), as seen in the highly social Meliponini (in the stingless bees) and Apini (in the corbiculates). Even in tropical climates, these societies must survive floral dearth periods. Low temperatures represent an additional challenge in temperate and high-altitude climates. In these environments, food storage capacity is likely critical in facilitating the transition from seasonal to perennial sociality.

The importance of food stores can be illustrated by comparing eusocial bumblebees and honey bees. Both bumblebee and honey bee colonies rely on stored pollen for brood development, and on nectar stores to feed the colony and to maintain brood temperatures above ambient. However, the energy reserves of a bumblebee colony span a timescale of days (Cartar & Dill, 1990; Heinrich, 2004), and the energy-based decisions foragers make vary from morning to afternoon (Cartar & Dill, 1990). Although they extend into the coldest climates of all bees, the lifespan of a temperate bumblebee colony is annual, with new queens hibernating over winter (Heinrich, 2004). In contrast, the stores of a temperate honey bee colony persist for months rather than days (Winston, 1991), and provide the energy capacity to thermoregulate over winter, allowing a perennial colony lifespan (Southwick, 1983; Southwick, *et al.*, 1990).

3.4.2 Biogeography

The bees form a diverse and cosmopolitan clade, extending into all continents except Antarctica (Michener, 2007). Ecological adaptation partially explains the geographic distributions of this specious taxon, which is estimated at over 20,000 species (Danforth, *et al.*, 2006). Different lineages may be found in temperate, xeric, and tropical environments, with some taxa occupying all three. The estimated geographical localities of the various bee taxa are given in Table 3.1. Biogeographically, there are numerous species of specialist bees in xeric temperate regions (e.g. 67 percent of

890 species in arid regions of California, Moldenke, 1979); these are primarily solitary or communal (Wcislo & Cane, 1996; Michener, 2007). In contrast, the Neotropical bee fauna is dominated by pollen generalists, and by social species, such as stingless bees (Michener, 1954; Michener, 2007; Roubik, 2006, 2012). Some of the most highly eusocial species, particularly *Apis* (honey bee) and *Bombus* (bumblebee) species, are found in temperate regions.

The bees (Apoidea) most likely originated in Africa (Michener, 1969; Danforth, *et al.*, 2006). The Melittidae considered most likely to be the earliest branch, are limited to Africa. From their African origin, bees have spread into all areas inhabitable by insects. The allodapine bees nicely illustrate the radiation of social bees (Michener, 1974; Tierney, *et al.*, 2008). Allodapines evolved initially in Africa, with branches subsequently extending into Asia and Australia (Michener, 1977; Tierney, *et al.*, 2008). Some allodapine clades are now widespread, ranging throughout Africa to Australia, whereas others are endemic to specific regions, particularly Australia. There is even a clade endemic to Madagascar.

In contrast to climatic effects on the radiation of warm-adapted species, the temperate-adapted bumblebees (*Bombus*) likely originated in the more temperate Old World, and potentially expanded during a global cooling period around 34 MYA. From these origins, they moved into northern and New World climates, and now represent a potential 250 species across the Northern Hemisphere and some areas of South America (Hines, 2008; Williams, *et al.*, 2008).The genus *Apis* (tribe Apini) includes up to ten recognized species, with the majority distributed through Asia, Africa, and surrounding areas (Arias & Sheppard, 2005), including: the Asiatic honey bee, *Apis cerana*, common to India, Japan and China,; the giant honey bee, *Apis* (*Megapis*) *dorsata*, which is found from India through Northern Asia; and the tiny dwarf honey bee, *Apis (Micrapis) florea*, which extends through the Arabian Peninsula into China, and some parts of Northern Africa (Hepburn & Radloff, 2011a, b). The distributions of other species, however, do not compare to that of the Western or European honey bee, *Apis mellifera,* which likely dispersed from Central Asia, possibly around 1 MYA (Arias & Sheppard, 1996), and now occurs either naturally or as feral populations in all continents except Antarctica (Winston, 1991).

3.4.3 Niches

Nest structure and location shape the primary ecological niches and the evolution of group living in bees. The connections between social evolution, ecological niche and biogeography have been discussed earlier. Nest location and thus nest architecture, itself, may also have influenced the diversity of bee social behavior. Most bees and spheciform wasps nest in the soil, with a vertical tunnel that sometimes branches, and in some taxa, lateral branches then led to cells. Eickwort & Kukuk (1990) presented data for halictid bees showing that lineages with communal behavior had longer branching and lateral tunnels than those with solitary or eusocial behavior. This variation connects with sociality if it is more difficult for a dominant female to monopolize access to cells dispersed in space than for clustered cells, as are found in many eusocial halictine bees. Comparable studies are not available for other groups, but in other halictines the overall nest architecture is the same for solitary and social species (Sakagami & Michener, 1962; see also Batra, 1966; Wcislo & Engel, 1996).

For *Apis*, nest architecture is clearly related to their advanced social behavior; comb is produced by collective behavior and allows for nest expansion as group size increases. The origins of this behavior are not clear, but the other corbiculate bees are also master builders, including the social orchid bee *Euglossa hyancinthina* that builds beautiful top-shaped nests of resin (Wcislo, *et al.*, 2012).

3.5 The Role of Evolutionary History in Shaping Sociality in Bees

Phylogenetically, roughly 20,000 species of bees constituting the Apiformes, plus some lineages of spheciform wasps (sometimes classified as Sphecidae), form a monophyletic taxon, the Apoidea (reviewed in Michener, 2007; Danforth, *et al.*, 2013). All major lineages of bees evolved by the mid- to late-Cretaceous, following the diversification of the flowering plants (Danforth, *et al.*, 2004; Hines, 2008; Cardinal & Danforth, 2013). The phylogenetic evidence suggests that the ancestral state for Apoidea was a resource specialist. Supporting this, the likely basal Melittidae tend to be floral specialists (oligolectic). Ancestrally, solitary females likely constructed ground nests, with tunnels leading to cells in which eggs were laid and provisioned with pollen and nectar.

As noted earlier, the transition from a solitary ancestor to eusociality (and back again to solitary living) has occurred repeatedly in the bees (Wheeler, 1928; Wilson, 1971; Michener, 1974; Cameron & Mardulyn, 2001), especially in the Xylocopinae (carpenter bees) and Halictidae (sweat bees) (Michener, 1985, 1990). Recent studies have attempted to answer the question of the number of independent origins for eusociality, using molecular sequence data coupled with mapping behavioral character states. These studies estimate that the number of independent transitions to eusocial behavior may be far fewer than previously believed: no more than four and possibly only two times in Halictidae (Danforth, *et al.*, 2003; Brady, *et al.*, 2006; Gibbs, *et al.*, 2012), once in the corbiculate Apidae (Cardinal & Danforth, 2011), and once in the common ancestor to Allodapani (Apidae) (Rehan, *et al.*, 2012). These studies also suggest that the numbers of times eusociality (or group living more generally) has been secondarily lost is greater than previously believed (Wcislo & Danforth, 1997), including at least twelve times in Halictinae (Danforth, *et al.*, 2003, 2006; Gibbs, *et al.*, 2012), once in corbiculate bees (Euglossini) (Cardinal & Danforth, 2011), and at least four times in xylocopine bees (Rehan, *et al.*, 2012).

Sociality did not necessarily drive phylogenetic radiations in bees. Although some of the more successful radiations of bees do indeed derive from eusocial clades, much more spectacular radiations have occurred in cleptoparasitic groups (e.g. Nomadinae, 1,286 species) and in primarily solitary clades (e.g. Megachilidae approximately 4000 species) (estimates of species numbers from Michener, 2007). In comparison to other taxa, bees present something of a paradox for social evolution. Although bees are often invoked as a key taxon for understanding the repeated evolution of sociality, they may in fact be more informative for teaching us about the constraining factors that impede the evolution of sociality, and about the conditions that may have precipitated its meltdown in some lineages.

II SOCIAL TRAITS

It is difficult to summarize the social traits for a group in which so many forms of social organization are found, and so we offer only a superficial representation of the diversity of bee social traits. Within this diversity, however, we continue to find general patterns of social strategies, from smaller egalitarian societies, to family groups, to eusociality. As already noted, these social traits – including individual and group longevity, fecundity, mating behavior, and even individual versus social complexity – ultimately trace back to the links between social strategy and ecology.

3.6 Traits of Social Species

3.6.1 Communication and Cognition

Almost by definition, social species must have mechanisms for communication. The diverse communication modalities used by bees include vibrational signaling (notably, they have no ears), chemical communication, and visual (i.e. behavioral) signals. A feature common to social animals is some mechanism to distinguish in-group members from out-group ones; in social insects this is generally a chemical-based recognition system (Fletcher & Michener, 1987). Recognition is likely based upon a stew of chemical sources, including cuticular hydrocarbons, Dufour's gland secretions, and wax (reviewed by Breed, 1998). Because data on cues used for nest recognition (as distinct from vision-based nest localization) are lacking for many lineages, the patterns of evolution in kin recognition signaling across the bees remains unclear. The Dufour's gland, located in the abdomen near the sting apparatus, may underlie one evolutionary pathway. The eusocial wasps analogously use mandibular gland secretions in recognition signaling (e.g. Dani, *et al.*, 1996), suggesting the signaling potential of glandular extracts may play an important role in recognition and thus sociality. As an interesting evolutionary comparison, Linsenmair (1987) reviewed evidence from a beautiful series of studies showing that fecal cues provide a means for nest recognition that is then co-opted for kin recognition in a social isopod.

The general question about the cognitive demands of sociality is not well answered for most weakly social bees. However, there is intriguing evidence that brain development is shaped by social context, even for facultatively social bees (Smith, *et al.*, 2010; Rehan, *et al.*, 2015). The brain mushroom bodies are involved in learning, memory and neural integration (Strausfeld, *et al.*, 1995; Farris, 2013). In *Megalopta* (Halictidae) and *Ceratina* (Apidae) some metrics of brain development indicate enhanced development of select brain regions in dominant females relative to subordinates (Smith, *et al.*, 2010; Rehan, *et al.*, 2015).

Independent of general allometric relationships between brain and body size (Eberhard & Wcislo, 2011), there is a theoretical expectation that brain size should increase with increasing group size and social complexity. This pattern seems to hold, however, only within the scope of social group sizes in which individuals can reasonably recognize each other. As social (eusocial) groups increase in size, individual workers theoretically should become more specialized (Fewell, *et al.*, 2009; Gronenberg & Riveros, 2009). In parallel, brain size should decrease in workers of highly eusocial species with extensive task specialization because a given worker has a smaller behavioral repertoire, and because social recognition becomes based on group membership, not individual identity (Gronenberg & Riveros, 2009). This hypothesis has found some support in bees, wasps, and ants (Gronenberg & Riveros, 2009; O'Donnell, *et al.*, 2015).

3.6.2 Lifespan and Longevity

In general, the lifespan of reproductively active bees follows those known for other insects, in that solitary individuals and workers generally have relatively short lives typical of other insects, while queens are long-lived. The extent of queen lifespan in eusocial species is dependent upon the level of eusociality and climate. However, the difference between queen and worker lifespan seems generally applicable to both temperate and tropical species (Wcislo, *et al.*, 1993; Yanega, 1997; Richards, 2000). In annual temperate eusocial species, such as bumblebees and some sweat bees (*Lasioglossum*), queen lifespan generally extends from emergence in one season to overwintering in diapause, to survival through the next season before succumbing in the next winter (reviewed in Michener, 1974; Richards, 2000). Queens of the Neotropical (but seasonally active) eusocial sweat bee *Lasioglossum umbripenne* (Halictidae) live about a year, while workers live roughly a month, and males live about 14 days (Wille & Orozco, 1970). In species with strongly perennial colonies, however, queen longevity can extend multiple years (Michener, 1974; Winston, 1991). The proximate mechanisms for queen longevity are being explored primarily for honey bee queens; their enhanced longevity and fecundity seem to be related to caste-specific differences in vitellogenin gene expression, and to mechanisms generating higher resistance to oxidative stress (Amdam, *et al.*, 2012; for facultatively social species, see Séguret, *et al.*, 2016).

The longevity of worker bees varies considerably, and is highly dependent upon season, forage availability (which ironically can produce higher forager mortality), and temperature. Indeed, in a three-year study of an *Osmia* bee, Bosch & Vicens (2006) state that both longevity and provisioning rate (influencing fecundity) are "strongly conditioned by stochastic factors". In tropical stingless bees (*Melipona*), workers live approximately 30 to 40 days (Roubik, 1982).

Honey bee workers, however, may survive several months overwinter, or less than a few weeks during periods of active colony growth (Winston, 1991).

3.6.3 Fecundity

Fecundity and type of sociality are interconnected. Most communal species include multiple reproductives, each of which can provision only a limited number of offspring; individual females of most species produce fewer than ten new adults (Michener, 1974, 2007). In contrast, most eusocial species are essentially family groups with a single matriarchal queen and multiple helpers, allowing a disconnect between the individual responsible for egg production and brood care. Success of advanced eusocial species depends in part upon group size, the queen's fecundity becomes orders of magnitude higher, from several to several millions (Michener, 1974, 2007). Their high fecundity is facilitated by the ability of female Hymenoptera to store sperm within a storage organ (the spermatheca). Sperm storage allows queens to mate with one or more males in a single or a few mating flights (Schlüns, *et al.*, 2005), minimizing risk, while maintaining the ability to produce a large number of offspring across a season or across years.

3.6.4 Age at First Reproduction

Age at first reproduction is a somewhat different concept for vertebrates and invertebrates. Bees, being holometabolous, develop from their pupal phase into adults that are capable of mating almost immediately. In social species, there is also the question of whether the concept of "reproduction" includes worker production, or is limited to production of new reproductively capable females and males. For annually living eusocial species production of worker and then reproductive brood is dependent upon the ability to produce two brood cycles in a single season (Hunt & Amdam, 2005). In perennial eusocial colonies, production of workers begins immediately, but colonies rarely produce reproductives until they have undergone a period of colony growth. The life cycle of the perennial eusocial colony is generally considered to have two phases: the ergonomic phase, as the colony grows from initiation to maturity; and the reproductive phase, in which mature colonies reproduce (Oster & Wilson, 1978; Houston, *et al.,* 1988). The length of the ergonomic phase, and the periodicity of reproduction (i.e. production of sexuals), varies with species and ecology.

In eusocial species, the question of reproduction can be posed both at the individual and group levels, because the colony actually reproduces when the queen produces new queens. In most eusocial species, colonies simultaneously produce new queens and males, which disperse to mate. The new queens then individually disperse to initiate a new colony of their own. However, Meliponine (stingless bee) and *Apis* (honey bee) colonies deviate from this norm. A honey bee colony reproduces (i.e. produces a daughter colony) when it swarms, a process of colony fission in which part of the colony leaves with the original queen and the other part stays in the natal nest with her newly mated daughter (reviewed by Winston, 1991; Seeley, 2009). In contrast, stingless bees, first establish a new nest site before the virgin queen flies out to a mating swarm of males (Roubik, 2006). As part of this difference, stingless bee colonies do not replace their standing queen to produce a swarm (reviewed in Roubik, 2006). However, swarms of *Apis* species usually include the colony's established queen, while the new queen (her daughter) remains with the original colony. Finally, honey bee colonies will produce new queens to replace a current queen as she ages and brood production decreases. Thus, for honey bees production of new queens may or may not generate a daughter colony (Winston, 1991). The two very different sequences suggest that honey bee and stingless bee swarming behaviors evolved independently.

3.6.5 Dispersal

Dispersal and mating are interwoven for bees, because most females disperse and mate simultaneously. Some communal and aggregative species show philopatry, with daughters either remaining near areas of their natal nests to construct their own, or returning to that area after overwintering (e.g. Yanega, 1990; Soucy, 2002). Philopatry has an impact on population structure and relatedness, and may be particularly important for eusocial and semisocial groups in which reproductive output may be limited to dominant females (e.g. Soucy, 2002; Ulrich, *et al.*, 2009) because philopatry increases average levels of relatedness.

3.7 Traits of Social Groups

3.7.1 Genetic Structure

Haplodiploidy generates two important effects on relatedness for Hymenoptera. First, it sets up a condition in which females sharing a mother and father in common are more highly related to their sisters than they would be to their daughters (Hamilton, 1964). Second, it generates asymmetries in relatedness, particularly between the sexes (Hamilton, 1972). From the perspective of a female bee, both her female and male offspring equally receive 50 percent of her genetic identity. Assuming no inbreeding, full sisters share an average of 75 percent of their allelic information in common by descent. Males, in contrast, produce only female offspring; because they have only one set of chromosomes, any daughters they produce have 100 percent of their genome. In terms of relatedness, males receive more value by producing offspring than in helping sisters, which share only 25 percent of alleles in common by descent.

What does relatedness within groups mean in terms of social evolution? This question has been such an on going focus that even in 1977, the late Howard Evans eloquently noted:

> *The literature relating to…[inclusive fitness and relatedness]… has recently become so vast that an erudite ant, were she able to grasp it at all, would surely become a hopeless psychotic. A termite, with her cellulose-digesting intestinal symbionts might better be able to handle it* (Evans, 1977).

The debate has not particularly softened since that time. We thus limit our discussion to a consideration of whether levels of relatedness within bee taxa generally support the hypothesis that relatedness may facilitate social evolution. The answer to this question, not surprisingly, depends upon social type. In general, relatedness among females within casteless social groups, particularly communal bee societies, is low, and in many cases approaches zero (Paxton, *et al.*, 1996; Moore & Kukuk, 2002; Kukuk, *et al.*, 2005). The costs and benefits of sociality in these taxa may be the most comparable to other cooperative, but non-eusocial, vertebrate and invertebrate taxa.

Measures of within-colony relatedness in primitively eusocial taxa provide a useful comparative test of the hypothesis that the early evolution of eusociality is facilitated by relatedness, with the associated prediction that monogamy should be the dominant reproductive strategy for primitively eusocial species (reviewed in Strassmann, 2001; Boomsma, *et al.*, 2011). Although mating and relatedness data are surprisingly limited given the diversity of bee species, the few examples tend generally (but not universally) to support this prediction (e.g. Kapheim, *et al.*, 2015). Estimated worker to worker relatedness in the primitively eusocial sweat bee, *Lasioglossum zephyrum* is approximately 0.70, not significantly different from 0.75 (Crozier, *et al.*, 1987), although with evidence of more complex genetic structure in some nests. Similarly, in the primitively eusocial *L. malachurum*, average mating number for queens is estimated at 1.13, suggesting a majority of queens are generally monandrous; however, a significant subset do mate multiply. Estimates of mate number in stingless bees (Meliponini, Peters, *et al.*, 1999), and in bumblebees (*Bombus*, Schmid-Hempel & Schmid-Hempel, 2000), also suggest that a strong majority of species are monandrous. Even in the exceptions, skew in mating frequency may point to monandry as the primary mating form. For example, queens of the polyandrous species, *Bombus hypnorum*, can mate with between 1 to 4 males, but have an effective mate number of 1.12, a similar effective mating number to many facultatively polyandrous ant species (as compiled by Strassmann, 2001).

Comparisons within facultatively eusocial/semisocial taxa suggest that relatedness, and concomitant indirect fitness gains, do not provide a complete argument for early social evolution in bees, as expected by theory (Hamilton, 1964), since ecological costs and benefits must be evaluated as well. In nests of *Ceratina australensis*, a few sisters (usually two) may cooperatively rear the brood of the dominant female. For nests with a non-reproductive secondary female, per capita reproduction (and thus the indirect fitness of the non-reproductive) is lower than that of females nesting alone. Secondary females retain full ability to reproduce, and so their inclusive fitness should be higher if solitary (Rehan, *et al.*, 2014). Thus, other factors beyond the indirect fitness benefits from relatedness must play a role in maintenance of facultative sociality for this species. These may involve, but are not limited to, costs of dispersal, predation, and/or parasitism for single female nests.

Although the females of most bee species are monandrous (reviewed in Paxton, 2005), mating frequencies by honey bee queens are exceptionally high (Page, 1980; Palmer & Oldroyd, 2000). This matches a similar trend to increasing polyandry with extreme social complexity in ants (reviewed in Oldroyd & Fewell, 2007). Several hypotheses have been tested around the functionality of extreme polyandry in the bees. Worker diversity may improve (1) colony functionality by facilitating genetically based division of labor (reviewed in Beshers & Fewell, 2001; Oldroyd & Fewell, 2007); and (2) colony resistance to disease by reducing the likelihood that a given pathogen will uniformly affect the whole colony (Baer & Schmid-Hempel, 2001; Tarpy, 2003; Seeley & Tarpy, 2007).

3.7.2 Group Structure, Breeding Structure and Sex Ratio

Breeding structure and sex ratios vary across the diversity of bee taxa, corresponding to their variation in social form (i.e. group structure). The breeding structure of communal species should be similar to solitary species, following Fisher's (1930) expectations of equal sex ratio investment. Studies of female investment in male versus female offspring (corrected for body mass) tend to support this expectation (Frohlich & Tepedino, 1986; Johnson, 1988). Across all bee species, females provide the offspring care. Thus, parental investment is female skewed, with the corresponding theoretical expectation that males compete for females (Bateman, 1948). Examples of male to mate competition are numerous; from the aggregative digger bees (Alcock, 1996), in which males wait for females to emerge and then fight over mating opportunities, to the bumblebees, in which males insert mating plugs into females to prevent subsequent males from mating, to the honey bee, in which the genitalia of males actually separate from their bodies during mating to form a massive plug that continues to pump in sperm as the female flies away.

In aggregations, "group size" essentially equals the size of the population, and is generally dictated by the population capacity of nesting sites. The group structure is essentially a series of individual nests aggregated around a suitable habitat. Again, as with other solitary species, sex ratios are expected to follow Fisherian expectations, and males to compete for females. As we move from solitary and aggregative to group living taxa, the nest or colony best represents the individual unit in any consideration of group structure. Thus, group living in bees can also be thought of as a condition in which populations can be subdivided into clearly distinguishable social groups. Nesting group sizes in communal bees are often quite small, which is somewhat unusual when compared to other social insect species. In a survey of group size in insects, Dornhaus, *et al.* (2012) found a larger proportion of bees than ants, wasps, or termites with reported group sizes fewer than ten individuals. Eusocial colony sizes become much larger, and can vary from several individuals (e.g. *Lasioglossum* and allodapine bees, Schwarz, *et al.*, 1998; Strohm & Bordon-Hauser, 2003) to tens of thousands of workers (in *Apis mellifera*, Winston, 1991). As a constant within this variation, however, only one or a few females are by definition actively engaged in reproduction (i.e. are queens, Michener, 1969; Wilson, 1971).

Because of their high reproductive skew, the concept of sex ratio in eusocial colonies becomes much more complex than for other group living bees. More females than males are generally produced by a eusocial colony, but most of these are sterile or functionally non-reproducing workers. The relative costs and relative values of male versus female sexuals also vary. At one end of the eusociality spectrum, the primitively eusocial sweat bee *Lasioglossum malachurum* produces nests with two successive worker broods within a season, totaling a dozen or so workers, followed by production of sexuals. Queens are approximately one and a half times the size of workers, which in turn are slightly larger than males. Thus, if equal numbers of males and reproductive females are produced, reproductive investment into males by mass is half that of females. The picture is complicated, however, in that colonies must produce a minimum of around four workers in the first worker brood cycle, with an additional second worker brood cycle to successfully produce reproductives at all (Strohm & Bordon-Hauser, 2003). Therefore, the consideration of costs of reproductive investment should include worker costs into the equation.

Sex investment becomes an even more complex equation in swarming reproducers (e.g. the honey bee). Colonies generally produce a few hundred or so males each season, and only a few new queens (Winston, 1991). Thus, investment into males seems on the surface to be much higher than investment into females. Colonies producing a new queen, however, must also fission. To do so, they need thousands of sterile workers to accompany the queen in the swarm in order to successfully found a new colony. Trivers & Hare (1976) argued that the high relatedness among females in eusocial hymenopteran colonies should generate a conflict in optimal sex ratios from the perspective of workers (female biased in a 3 to 1 ratio) versus queens (equal investment). Do colonies deviate from equal investment to male versus female reproduction? Empirical measures of reproductive investment in the primitively eusocial species *Halictus rubricundus*, (Yanega, 1996) and *Lasioglossum laevissimum* (Packer & Owen, 1994) showed female-biased sex ratio, of approximately two females per male; intermediate between queen and worker preferences. In an experimental test of queen-worker sex ratio preference, Mueller (1991) removed queens from colonies of the primitively eusocial species, *Augochlorella striata*,, allowing workers to control reproduction. Queen right colonies produced a more female-biased sex ratio than did queenless colonies, suggestive of queen–worker reproductive conflict.

III SOCIAL SYNTHESIS

3.8 A Summary of Bee Sociality

Bees are one of the most studied but perhaps least thoroughly understood of the social taxa. They can be distinguished as a clade by their phenomenal diversity in social behavior and ecology. Observers watching orchid bees in a tropical rainforest, ground nesting bees in a desert, or bumblebees in an alpine meadow would come to very different conclusions about what it means to be a social bee. Yet, fascinatingly, this diverse monophyletic clade is generally limited to a specific form of nutrition (floral resources, primarily pollen and nectar), and constrained seasonally by ectothermy, by resource availability, and energetically by flight demands. Despite this, bees display a biogeographic range and associated diversity of social types that matches or exceeds all other social taxa.

The vast majority of bee species are primarily or entirely solitary, even though behavioral antecedents for sociality seem to be widespread at low frequencies (Wcislo, 1997; Holbrook, *et al.*, 2009; Flores-Prado, 2012). Despite their phylogenetic rarity, the relatively few obligately eusocial clades dominate ecologically. Honey bees (*Apis*) dominate most habitats where they occur, and *Apis mellifera* (either European or Africanized subspecies) has invasively colonized most of the New World (Winston, 1992; Whitfield, *et al.*, 2006). In the world's tropics, stingless bees (Meliponini) are exceptionally diverse and abundant and are ecologically dominant (Roubik, 2012).

Group formation in bees probably occurs for many of the same behavioral advantages as for other taxa: primarily improved offspring care, offspring defense, and resource competition. Attributes more specific to bees include adaptation for floral resources which are potentially widely available; the ability to construct nests that flexibly allow expansion; and a haplodiploid genetic system with high sibling relatedness. Although all of these factors likely facilitated the transition to sociality, they may not be specific drivers. As an example, pollen specialization seems to impede social evolution; however, variation in resource availability does not seem to explain the variation in social behavior seen among generalist bees. Other factors, such food storage and improved modulation of the physical environment, become useful to derived social taxa as they grow larger and more complex, but are unlikely to explain the transition to sociality itself. Despite these issues – or perhaps because of them – the bees offer one of the richest natural history contexts for studying social evolution, and provide an important testing ground for dissecting the mechanisms and selection forces underlying social behavior.

3.9 Comparative Perspectives on Bee Sociality

Sociality has evolved and been evolutionary lost repeatedly in bees. Their sociality cannot be explained by a specific evolutionary or ecological breakthrough, as seen in the ants (Chapter 2) and termites (Chapter 5). Termites, for example, which live in a typically indigestible food for animals, may be socially constrained by the costs of dispersal and a need for transmission of gut symbionts (reviewed in Thorne, 1997; Chapter 5). In contrast, bees experience diverse ecological conditions in their food supply and climate, and express a diversity of life history strategies.

Part of the difference in sociality is also phylogenetic. In the ants, the transition to eusociality likely occurred before their diversification (Wilson & Hölldobler, 2005; Moreau, *et al.*, 2006). Most bees, by contrast, remain solitary. Advanced eusociality in bees is thus phyletically much more restricted than in ants or termites. Some taxa, such as honey bees, stingless bees, and bumblebees, have crossed the so-called "point of no return" and have never evolved secondarily solitary species; however, no halictid species have crossed this threshold (Wcislo & Danforth, 1997). What factors have influenced or impeded the evolution of group living in many clades of bees? What clade-specific biological factors have promoted the evolution of eusociality in some lineages, and a different form of group living in others? The bees represent an ecological and natural history frontier for social evolution, the vast majority of which remains to be discovered.

3.10 Concluding Remarks

Phylogenetically and ecologically, bees are essentially a very large and very successful group of pollen-feeding wasps. Their monophyletic origin from a sphecoid wasp ancestor was followed by a rapid and highly successful radiation, with nearly 20,000 identified species (Michener, 2007). The success of the bees is demonstrated by their cosmopolitan representation. They have successfully found ways to live in the different ecologies of tropical, xeric, and desert environments, as well as evolving social adaptations to face the seasonal climatic challenges of temperate regions.

The bees represent arguably the most richly diverse array of social systems within any single animal taxon. At first glance, this level of social diversity is a puzzle. Bees as a group retain several key ecological similarities, especially in their use of floral resources, and to a lesser degree in their nesting biology (central place foragers utilizing constructed nests). Physiologically females all retain flight, which imposes biomechanical limits on body allometry and energy use. Thus, their diversity in behavior and social type must ultimately be connected to the unique problems that different environments impose, and that sociality can solve. A great challenge is to understand the extent to which the diversity in social forms represents solutions to diverse ecological problems.

Bees represent an open frontier in social biological investigation. They represent new worlds of sociality waiting to be discovered. They also provide tractable natural contexts in which to directly compare the costs and benefits of social versus solitary living. This accessibility for natural behavioral studies is coupled – at least for some species – with a depth of genomic, comparative phylogenetic, and physiological information. This provides a platform from which to ask and answer questions of bee sociality across all levels of analysis. Thus, bees represent one of the most promising of all taxa in which to ask the fundamental question of social evolution: why (and why not) be social?

References

Abrams, J. & Eickwort, G. C. (1981) Nest switching and guarding by the communal sweat bee *Agapostemon virescens* (Hymenoptera, Halictidae). *Insectes Sociaux*, **28**, 105–116.

Alcock, J. (1980) Natural selection and the mating systems of solitary bees. *American Scientist*, **68**, 146–153.

(1996) The relation between male body size, fighting, and mating success in Dawson's burrowing bee, *Amegilla dawsoni* (Apidae, Apinae, Anthophorini). *Journal of Zoology*, **239**, 663–674.

Alexander, R. D. (1974) The evolution of social behavior. *Annual Review of Ecology and Systematics*, **5**, 325–383.

Alexander, R. D., Noonan, K. M., & Crespi, B. J. (1991) The evolution of eusociality. *In*: P. W. Sherman, Jarvis, J. U. M., & Alexander, R. D. (eds.) *The Biology of the Naked Mole Rat*. Princeton, NJ: Princeton University Press, pp. 3–44.

Amdam, G. V., Fennern, E., & Havukainen, H. (2012) Vitellogenin in honey bee behavior and lifespan. *In*: Galizia, G., Eisenhardt, D., & Giurfa, M. (eds.) *Honeybee Neurobiology and Behavior*. Netherlands: Springer, pp. 17–29.

Anderson, K. E., Sheehan, T. H., Eckholm, B. J., Mott, B. M., & Degrandi-Hoffman, G. (2011) An emerging paradigm of colony health: Microbial balance of the honey bee and hive (*Apis mellifera*). *Insectes Sociaux*, **58**, 431–444.

Arias, M. C. & Sheppard, W. S. (1996) Molecular phylogenetics of honey bee subspecies (*Apis mellifera* L.) inferred from mitochondrial DNA sequences. *Molecular Phylogenetics and Evolution*, **5**, 557–566.

(2005) Phylogenetic relationships of honey bees (Hymenoptera: Apinae: Apini) inferred from nuclear and mitochondrial DNA sequence data. *Molecular Phylogenetics and Evolution*, **37**, 25–35.

Ayasse, M. & Paxton, R. J. (2002) Brood protection in social insects. *In*: Hilker, M. & Meiners, T. (eds.) *Chemoecology of Insects Eggs and Egg Deposition*. Berlin: Blackwell, pp. 117–148.

Ayasse, M., Paxton, R. J. & Tengö, J. (2001) Mating behavior and chemical communication in the order Hymenoptera. *Annual Review of Entomology*, **46**, 31–78.

Baer, B. & Schmid-Hempel, P. (2001) Unexpected consequences of polyandry for parasitism and fitness in the bumblebee, *Bombus terrestris. Evolution*, **55**, 1639–1643.

Bateman, A. J. (1948) Intra-sexual selection in *Drosophila. Heredity*, **2**, 249–368.

Batra, S. W. T. (1966) Social behavior and nests of some nomiine bees in India (Hymenoptera, Halictidae). *Insectes Sociaux*, **13**, 145–154.

Beshers, S. N. & Fewell, J. H. (2001) Models of division of labor in social insects. *Annual Review of Entomology*, **46**, 413–440.

Boomsma, J. J., Beekman, M., Cornwallis, C. K., Griffin, A. S., Holman, L., *et al.* (2011) Only full-sibling families evolved eusociality. *Nature*, **471**, E4–E5.

Bosch, J. & Vicens, N. (2006) Relationship between body size, provisioning rate, longevity and reproductive success in females of the solitary bee *Osmia cornuta. Behavioral Ecology and Sociobiology*, **60**, 26–33.

Brady, S. G., Sipes, S., Pearson, A., & Danforth, B. N. (2006) Recent and simultaneous origins of eusociality in halictid bees. *Proceedings of the Royal Society of London B*, **273**, 1643–1649.

Breed, M. D. (1998) Chemical cues in kin-recognition: Criteria for identification, experimental approaches, and the honey bee as an example. *In*: Vander Meer, R. K., Breed, M. D., Espelie K. E., & Winston M. L. (eds.) *Pheromone Communication in Social Insects.* Boulder: Westview Press, pp. 57–78.

Breed, M. D., Guzmán-Novoa, E., & Hunt, G. J. (2004) Defensive behavior of honey bees: Organization, genetics, and comparisons with other bees. *Annual Reviews in Entomology*, **49**, 271–298.

Bromley, S. W. (1948) Honey-bee predators. *Journal of the New York Entomological Society*, **56,** 195–199.

Buchwald, R. & Breed, M. D. (2005) Nestmate recognition cues in a stingless bee, Trigona fulviventris. *Animal Behaviour*, **70**, 1331–1337

Cameron, S. A. & Mardulyn P. (2001) Multiple molecular data sets suggest independent origins of highly eusocial behavior in bees (Hymenoptera: Apinae). *Systematic Biology*, **50**, 194–214.

Cardinal, S. & Danforth, B. N. (2011) The antiquity and evolutionary history of social behavior in bees. *PLoS ONE*, **6**, e21086.

(2013) Bees diversified in the age of eudicots. *Proceedings of the Royal Society of London B*, **280**, 20122686.

Carlin, N. F. & Frumhoff, P. C. (1990) Nepotism in the honey bee. *Nature*, **346**, 706–707.

Cartar, R. V. & Dill, L. M. (1990) Colony energy requirements affect the foraging currency of bumblebees. *Behavioral Ecology and Sociobiology*, **27**, 377–383.

Châline, N., Martin, S. J., & Ratnieks, F. L. (2005) Absence of nepotism toward imprisoned young queens during swarming in the honey bee. *Behavioral Ecology*, **16**, 403–409.

Costa, J. T. & Fitzgerald, T. D. (2005) Social terminology revisited: Where are we ten years later? *Annales Zoologici Fennici*, **42,** 559–564.

Crozier, R. H., Smith, B. H., & Crozier, Y. C. (1987) Relatedness and population structure of the primitively eusocial bee *Lasioglossum zephyrum* (Hymenoptera: Halictidae) in Kansas. *Evolution*, **41**, 902–910.

Danforth, B. N. (1991a) Female foraging and intranest behavior of a communal bee, Perdita portalis (Hymenoptera: Andrenidae). *Annals of the Entomological Society of America*, **84**, 537–548.

(1991b) The morphology and behavior of dimorphic males in Perdita portalis (Hymenoptera: Andrenidae). *Behavioral Ecology and Sociobiology*, **29**, 235–247.

(2002) Evolution of sociality in a primitively lineage of bees. *Proceedings of the National Academy of Sciences USA*, **99**, 286–290.

Danforth, B. N., Conway, L. & Ji, S. (2003) Phylogeny of eusocial *Lasioglossum* reveals multiple losses of eusociality within a primitively eusocial clade of bees (Hymenoptera: Halictidae). *Systematic Biology*, **52**, 23–36.

Danforth, B. N., Brady, S. G., Sipes, S. D., & Pearson, A. (2004) Single-copy nuclear genes recover Cretaceous-age divergences in bees. *Systematic Biology*, **53**, 309–326.

Danforth, B. N., Sipes, S., Fang, J., & Brady, S. G. (2006) The history of early bee diversification based on five genes plus morphology. *Proceedings of the National Academy of Sciences USA*, **103**, 15118–15123.

Danforth, B. N., Cardinal, S., Praz, C., Almeida, E. A. B., & Michez, D. (2013) The impact of molecular data our understanding of bee phylogeny and evolution. *Annual Review of Entomology*, **58**, 57–78.

Dani, F. R., Fratini, S., & Turillazzim, S. (1996) Behavioural evidence for the involvement of Dufour's gland secretion in nestmate recognition in the social wasp *Polistes dominulus* (Hymenoptera: Vespidae). *Behavioral Ecology and Sociobiology*, **38**, 311–319.

Dew, R. M., Tierney, S. M., & Schwarz, M. P. (2015) Social evolution and casteless societies: Needs for new terminology and a new evolutionary focus. *Insectes Sociaux*, **1**, 5-14.

Dornhaus, A. & Chittka, L. (1999) Insect behaviour: Evolutionary origins of bee dances. *Nature*, **401**, 38.

Dornhaus, A., Powell, S., & Bengston, S. (2012) Group size and its effects on collective organization. *Annual Review of Entomology*, **57**, 123–141.

Dunn, T. & Richards, M. H. (2003) When to bee social: Interactions among environmental constraints, incentives, guarding, and relatedness in a facultatively social carpenter bee. *Behavioral Ecology*, **14**, 417–424.

Dyer, F. C. (2002) The biology of the dance language. *Annual Review of Entomology*, **47**, 917–949.

Eberhard, W. G. & Wcislo, W. T. (2011) Grade changes in brain–body allometry: Morphological and behavioural correlates of brain size in miniature spiders, insects and other invertebrates. *Advances in Insect Physiology*, **40**, 155.

Eickwort, G. C. & Kukuk, P. F. (1990) The relationship between nest architecture and sociality in halictine bees. *In*: Veeresh, G. K., Mallik, B. & Viraktamath, C. A. (eds.) *Social Insects and the Environment*. New Dehli: Oxford & IBH Publishing Co. pp. 664–665.

Eickwort, G. C., Eickwort, J. M., Gordon, J., Eickwort, M. A., & Wcislo, W. T. (1996) Solitary behavior in a high-altitude population of the social sweat bee *Halictus rubicundus* (Hymenoptera: Halictidae). *Behavioral Ecology and Sociobiology*, **38**, 227–233.

Engel, P. & Moran, N. A. (2013) The gut microbiota of insects: Diversity in structure and function. *FEMS Microbiology Reviews*, **37**, 699–735.

Evans, H. E. (1977) Commentary: Extrinsic versus intrinsic factors in the evolution of insect sociality. *Bioscience*, **27**, 613–617.

Farris, S. M. (2013) Evolution of complex higher brain centers and behaviors: Behavioral correlates of mushroom body elaboration in insects. *Brain, Behavior and Evolution*, **82**, 9–18.

Fewell, J. H. (2003) Social insect networks. *Science*, **301**, 1867–1870.

Fewell, J. H. & Winston, M. L. (1992) Colony state and regulation of pollen foraging in the honey bee, *Apis mellifera* L. *Behavioral Ecology and Sociobiology*, **30**, 387–393.

Fewell, J. H., Schmidt, S. K., & Taylor, T. (2009) Division of labor in the context of complexity. *In*: Gadau, J. & Fewell, J. H. (eds.) *Organization of Insect Societies: From Genome to Sociocomplexity.* Cambridge, MA: Harvard University Press, pp. 483–502.

Field, J., Paxton, R. J., Soro, A., & Bridge, C. (2010) Cryptic plasticity underlies a major evolutionary transition. *Current Biology*, **20**, 2028–2031.

Fisher, R. A. (1930) *The Genetical Theory of Natural Selection.* Oxford: Clarendon Press.

Fletcher, D. J. C. & Michener, C. D. (eds.) (1987) *Kin Recognition in Animals.* Chichester: Wiley.

Flores-Prado, L. (2012) Evolución de la sociabilidad en hymenopteras: Rasgos conductuales vinculados a niveles sociales y precursors de sociabilidad en especies solitarias. *Revista Chilena de Historia Natural*, **85**, 245–266.

Frohlich, D. R. & Tepedino, V. J. (1986) Sex ratio, parental investment, and interparent variability in nesting success in a solitary bee. *Evolution*, **40,** 142–151.

Gadagkar, R. (1991) Demographic predisposition to the evolution of eusociality: A hierarchy of models. *Proceedings of the National Academy of Sciences USA*, **88**, 10993–10997.

Gerling, D., Velthuis, H. H. W., & Hefetz, A. (1989) Bionomics of the large carpenter bees of the genus Xylocopa. *Annual Review of Entomology*, **34**, 163–190.

Gibbs, J., Brady, S. G., Kanda, K., & Danforth, B. N. (2012) Phylogeny of halictine bees supports a shared origin of eusociality for *Halictus* and *Lasioglossum* (Apoidea: Anthophila: Halictidae). *Molecular and Phylogenetic Evolution*, **65**, 926–939.

Gronenberg, W. & Riveros, A. J. (2009) Social brains and behavior: Past and present. *In*: Gadau, J. & Fewell, J. H. (eds.) *Organization of Insect Societies: From Genome to Sociocomplexity.* Cambridge: Harvard University Press, pp. 377–401.

Grüter, C., Kärcher, M. H., & Ratnieks, F. L. W. (2011) The natural history of nest defence in a stingless bee, *Tetragonisca angustula* (Latreille) (Hymenoptera: Apidae), with two distinct types of entrance guards. *Neotropical Entomology*, **40**, 55–61.

Hamilton, W. D. (1964) The genetical evolution of social behaviour. *II. Journal of Theoretical Biology*, **7**, 17–52.

(1972) Altruism and related phenomena, mainly in social insects. *Annual Review of Ecology and Systematics*, **3,** 193–232.

Harrison, J. F., Woods, H. A., & Roberts, S. P. (2012) *Ecological and Environmental Physiology of Insects.* Oxford: Oxford University Press.

Heinrich, B. (1985) The social physiology of temperature regulation in honeybees. *In*: Hölldobler, B. & Lindauer, M. (eds.) *Experimental Behavioral Ecology and Sociobiology.* Sunderland, MA: Sinauer, pp. 393–406.

(2004) *Bumblebee Economics.* Cambridge, MA: Harvard University Press.

Hepburn, H. R. & Radloff, S. E. (2011a) Biogeography of the dwarf honeybees, *Apis andreniformis* and Apis florea. *Apidologie*, **42**, 293–300.

(2011b) *Honeybees of Asia.* Berlin: Springer.

Hines, H. M. (2008) Historical biogeography, divergence times, and diversification patterns of bumblebees (Hymenoptera: Apidae: *Bombus*). *Systematic Biology*, **57**, 58–75.

Hogendoorn, K. & Velthuis, H. H. W. (1999) Task allocation and reproductive skew in social mass provisioning carpenter bees in relation to age and size. *Insectes Sociaux*, **46,** 198–207.

Holbrook, C. T., Clark, R. M., Jeanson, R., Bertram, S. M., Kukuk, P. F., *et al.* (2009) Emergence and consequences of division of labor in associations of normally solitary sweat bees. *Ethology*, **115**, 301–310.

Houston, A., Schmid-Hempel, P., & Kacelnik, A. (1988) Foraging strategy, worker mortality, and the growth of the colony in social insects. *The American Naturalist*, **131**, 107–114.
Hunt, J. H. (2007) *The Evolution of Social Wasps*. Oxford: Oxford University Press.
Hunt, J. H. & Amdam, G. V. (2005) Bivoltinism as an antecedent to eusociality in the paper wasp genus *Polistes*. *Science*, **308**, 264–267.
Jaramillo, C. & Cárdenas, A. (2013) Global warming and neotropical rainforests: A historical perspective. *Annual Review of Earth and Planetary Sciences*, **41**, 741–766.
Jeanson, R., Kukuk, P. F., & Fewell, J. H. (2005) Emergence of division of labour in halictine bees: Contributions of social interactions and behavioural variance. *Animal Behaviour*, **70**, 1183–1193.
Johnson, M. D. (1988) The relationship of provision weight to adult weight and sex ratio in the solitary bee, Ceratina calcarata. *Ecological Entomology*, **13**, 165–170.
Kapheim, K. M., Nonacs, P., Smith, A. R., Wayne, R. K., & Wcislo, W. T. (2015) Kinship, parental manipulation and evolutionary origins of eusociality. *Proceedings of the Royal Society of London B*, **282**, 20142886.
Ken, T., Hepburn, H. R., Radloff, S. E., Yusheng, Y., Yiqiu, L., *et al.* (2005) Heat-balling wasps by honeybees. *Naturwissenschaften*, **92**, 492–495.
Kocher, S. D. & Paxton, R. J. (2014) Comparative methods offer powerful insights into social evolution in bees. *Apidologie*, **45**, 289–305.
Kukuk, P. F. & Schwarz, M. (1987) Intranest behavior of the communal sweat bee Lasioglossum (Chilalictus) erythrurum (Hymenoptera: Halictidae). *Journal of the Kansas Entomological Society*, **60**, 58–64.
Kukuk, P. F., Bitney, C. & Forbes, S. H. (2005) Maintaining low intragroup relatedness: Evolutionary stability of nonkin social groups. *Animal Behaviour*, **70**, 1305–1311.
Lenoir, A., d'Ettorre, P., Errard, C., & Hefetz, A. (2001) Chemical ecology and social parasitism in ants. *Annual Review of Entomology*, **46**, 573–599.
Lin, N. & Michener, C. D. (1972) Evolution of sociality in insects. *Quarterly Review of Biology*, **47**, 131–159.
Linsenmair, K. E. (1987) Kin recognition in subsocial arthropods, in particular the desert isopod *Hemilepistus reaumuri*. *In*: Fletcher, D. J. C. & Michener, C. D. (eds.) *Kin Recognition in Animals*. New York: John Wiley, pp. 121–208.
Linsley, E. (1958) The ecology of solitary bees. *Hilgardia*, **27**, 543–599.
Malyshev, S. I. (1968) *Genesis of the Hymenoptera and the Phases of their Evolution*. Republished by Springer Online (2012).
McFrederick, Q., Wcislo, W., Hout, M., Mueller, U. 2014. Host developmental stage, not host sociality, affects bacterial community structure in socially polymorphic bee. *FEMS Microbiology Ecology* 88: 398–406.
Michener, C. D. (1954) Bees of Panama. *Bulletin of the American Museum Natural History*, **104**, 1–176.
(1958) The evolution of social behavior in bees. *Proceedings of the 10th International Congress of Entomology*, **2**, 441–447.
(1964) Evolution of the nests of bees. *American Zoologist*, **4**, 227–239.
(1969) Comparative social behavior of bees. *Annual Review of Entomology*, **14**, 299–342.
(1974) *The Social Behavior of the Bees: A Comparative Study*. Cambridge, MA: Harvard University Press.
(1977) Discordant evolution and the classification of allodapine bees. *Systematic Zoology*, **26**, 32–56.

(1985) From solitary to eusocial: Need there be a series of intervening species? *In*: Holldobler, B. & Lindauer, M. (eds.) *Experimental Behavioral Ecology and Sociobiology*. Stuttgart: Fischer, pp. 293–305.

(1990) Reproduction and castes in social halictine bees. *In*: Engels, W. (ed.) *Social Insects*. Berlin: Springer, pp. 77–121

(2007) *The Bees of the World, 2nd Edition*. Baltimore: Johns Hopkins University Press.

Minckley, R. L., Wcislo, W. T., Yanega, D., & Buchmann, S. L. (1994) Behavior and phenology of a specialist bee (*Dieunomia*) and sunflower (*Helianthus*) pollen availability. *Ecology*, **75,** 1406–1419.

Moldenke, A. R. (1979) Host–plant coevolution and the diversity of bees in relation to the flora of North America. *Phytology*, **43**, 357–419

Moore, A. J. & Kukuk, P. F. (2002) Quantitative genetic analysis of natural populations. *Nature Reviews Genetics*, **3**, 971–978.

Moran, N. A. (2015) Genomics of the honey bee microbiome. *Current Opinion in Insect Science*, **10**, 22–28.

Moreau, C. S., Bell, C. D., Vila, R., Archibald, S. B., & Pierce, N. E. (2006) Phylogeny of the ants: Diversification in the age of angiosperms. *Science*, **312**, 101–104.

Mueller, U. G. (1991) Haplodiploidy and the evolution of facultative sex ratios in a primitively eusocial bee. *Science*, **254**, 442–444.

Müller, H. (1872) Anwendung der Darwinschen Lehre auf Bienen. *Verhhandlungen des naturhistorischen Vereines der preussischen Rheinlande und Westphalens*, **29**, 1–96.

Nieh, J. C. (1999) Stingless-bee communication. *American Scientist*, **87**, 428–435.

Nieh, J. C., Kruizinga, K., Barreto, L. S., Contrera, F. A. L., & Imperatriz-Fonseca, V. L. (2005) Effect of group size on the aggression strategy of an extirpating stingless bee, *Trigona spinipes*. *Insectes Sociaux*, **52**, 147–154.

Noll, F. B., Zucchi, R., Jorge, J. A., & Mateus, S. (1996) Food collection and maturation in the necrophagous stingless bee, *Trigona hypogea* (Hymenoptera: Meliponinae). *Journal of the Kansas Entomological Society*, **69**, 287–293.

O'Donnell, S., Bulova, S. J., DeLeon, S., Khodak, P., Miller, S., *et al.* (2015) Distributed cognition and social brains: Reductions in mushroom body investment accompanied the origins of sociality in wasps (Hymenoptera: Vespidae). *Proceedings of the Royal Society of London B*, **282**, 20150791

Oldroyd, B. P. & Fewell, J. H. (2007) Genetic diversity promotes homeostasis in insect colonies. *Trends in Ecology and Evolution* **22**, 408–413.

O'Neil, K. M. (2001) *Solitary Wasps. Behavior and Natural History*. Ithaca: Comstock Publishing Associates.

Oster, G. F. & Wilson, E. O. (1978) *Caste and Ecology in the Social Insects*. Princeton: Princeton University Press.

Packer, L. & Owen, R. E. (1994) Relatedness and sex ratio in a primitively eusocial halictine bee. *Behavioral Ecology and Sociobiology*, **34**, 1–10.

Page, R. E. (1980) The evolution of multiple mating behavior by honey bee queens (*Apis mellifera* L.). *Genetics*, **96**, 263–273.

Page, R. E., Robinson, G. E., & Fondrk, M. K. (1989) Genetic specialists, kin recognition and nepotism in honey-bee colonies. *Nature*, **338**, 576–579.

Palmer, K. A. & Oldroyd, B. P. (2000) Evolution of multiple mating in the genus Apis. *Apidologie*, **31**, 235–248.

Pankiw, T., Page, R. E., & Fondrk, M. K. (1998) Brood pheromone stimulates pollen foraging in honey bees (*Apis mellifera*). *Behavioral Ecology and Sociobiology*, **44**, 193–198.

Paxton, R. J. (2005) Male mating behaviour and mating systems of bees: An overview. *Apidologie*, **36**, 145–156.

Paxton, R. J., Thorén, P. A., Tengö, J., Estoup, A., & Pamilo, P. (1996) Mating structure and nestmate relatedness in a communal bee, *Andrena jacobi* (Hymenoptera, Andrenidae), using microsatellites. *Molecular Ecology*, **5**, 511–519.

Paxton, R. J., Kukuk, P. F., & Tengö, J. (1999) Effects of familiarity and nestmate number on social interactions in two communal bees, *Andrena scotica* and *Panurgus calcaratus* (Hymenoptera, Andrenidae). *Insectes Sociaux*, **46**, 109–118.

Peters, J. M., Queller, D. C., Imperatriz-Fonseca, V. L., Roubik, D. W. & Strassmann, J. E. (1999) Mate number, kin selection and social conflicts in stingless bees and honeybees. *Proceedings of the Royal Society of London B*, **266**, 379–384.

Plateaux-Quénu, C., Plateaux, L., & Packer, L. (2000) Population-typical behaviours are retained when eusocial and non-eusocial forms of *Evylaeus albipes* (F.) (Hymenoptera, Halictidae) are reared simultaneously in the laboratory. *Insectes Sociaux*, **47**, 263–270.

Prager, S. M. (2014) Comparison of social and solitary nesting carpenter bees in sympatry reveals no advantage to social nesting. *Biological Journal of the Linnean Society*, **113**, 998–1010.

Queller, D. C. & Strassmann, J. E. (1998) Kin selection and social insects. *Bioscience,* **48**, 165–175.

Ratnieks, F. L. & Visscher, P. K. (1989) Worker policing in the honeybee. *Nature*, **342**, 796–797.

Rehan, S. M., Leys, R., & Schwarz, M. P. (2012) A mid-Cretaceous origin of sociality in Xylocopine bees with only two origins of true worker castes indicates severe barriers to eusociality. *PLoS ONE*, **7**, e34690.

Rehan, S. M., Richards, M. H., Adams, M., & Schwarz, M. P. (2014) The costs and benefits of sociality in a facultatively social bee. *Animal Behaviour*, **97**, 77–85.

Richards, M. H. (2000) Evidence for geographic variation in colony social organization in an obligately social sweat bee, *Lasioglossum malachurum* Kirby (Hymenoptera; Halictidae). *Canadian Journal of Zoology*, **78**, 1259–1266.

Rosenheim, J. A. (1990) Density-dependent parasitism and the evolution of aggregated nesting in the solitary Hymenoptera. *Annals of the Entomological Society of America*, **83**, 277–286.

Roubik, D. W. (1982) Seasonality in colony food storage, brood production and adult survivorship: Studies of Melipona in tropical forest (Hymenoptera: Apidae). *Journal of the Kansas Entomological Society*, **55**, 789–800.

(1989) *Ecology and Natural History of Tropical Bees.* Cambridge: Cambridge University Press.

(2006) Stingless bee nesting biology. *Apidologie*, **37**, 124–143.

(2012) Ecology and Social Organization of Bees. *In*: *eLS*. Chichester: John Wiley & Sons Ltd, www.els.net.

Roubik, D. W. & Ackerman, J. D. (1987) Long-term ecology of euglossine orchid-bees (Apidae: Euglossini) in Panama. *Oecologia*, **73**, 321–333.

Sakagami, S.F. & Michener, C. D. (1962) *The Nest Architecture of the Sweat Bees (Halictinae): A Comparative Study of Behavior.* Lawrence, Kansas: University of Kansas Press.

Schmickl, T. & Crailsheim, K. (2004) Inner nest homeostasis in a changing environment with special emphasis on honey bee brood nursing and pollen supply. *Apidologie*, **35**, 249–263.

Schmid-Hempel, P. (1998) *Parasites in Social Insects*. Princeton: Princeton University Press.

Schmid-Hempel, R. & Schmid-Hempel, P. (2000) Female mating frequencies in *Bombus* spp. from Central Europe. *Insectes Sociaux*, **47**, 36–41.

Schlüns, H., Moritz, R. F., Neumann, P., Kryger, P., & Koeniger, G. (2005) Multiple nuptial flights, sperm transfer and the evolution of extreme polyandry in honeybee queens. *Animal Behaviour*, **70**, 125–131.

Schürch, R., Accleton C., & Field J. (2016) Consequences of a warming climate for social organisation in sweat bees. *Behavioral Ecology and Sociobiology*, **70**, 1131–1139.

Schwarz, M. P., Bull, N. J., & Hogendoorn, K. (1998) Evolution of sociality in the allodapine bees: A review of sex allocation, ecology and evolution. *Insectes Sociaux*, **45**, 349–368.

Schwarz, M. P., Tierney, S. M., Zammit, J., Schwarz, P. M., & Fuller, S. (2005) Brood provisioning and colony composition of a Malagasy species of Halterapis: Implications for social evolution in the allodapine bees (Hymenoptera: Apidae: Xylocopinae). *Annals of the Entomological Society of America*, **98**, 126–133.

Schwarz, M. P., Richards, M. H. & Danforth, B. N. (2007) Changing paradigms in insect social evolution: Insights from halictine and allodapine bees. *Annual Review of Entomology*, **52**, 127–150.

Schwarz, M. P., Tierney, S. M., Rehan, S. M., Chenoweth L. B., & Cooper, S. J. B. (2011) The evolution of eusociality in allodapine bees: Workers began by waiting. *Biological Letters*, **7**, 277–280.

Seeley, T. D. (1994) Honey bee foragers as sensory units of their colonies. *Behavioral Ecology and Sociobiology*, **34**, 51–62.

(1997) Honey bee colonies are group-level adaptive units. *The American Naturalist*, **150**, S22-S41.

(2009) *The Wisdom of the Hive: The Social Physiology of Honey Bee Colonies*. Cambridge, MA: Harvard University Press.

Seeley, T. D. & Tarpy, D. R. (2007) Queen promiscuity lowers disease within honeybee colonies. *Proceedings of the Royal Society of London B*, **274**, 67–72.

Séguret A., Bernadou A., & Paxton R. J. 2016. Facultative social insects can provide insights into the reversal of the longevity/fecundity trade-off across the eusocial insects. *Current Opinion in Insect Science* **16**, 95–103.

Shorter, J. R. & Rueppell, O. (2012) A review on self-destructive defense behaviors in social insects. *Insectes Sociaux*, **59**, 1–10.

Smith, A. R., Kapheim, K. M., O'Donnell, S. & Wcislo, W. T. (2009) Social competition but not subfertility leads to a division of labour in the facultatively social sweat bee *Megalopta genalis* (Hymenoptera: Halictidae). *Animal Behaviour*, **78**, 1043–1050.

Smith, A. R., Seid, M. A., Jimenez, L. & Wcislo, W. T. (2010) Socially induced brain development in the mushroom bodies of a facultatively social sweat bee Megalopta genalis. *Proceedings of the Royal Society Series B*, 277, 2157–2163.

Soro, A., Field, J., Bridge, C., Cardinal, S. C. & Paxton, R. J. (2010) Genetic differentiation across the social transition in socially polymorphic sweat bee, *Halictus rubicundus*. *Molecular Ecology*, **19**, 3351–3363.

Soucy, S. L. (2002) Nesting biology and socially polymorphic behavior of the sweat bee Halictus rubicundus (Hymenoptera: Halictidae). *Annals of the Entomological Society of America*, **95**, 57–65.

Soucy, S. L. & Danforth, B. N. (2002) Phylogeography of the socially polymorphic sweat bee *Halictus rubicundus* (Hymenoptera: Halictidae). *Evolution*, **56**, 330–341.

Southwick, E. E. (1983) The honey bee cluster as a homeothermic superorganism. *Comparative Biochemistry and Physiology A*, **75**, 641–645.

Southwick, E. E., Roubik, D. W., & Williams, J. M. (1990) Comparative energy balance in groups of Africanized and European honey bees: Ecological implications. *Comparative Biochemistry and Physiology A*, **97**, 1–7.

Starks, P. T. & Gilley, D. C. (1999) Heat shielding: A novel method of colonial thermoregulation in honey bees. *Naturwissenschaften*, **86**, 438–440.

Starks, P. T., Johnson, R. N., Siegel, A. J., & Decelle, M. M. (2005) Heat shielding: A task for youngsters. *Behavioral Ecology*, **16**, 128–132.

Stow, A., Briscoe, D., Gillings, M., Holley, M., Smith, S., *et al.* (2007) Antimicrobial defences increase with sociality in bees. *Biological Letters*, **3**, 422–424.

Strassmann, J. (2001) The rarity of multiple mating by females in the social Hymenoptera. *Insectes Sociaux*, **48**, 1–13.

Strausfeld, N. J., Buschbeck, E. K., & Gomez, R. S. (1995) The arthropod mushroom body: Its functional roles, evolutionary enigmas and mistaken identities. *In*: Breidbach, O. & Kutsch, W. (eds.) *The Nervous Systems of Invertebrates: An Evolutionary and Comparative Approach.* Basel: Birkhäuser, pp. 349–381.

Strohm, E. & Bordon-Hauser, A. (2003) Advantages and disadvantages of large colony size in a halictid bee: The queen's perspective. *Behavioral Ecology*, **14**, 546–553.

Tarpy, D. R. (2003) Genetic diversity within honeybee colonies prevents severe infections and promotes colony growth. *Proceedings of the Royal Society of London B*, **270**, 99–103.

Tarpy, D. R., Gilley, D. C., & Seeley, T. D. (2004) Levels of selection in a social insect: A review of conflict and cooperation during honey bee (Apis mellifera) queen replacement. *Behavioral Ecology and Sociobiology*, **55**, 513–523.

Thorne, B. L. (1997) Evolution of eusociality in termites. *Annual Review of Ecology and Systematics*, **28**, 27–54.

Tierney, S. M., Smith, J. A., Chenoweth, L., & Schwarz, M. P. (2008). Phylogenetics of allodapine bees: A review of social evolution, parasitism and biogeography. *Apidologie*, **39**, 3–15.

Trivers, R. L. & Hare, H. (1976) Haploidploidy and the evolution of the social insect. *Science*, **191**, 249–263.

Ulrich, Y., Perrin, N., & Chapuisat, M. (2009) Flexible social organization and high incidence of drifting in the sweat bee, *Halictus scabiosae*. *Molecular Ecology*, **18**, 1791–1800.

von Frisch, K. (1967) *The Dance Language and Orientation of Bees*. Cambridge, MA: Harvard University Press

Wcislo, W. T. (1987) The role of learning in the mating biology of a sweat bee Lasioglossum zephyrum (Hymenoptera: Halictidae). *Behavioral Ecology and Sociobiology*, **20**, 179–185.

(1992) Attraction and learning in mate-finding by solitary bees, *Lasioglossum* (Dialictus) *figueresi* Wcislo and *Nomia triangulifera* Vachal (Hymenoptera: Halictidae). *Behavioral Ecology and Sociobiology*, **31**, 139–148.

(1997) Are behavioral classifications blinders to studying natural variation?. *In:* Choe, J. C. & Crespi, B. J. (eds.) *The Evolution of Social Behavior in Insects and Arachnids*. Cambridge: Cambridge University Press, pp. 8-13.

(2005) Social labels: We should emphasize biology over terminology and not vice versa. *Annales Zoologici Fennici*, **42**, 565–568.

Wcislo, W. T. & Cane, J. H. (1996) Floral resource utilization by solitary bees (Hymenoptera: Apoidea) and exploitation of their stored foods by natural enemies. *Annual Review of Entomology*, **41**, 257–286.

Wcislo, W. T. & Danforth, B. N. (1997) Secondarily solitary: The evolutionary loss of social behavior. *Trends in Ecology and Evolution*, **12**, 468–474.

Wcislo, W. T. & Engel, M. S. (1996) Social behavior and nest architecture of nomiine bees (Hymenoptera: Halictidae; Nomiinae). *Journal of the Kansas Entomological Society*, **69**, 158–167.

Wcislo, W. T. & Tierney, S. M. (2009) The evolution of communal behavior in bees and wasps: An alternative to eusociality. *In*: Gadau, J. & Fewell, J. H. (eds.) *Organization of Insect Societies: From Genome to Sociocomplexity*, Cambridge, MA: Harvard University Press, pp. 148–169.

Wcislo, W. T., Wille, A. & Orozco, E. (1993) Nesting biology of tropical solitary and social sweat bees, *Lasioglossum (Dialictus) figueresi* Wcislo and *L.*(D.*) aeneiventre* (Friese) (Hymenoptera: Halictidae). *Insectes Sociaux*, **40**, 21–40.

Wcislo, D., Vargas, G., Ihle, K., & Wcislo, W. (2012) Nest construction behavior by the orchid bee *Euglossa hyacinthina*. *Journal of Hymenoptera Research*, **29**, 15–20.

Weidenmüller, A., Kleineidam, C., & Tautz, J. (2002) Collective control of nest climate parameters in bumblebee colonies. *Animal Behaviour*, **63**, 1065–1071.

Wheeler, W. M. (1928) *The Social Insects Their Origin And Evolution*. London: Kegan Paul Trench Trubner and Co Ltd.

Wheeler, D. E. (1986) Developmental and physiological determinants of caste in social Hymenoptera: Evolutionary implications. *The American Naturalist*, **128**, 13–34.

Whitfield, C. W., Behura, S. K., Berlocher, S.H., Clark, A. G., Johnston, J. S., *et al.* (2006) Thrice out of Africa: Ancient and recent expansions of the honey bee, *Apis mellifera*. *Science*, **314**, 642–645.

Wille, A. & Orozco, E. (1970). The life cycle and behavior of the social bee *Lasioglossum* (Dialictus) *umbripenne* (Hymenoptera: Halictidae). *Revista de Biologia Tropical* **17**, 199–245.

Williams, P., Cameron, S. A., Hines, H. M. Cederberg, B., & Rasmont, P. (2008) A simplified subgeneric classification of the bumblebees (genus Bombus). *Apidologie* **39**: 46–74.

Wilson, E. O. (1971) *The Insect Societies*. Cambridge, MA: Harvard University Press.

Wilson, E. O. & Hölldobler, B. (2005) The rise of the ants: A phylogenetic and ecological explanation. *Proceedings of the National Academy of Sciences USA*, **102**, 7411–7414.

Winston, M. L. (1991) *The Biology of the Honey Bee*. Cambridge: Harvard University Press.

(1992) The biology and management of Africanized honey bees. *Annual Review of Entomology*, **37**, 173–193.

Winston, M. L. & Michener, C. D. (1977) Dual origin of highly social behavior among bees. *Proceedings of the National Academy of Sciences USA*, **74**, 1135–1137.

Yagi, N. & Hasegawa, E. (2012) A halictid bee with sympatric solitary and eusocial nests offers evidence for Hamilton's rule. *Nature Communications*, **3**, 939.

Yanega, D. (1990) Philopatry and nest founding in a primitively social bee, *Halictus rubicundus*. *Behavioral Ecology and Sociobiology*, **27**, 37–42.

(1992) Does mating determine caste in sweat bees? (Hymenoptera: Halictidae). *Journal of the Kansas Entomological Society*, **65**, 231–237.

(1996) Sex ratio and sex allocation in sweat bees (Hymenoptera: Halictidae). *Journal of the Kansas Entomological Society*, 69, 98–115.

(1997) Demography and sociality in halictine bees (Hymenoptera: Halictidae). *In*: Choe, J. & Crespi, B. J. (eds.) *The Evolution of Social Behavior in Insects and Arachnids*. Cambridge, MA: Harvard University Press, pp. 293–315.

4 Sociality in Wasps

James H. Hunt and Amy L. Toth

Overview

Wasps encompass solitary, communal, and facultative, obligate, and swarm-founding social species and are important model organisms for study of the origin and elaboration of insect sociality. Common names for social species are hover wasps, paper wasps, yellowjackets, hornets, and swarm-founding wasps. Excepting a few communal species, all social wasps are in a single family, Vespidae. Social wasps occur worldwide except in extreme dry or cold climates. Nourishment dynamics and dominance interactions shape intra-colony social structure. Communication can be chemical, vibrational, or visual. Differentiation of egg-layers and workers can occur among adults or larvae via differential feeding, dominance, and corresponding changes in gene expression. Some species have definitive queen and worker castes determined during larval development. Most colonies are comprised of related individuals, but workers may care for unrelated individuals. The diversity of social forms makes wasps one of the most informative taxa for integrative and comparative studies of ecological and genetic drivers of cooperative behavior and the evolution of insect sociality.

I SOCIAL DIVERSITY

4.1 How Common is Sociality in Wasps?

Order Hymenoptera has about 115,000 described species out of an estimated one million. Social wasps occur in only three of more than 37 families of stinging wasps (Aculeata): Pompilidae, Sphecidae, and Vespidae. All ants and some bees are social, but most Hymenoptera are solitary. Social wasps comprise a minuscule fraction of Hymenoptera.

We sincerely thank two anonymous reviewers who responded to an early draft with detailed and very helpful critiques. Improvements reflecting their substantial knowledge and experience exist throughout the chapter. Raghavendra Gadagkar participated in early discussions and contributed to identifying and organizing topics. We thank the editors, Dustin Rubenstein and Patrick Abbot, for text reviews and for finding a delicate balance between encouragement and prodding in our progress toward completion, although we all agree that a bit more prodding was called for.

4.2 Forms of Sociality in Wasps

All wasp societies are centered on nests within which larvae are reared (Wenzel, 1991). Few Pompilidae construct nests (Evans & Yoshimoto, 1962). All Sphecidae and Vespidae construct nests or use pre-existing nest-like cavities, but most Sphecidae and Vespidae are solitary. A few wasp species have communal sociality in which several same-generation females interact among themselves on a shared nest, but their life history is otherwise the same as solitary wasps. Wasp sociality is otherwise founded on solitary life with maternal care (i.e. subsociality) and encompasses facultative, obligate, and swarm-founding sociality (Tables 4.1 and 4.2), all of which occur only in Vespidae.

Vespidae has six currently recognized subfamilies (Carpenter, 1982), although molecular phylogenetic analysis points to "zethines" as a seventh subfamily, Zethinae (Hines, *et al.*, 2007) (Figure 4.1). Figure 4.2 uses the phylogeny of Hines, *et al.* (2007) to illustrate the forms of sociality in the subfamilies of Vespidae. Solitary life histories characterize Euparagiinae (1 genus, 10 species, Carpenter & Kimsey, 2009), Masarinae, (19 genera, 250 species [including one communal species], Gess, 1996), and most Eumeninae including "zethines" (180 genera, more than 3,000 species, Carpenter, 1986). Subfamilies Stenogastrinae, Vespinae, and Polistinae (shown divided into its four tribes [names ending . . .ini]) are comprised entirely of social species. All Stenogastrinae (stenogastrines) have facultative sociality. Vespinae (vespines) have obligate sociality, excepting the swarm-founding genus *Provespa*. Polistinae (polistines) is comprised of paper wasps, which have obligate sociality, and three independent lineages of swarm-founding wasps: Epiponini (epiponines), *Polybioides*, and *Ropalidia* subgenus *Icarielia*. Reproductive and worker castes exist only among females, which are the focus of discussion unless males are specifically mentioned.

Although phylogenetic analysis of combined morphological, behavioral and molecular data yields a single origin of sociality (Pickett & Carpenter, 2010; Piekarski, *et al.*, 2014), four molecular phylogenetic studies that exclude phenotype data (Schmitz & Moritz, 1998, 2000; Hines, *et al.*, 2007; Piekarski, *et al.*, 2014) and four re-analyses of the Hines, *et al.* (2007) data by Pickett & Carpenter (2010) show Stenogastrinae to be a separate clade from Polistinae and Vespinae, implying that sociality evolved twice in Vespidae.

4.2.1 Communal Sociality

Communal nesting occurs in a few species of Crabronidae, Pompilidae, Sphecidae, and subfamily Eumeninae of Vespidae. Two or more same-generation adults construct a nest of aggregated brood cells where females interact, but each produces its own offspring. Communal societies lack castes and often contain non-relatives (Wcislo & Tierney, 2009). Possible benefits include passive or active nest defense and shared nest construction costs. Possible costs include nest cell usurpation and theft of prey items from nestmates' nest cells.

Table 4.1 Grades of wasp sociality and example taxa. Taxa and references in the table are a small sample of a large literature. Evans & West-Eberhard (1970), Spradbery (1973), and Hunt (2007) review both solitary and social wasps. O'Neill (2001) reviews solitary wasps. Spradbery (1973) and Ross & Matthews (1991) review social wasps. Turillazzi (1991, 1996, 2012) reviews Stenogastrinae. Gadagkar (1991a) reviews *Polistes, Belonogaster, Parapolybia* and independent-founding *Ropalidia*. Turillazzi (1996) reviews *Belonogaster*. Gadagkar (1996, 2001) treats a single species, *Ropalidia marginata*, in depth. Reeve (1991) and Turillazzi & West-Eberhard (1996) review *Polistes*. Edwards (1980), Matsuura & Yamane (1990), Matsuura (1991), Greene (1991), and Archer (2012) review Vespinae. Jeanne (1991b) reviews the swarm-founding genera *Ropalidia, Polybioides, Provespa*, and polistine tribe Epiponini.

Grade of sociality and example taxa	References
Solitary	O'Neill, 2001
Vespidae	
Euparagiinae	Clement & Gressell (1968); Trostle & Torchio (1986)
Masarinae	Gess (1996)
majority of Eumeninae	Cowan (1991)
Communal	Wcislo & Tierney (2009)
Pompilidae	
Auplopus semialatus	Wcislo, *et al.* (1988)
Sphecidae	
Cerceris australis	Evans & Hook (1982)
Vespidae	
Masarinae	
Trimeria howardi	Zucchi, *et al.* (1976)
Eumeninae	
Zethus miniatus	West-Eberhard (1987a)
Communal with Overlapping Generations	
Vespidae	
Eumeninae	
Montezumia cortesioides	West-Eberhard (2005)
Variably Solitary to Social	
Sphecidae	
Pemphredoninae	
Microstigmus comes	Lucas, *et al.* (2011)
Arpactophilus mimi	Matthews & Naumann (1988)
Spilomena subterranea	McCorquodale & Naumann (1988)
Facultative Sociality	
Vespidae	
Stenogastrinae	Field (2008); Turillazzi (1991, 2012)
Eustenogaster species	Hansell (1987)
Obligate Sociality without Queen/Worker Dimorphism	
Vespidae	Ross & Matthews (1991)
Polistinae	Turillazzi & West-Eberhard (1996)
Polistini	
Polistes (200+ species)	Hunt (2007); Reeve (1991)
Mischocyttarini	
Mischocyttarus drewseni	Jeanne (1972)

Table 4.1 (*cont.*)

Grade of sociality and example taxa	References
Ropalidiini	
Ropalidia marginata	Gadagkar (2001)
Belonogaster petiolata	Keeping (1992)
Parapolybia varia	Yamane (1985)
Obligate Sociality with Queen/Worker Dimorphism	
Vespidae	
Polistinae	
Polistini	
Polistes olivaceous?	Alam (1958); Kundu (1967)
Ropalidiini	
Ropalidia ignobilis	Wenzel (1992)
Vespinae	Archer (2012); Matsuura & Yamane (1990)
Vespa crabro	Archer (1993)
Vespula consobrina	Akre, *et al.* (1982)
Dolichovespula maculata	Akre & Myhre (1992)
Swarm-Founding Sociality	
Vespidae	Jeanne (1991a); West-Eberhard (1982)
Polistinae	
Epiponini	
Polybia occidentalis	Hunt, *et al.* (1987); Jeanne (1986)
Ropalidiini	
Ropalidia montana	Jeanne & Hunt (1992); Yamane, *et al.* (1983)
Polybioides tabidus	Turillazzi, *et al.* (1994)
Vespinae	
Provespa anomala	Matsuura (1999)

Subterranean *Cerceris australis* (Crabronidae) nests can reach 100 nest cells with two or three overlapping generations (Evans & Hook, 1982). Female *Auplopus semialatus* (Pompilidae) can cooperatively construct nest cells, defend against parasitoid wasps, and tolerate one another, but only in the absence of available prey (Wcislo, *et al.*, 1988). When spider prey arrive at the nest, aggression ensues and spiders are taken from other nest cells; therefore these wasps act as communal cleptoparasites (Wcislo, *et al.*, 1988).

In Sphecidae, multiple *Arpactophilus mimi* females construct nests in abandoned nest cells of mud dauber wasps. Limited availability of unoccupied mud dauber nest cells may be a factor in *A. mimi* sociality (Matthews & Naumann, 1988). *Spilomena subterranea* varies from solitary to shared nesting with an average of 2.5 females and 0.8 males (McCorquodale & Naumann, 1988). Sociality varies both within and among *Microstigmus* species. About half of *M. comes* nests contain two or more females (Matthews, 1968) that cooperatively mass provision nest cells with Collembola. Genetic data indicate egg laying by only one wasp (Ross & Matthews, 1989a,b). More than one wasp (maximum six) occupied 37 of 58 nests of *M. nigrophthalmus* (Lucas, *et al.*, 2011). *M. nigrophthalmus* nest cells are progressively provisioned, and nests are usually inherited and re-used through several successive generations (Melo, 2000). Thus, *Microstigmus* possesses the most developed sociality outside of Vespidae.

Table 4.2 Life history characteristics and their fitness benefits for grades of wasp sociality.

Grade of sociality	Fitness benefits
Solitary; Communal; Communal with Overlapping Generations; Variably Solitary to Social	
Mass Provisioning	
Construct cell to contain a single offspring; cells may be separate or contiguous in a nest; oviposit in empty cell; provision cell with multiple prey items that have been stung and lightly anesthetized	Multiple prey items obviate need to find single prey item of sufficient mass to rear wasp larva; anesthetization ensures that prey items remain alive and do not decay prior to being eaten by the wasp larva
Truncated Progressive Provisioning	
Construct single cell, lay egg, wait for egg to hatch; place prey items that have been mandibulated until flaccid into cell slowly as the larva eats them and grows, then fill cell quickly with prey and seal it in the later stages of larval growth	Prey items that have been mandibulated until flaccid reduce risk that prey items will introduce endoparasitoids in the prey item into the nest cell where they might kill the wasp larva; mandibulation of the prey item kills it; therefore it will not remain fresh if the cell was mass provisioned, and thus progressive provisioning is necessary
Fully Progressive Provisioning	
Progressively provision single nest cell with prey items that have been mandibulated until flaccid before sealing cell when larva has completed growth and is ready to pupate	Further reduces risk of introducing endoparasitoids in prey items
Facultative Sociality	
Nests founded by single wasp or small group; nest cells may or may not be contiguous; multiple cells open simultaneously and can contain larvae of all developmental stages from egg to fully grown; often but not always multiple adults are at a nest; adults emerge from pupation while larvae are present in other nest cells; indirectly feed multiple larvae simultaneously using prey items that have been lightly minced in the mandibles; feed on clear liquid, probably produced by larvae with large salivary glands; share liquids among adults via mouth- to-mouth trophallaxis; new adults may or may not remain at natal nest	Multi-cell nest accommodates larvae that span ages from egg to fully grown, enabling continuous nest enlargement rather than rather than step-like as in communal wasps; minced prey items obviate risk of introducing parasitoids in prey items; indirect feeding lets larva consume prey item over short time span; multiple adults enable partitioning into maternal reproductive and allomaternal worker roles; emergence of offspring while larvae are present in other nest cells provides opportunity for allomaternal care; clear liquid may be larval saliva and, if so, nourishing to adults that consume it; trophallaxis distributes nourishment among adults; adults have diverse reproductive options
Obligate Sociality	
Nests founded by single wasp or small group; contiguous multiple cells open simultaneously containing larvae of all developmental stages from egg to fully grown; directly feed multiple larvae simultaneously using prey items that have been thoroughly malaxated in the mandibles; imbibe hemolymph from malaxated prey; drink nutritious larval saliva directly from larvae; share liquids among adults via mouth to mouth trophallaxis; first-emerged	Multi-cell nest accommodates larvae that span ages from egg to fully grown, enabling continuous nest enlargement rather than rather than step-like as in communal wasps; malaxation obviates risk of introducing endoparasitoids from prey items; malaxated prey would decay quickly, which is obviated by direct feeding; the prey hemolymph imbibed during malaxation is a source of rich nourishment; larval saliva is a source of rich

Table 4.2 (*cont.*)

Grade of sociality	Fitness benefits
offspring typically remain at natal nest and engage in allomaternal behavior; foundress(es) or other egg laying wasps cannot produce reproductive offspring without allomaternal care by wasps that emerge at the nest; Vespinae have developmental switch point leading to distinct worker and gyne phenotypes; colony lifetime is determinate – annual in temperate zones, longer term in tropics	nourishment; trophallaxis distributes nourishment among adults; allomaternal behavior of first-emerged adults at their natal nest contributes to accelerated egg laying by queen or other egg-layer, accelerated nest expansion, and enhanced nourishment of nestmate larvae that will become reproductives
Swarm-founding Sociality	
Colonies of multiple adults containing reproductive(s) and workers; reproductive units are swarms containing reproductive(s) and workers; never a life cycle stage of reproductive(s) without workers; feeding and nourishment patterns as in obligate sociality	Continuous presence of workers supports higher number of reproductive offspring produced by colony reproductive(s), however reproductive units are new swarms rather than new individual reproductive offspring

Communal Vespidae include *Trimeria howardi* (Masarinae) (Zucchi, et al., 1976) and three Eumeninae, including "zethines": *Zethus miniatus* (West-Eberhard, 1987a), *Xenorhynchium nitidulum* (West-Eberhard, 1987b), and *Montezumia cortesioides* (West-Eberhard, 2005). There may be other communal eumenines (West-Eberhard, 2005). *M. cortesioides* reuse brood cells on their shared nest, compete strongly for empty cells, and steal prey from open cells (West-Eberhard, 2005). Communal eumenines progressively provision their larvae (West-Eberhard, 2005), a life history trait also found in all Vespidae with more complex sociality.

4.2.2 Facultative Sociality

Found exclusively in Stenogastrinae, the hover wasps, most or all facultatively social females, mate and can lay male (unfertilized) and female (fertilized) eggs. Nests are typically small, reaching about 100 cells at most, with an average of two to four females and thirteen the known maximum (Turillazzi, 2012). Some species average fewer than two females per colony, thus they can be called "social" for only brief periods of their colony cycle (Turillazzi, 2012). Adult interactions include sharing foraged prey items (often via theft from foragers), mouth-to-mouth exchange of nourishment liquids (trophallaxis), dominance, and fighting. All show maternal and/or allomaternal (i.e. worker) larval care. Multiple females often have developed ovaries, but a single dominant or principal egg layer typically lays all the eggs. Reproductive options include independent nest founding, usurping a nest from its foundress, joining a single female at a newly founded nest, joining a multi-female association, adopting an orphan nest containing older brood, or working at the natal nest awaiting ascent in the dominance hierarchy to become the egg layer. Queen turnover is high, nest-sharing adults are often unrelated, and care of unrelated larvae is common. Non-reproducing worker status for a

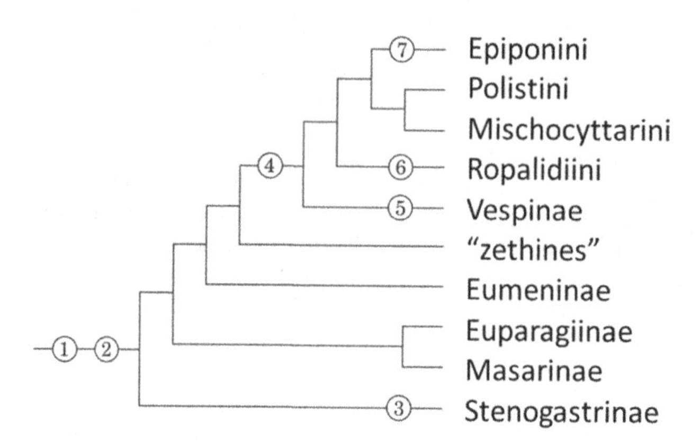

Figure 4.1 Subfamilies of Vespidae and Tribes of Polistinae from a 4-gene molecular phylogeny. Eumeninae *sensu stricto* and "zethines" taken together have traditionally been considered to comprise the subfamily Eumeninae (*sensu lato*). Subfamily Polistinae is subdivided into its four tribes (names ending in . . . ini). Major traits salient to sociality are placed on numbered branches where they first appear. **1**: Hymenoptera [H] Suborder Aculeata [A] – haplodiploidy [H], monandry [H], emergence from pupation with undeveloped ovaries [A], constraint by the wasp waist to obtain liquid nourishment [A]. **2**: (Vespidae) Females solitary, nesting behavior, oviposition precedes larval provisioning, nest cells constructed and provisioned sequentially, one offspring per nest cell, larvae provisioned with intact prey. **3**: (Stenogastrinae) Solitary foundress or co-foundresses, multiple larvae provisioned simultaneously until pupation, larvae progressively provisioned with minced or lightly malaxated prey, provisioning indirect (not mouth to mouth), adults lap saliva from ventral portion of larvae (not mouth-to-mouth). **4**: (Vespinae + Polistinae) Solitary foundress, multiple larvae provisioned simultaneously until pupation, provisioning direct (mouth-to-mouth) with thoroughly malaxated prey, adults at the nest drink prey hemolymph during malaxation and larval saliva (mouth-to-mouth). **5**: (Vespinae) Morphologically discrete worker and queen castes, most species with obligate sociality, one single-queen swarm-founding genus. **6**: (Ropalidiini) Some species with obligate sociality, other species are multi-queen swarm-founding. **7**: (Epiponini) Multi-queen swarm-founding colonies. Details and additional traits are given in Hunt (1999, 2007).

full lifetime characterizes failure to ascend the dominance hierarchy and become the dominant egg-layer (Field, *et al.*, 1999; Shreeves & Field, 2002).

4.2.3 Obligate Sociality

Vernacular names for wasps with obligate sociality (usually called independent founding) are paper wasp for independent-founding Polistinae, hornet for *Vespa*, and yellowjacket for *Vespula* and *Dolichovespula* (sometimes called blackjackets). In obligate sociality, a solitary foundress, albeit of a social species, is *de facto* a solitary wasp that constructs the nest, lays the eggs, and forages for nourishment for herself and her larvae before her first adult offspring emerge (Hunt, 2007). First-emerged offspring remain at their natal nest and direct allomaternal (worker) behavior toward larvae that are not their own. This occurs before the time in their lives they would care for their own larvae; therefore, it is heterochronic expression of maternal behavior (Linksvayer & Wade, 2005). The foundress diminishes her work and increases egg laying as workers assume larval care. The foundress becomes queen, and larvae that workers care for become her reproductive

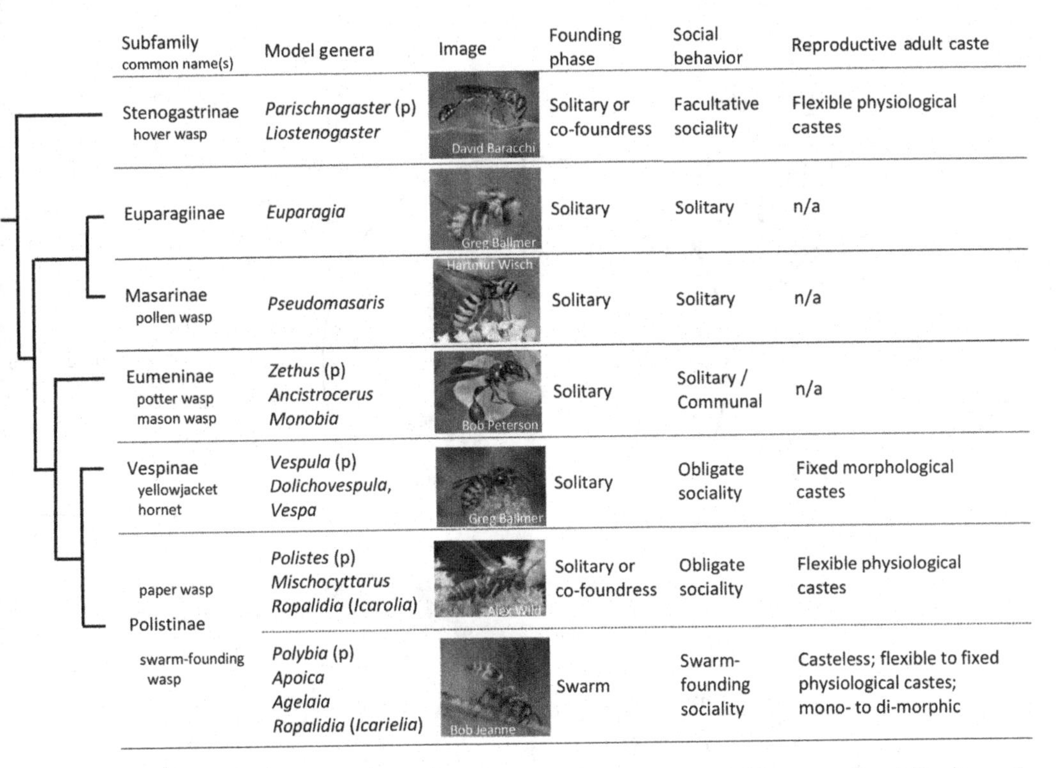

Subfamily common name(s)	Model genera	Image	Founding phase	Social behavior	Reproductive adult caste
Stenogastrinae hover wasp	*Parischnogaster* (p) *Liostenogaster*	David Baracchi	Solitary or co-foundress	Facultative sociality	Flexible physiological castes
Euparagiinae	*Euparagia*	Greg Ballmer	Solitary	Solitary	n/a
Masarinae pollen wasp	*Pseudomasaris*	Hartmut Wisch	Solitary	Solitary	n/a
Eumeninae potter wasp mason wasp	*Zethus* (p) *Ancistrocerus* *Monobia*	Bob Peterson	Solitary	Solitary / Communal	n/a
Vespinae yellowjacket hornet	*Vespula* (p) *Dolichovespula,* *Vespa*	Greg Ballmer	Solitary	Obligate sociality	Fixed morphological castes
Polistinae paper wasp	*Polistes* (p) *Mischocyttarus* *Ropalidia* (*Icarolia*)	Alex Wild	Solitary or co-foundress	Obligate sociality	Flexible physiological castes
Polistinae swarm-founding wasp	*Polybia* (p) *Apoica* *Agelaia* *Ropalidia* (*Icarielia*)	Bob Jeanne	Swarm	Swarm-founding sociality	Casteless; flexible to fixed physiological castes; mono- to di-morphic

Figure 4.2 Vespidae phylogenetic tree showing model genera and features of sociality for each subfamily. Almost all species of Vespinae are characterized by solitary foundresses, except swarm-founding with a single queen in *Provespa*. Polistinae is divided into paper wasps and swarm-founding wasps. The solitary founding life history, including co-founding, is called independent founding. Swarm-founding with multiple queens evolved three times in Polistinae: Epiponini in the Western Hemisphere (e.g. *Polybia, Apoica, Agelaia*) and, in the Eastern Hemisphere,, *Polybioides* and *Ropalidia* subgenus *Icarielia*. Swarm-founding therefore evolved independently four times in Vespidae. In the figure, (p) signifies photograph.

offspring. Because the queen's production of reproductive offspring requires workers, sociality is obligate. Obligately social polistines lack morphological castes, and all females can be inseminated and lay eggs; therefore, they are sometimes called "primitively eusocial." Obligately social vespines are sometimes called "advanced eusocial" because queens are discretely larger than workers and can mate and store sperm, whereas workers cannot (Gotoh, *et al.*, 2008). These traits phylogenetically separate Vespinae from Polistinae, but their obligate sociality life history is the same.

There are exceptions to obligate sociality. Some single foundress *Polistes biglumis* produce reproductive offspring as the first and only brood (Fucini, *et al.*, 2009). *Ropalidia formosa* nests with only two wasps are "nearly solitary" (Wenzel, 1987a). Several Polistinae and Vespinae are obligate social parasites that evict the foundress of another species from her nest at about the time her first offspring emerge, and the foundress' workers thereafter rear reproductive offspring of the social parasite, which has no worker offspring (Cervo & Dani, 1996; Greene, 1991).

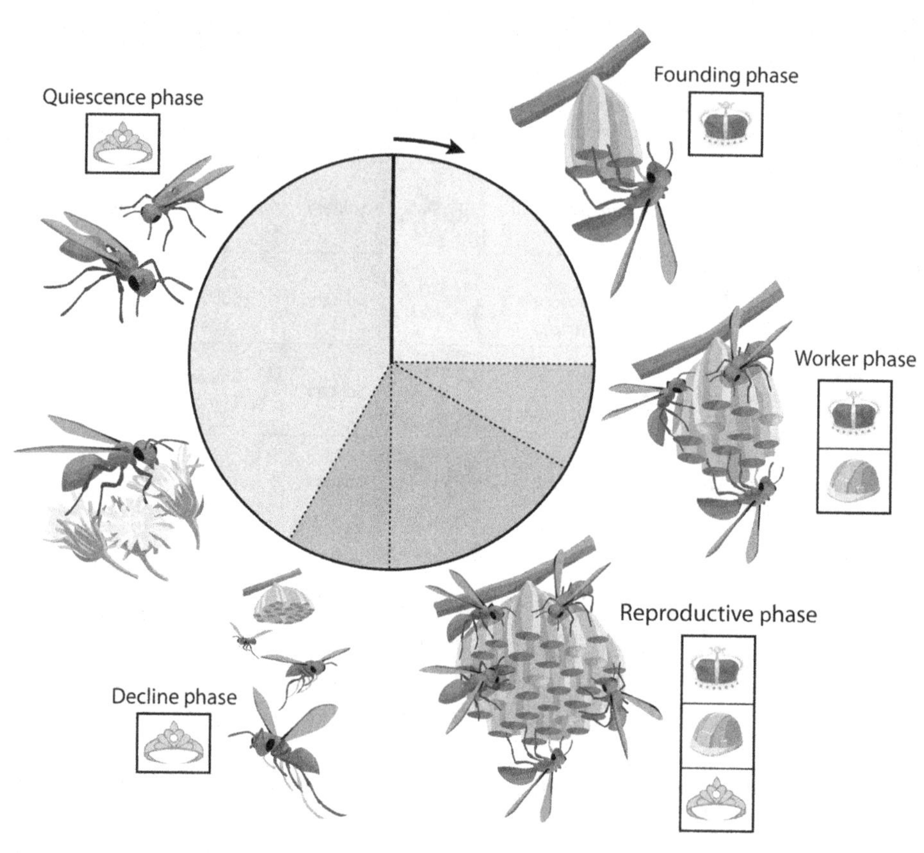

Figure 4.3 Colony cycle of an annual paper wasp, exemplified by *Polistes metricus*. *Founding Phase*. The annual cycle begins (arrow) when a female emerges from quiescence with undeveloped ovaries. Feeding at flowers enables ovary development, which precedes nest founding. Oviposition occurs immediately upon initiation of each nest cell by the now-maternal foundress (crown icon) that performs all maternal behaviors in addition to oviposition, including foraging, feeding larvae, and nest construction. At this stage she is *de facto* a solitary wasp. *Worker Phase*. The first female offspring to emerge in early summer are workers (hard hat icon) that forage, feed larvae, and construct the expanding nest. The foundress transitions into a true social queen that limits her activities to oviposition and feeding larvae using foods brought to the nest by workers. *Reproductive Phase*. Female offspring emerging in mid-summer are non-working future foundresses called gynes (tiara icon). Males, not shown, emerge synchronously with gynes. *Decline Phase*. In late summer the queen and workers die. Gynes and males depart the nest, which now is empty and unattended. *Quiescence Phase*. Gynes and males mate and feed at fall flowers. With the onset of cooler weather, males die, and gynes enter a torpor-like quiescence in sheltered concealment. Workers live, at most, only from their emergence until the end of the reproductive phase. A gyne from one year can become a foundress and then queen the following year, thus the tiara to crown transition occurs in individuals. Figure reprinted from Hunt, *et al.* (2011).

Figure 4.3 shows the annual colony cycle of *Polistes metricus*, a temperate zone single-foundress paper wasp with obligate sociality. A solitary foundress constructs the nest and rears her first offspring. Workers emerge physiologically primed for ovary development and reproduction (Bohm, 1972; Mead, *et al.*, 1995), and when they encounter larvae in other nest cells, those larvae act as releaser cues for allomaternal behavior. Non-working females called gynes emerge in mid-to-late portions of the active phase. Gynes have characteristics of reproductive diapause (Deleurance, 1949, 1952; Hunt & Amdam, 2005; Hunt, *et al.*, 2007, 2010) and never exhibit ovary development during the year in which they emerge. Instead, they mate, overwinter, and become foundresses and, after the first workers emerge, colony queens in the following year. Paper wasp colonies lasting longer than one year occur in moist, aseasonal tropics. Phylogeographic cladistics analysis places the origin of *Polistes* in Southeast Asia (Santos, *et al.*, 2014).

Most *Vespa*, *Vespula*, and *Dolichovespula* have the same life cycle as temperate zone paper wasps. Some *Vespa* living in aseasonal tropics of southeast Asia have colonies lasting longer than one year, and some tropical hornets have multiple foundress colonies formed when several inseminated gynes join a foundress on its nest prior to worker emergence (Matsuura, 1991; Kojima, *et al.*, 2001). Some yellowjackets occasionally have colonies that last longer than one year and produce gynes that leave the nest, mate, and return as supernumerary queens (Greene, 1991; Vetter & Vischer, 1997; Pickett, *et al.*, 2001).

4.2.4 Swarm-founding Sociality

Swarm founding is a major evolutionary transition *sensu* Maynard Smith & Szathmáry (1995). Reproductive propagules are swarms containing a single queen (*Provespa*, Matsuura, 1999) or multiple queens (all others) and a larger number of workers that leave their natal colony and establish a new nest elsewhere. Queens are exclusively egg-layers. Workers perform all other roles. Swarm-founding Polistinae are the Paleotropical genera *Polybioides* and *Ropalidia* subgenus *Icarielia* and the Neotropical tribe Epiponini, which contains nineteen genera (Carpenter, 2004).

4.3 Why Wasps Form Social Groups

Several ecological and life history factors contribute to the success of social groups in wasps. Social nesting can provide benefits including improved food acquisition, defense against predators and parasites, regulation of nest homeostasis, and distributed work in the care of young.

4.3.1 Resource Acquisition and Use

Resource acquisition by nesting wasps is limited by a wasp's foraging range. Most solitary species take only a narrow range of prey items, e.g. a single species, genus, or family or, in mud daubers, spiders in a narrow size range. Social wasps take a broader taxonomic diversity and size range of prey. This evolutionarily significant correlate of sociality enables social wasps to collect more prey biomass than solitary wasps and rear more offspring.

A challenge for all predatory wasps is that intact prey may contain parasitoid eggs or larvae that can then consume the wasp larva. Social wasps and a few related solitary species have adapted to the parasitoid problem by progressively provisioning their larvae through the full course of larval development with prey items that have been thoroughly pre-processed. Prey are minced into small fragments by Stenogastrinae or thoroughly malaxated (i.e. kneaded using the mandibles into an unrecognizable mass) by Polistinae and Vespinae, thereby killing any parasitoid eggs or larvae in the prey.

4.3.2 Predator Avoidance

Foraging solitary wasps cannot defend nests against cleptoparasites (usually other wasp species) that enter nest cells and lay an egg that becomes a larva that feeds on provisions gathered by the nesting wasp. Multiple adults on social wasp nests can defend against cleptoparasites and parasitoids that attempt to inject eggs into larvae. Nest predation by birds and by hornets in the Paleotropics are major causes of colony failure. Social wasps have diverse of visual behaviors to deter predators (Akre, 1982; O'Donnell, *et al.*, 1997). Some epiponines can spray highly irritating venom into vertebrate predators' eyes (Hunt, *et al.*, 2001b). Vespinae and many swarm founding wasps have painful stings and aggressive nest defense.

As defense against ants, some Stenogastrinae use a gelatinous glandular secretion applied in rings around the narrow exposed plant rootlet to which their nest is attached (Turillazzi & Pardi, 1981; Turillazzi, 1987; Sledge, *et al.*, 2000). Paper wasp nests hang from a slender stalk (nest petiole) that adults rub with the underside of their abdomen, depositing a non-gelatinous glandular secretion with ant repellent properties (Jeanne, 1970; Smith, *et al.*, 2001). Enclosed nests of swarm-founding species, many of which lack the gland, may be an adaptive defense against ant predation (Smith, *et al.*, 2001).The small epiponine *Parachartergus colobopterus* can spray venom that may repel ants (Jeanne & Keeping, 1995). Ant predation on wasps' brood probably plays a major role in short nest durations of many tropical species (Jeanne, 1991a).

Polistes single foundress colonies generally have higher failure rates than co-foundress colonies (Gibo, 1978; Reeve, 1991; Field, *et al.*, 1998a; Tibbetts & Reeve, 2003), indicating that predation pressure can favor multiple foundress associations (Strassmann, *et al.*, 1988; Tindo, *et al.*, 2008).

4.3.3 Homeostasis

Social wasps have diverse adaptations to deal with environmental variation (Seeley & Heinrich, 1981; Jones & Oldroyd, 2007). Unenclosed paper wasp nests provide limited ability to regulate nest temperature, which probably restricts paper wasp distributions to temperate and tropical zones. Possible adaptations to nesting in more northernly latitudes include nesting in sun-warmed sites (Jeanne & Morgan, 1992) and constructing a "functional envelope" of elongated empty cells at the nest periphery (Yamane &

Kawamichi, 1975). Cooling the nest and brood during elevated ambient temperatures is the apparent function of an adult fanning its wings while standing on the nest, often also expelling water droplets to foster evaporative cooling (Pardi, 1948; Giannotti, 1999). Even though it is unknown whether any paper wasp maintains an elevated brood temperature (Weiner, *et al.*, 2010), social regulation of colony temperature can be achieved to some degree.

Vespinae have a multi-layered nest envelope and can maintain the brood chamber at or near constant temperature if colony biomass and worker activity generate heat that exceeds heat loss, but thermoregulatory ability is poor at small initial and declining colony sizes (Gibo, *et al.*, 1974a,b; Martin, 1988, 1992). Hornet foundresses can warm their first brood by curling their body around the nest pedicel (a larger attachment than paper wasps' nest petiole) within the initial single-layer envelope (Makino & Yamane, 1980). Hornet workers in empty cells adjacent to pupae can pump the abdomen to generate heat (Ishay, 1973). *Vespa, Vespula* and *Dolichovespula* fan at the nest entrance during high ambient temperatures (Greene, 1991; Ishay, 1973; Riabinin, *et al.*, 2004). Brood chamber temperature regulation probably plays a role in yellowjackets' occurrence in high latitudes.

4.3.4 Mating

Mating in social wasps takes place separately from colony life. Male Stenogastrinae perform "patrolling" flights in open areas that may involve pheromone emission (Turillazzi, 1983; Turillazzi & Francescato, 1990; Beani & Turillazzi, 1994). Females may come to patrolling aggregations for mating (Pagden, 1962; Turillazzi, 1983). *Polistes* occasionally mate on nests (Kundu, 1967; West Eberhard, 1969; Hook, 1982; O'Donnell, 1994), but most mating occurs away from nests (West Eberhard, 1969; Kasuya, 1981a; Reed & Landolt, 1991). *Belonogaster petiolata* males may attempt mating on the nest (Keeping, *et al.*, 1986). Some *Polistes* defend territories that contain no resources of value (Beani, *et al.*, 1992; Polak, 1993) or are near nesting or hibernation sites (West Eberhard, 1969; Kasuya, 1981a; Post & Jeanne, 1983). Larger *P. canadensis* males hold territories for more days than smaller males, and all territorial males are larger than patrolling males (Polak, 1993). Territoriality and either patrolling or searching female feeding sites are alternative mating strategies in *P. canadensis* and *P. fuscatus* (Post & Jeanne, 1983; Polak, 1993). Males of other species patrol common routes, stopping at collective perches (Beani, *et al.*, 1992). Male aggregations that can be leks assemble near nests, hibernacula, or environmental features such as hilltops or other distinctive features (Reed & Landoldt, 1991; Beani, *et al.*, 1992). Males often rub the underside of the abdomen, mandibles, or legs on frequently-used perches, indicating scent marking (Wenzel, 1987b; Beani & Calloni, 1991; Reed & Landolt, 1991).

Most solitary wasps are singly-inseminated (Thornhill & Alcock, 1983), as are most social wasps, including *Polistes* species, the swarm-founders *Parachartergus colobopterus, Brachygastra mellifica* and *Polybioides tabidus*, the hornet *Vespa crabro*, and the yellowjacket genus *Dolichovespula* (Strassmann, 2001). Double or triple inseminations in *Ropalidia marginata, Polistes biglumis, Vespa crabro*, and *Dolichovespula* spp. (Muralidharan, *et al.*, 1986; Seppä, *et al.*, 2011) indicate that a physical constraint to multiple mating is unlikely (Strassmann, 2001). Multiple mating is a derived trait (Foster, *et al.*, 1999) that in *Vespula* is concomitant with colonies that often reach thousands of offspring from a single queen.

4.3.5 Offspring Care

Most eumenines mass (fully) provision nest cells with intact prey and have no larval contact after cell closure. Progressive provisioning with intact prey characterizes some eumenines, including the three communal species, with cell closure shortly before a larva completes development. Stenogastrinae place minced prey near the mouthparts of larvae that then feed on it – a mode called indirect provisioning. Polistinae and Vespinae provision larvae directly: Adults thoroughly malaxate prey and hold it against a larva's mouthparts. During malaxation the adult imbibes prey hemolymph (Hunt, 1984) that is subsequently regurgitated and fed to larvae. Progressively provisioning larvae until pupation with processed prey is a central element of wasp sociality.

4.4 The Role of Ecology in Shaping Sociality in Wasps

Social wasps are cosmopolitan and occur in a wide variety of habitat types. Temperate species with obligate sociality typically have annual colonies. Many tropical species with obligate sociality have colonies that last longer than one year. Tropical wasps attain the largest species-typical colony sizes. Yamane (1996) reviews ecological factors affecting the *Polistes* life cycle. Strassmann & Queller (1989) review ecological factors affecting the evolution of wasp sociality.

4.4.1 Habitat and Environment

Social wasps live in vegetated deserts, arid scrublands, dry forests, mesic forests, rainforests, high latitude/altitude coniferous forests, and arctic tundra. Paper wasps nesting in vegetated deserts can experience temperatures as high as 45°C; those nesting in high latitudes can experience temperatures as low as 0°C. Sheehan, *et al.* (2015) give an in-depth investigation of environmental variables that affect the frequency and biogeographic distribution of multiple foundress *Polistes* colonies. Some *Polistes* and *Mischocyttarus* gynes migrate to higher elevation, cooler sites during the lowland dry season and return as foundresses when the wet season begins (Hunt, *et al.*, 1999; Manzanilla, *et al.*, 2000; Gobbi, *et al.*, 2006). Swarm-founding sociality generally requires an equable year-round climate, but some Epiponini nest in regions with strong wet/dry seasonality and diminish or suspend brood rearing during the dry season. Others migrate seasonally to cooler, moister, sometimes higher elevation sites, remain clustered there without constructing a nest, then return and establish a new nest (Naumann, 1970; Hunt, *et al.*, 2001a).

4.4.2 Biogeography

Asynchronous and aseasonal colony cycles of Stenogastrinae reflect their distribution exclusively in lowland wet tropics of the Indo-Pacific. Paper wasps are distributed worldwide in temperate and tropical zones: cosmopolitan *Polistes*, Western Hemisphere *Mischocyttarus*, East Asian *Parapolybia*, sub-Saharan *Belonogaster*, and *Ropalidia* subgenus *Icarolia* in Afrotropical, Indomalayan and Australasian regions. Swarm-founding wasps exist only in tropical and, in a few species of Epiponini, lower subtropical latitudes (e.g. Sugden & McAllen, 1994). Vespinae are distributed primarily in temperate and boreal latitudes, although some *Dolichovespula* extend well into high latitudes and some *Vespa* occur in tropical lowlands of southern and southeastern Asia (Matsuura & Yamane, 1990). The lowest latitude of a non-adventive Western Hemisphere vespine is the yellowjacket *Vespula squamosa* in highland forests of Honduras (Hunt, *et al.*, 2001b).

4.4.3 Niches

Social wasps take a wide range of prey and occupy a broad predatory niche. Ecological impact of social wasp predation on other insects and arachnids can sometimes be substantial. In an extreme example, *Vespula vulgaris* and *V. germanica* introduced into New Zealand reach high densities based on abundant honeydew from scale insects and take over 99 percent of available Lepidoptera larvae and spiders (Harris, 1991; Beggs, 2001). Epiponine wasps could be the most significant insect predators of other insects in the Neotropics. Masarinae pollinate flowers as they gather pollen and nectar (Gess, 1996). Paper wasps visit flowers for nectar but are inefficient pollinators.

4.5 The Role of Evolutionary History in Shaping Sociality in Wasps

Hunt (1999) maps traits of morphology and behavior relevant to wasp social evolution onto a phylogeny of Hymenoptera. The phylogenetic tree in Figure 4.1 similarly provides a framework to map evolutionary changes from ancestral to derived states for traits of anatomy, reproduction, and behavior relevant to sociality in Vespidae (Hunt, 2007). Numbered traits on one branch are novel adaptations in the shared common ancestor of all taxa, both solitary and social, on more distal branches. Haplodiploidy enables wasps to choose offspring sex by fertilizing eggs (producing females) or not (producing males), an important ancestral trait for producing only females, which work, during periods of colony growth. Monandry (single insemination) is ancestral. Emergence from pupation with undeveloped ovaries necessitates proteinaceous nourishment prior to oviposition, but the thread waist constrains possible sources to liquids only. Nesting protects larvae. Oviposition in an empty nest cell enables progressive larval provisioning from eclosion from the egg to pupation. Mincing (Stenogastrinae) or malaxating (Polistinae) prey items enables provisioning larvae with easily ingested prey tissue and eliminates risk of introducing parasitoid eggs or larvae. Larval saliva provides a source of proteinaceous nourishment at the nest. It is the key trait in social wasp evolution that attracts newly-emerged females with active reproductive physiology but undeveloped ovaries to remain at their natal nest (Roubaud, 1916; Hunt, 2007), thereby establishing an overlap of generations and context for allomaternal care. Traits at numbered branches near social taxa have higher salience for sociality than traits at more basal branches; however, sociality in Vespidae incorporates a synthesis of all traits shown in Figure 4.1 (Hunt, 1999, 2007).

II SOCIAL TRAITS

The defining feature of social wasps is the existence of caste-containing societies with highly fecund queens and non-reproducing workers. Much work has focused upon understanding behavioral traits associated with sociality, including visual, chemical, and vibrational communication. Similar attention has been given to traits of development, physiology, genetics associated with the evolution of castes, and the evolution of increased female lifespan, especially in queens.

4.6 Traits of Social Species

4.6.1 Cognition and Communication

Nestmate recognition is ubiquitous in social wasps. Individuals must be assessed to either "belong or not belong" to a colony. Aggressive rejection of non-nestmates is based on odors, especially cuticular hydrocarbons (CHCs), acquired from the nest and nestmates (reviewed in Richard & Hunt, 2013). *Polistes* nestmates have a CHC profile distinct from that of other colonies, providing a likely cue to group membership. Young individuals may have a more generic CHC profile and are more likely to be accepted by other colonies or even drift between nests, which appears to be relatively common (Kasuya, 1981b; Tsuchida & Itô, 1987; Gamboa, 2004). In *Polistes dominula* [dominulus] and *Vespa crabro*, CHCs differ between a foundress' offspring and those of a usurping queen or following within-colony queen turnover (Dani, *et al.*, 2004). Workerless obligate social parasite *Polistes* species adopt the CHC odor of the *P. dominula* colony they invade and usurp (Sledge, *et al.*, 2001a; Lorenzi, *et al.*, 2004). CHC profiles differ among *P. dominula* co-foundresses that differ in dominance status, and profiles can dynamically change after an individual ascends in dominance rank (Sledge, *et al.*, 2001b), although there is no clear "signature of" CHC dominance status (Monnin, 2006; Toth, *et al.*, 2014). *Polistes fuscatus* can recognize some non-nestmate kin, presumably due to genetically-based differences in cuticular odor (Gamboa, *et al.*, 1987, Gamboa 1988).

Several *Polistes* species have variable facial color patterns (Tibbetts, 2004), which are also found in at least two species of Stenogastrinae (Baracchi, *et al.*, 2013). In *Polistes dominula*, coloration of the clypeus (facial region between the eyes and above the mandibles) is yellow or yellow with black spots or splotches in varied sizes and shapes (Tibbetts & Dale, 2004). Laboratory studies of *P. dominula* in its invasive range in the United States found that wasps with more broken black patterning tend to have better fighting ability and are dominant in staged contests between wasp pairs (Tibbetts, 2006; Tibbetts & Lindsay, 2008). Greater amounts of black and broken patterning are associated with higher juvenile hormone titers and better nourishment during development, suggesting that clypeal markings indicate reproductive potential (Tibbetts & Curtis, 2007; Tibbetts, 2010). Field studies of *P. dominula* in its native range in Italy and Spain, however, showed no relationship between facial color patterns and size, probability of surviving overwintering, social rank in spring associations, health status, or reproductive success (Cervo, *et al.*, 2008, Green, *et al.*, 2013). Some species, such as *P. fuscatus*, may use highly variable facial color patterns for individual recognition (Tibbetts, 2002; Sheehan & Tibbetts, 2008). Not all *Polistes* species have variable facial color patterns, and visually based individual recognition in *Polistes* likely depends upon the social and nesting context of each species (Greene & Field, 2011; Tibbetts, *et al.*, 2011).

Vibrational signals (reviewed in Hunt & Richard, 2013) occur in nine genera (Brennan, 2007) but are likely under-documented and under-appreciated for their roles in colony life. Three main vibratory signals in *Polistes* are antennal drumming, abdomen wagging, and lateral vibrations of the entire body. All play roles in adult-larva interactions and are performed most often by foundresses in pre-worker emergence nests (Cummings, *et al.*, 1999; Suryanarayanan & Jeanne, 2008; Suryanarayanan, *et al.*, 2011a). Antennal drumming is a maternal manipulation (Alexander, 1974) with an indirect genetic effect on larvae that reduces their reproductive capacity and increases their likelihood of becoming a worker following emergence as adults (Suryanarayanan, *et al.*, 2011a,b; Jandt, *et al.*, 2014).

4.6.2 Lifespan and Longevity

Colony lifespans vary widely in social wasps, with high nest failure rates in many species. Nest failure rates for Stenogastrinae are high (Turillazzi, 2012). The colony cycle for *Parischnogaster nigricans serrei* was 7 to 8 months for three nests (Turillazzi, 2012), although mature *P. nigricans serrei* colonies can last indefinitely (Turillazzi, 1985). Development time from egg to adult emergence can be extremely long – an average of 100 days in *Liostenogaster flavolineata* (Turillazzi, 2012).

In Vespinae, colony failures prior to gyne emergence were 83 percent in *V. crabro*, 75 percent in *V. analis*, and 69 percent in *V. tropica* (Matsuura & Yamane, 1990). Yellowjacket colonies' survival until the end of the colony cycle was 5.5 percent for *Vespula maculifrons* (MacDonald & Matthews, 1981), a combined 29 percent and 54 percent in successive years for *Dolichovespula maculata* and *D. arenaria* nests on buildings (Pallett, 1984), and 63 percent for *Vespula vulgaris* (Archer, 1981).

The epiponine *Parachartergus colobopterus* has log-linear colony survivorship averaging 347 days, 10 percent remaining at 600 days, and 0.5 percent of colonies reaching the maximum 1,500 days (Strassmann, *et al.*, 1997a). Given that a swarm-founding wasp colony can maintain integrity by absconding from its current nest and founding a new nest elsewhere, colonies can arguably be said to approach immortality if they continue to produce new adults and daughter colonies. The epiponines *Chartergus chartarius*, *Epipona tatua*, and *Polybia liliacea* have durable nest exteriors (perhaps to exclude bird or bat predation: Jeanne, 1991a) and can last for years. *Polybia scutellaris* nests can last 25–30 years (Vesy-Fitzgerald, 1938; Richards, 1978).

Solitary wasps' life spans are typically short. The mud dauber *Sceliphron assimile* has log-linear survivorship of nesting females, reflecting stochastic mortality events, with a maximum 45 to 50 day reproductive longevity (Hunt, 1993). Adult longevity in social wasps may be much shorter than colony lifespans and varies widely depending on species, caste, and climate (Toth, *et al.*, 2016). In general, social wasps show the typical reversal of the longevity-fecundity tradeoff also found in other social insects: queens are not only more fecund but also are longer lived than workers. Queen-worker longevity differences are less pronounced in aseasonal tropical species compared to seasonal or temperate species, in which queens usually live for one year, whereas workers may only live for a few weeks. Across species, there is also a trend for large-colony species to have shorter worker longevities than small-colony species (Toth, *et al.*, 2016). Shorter worker lifespans may be selected for in more derived forms of sociality. The "disposable caste/soma" idea posits that colonies should reduce investment in maintenance and lifespan of non-reproductive members with high extrinsic mortality (Lucas & Keller, 2014).

4.6.3 Fecundity

The solitary mud dauber *Sceliphron assimile* averages 1.6 female offspring with a maximum of 26 (Hunt, 1993). In wasps with obligate sociality, mortality can be extremely high for foundresses and small colonies, but the few colonies that reach the gyne production stage can produce many reproductive offspring. It can be said that solitary wasps have high probability of low fecundity, whereas social wasps have low probability of high to very high fecundity. A small number of queens produce the majority of foundresses in the next generation, a demographic factor with a direct fitness effect that would have played a major role in the evolution of obligate sociality (Hunt, 2007).

Fecundity in the facultatively social Stenogastrinae would be low, but movement between nests, usurpation, etc., inhibit taking robust data. *Polistes* colonies rarely exceed 100 contemporaneous adults (Pickett, *et al.*, 2001), but a nest of *Polistes olivaceous* in India had 1,546 nest cells (Alam, 1958), and a *P. annularis* nest in Kansas produced at least 1,500 gynes (Wenzel, 1989). Due to high colony failure rates, maximum colony size is far from typical for any *Polistes* species and at best provides a rough estimate of maximum gyne production. Tropical paper wasps with long-lived colonies can have within-colony queen replacements (serial polygyny), which may be common. Queen durations ranged from 7 to 236 days in *Ropalidia marginata* (Gadagkar, *et al.*, 1993) and 32 to 88 days in *Mischocyttarus drewseni* (Jeanne, 1972).

Typical yellowjacket colonies in the U.S. can reach sizes of 8,000 to nearly 600,000 (MacDonald & Matthews, 1981; Pickett, *et al.*, 2001). Colonies in the southeastern United States and California that overwinter and become multi-queen reach very large sizes (Ratnieks, *et al.*, 1996; Pickett, *et al.*, 2001). Gyne production in Vespinae can reflect species-typical colony sizes, e.g. a few tens of gynes in *Vespa tropica* vs. an estimated 1,000 in *V. simillima* (Matsuura & Yamane, 1990). Although colony sizes and gyne productivities are noteworthy, effective fecundity is the number of gynes in one year that produce successful colonies the following year. In England, *Vespula vulgaris* and *Dolichovespula sylvestris* gynes had 99.9 and 99.6 percent cumulative mortalities resulting in per-colony means of 1.1 and 0.9 successful colonies the following year (Archer, 1984).

Swarm-founding wasp colony sizes are species-typical, ranging from a few tens in some Epiponini to tens of thousands. The largest colonies are those of *Agelaia vicina* in Brazil, reaching an estimated 7.5 million nest cells and one million adults (Zucchi, *et al.*, 1995). Swarm-founding effective fecundity is not colony size, however, but is instead the number of a colony's successful reproductive swarms, which has never been quantified.

4.6.4 Age at First Reproduction

Given that all female Stenogastrinae are potentially reproductive, age at first reproduction could be age at first oviposition but would more accurately be age at oviposition for the first successful adult emergence. Age at first reproduction for independent-founding Polistinae and Vespinae with an annual life cycle could be the age as timed from the end of diapause to first oviposition. However, first offspring are usually workers, therefore age at first reproduction could be queen age at oviposition of the first reproductive offspring. In a more abstract view, queens producing gynes in a region with the same year-to-year environmental conditions would have been gynes in the preceding year and have an average calendar age at first reproduction of one year.

If the queen should be lost in a multi-queen paper wasp colony, the most dominant co-foundresses becomes queen in *Polistes dominula* (*P. gallicus* in Pardi, 1948) and *P. canadensis* (*P. erythrocephalus* in West Eberhard, 1969). In species with annual colony cycles, if the queen of a *Polistes* colony should die with no replacement co-foundress, the oldest or one of the older workers can become queen and reproduce in its first year in *Polistes dominula* (Pardi, 1948), *P. exclamans* (Strassmann & Meyer, 1983), and *P. instabilis* (Hughes & Strassman, 1988). In *P. exclamans*, such queens can mate with early males and produce females (Strassmann, 1981a,c). *Ropalidia revolutionalis* replacement queens appear in sequence in a dominance hierarchy of co-foundresses followed by serial replacement in a dominance hierarchy of workers (Henshaw, *et al.*, 2004). *Ropalidia marginata* replacement queens do not fit the pattern of other paper wasps in that they are not the dominant wasp at the time of queen loss (Gadagkar, 1991a, 2001). In tropical species with colonies persisting for extended periods, the replacement queen is young and has done little foraging in *P. canadensis* (West Eberhard, 1969) and *Mischocyttarus drewseni* (Jeanne, 1972). *Parapolybia indica* is a temperate zone paper wasp in which replacement queens can have older as well as younger sisters (Suzuki, 2003). Young offspring are the potential queens in the epiponines *Metapolybia aztecoides* (West-Eberhard, 1978), *Parachartergus colobopterus* (Strassmann, *et al.*, 1997a), and *Polybioides tabidus* (Henshaw, *et al.*, 2000). This could be the case in swarm-founding wasps generally.

4.6.5 Dispersal

Gynes of wasps with obligate sociality often establish nests in the same area as or even close to their natal nest. This pattern, called philopatry, constitutes low or no dispersal. Philopatry is common in *Polistes* (Reeve, 1991) and occurs in *Mischocyttarus mexicanus* (Litte, 1977). Philopatric co-foundresses can often be from the same nest of the previous year (Klahn, 1979). In tropical India, female *P. stigma* leave the nest about one week following emergence and show no philopatric tendency (Suzuki & Ramesh, 1992). This, in combination with single foundresses, is an apparent adaptation to nest predation by hornets, which repeatedly forage in sites where they previously found *P. stigma* colonies.

Wasp swarms move slowly in loose associations rather than quickly in compact formations as do honey bees (Jeanne, 1975). Individuals move from landmark to landmark on which preceding wasps applied a pheromone marker by rubbing with secretory glands on the underside of the gaster (Jeanne, 1975; Naumann, 1975). *Apoica pallens* apparently coordinates swarm movement using an aerial pheromone released by elevating the gaster and exposing glands between its terminal segments (Hunt, *et al.*, 1995). Seasonal colony relocation between habitats could involve as-yet unmeasured distances of kilometers, but between-habitat migration does not constitute dispersal *per se*.

4.6.6 Body Size

Female solitary wasps are often larger than males, but wasps of each sex are usually monomorphic. Stenogastrinae are monomorphic (Turillazzi, 1991). Most paper wasps have no discrete queen/worker difference, although there can be a range of body sizes (Haggard & Gamboa, 1980). Discrete queen-worker difference occurs in the paper wasps *Belonogaster petiolata* (Keeping, 2000, 2002), *Ropalidia ignobilis* (Wenzel, 1992), and perhaps *Polistes olivaceous* (Alam, 1958; Kundu, 1967). Vespinae are dimorphic, with queens larger than workers (Matsuura & Yamane, 1990; Spradbery, 1993; Archer, 2012).

The swarm-founding *Provespa* species have queen-worker dimorphism as in other Vespinae. The two swarm-founding ropalidiine genera, *Polybioides* and *Ropalidia* subgenus *Icarielia*, have discrete queen-worker dimorphism (Yamane, *et al.*, 1983; Jeanne & Hunt, 1992; Turillazzi, *et al.*, 1994; Fukuda, *et al.*, 2003). The polistine tribe Epiponini has the greatest diversity of queen-worker dimorphism (O'Donnell, 1998b; Shima, *et al.*, 1998), with four distinguishable categories: monomorphic with reproductives having few developing ova, monomorphic with readily identifiable queens having many developing ova, queens larger but mostly the same shape as workers, and queens shaped differently from workers (Noll, *et al.*, 2004; Noll & Wenzel, 2008). The latter two categories are based on differences in larval development, with quantity and/or quality of nourishment playing a major role (O'Donnell, 1998b).

4.7 Traits of Social Groups

4.7.1 Genetic Structure

Like all Hymenoptera, social wasps are haplodiploid, a sex-determining system in which females are diploid and derived from fertilized eggs, whereas males are haploid and derived from unfertilized eggs. Like solitary wasps (Thornhill & Alcock, 1983), most social Vespidae are singly mated (Strassmann, 2001). Colonies with a singly-mated queen and "supersister" workers (i.e. daughters from a singly-inseminated mother) are thus predicted to have a mean sister-sister relatedness of 0.75. Despite this prediction, realized relatedness rarely reaches 0.75, instead varyinh widely due to a diversity of factors, including multiply-mated queens, multiple queens, egg-laying workers, or nest adoption or usurpation (Ross & Carpenter, 1991). Measured and estimated relatedness for *Polistes* colonies ranges from 0.34 to 0.80 (Ross & Carpenter 1991), and it also ranges widely in swarm-founding wasps (e.g. 0.10 to 0.65 in *Polybia* spp., Queller, *et al.*, 1988). Although related workers typify social species, numerous examples of unrelated conspecific workers exist: in facultatively social Stenogastrinae (Turillazzi, 1991, 2012; Landi, *et al.*, 2003), following nest usurpation by facultative social parasites in Vespinae (Sakagami & Fukushima, 1957; Greene, 1991; Matsuura, 1991; Archer, 2012), and in *Polistes* by adoption of an empty nest or usurpation by a facultative social parasite (Yoshikawa, 1955; Klahn, 1988; Makino, 1989; Makino & Sayama, 1991; Nonacs & Reeve, 1993; Starks, 1998; Hunt, 2009).

Polistes co-foundress associations typically range from 2 to 10 with an average between 2 and 5 (Liebert & Starks, 2006; Queller, *et al.*, 2000). Oviposition is almost exclusively by a single dominant (Field, *et al.*, 1998b; Queller, *et al.*, 2000; Liebert & Starks, 2006; Nonacs, *et al.*, 2006). The proportion of related co-foundresses in *P. dominula* is 65 to 85 percent (Queller, *et al.*, 2000; Leadbeater, *et al.*, 2011) and 77–93 percent in *P. fuscatus* (Klahn, 1979). Unrelated co-foundresses can be a non-negligible proportion of the population, and there are rare cases of two species co-founding (Snelling, 1952; Hunt & Gamboa, 1978; O'Donnell & Jeanne, 1991). Therefore, relatedness is not required for co-founding. Learned olfactory cues from cuticular hydrocarbons can potentially be a proxy for identifying individuals that are *usually* kin (Klahn & Pilgrim 1985; Gamboa, *et al.*, 1986).

There is disagreement on the importance of relatedness in the evolution of sociality (e.g. Foster, *et al.*, 2006; Hunt 2007; Abbot, *et al.*, 2010; Nowak, *et al.*, 2010; Wilson & Hölldobler, 2005). Social wasps provide evidence both for and against the role of relatedness in group formation and integration. Although a review of the controversy is beyond the scope of this chapter, social wasps have provided tests of some key predictions of kin selection. Ancestral single mating by queens (Strassmann, 2001), absence of male production by workers, queen mating frequency in vespines, and sex ratios in several polistines (reviewed by Bourke, 2005) support predictions of kin selection theory. Low intra-colony relatedness does not support kin selection predictions. Alternative models to inclusive fitness focus on individual fitness (Nowak, *et al.*, 2010) or group selection (Wilson & Hölldobler, 2005). Wilson & Hölldobler (2005) and Wilson (2008) suggest that relatedness is not a driver of the evolution of sociality, but an evolutionary role for relatedness among colony members arises as a consequence rather than cause of sociality.

4.7.2 Group Structure, Breeding Structure, and Sex Ratio

The fundamental breeding structure of facultative and obligate sociality is independent founding by a singly-inseminated female (Strassmann, 2001). Swarm-founding wasp colonies, except those of *Provespa*, have multiple queens, among which single insemination is the apparent norm (Strassmann, 2001), and most or all queens lay eggs. Queen number in newly-founded colonies diminishes before the emanation of new swarms containing many young queens, resulting in a pattern called cyclical oligogyny (Strassmann, *et al.*, 1992; Queller, *et al.*, 1993; Henshaw, *et al.*, 2000, 2004; Nascimento, *et al.*, 2004).

Haplodiploid sex determination enables natural selection for sex ratio bias, which is widespread and diverse throughout Hymenoptera. Selection for sex ratio bias in social wasps resulted in primarily female offspring as the earliest brood (protogyny). Because females have maternal behavioral repertoires and because males do not work, protogyny yields an early brood of potentially allomaternal workers rather than non-working, energy-consuming males.

Yoshikawa, *et al.* (1969) report a *Polistes*-like dominance hierarchy in a colony of a *Parischnogaster* species. In *P. jacobsoni*, a dominant female is easily discernable by her aggressive behavior and nestmates' submissive postures and behaviors, and the dominant had the greatest ovary development in five of six nests (Turillazzi, 1988). Dominance interactions in *Liostenogaster vechti* are rare (Turillazzi, 1988).

Dominance hierarchies exist in *Polistes*, *Mischocyttarus*, *Belonogaster*, and *Ropalidia* (Pardi, 1948; Tindo, *et al.*, 1997; O'Donnell, 1998a; Sumana & Gadagkar, 2001). Dominance interactions occur among co-foundresses during nest founding, between queens and workers after worker emergence, and between gynes in pre-hibernation aggregations (Dapporto, *et al.*, 2006). *Polistes* dominance interactions include physical aggression; ritualized biting, grappling, and stinging; and signal behaviors including open mandible lunging, posturing, wagging the abdomen, and placing the body on top of another individual. Dominance is established rapidly within a few minutes of interactions between two individuals. Dominant individuals typically retain dominant status by occasional signal behaviors and oophagy (i.e. egg-eating) of any eggs laid by subordinates (Jandt, *et al.*, 2014). Deleurance (1948) removed the developed ovaries of dominant *Polistes*, but when an ovariectomized dominant was returned to the nest she retained her dominant status. Dominance hierarchies are somewhat fluid in respect to newly-emerged lower ranked individuals or death of higher ranked individuals, including the queen (West Eberhard, 1969; Jeanne, 1972).

In most paper wasps, dominant queens are aggressive, and tasks are partitioned by queen-to-worker top-down control (e.g. *Polistes fuscatus*, Reeve & Gamboa, 1983, 1987). Colony activity is initiated by workers, not queens, in *Polistes instabilis* and *P. dominula* (Jha, *et al.*, 2006). Dominant individuals spend more time on the nest and initiate new nest cells, whereas subordinates typically do more foraging, extending nest cells, and brood care (Gadagkar, 1980; Gadagkar & Joshi, 1983). *Ropalidia marginata* has three behavioral worker castes called sitters, fighters, and foragers (Gadagkar & Joshi, 1983). *R. marginata* queens are docile and subordinate or may not participate in dominance interactions (Chandrashekar & Gadagkar, 1991; Gadagkar, 2001), but following queen loss the successor queen quickly expresses dominance behaviors toward other colony members that continue diminishingly until her queen status is established and she becomes docile and subordinate. Dominance interactions among *R. marginata* workers regulate foraging rather than reinforce queen status (Sumana & Gadagkar, 2001), and *R. marginata* appears to be approaching colony-level self-organization of work without top-down control (Premnath, 1995).

Age polyethism is not well developed in wasps with obligate sociality (Jeanne, 1991a) but occurs in some vespines (Kim, *et al.*, 2012). *Ropalidia marginata* has a honey bee-like age polyethism in which task allocation is more strongly linked to a worker's relative age among nestmates than to age since emergence from pupation (Naug & Gadagkar, 1998; Agrahari & Gadagkar, 2004). Swarm-founding Polistinae show the strongest age polyethism in Vespidae (Jeanne, 1991b). Young *Polybia occidentalis* build nests and care for brood, intermediate age individuals maintain and defend the nest, and older individuals are foragers (Jeanne, *et al.*, 1988). Nest building and repair in *P. occidentalis* is mediated by worker-worker interactions among three specialized worker groups – wood-pulp foragers, water foragers, and builders – that coordinate and regulate activities in a self-organized manner (Jeanne, 1986). The system works efficiently, is capable of increasing or decreasing total building effort, and can self-correct in response to unexpected perturbations that cause mismatched efforts of the three groups (Jeanne, 1986).

Several vespine species have dominance interactions among post-hibernation gynes, including differential food intake, biting, and stinging (Matsuura, 1984; Greene, 1991). Overt aggression is rare in vespine colonies with a functional queen. Instead, queen reproductive dominance is typically established and maintained by chemical means, likely involving the production of ovary-inhibiting pheromones (Greene, 1991). Many vespine workers can undergo ovary development and lay male-producing eggs if a colony loses its queen, but the presence of a few egg-laying workers in *Dolichovespula* queen-right colonies (Foster, *et al.*, 2001) may reflect suppression of workers' ovary development by among-worker dominance interactions (Greene, 1991).

Queens' status in swarm-founding wasp evolution moved from dominance behavior to dominance signals (e.g. *Metapolybia aztecoides*, West-Eberhard, 1978) and/or queen pheromones (Kocher & Grozinger, 2011). Pheromones are presumed to be the primary modality by which vespines suppress worker reproduction (Greene, 1991). The paper wasp *Ropalidia marginata* has a queen pheromone (Bhadra, *et al.*, 2010). If sought, queen pheromones will doubtlessly be found in other paper wasps and Epiponini.

III SOCIAL SYNTHESIS

4.8 A Summary of Wasp Sociality

All stinging Hymenoptera (suborder Aculeata) are known as wasps, excepting the major radiations known as ants and bees. Most wasps are solitary: each female, acting alone, produces her offspring, all of which are reproductives. Social wasps comprise a minuscule fraction of Aculeata. Vespidae encompasses the full range of grades of insect sociality: solitary, communal, and facultative, obligate, and swarm-founding sociality. Everything from the origins of worker behavior to the elaboration of sociality into complex forms can be studied within this family.

4.9 Comparative Perspectives on Wasp Sociality

Together with all termites, all ants, and some bees, wasps are considered one of the four main social insect taxa. The sociality of all four is characterized by species that fulfill the classical defining criteria for eusociality. Wasps, ants, and bees are Hymenoptera in suborder Aculeata. There have been multiple origins of sociality within Aculeata, especially among bees (Michener, 1974). Excepting communal sociality, the evolution of Hymenopteran sociality is most accurately viewed as multiple parallel evolutions of mother-daughter social structures arising from solitary nesting with subsocial maternal care. There is, therefore, a suite of traits that are similar in all social Hymenoptera, including anatomy, physiology, behavior, and life history (Wilson, 1971). In addition, many of the same metabolic pathways and molecular functions, but not exactly the same genes, are associated with caste differences in paper wasps, bees, and ants (Berens, *et al.*, 2015). These data lend support to the notion that there is a conserved "toolkit" of gene networks related to caste evolution across social Hymenoptera (Toth, *et al.*, 2010).

The shared common ancestor of the four tribes of corbiculate bees (Chapter 3) would have had been social with a mother-daughter colony structure. Most orchid bees (Euglossini) have reverted to solitary life, but some are communal, and some are weakly social (Cardinal & Danforth, 2011). Bumble bees (Bombini), excepting the obligate social parasite clade *Psythyrus*, have obligate sociality identical to that of Vespinae. All species of *Apis* (Apini) are single-queen and swarm-founding. The honey bee, *Apis mellifera*, has developmental, morphological, behavioral, and reproductive differences between workers and queens (Michener, 1974) that are analogous to some swarm-founding wasps. Also analogously to Polistinae and Vespinae, caste determination in *A. mellifera* involves specific gene activation in both workers and queens

(Evans & Wheeler, 1999). Stingless bees (Meliponini), excepting a few polygynous species (Velthuis, *et al.*, 2001) are several genera of swarm-founding, single-queen species (Michener, 1974). The three social tribes of corbiculate bees therefore reflect the matrifilial sociality of their common ancestor, as is the case with the ancestral matrifilial sociality of Polistinae and Vespinae.

Hymenoptera have complete metamorphosis (holometabolism) comprised of larva to pupa to adults, whereas termites (Chapter 5) have incomplete metamorphosis (hemimetabolism) and pass through a series of molts, with each instar having increasingly adult-like features before reaching the adult reproductive form with wings. All termites have a soldier caste, which is not the case for social wasps. Colonies are initiated not by a single foundress but instead by a male/female reproductive pair, and both male and female termites work. Immature soldiers and workers of "one piece" termites can become supplementary or replacement reproductives, whereas "multiple piece" termites have a discrete worker caste incapable of reproduction. Despite these significant differences between social wasps and termites, there are also significant similarities. Termite sociality is based upon a subsocial ancestor (Nalepa, 2015). Nalepa (2015) lays out a scenario in which a wood roach-like ancestor living in a nest chamber from which immatures did not disperse could have had a second brood, with the immature first brood providing heterochronic alloparental care to the second brood in the same social structure as one piece termites and wasps with obligate sociality. From this initial structure of two overlapping generations, termites evolved a suite of morphological, behavioral, and physiological traits that made their sociality obligate (Nalepa, 2015). In addition to similarity in evolutionary history, striking overlap occurs between termites and social Hymenoptera in genes controlling caste and the shift of parental care to offspring via heterochrony, suggesting co-option of similar developmental mechanisms (Howard & Thorne, 2011).

Thrips, aphids (Chapter 6), and snapping shrimp (Chapter 8) are also hemimetabolous. Their worker behavior consists of defending a plant gall or marine sponge rather than brood care, thus it is fundamentally different from wasp sociality. Nonetheless, a significant analogy among all social species is occupancy of a nest or nest-like structure.

Like some social vertebrates and despite fundamental differences between insects and vertebrates, simple wasp societies are characterized by a combination of conflict (i.e. dominance hierarchies) and cooperation (e.g. group defense, care of young, colony co-founding) (Brockmann, 1997). In particular, *Polistes* and other small-colony social wasps are structured by dominance hierarchies that show striking similarities to hierarchically-structured vertebrate societies, including those of coopratively breeding birds, mammals, and even humans (Crespi & Yanega, 1995).

4.10 Concluding Remarks

Wasps span the full range of insect sociality: solitary, communal, and facultative, obligate, and swarm-founding sociality. Although a few communal species occur in

Pompilidae and Sphecidae, all other social wasps are in Vespidae, a family containing all grades of sociality, from solitary to swarm-founding. Great diversity exists in social wasp natural history, ecology, behavior, and colony structure. Molecular tools to investigate phylogenetics and relatedness are well established, and molecular tools to investigate mechanisms of development, physiology, and gene expression are being rapidly developed. Since the 1970s, there has been a narrow focus on the role of relatedness in the evolution of wasp sociality. A more comprehensive view of wasp sociality is emerging from mechanistic approaches to study the evolution of sociality including maternal control and its effects on development, behavior, and reproduction; reproductive heterochrony at behavioral, physiological and gene expression levels; and insights being derived from "toolkits" of shared genes and physiological pathways that play roles in the multiple origins and parallel evolutions of insect sociality. Although wasps provide the best context to study the evolution of insect sociality, there is a paucity of studies on physiology and genetics of several groups, especially Eumeninae and Stenogastrinae (Jandt & Toth, 2015). We suggest fruitful areas for expanding future study of wasp sociality will include comparative genomics, epigenetics, and indirect genetic effects.

References

Abbot, P., Withgott, J. H., & Moran, N. A. (2001) Genetic conflict and conditional altruism in social aphid colonies. *Proceedings of the National Academy of Sciences USA*, **98**, 12086–12071.

Agrahari, M. & Gadagkar, R. (2004) Hard working nurses rather than over-aged nurses permit *Ropalidia marginata* to respond to the loss of young individuals. *Insectes Sociaux* **51**, 306–307.

Akre, R. D. (1982) Social wasps. *In*: H. R. Hermann (ed.) *Social Insects* Vol. IV, New York: Academic Press, pp. 1–105.

Akre, R. D. & Myhre, E. A. (1992) Nesting biology and behavior of the baldfaced hornet, *Dolichovespula maculata* (L.) (Hymenoptera: Vespidae) in the Pacific Northwest. *Melanderia*, **48**, 1–33.

Alam, S. M. (1958) Some interesting revelations about the nest of *Polistes hebroeus* Fabr. (Vespidae, Hymenoptera) - the common yellow wasp of India. *Proceedings of the Zoological Society of Calcutta*, **11**, 113–122.

Alexander, R. D. (1974) The evolution of social behavior. *Annual Review of Ecology and Systematics*, **5**, 325–383

Archer, M. E. (1981) Successful and unsuccessful development of colonies of *Vespula vulgaris* (Linn.) (Hymenoptera: Vespidae). *Ecological Entomology*, **6**, 1–10.

(1984) Life and fertility tables for the wasp species *Vespula vulgaris* and *Dolichovespula sylvesteris* (Hymenoptera: Vespidae) in England. *Entomologia Generalis*, **9**, 181–188.

(1993) The life-history and colonial characteristics of the Hornet, *Vespa crabro* L. (Hym., Vespinae). *Entomologist's Monthly Magazine*, **124**, 117–122.

(2012) *Vespine Wasps of the World: Behaviour, Ecology & Taxonomy of the Vespinae.* Manchester: Siri Scientific Press.

Baracchi, D., Petrocelli, I., Cusseau, G., *et al.* (2013) Facial markings in the hover wasps: Quality signals and familiar recognition cues in two species of Stenogastrinae. *Animal Behaviour*, **85**, 302–212.

Beani, L. & Calloni, C. (1991) Male rubbing behavior and the hypothesis of pheromonal release in polistine wasps (Hymenoptera: Vespidae). *Ethology Ecology & Evolution, Special Issue* **1**, 51–54.

Beani, L. & Turillazzi, S. (1994) Aerial patrolling and the stripes-display in males of *Parischnogaster mellyi* (Hymenoptera Stenogastrinae).. *Ethology Ecology & Evolution*, **6**, *Supplement* 1, 43–46.

Beani, L., Cervo, R., Lorenzi, C. M., & Turillazzi, S. (1992) Landmark-based mating systems in 4 *Polistes* species (Hymenoptera, Vespidae). *Journal of the Kansas Entomological Society* **65**, 211–217.

Beggs, J. (2001) The ecological consequences of social wasps (*Vespula* spp.) invading an ecosystem that has an abundant carbohydrate source. *Biolgical Conservation* **99**, 17–28.

Berens, A. J., Hunt, J. H., & Toth, A. L. (2015) Comparative transciptomics of convergent evolution: Different genes but conserved pathways underlie caste phenotypes across lineages of eusocial insects. *Molecular Biology and Evolution*, **32**, 690–703.

Bhadra, A., Mitra, A., Sujata, A. D., *et al.* (2010) Regulation of reproduction in the primitively eusocial wasp *Ropalidia marginata*: On the trail of the queen pheromone. *Journal of Chemical Ecology*, **36**, 424–431.

Bohm, M. K. (1972) Effects of environment and juvenile hormone on ovaries of the wasp, *Polistes metricus*. *Journal of Insect Physiology*, **18**, 1875–1883.

Bourke, A. F. G. (2005) Genetics, relatedness and social behaviour in insect societies. *Symposium - Royal Entomological Society of London*, **22**, 1–30.

Brennan, B. J. (2007) Abdominal wagging in the social paper wasp *Polistes dominulus*: Behavior and substrate vibrations. *Ethology*, **113**, 692–702.

Brockmann, H. J. (1997) Cooperative breeding in wasps and vertebrates: The role of ecological constraints. *In*: Choe, J. C., & Crespi, B. J. (eds.) *Social Behavior in Insects and Arachnids*, Cambridge University Press, pp. 347–371.

Bunn, D. S. (1986) The nesting cycle of the hornet *Vespa crabro* L. (Hym., Vespidae). *Entomologists Monthly Magazine*, **124**, 117–122.

Cardinal, S., & Danforth, B. N. (2011) The antiquity and evolutionary history of social behavior in bees. *PLoS ONE*, **6**, e21086.

Carpenter, J. M. (1982) The phylogenetic relationships and natural classification of the Vespoidea (Hymenoptera). *Systematic Entomology*, **7**, 11–38.

(1986) A synonymic generic checklist of the Eumeninae (Hymenoptera: Vespidae). *Psyche*, **93**, 61–90.

(2004) Synonymy of the genus *Marimbonda* Richards, 1978, with *Leipomeles* Möbius, 1856 (Hymenoptera: Vespidae; Polistinae), and a new key to the genera of paper wasps of the New World. *American Museum Novitates*, **3465**, 1–16.

Carpenter, J. M. & Kimsey, L. S. (2009) The genus *Euparagia* Cresson (Hymennoptera: Vespidae; Euparagiinae). *American Museum Novitates*, **3643**, 1–11.

Cervo, R. & Dani, F. R. (1996) Social parasitism and its evolution in *Polistes*. *In*: S. Turillazzi, S. & West-Eberhard, M. J. (eds.) *Natural History and Evolution of Paper-wasps*, Oxford: Oxford University Press, pp. 98–112.

Cervo, R., Dapporto, L., Beani, L., Strassmann, J. E & Turillazzi, S. (2008) On status badges and quality signals in the paper wasp *Polistes dominulus*: Body size, facial colour patterns and herarchical rank. *Proceeding of the Royal Society of London B*, **275**, 1189–1196.

Chandrashekara, K. & Gadagkar, R. (1991) Behavioural castes, dominance and division of labour in a primitively eusocial wasp. *Ethology*, **87**, 269–283.

Clement, S. L. & Grissell, E. E. (1968) Observatons of the nesting habits of *Eupargia scutellaris* Cresson. *The Pan-Pacific Entomologist*, **44**, 34–37.

Cowan, D. P. 1991. The solitary and presocial Vespidae. In K. G. Ross and R. W. Matthews (eds.), *The Social Biology of Wasps*, pp. 33–73. Comstock Publishing Associates, Cornell University Press, Ithaca.

Crespi, B. J. & Yanega, D. (1995) The definition of eusociality. *Behavioral Ecology*. 6:109–115.

Cummings, D. L. D., Gamboa, G. J., & Harding, B. J. (1999) Lateral vibrations by social wasps signal larvae to withold salivary secretions (*Polistes fuscatus*, Hymenoptera: Vespidae). *Journal of Insect Behavior*, **12**, 465–473.

Dani, F. R., Foster, K. R., Zacchi, F., *et al.* (2004) Can cuticular lipids provide sufficient information for within-colony nepotism in wasps? *Proceeding of the Royal Society of London B*, **271**, 745–753.

Dapporto, L., Palagi, E., Cini, A., & Turillazzi, S. (2006) Prehibernating aggregations of *Polistes dominulus*: An occasion to study early dominance assessment in social insects. *Naturwissenschaften*, **93**, 321–324.

Deleurance, É.-P. (1948) Le comportement reproducteur est indépendant de la presence des ovaries chez *Polistes* (Hyménoptères Véspides). *Comptes rendus hebdomadaires des Séances de l'Académie des Sciences Paris*, **227**, 866–867.

(1949) Sur le déterminisme de l'apparition des ouvrières et des fondatrices-filles chez les *Polistes* (Hyménoptères: Vespides). *Comptes rendus hebdomadaires des Séances de l'Académie des Sciences Paris*, **229**, 303–304.

(1952) Le polymorphisme sociale et son déterminisme chez les guêpes. *Colloques internationaux du Centre National de la Recherche scientifique*, **34**, 141–155.

Edwards, R. (1980) *Social Wasps: Their Biology and Control*. East Grinstead: Rentokill Limited.

Evans, H. E. & Hook, A. W. (1982) Communal nesting in the digger wasp *Cerceris australis* (Hymenoptera: Sphecidae). *Australian Journal of Zoology*, **30**, 557–568.

Evans, H. E. & West-Eberhard, M. J. (1970) *The Wasps*. Ann Arbor, MI: University of Michigan Press.

Evans, H. E. & Yoshimoto, C. M. (1962) The ecology and nesting behavior of the Pompilidae of the northeastern United States. *Miscellaneous Publications of the Entomological Society of America*, **3**, 66–119.

Evans, J. D. & Wheeler, D. E. (1999) Differential gene expression between developing queens and workers in the honey bee, *Apis mellifera. Proceedings of the National Academy of Sciences USA*, **96**, 5575–5580.

Felippotti, G. T., Tanaka, G. M. Jr., Noll, F. B., & Wenzel J. W. (2009) Discrete dimorphism among castes of the bald-faced hornet *Dolichovespula maculata* (Hymenoptera: Vespidae) in different phases of the colony cycle. *Journal of Natural History*, **43**, 2481–2490.

Field, J. (2008) The ecology and evolution of helping in hover wasps (Hymenoptera: Stenogastrinae). *In*: Korb, J. & Heinze, J. (eds.) *Ecology of Social Evolution*, pp. 85–107. Springer, Heidelberg.

Field, J., Foster, W., Shreeves, G., & Sumner, S. (1998a) Ecological constraints on independent nesting in facultatively eusocial hover wasps. *Proceedings of the Royal Society of London B*, **265**, 973–977.

Field, J., Solís, C. R., Queller, D. C., & Strassman, J. E. (1998b) Social and genetic structure of paper wasp cofoundress associations: Tests of reproductive skew models. *The American Naturalist*, **151**, 545–563.

Field, J., Shreeves, G., & Sumner, S. (1999) Group size, queuing and helping decisions in facultatively eusocial hover wasps. *Behavioral Ecology and Sociobiology*, **45**, 378–385.

Foster, K. R., Seppä, P., Ratnieks, F. L.W., & Thorén, P. A. (1999) Low paternity in the hornet *Vespa crabro* indicates that multiple mating by queens is derived in vespine wasps. *Behavioral Ecology and Sociobiology*, **46**,252–257.

Foster, K. R., Ratnieks, F. L. W., Gyllenstrand, N., & Thorén, P. A. (2001) Colony kin structure and male production in *Dolichovespula* wasps. *Molecular Ecology*, **10**, 1003–1010.

Foster, K. R., Wenseleers, T., & Ratnieks, F. L. W. (2006) Kin selection is the key to altruism. *Trends in Ecology and Evolution*, **21**, 57–60.

Fucini, S., Di Bona, V., Mola, F., Piccaluga, C., & Lorenzi, M. C. (2009) Social wasps without workers: Geographic variation of caste expression in the paper wasp *Polistes biglumis*. *Insectes Sociaux*, **56**, 347–358.

Fukuda, H., Kojima, J., & Jeanne, R. L. (2003) Colony specific morphological caste differences in an old world, swarm-founding polistine, *Ropalidia romandi* (Hymenoptera: Vespidae). *Entomological Science*, **6**, 37–47.

Gadagkar, R. (1980) Dominance hierarchy and division of labour in the social wasp, *Ropalidia marginata* (Lep.) (Hymenoptera: Vespidae). *Current Science*, **49**, 772–775.

(1991a) *Belonogaster, Mischocyttarus, Parapolybia*, and independent-founding *Ropalidia*. *In*: Ross, K. G. & Matthews, R. W. (eds.) *The Social Biology of Wasps*. Ithaca, NY: Comstock Publishing Associates, Cornell University Press, pp. 149–190.

(1996) The evolution of eusociality, including a review of the social status of *Ropalidia marginata*. *In*: Turillazzi, S. & West-Eberhard, M. J. (eds.) *Natural History and Evolution of Paper-wasps*. Oxford: Oxford University Press, 248–271.

(2001) *The Social Biology of Ropalidia marginata - Toward Understanding the Evolution of Eusociality*. Cambridge, MA: Harvard University Press.

Gadagkar, R. & Joshi, N. V. (1983) Quantitative ethology of social wasps: Time-activity budgets and caste differentiation in *Ropalidia marginata* (Lep.) (Hymenoptera: Vespidae). *Animal Behaviour*, **31**, 26–31.

Gadagkar, R., Chandrashekara, K., Chandran, S., & Bhagavan, S. (1993) Serial polygyny in the primitively eusocial wasp *Ropalidia marginata*: Implications for the evolution of sociality. *In*: Keller, L. (ed.) *Queen Number and Sociality in Insects*, Oxford: Oxford University Press, pp. 188–214.

Gamboa, G. J. (1988) Sister, ant-niece, and cousin recognition by social wasps. *Behavior Genetics*, **18**, 409–423.

(2004) Kin recognition in eusocial wasps. *Annales Zoologici Fennici*, **41**, 789–808.

Gamboa, G. J., Reeve, H. K., Ferguson, I. D., & Wacker, T. L. (1986) Nestmate recognition in social wasps: The origin and acquisition of recognition odours. *Animal Behaviour*, **34**, 685–695.

Gamboa, G. J., Klahn, J. E., Parman, A. O., & Ryan, R. E. (1987) Discrimination between nestmate and non-nestmate kin by social wasps (*Polistes fuscatus*, Hymenoptera: Vespidae). *Behavioral Ecology and Sociobiology*, **21**, 125–128.

Gess, S. K. (1996) *The Pollen Wasps: Ecology and Natural History of the Masarinae*. Cambridge, MA: Harvard University Press.

Giannotti, E. (1999) Social organization of the eusocial wasp *Mischocyttarus cerberus styx* (Hymenoptera: Vespidae). *Sociobiology*, **33**, 325–338.

Gibo, D. L. (1978) The selective advantage of foundress associations in *Polistes fuscatus* (Hymenoptera: Vespidae): A field study of the effects of predation on productivity. *Canadian Entomologist*, **110**, 519–540.

Gibo, D. L., Yarascavitch, R. M., & Dew, H. E. (1974a) Thermoregulation in colonies of *Vespula arenaria* and *Vespula maculata* (Hymenoptera-Vespidae) under normal conditions and under cold stress. *Canadian Entomologist*, **106**, 503–507.

Gibo, D. L., Dew, H. E., & Hajduk, A. S. (1974b) Thermoregulation in colonies of *Vespula arenaria* and *Vespula maculata* (Hymenoptera-Vspidae). 2. Relation between colony biomass and calorie production. *Canadian Entomologist*, **106**, 873–879.

Gobbi, N., Noll, F. B, & Penna, M. A. H. (2006) "Winter" aggregations, colony cycle, and seasonal phenotypic change in the paper wasp *Polistes versicolor* in Brazil. *Naturwissenschaften*, **93**, 487–494.

Gotoh, A., Billen, J., Hashim, R., & Ito, F. (2008) Comparison of spermatheca morphology between reproductive and non-reproductive females in social wasps. *Arthropod Structure and Development*, **37**, 199–209.

Green, J. P. & Field, J. (2011) Inter-population variation in status signalling in the paper wasp *Polistes dominulus*. *Animal Behaviour*, **81**, 205–209.

Green, J. P., Leadbeater, E., Carruthers, J. M., *et al.* (2013) Clypeal patterning in the paper wasp *Polistes dominulus*: No evidence of adaptive value in the wild. *Behavioral Ecology*, **24**, 623–633.

Greene, A. (1991) *Dolichovespula* and *Vespula*. *In*: Ross, K. G., & Matthews, R. W. (eds.) *The Social Biology of Wasps*. Ithaca, NY: Comstock Publishing Associates, Cornell University Press, pp. 263–305.

Haggard, C. M. & Gamboa, G. J. (1980) Seasonal variation in body size and reproductive condition of a paper wasp, *Polistes metricus* (Hymenoptera: Vespidae). *The Canadian Entomologist*, **112**, 239–248.

Hansell, M. H. (1987) Elements of eusociality in colonies of *Eustenogaster calyptodoma* (Sakagami & Yoshikawa) (Stenogastrinae, Vespidae). *Animal Behaviour*, **35**, 131–141.

Harris, R. J. (1991) Diet of the wasps *Vespula vulgaris* and *V. germanica* in honeydew beech forest of the South Island, New Zealand. *New Zealand Journal of Zoology*, **18,** 159–170.

Heinrich, B. (1984) Strategies of thermoregulation and foraging in two vespid wasps, *Dolichovespula maculata* and *Vesupla vulgaris*. *Journal of Comparative Physiology B*, **154**, 175–180.

Henshaw, M. T., Strassmann, J. E., & Queller, D. C. (2000) The independent origin of a queen number bottleneck that promotes cooperation in the African swarm-founding wasp, *Polybioides tabidus*. *Behavioral Ecology and Sociobiology*, **48**, 478–483.

Henshaw, M. T., Robson, S. K. A., & Crozier, R. H. (2004) Queen number, queen cycling and queen loss: The evolution of complex multiple queen societies in the social wasp genus *Ropalidia*. *Behavioral Ecology and Sociobiology*, **55**, 469–476.

Hines, H. M., Hunt, J. H., O'Connor, T. K., Gillespie, J. J., & Cameron, S. A. (2007) Multigene phylogeny reveals eusociality evolved twice in vespid wasps. *Proceedings of the National Academy of Sciences USA*, **104**, 3295–3299.

Hook, A. (1982) Observations on a declining nest of *Polistes tepidus* (F.) (Hymenoptera: Vespidae). *Journal of the Australian Entomological Society*, **21**, 277–278.

Howard, K. J. & Thorne, B. L. (2011) Eusocial evolution in termites and Hymenoptera. *In*: Bignell, D. E., Roisin, Y., & Lo, N. (eds.) *Biology of Termites: A Modern Synthesis*, Dordrecht: Springer, pp. 97–132.

Hughes, C. R. & Strassmann, J. E. (1988) Age is more important than size in determining dominance among workers in the primitively eusocial wasp, *Polistes instabilis. Behaviour*, **107**, 1–14.

Hunt, J. H. (1984) Adult nourishment during larval provisioning in a primitively social paper wasp, *Polistes metricus* Say. *Insectes Sociaux*, **31**, 452–460.

(1991) Nourishment and the evolution of the social Vespidae. *In*: Ross, K. G. & Matthews, R. W. (eds.) *The Social Biology of Wasps*, Ithaca NY: Comstock Publishing Associates, Cornell University Press, pp. 426–450.

(1993) Survivorship, fecundity, and recruitment in a mud dauber wasp, *Sceliphron assimile* (Hymenoptera: Sphecidae). *Annals of the Entomological Society of America*, **86**, 51–59.

(1994) Nourishment and evolution in wasps *sensu lato*. *In*: Hunt, J. H. & Nalepa, C. A. (eds.) *Nourishment and Evolution in Insect Societies*. Boulder CO: Westview Press, pp. 211–244.

(1999) Trait mapping and salience in the evolution of eusocial vespid wasps. *Evolution*, **53**, 225–237.

(2007) *The Evolution of Social Wasps*. New York: Oxford University Press.

(2009) Interspecific adoption of orphaned nests by *Polistes* paper wasps (Hymenoptera: Vespidae). *Journal of Hymenoptera Research*, **18**, 136–139.

Hunt, J. H. & Amdam, G. V. (2005) Bivoltinism as an antecedent to eusociality in the paper wasp genus *Polistes*. *Science*, **308**, 264–267.

Hunt, J. H. & Gamboa, G. J. (1978) Joint nest use by two paper wasp species. *Insectes Sociaux*, **25**, 373–374.

Hunt, J. H. & Richard, F.-J. (2013) Intracolony vibroacoustic communication in social insects. *Insectes Sociaux*, **60**, 403–417.

Hunt, J. H., Jeanne, R. L., Baker, I., & Grogan, D. E. (1987) Nutrient dynamics of a swarm-founding social wasp species, *Polybia occidentalis*(Hymenooptera: Vespidae). *Ethology*, **75**, 291–305.

Hunt, J. H., Brown, P. A., Sago, K. M., & Kerker, J. A. (1991) Vespid wasps eat pollen (Hymenoptera: Vespidae). *Journal of the Kansas Entomological Society*, **64**, 127–130.

Hunt, J. H., Jeanne, R. L., & Keeping, M. G. (1995) Observations on *Apoica pallens*, a nocturnal Neotropical social wasp (Hymenoptera: Vespidae, Polistinae, Epiponini). *Insectes Sociaux*, **42**, 223–236.

Hunt, J. H., Brodie, R. J., Carithers, T. P., Goldstein, P. Z., & Janzen, D. H. (1999) Dry season migration by Costa Rican lowland paper wasps to high elevation cold dormancy sites. *Biotropica*, **31**, 192–196.

Hunt, J. H., O'Donnell, S., Chernoff, N., & Brownie, C. (2001a) Observations on two Neotropical swarm-founding wasps, *Agelaia yepocapa* and *A. panamaensis* (Hymenoptera: Vespidae). *Annals of the Entomological Society of America*, **94**, 555–562.

Hunt, J. H., Cave, R. D., & Borjas, G. R. (2001b). First records from Honduras of a yellowjacket wasp, *Vespula squamosa* (Drury) (Hymenoptera: Vespidae, Vespinae). *Journal of the Kansas Entomological Society*, **74**, 118–119.

Hunt, J. H., Kensinger, B. A., Kossuth, J., *et al.* (2007) A diapause pathway underlies the gyne phenotype in *Polistes* wasps, revealing an evolutionary route to caste-containing insect societies. *Proceedings of the National Academy of Sciences USA*, **104**, 14020–14025.

Hunt, J. H., Wolschin, F., Henshaw, M. T., Newman, T. C., Toth, A. L. & Amdam, G. V. (2010) Differential gene expression and protein abundance evince ontogenetic bias toward castes in a primitively eusocial wasp. *PLoS ONE*, 5, e10674.

Hunt, J. H., Mutti, N. S., Havukainen, H., Henshaw, M. T., & Amdam, G. V. (2011) Development of an RNA interference tool, characterization of its target, and an ecological test of caste differentiation in the eusocial wasp *Polistes*. *PLoS ONE*, 6, e26641.

Ishay, J. S. (1973) Thermoregulation by social wasps: Behavior and pheromones. *Transactions of the New York Academy of Sciences*, **35**, 447–462.

Jandt, J. M. & Toth, A. L. (2015) Chapter 3 - Physiological and genomic mechanisms of social organization in wasps (Family: Vespidae). *Advances in Insect Physiology*, **48**, 95–130.

Jandt, J. M. Tibbetts, E. A., & Toth, A. L. (2014) *Polistes* paper wasps: A model genus for the study of social dominance hierarchies. *Insectes Sociaux*, **61**, 11–27.

Jeanne, R. L. (1970) Chemical defence of Brood by a social wasp. *Science*, **168**, 1465–1466.

(1972) Social biology of the neotropical wasp *Mischocyttarus drewseni*. *Bulletin of the Museum of Comparative Zoology, Harvard University*, **144**, 63–150.

(1975) Behavior during swarm movement in *Stelopolybia areata* (Hymenoptera: Vespidae). *Psyche*, **82**, 259–264.

(1986) The organization of work in *Polybia occidentalis*: The costs and benefits of specialization in a social wasp. *Behavioral Ecology and Sociobiology*, **19**, 333–341.

(1991a) The swarm-founding Polistinae. *In*: Ross, K. G., & Matthews, R. W. (eds.) *The Social Biology of Wasps*. Ithaca NY: Comstock Publishing Associates, Cornell University Press, pp. 191–231.

(1991b) Polyethism. *In*: Ross, K. G. & Matthews, R. W. (eds.) *The Social Biology of Wasps*. Ithaca NY: Comstock Publishing Associates, Cornell University Press, pp. 389–425.

Jeanne, R. L. & Hunt, J. H. (1992) Observations on the social wasp *Ropalidia montana* from peninsular India. *Journal of Biosciences*, **17**, 1–14.

Jeanne, R. L. & Keeping, M. G. (1995) Venom spraying in *Parachartergus colobopterus*: A novel defensive behavior in a social wasp (Hymenoptera: Vespidae). *Journal of Insect Behavior*, **8**, 433–442.

Jeanne, R. L. & Morgan, R. C. (1992) The influence of temperature on nest site choice and reproductive strategy in a temperate zone *Polistes* wasp. *Ecological Entomology*, **17**, 135–141.

Jeanne, R. L., Downing, H. A., & Post, D. C. (1988) Age polyethism and individual variation in *Polybia occidentalis*, an advanced eusocial wasp. *In*: Jeanne, R. L. (ed.) *Interindividual Behavioral Variability in Social Insects*. Boulder CO: Westview Press, pp. 323–357.

Jha, S., Casey-Ford, R.-G., Pedersen, *et al.* (2006) The queen is not a pacemaker in the small colony wasps *Polistes instabilis* and *P. dominulus*. *Animal Behaviour* **71**, 1197–1203.

Jones, J. C. & Oldroyd, B. P. (2007) Nest thermoregulation in social insects. *Advances in Insect Physiology*, **33**, 153–191.

Kasuya, E. (1981a) Male mating territory in a Japanese paper wasp, *Polistes jadwigae* Dalla Torre (Hymenoptera, Vespidae). *Kontyû*, **49**, 607–614.

(1981b) Internidal drifting of workers in the Japanese paper wasp *Polistes chinensis antennalis* (Vespidae; Hymenoptera). *Insectes Sociaux*, **28**, 343–346.

Keeping, M. G. (1992) Social organization and division of labor in colonies of the polistine wasp, *Belonogaster petiolata*. *Behavioral Ecology and Sociobiology*, **19**, 333–341.

(2000) Morpho-physiological variability and differentiation of reproductive roles among foundresses of the primitively eusocial wasp, *Belonogaster petiolata* (DeGeer) (Hymenoptera, Vespidae). *Insectes Sociaux*, **47**, 47–154.

(2002) Reproductive and worker castes in the primitively eusocial wasp *Belonogaster petiolata* (DeGeer) (Hymenoptera: Vespidae): Evidence for pre-imaginal differentiation. *Journal of Insect Physiology*, **48**, 867–879.

Keeping, M. G., Lipschitz, D., & Crewe, R. (1986) Chemical mate recognition and release of male sexual behavior in polybiine wasp, *Belonogaster petiolata* (Degeer) (Hymenoptera: Vespidae). *Journal of Chemical Ecology*, **12**, 773–779.

Kim, B., Kim, K. W., & Choe, J. C. (2012) Temporal polyethism in Korean yellowjacket foragers, *Vespula koreensis*. *Insectes Sociaux*, **59**, 263–268.

Klahn, J. E. (1979) Philopatric and nonphilopatric foundress associations in the social wasp *Polistes fuscatus*. *Behavioral Ecology and Sociobiology*, **5**, 417–424.

(1988) Intraspecific comb usurpation in the social wasp *Polistes fuscatus*. *Behavioral Ecology and Sociobiology*, **23**, 1–8.

Klahn, J. & Pilgrim, D. (1985) Kin recognition and brood tolerance in the paper wasp *Polistes fuscatus* (Hymenoptera: Vespidae). *Journal of the Kansas Entomological Society*, **58**, 567–568.

Kocher, S. D. & Grozinger, C. M. (2011) Cooperation, conflict, and the evolution of queen pheromones. *Journal of Chemical Ecology*, **37**, 1263–1275.

Kojima, J., Hartini, S., Noerdjito, W. A., *et al.* (2001) Descriptions of pre-emergence nests and mature larvae of *Vespa fervida*, with a note on a multiple-foundress colony of *V. affinis* in Sulawesi (Hymenoptera: Vespidae; Vespinae). *Entomological Science*, **4**, 355–360.

Kundu, H. L. (1967) Observations on *Polistes hebraeus* (Hymenoptera). Birla Institute of Technological Science, *Journal (Pilani)*, **1**, 152–161.

Landi, M., Queller, D. C., Turillazzi, S., & Strassmann, J. E. (2003) Low relatedness and frequent queen turnover in the stenogastrine wasp *Eustenogaster fraterna* favor the life insurance over the haplodiploid hypothesis for the origin of eusociality. *Insectes Sociaux*, **50**, 262–267.

Leadbeater, E., Carruthers, J. M., Green, J. P., Rosser, N. S., & Field, J. (2011) Nest inheritance is the missing source of direct fitness in a primitively eusocial insect. *Science*, **333**, 874–876.

Liebert, A. E. & Starks, P. T. (2006) Taming of the skew: Transactional models fail to predict reproductive partitioning in the paper wasp *Polistes dominulus*. *Animal Behaviour*, **71**, 913–923.

Linksvayer, T. A. & Wade, M. J. (2005) The evolutionary origin and elaboration of sociality in the aculeate Hymenoptera: Maternal effects, sib-social effects, and heterochrony. *Quarterly Review of Biology*, **80**, 317–336.

Litte, M. (1977) Behavioral ecology of the social wasp, *Mischocyttarus mexicanus*. *Behavioral Ecology and Sociobiology*, **2**, 229–246.

Lorenzi, M. C., Cervo, R., Zacchi, F., Turillazzi, S., & Bagnères, A. G. (2004) Dynamics of chemical mimicry in the social parasite wasp *Polistes semenowi* (Hymenoptera: Vespidae). *Parasitology*, **129**, 643–651.

Lucas, E. R. & Keller, L. (2014) Ageing and somatic maintenace in social insects. *Current Opinion in Insect Science*, **5**, 31–36.

Lucas, E. R., Martins, R. P, Zanette, L. R. S., & Field, J. (2011) Social and genetic structure in colonies of the social wasp *Microstigmus nigrophthalmus*. *Insectes Sociaux*, **58**, 107–114.

MacDonald, J. F. & Matthews, R. W. (1981) Nesting biology of the eastern yellowjacket, *Vespula maculifrons* (Hymenoptera: Vespidae). *Journal of the Kansas Entomological Society*, **54**, 433–457.

Makino, S. (1989) Usurpation and nest rebuilding in *Polistes riparius*: Two ways to reproduce after the loss of the original nest (Hymenoptera: Vespidae). *Insectes Sociaux*, **36**, 116–128.

Makino, S. & Sayama, K. (1991) Comparison of intraspecific nest usurpation between two haplometrotic paper wasp species (Hymenoptera: Vespidae: *Polistes*). *Journal of Ethology*, **9**, 121–128.

Makino, S. & Yamane, S. (1980) Heat production by the foundress of *Vespa simillima*, with description of its embryo nest (Hymenoptera: Vespidae). *Insecta Matsumurana*, **19**, 89–101.

Manzanilla, J., de Sousa, L., & Sánchez, D. (2000) High densities of *Polistes versicolor versicolor* (Oliver 1791) (Hymenoptera: Vespidae) at Cerro La Laguna, Turimiquire massif, Anzoátegui State, Venezuela. *Boletín de Entomología Venezolana*, **15**, 245–248.

Martin, J. S. (1988) Thermoregulation in *Vespa simillima xanthoptera* (Hymenoptera, Vespidae). *Kontyu*, **56**, 674–677.

(1992) Nest thermoregulation in *Vespa affinis* (Hymenoptera, Vespidae). *Japanese Journal of Entomology*, **60**, 483–486.

Matsuura, M. (1984) Comparative biology of the five Japanese species of the genus *Vespa* (Hymenoptera, Vespidae). *Bulletin of the Faculty of Agriculture, Mie University*, **69**, 1–131.

(1991) *Vespa* and *Provespa*. *In*: Ross, K. G. & Matthews, R. W. (eds.) *The Social Biology of Wasps*. Ithaca NY: Comstock Publishing Associates, Cornell University Press, pp. 232–262.

(1999) Size and composition of swarming colonies in *Provespa anomala* (Hymenoptera, Vespidae), a nocturnal social wasp. *Insectes Sociaux*, **46**, 219–223.

Matsuura, M. & Yamane, S. (1990) *Biology of the Vespine Wasps*. Berlin: Springer-Verlag.

Matthews, R. W. (1968) *Microstigmus comes*: Sociality in a sphecid wasp. *Science*, **160**, 787–788.

Matthews, R. W. & Naumann, I. D. (1988) Nesting biology and taxonomy of *Arpactophilus mimi*, a new species of social sphecid (Hymenoptera: Sphecidae). *Australian Journal of Zoology*, **36**, 585–597.

Maynard Smith, J. & Szathmáry, E. (1995) *The Major Transitions in Evolution*. Oxford: W. H. Freeman / Spektrum.

McCorquodale, D. B. & Naumann, I. D. (1988) A new Australian species of communal ground nesting wasp in the genus *Spilomena* Shuckard (Hymenoptera: Sphecidae: Pemphredoninae). *Journal of the Australian Entomological Society*, **27**, 221–231.

Mead, F., Habersetzer, C., Gabouriaut, D., & Gervet, J. (1995) Nest-founding behavior induced in the first descendants of *Polistes dominulus* Christ (Hymenoptera, Vespidae) colonies. *Insectes Sociaux*, **42**, 385–396.

Melo, G. A. R. (2000) Comportamento social em vespas da família Sphecidae (Hymenoptera, Apoidea). *In*: Martins, R. P., Lewinsohn, T. M., & Barbeitos, M. S. (eds.) *Ecologia e Comportamento de Insects*. Rio de Janeiro: *Oecologia Brasiliensis*, pp. 85–130.

Michener, C. D. (1974) *The Social Behavior of the Bees: A Comparative Study*. Cambridge, MA: Belknap Press of Harvard University Press.

Monnin, T. (2006) Chemical recognition of reproductive status in social insects. *Annales Zoologici Fennici*, **43**, 515–530.

Muralidharan, K., Shaila, M. S., & Gadagkar, R. (1986) Evidence for multiple mating in the primitively eusocial wasp *Ropalidia marginata* (Lep.) (Hymenoptera: Vespidae). *Journal of Genetics*, **153**, 153–158.

Nalepa, C. A. (2015) Origin of termite eusociality: Trophallaxis integrates the social, nutritional, and microbial environments. *Ecological Entomology*, **40**, 323–335.

Nascimento, F. S., Tannure-Nacimento, I. C., & Zucchi, R. (2004) Behavioral mediators of cyclical oligogyny in the Amazonian swarm-founding wasp *Asteloeca ujhelyii*. *Insectes Sociaux*, **51**, 17–23.

Naug, D. & Gadagkar, R. (1998) The role of age in temporal polyethism in a primitively eusocial wasp. *Behavioral Ecology and Sociobiology*, **42**, 37–47.

Naumann, M. G. (1970) *The nesting behavior of* Protopolybia pumila *in Panama (Hymenoptera: Vespidae).* Ph.D. dissertation, University of Kansas.

(1975) Swarming behavior: Evidence for communication in social wasps. *Science*, **189**, 642–644.

Noll, F. B. & Wenzel, J. W. (2008) Caste in the swarming wasps: 'queenless' societies in highly social insects. *Biological Journal of the Linnean Society*, **93**, 509–522.

Noll, F. B., Wenzel, J. W., & Zucchi, R. (2004) Evolution of caste in Neotropical swarm-founding wasps (Hymenoptera: Vespidae; Epiponini). *American Museum Novitates*, **3467**, 1–24.

Nonacs, P. & Reeve, H. K. (1993) Opportunistic adoption of orphaned nests in paper wasps as an alternative reproductive strategy. *Behavioural Processes*, **30**, 47–59.

Nonacs, P., Liebert, A. E., & Starks, P. T. (2006) Transactional skew and assured fitness return models fail to predict patterns of cooperation in wasps. *The American Naturalist.* **167**, 467–480.

Nowak, M. A., Tarnita, C. E., & Wilson, E. O. (2010) The evolution of eusociality. *Nature*, **466**, 1057–1062.

O'Donnell, S. (1994) Nestmate copulation in the Neotropical eusocial wasp *Polistes instabilis* De Saussure (Hymenoptera: Vespidae). *Psyche*, **101**, 33–36.

(1998a) Dominance and polyethism in the eusocial wasp *Mischocyttarus mastigophorus* (Hymenoptera: Vespidae). *Behavioral Ecology and Sociobiology*, **43**, 327–331.

(1998b) Reproductive caste determination in eusocial wasps (Hymenoptera: Vespidae). *Annual Review of Entomology*, **43**, 323–346.

O'Donnell, S. & Jeanne, R. L. (1991) Interspecific occupation of a tropical social wasp colony (Hymenoptera: Vespidae: *Polistes*). *Journal of Insect Behavior*, **4**, 397–400.

O'Donnell, S., Hunt, J. H., & Jeanne, R. L. (1997) Gaster-flagging during colony defense in Neotropical swarm-founding wasps (Hymenoptera: Vespidae, Epiponini). *Journal of the Kansas Entomological Society*, **70**, 175–180.

O'Neill, K. M. (2001) *Solitary Wasps: Behavior and Natural History*. Ithaca NY: Cornell University Press.

Pagden, H. T. (1962) More about *Stenogaster*. *Malayan Nature Journal* **12**, 131–148.

Pallett, M. J. (1984) Nest site selection and survivorship of *Dolichovespula arenaria* and *Dolichovespula maculata* (Hymenoptera: Vespidae). *Canadian Journal of Zoology*, **62**, 1268–1272.

Pardi, L. (1948) Dominance order in *Polistes* wasps. *Physiological Zoology*, **21**, 1–13.

Peters, J. M., Queller, D. C., Strassmann, J. E., & Solís, C. R. (1995) Maternity assignment and queen replacement in a social wasp. *Proceeding of the Royal Society of London B*, **260**, 7–12.

Pickett, K. M. & Carpenter, J. M. (2010) Simultaneous analysis and the origin of eusociality in the Vespidae (Insecta: Hymenoptera). *Arthropod Systematics & Phylogeny*, **68**, 3–33.

Pickett, K. M., Osborne, D. M., Wahl, D., & Wenzel, J. W. (2001) An enormous nest of *Vespula squamosa* from Florida, the largest social wasp nest reported from North America, with notes on colony cycle and reproduction. *Journal of the New York Entomological Society*, **109**, 408–415.

Piekarski, P., Longair, R., & Rogers, S. (2014) Monophyly of eusocial wasps (Hymenoptera: Vespidae): Molecules and morphology tell opposing histories. *Journal of Undergraduate Research in Alberta*, **4**, 11–14.

Polak, M. (1993) Competition for landmark territories among male *Polistes canadensis* (L.. (Hymenoptera: Vespidae): Large-size advantage and alternative mate-aquisition tactics. *Behavioral Ecology*, **4**, 325–331.

Post, D. C. & Jeanne, R. L. (1983) Male reproductive behavior of the social wasp *Polistes fuscatus* (Hymenoptera: Vespidae). *Zeitschrift für Tierpsychologie*, **62**, 157–171.

Premnath, S., Sinha, S., & Gadagkar, R. (1995) Dominance relationship in the establishment of reproductive division of labour in a primitively eusocial wasp (*Ropalidia marginata*). *Behavioral Ecology and Sociobiology*, **39**, 125–132.

Queller, D. C. (1989) The evolution of eusociality: Reproductive head starts of workers. *Proceedings of the National Academy of Sciences USA*, **86**, 3224–3226.

(1996) The origin and maintenance of eusociality: The advantage of extended parental care. *In*: Turillazzi, S. & West-Eberhard, M. J. (eds.) *Natural History and Evolution of Paper-wasps*, Oxford: Oxford University Press, pp. 218–234.

Queller, D. C., Strassman, J. E., & Hughes, C. R. (1988) Genetic relatedness in colonies of tropical wasps with multiple queens. *Science*, **242**, 1155–1157.

Queller, D. C., Hughes, C. R., & Strassmann, J. E. (1990) Wasps fail to make distinctions. *Nature*, **344**, 388.

Queller, D. C., Negrón-Sotomayor, J. A., Strassmann, J. E., & Hughes, C. R. (1993) Queen number and genetic relatedness in a neotropical wasp, *Polybia occidentalis*. *Behavioral Ecology*, **4**, 7–13.

Queller, D. C., Zacchi, F., Cervo, R. Turillazzi, S., Henshaw, M. T., *et al.* (2000) Unrelated helpers in a social insect. *Nature*, **405**, 784–787.

Ratnieks, F. L. W., Vetter, R. S., & Visscher, P. K. (1996) A polygynous nest of *Vespula pensylvanica* from California with a discussion of possible factors influencing the evolution of polygyny in *Vespula*. *Insectes Sociaux*, **43**, 401–410.

Reed, H. C. & Landolt, P. J. (1991) Swarming of paper wasp (Hymenoptera, Vespidae) sexuals at towers in Florida. *Annals of the Entomological Society of America*, **84**, 628–635.

Reeve, H. K. (1991) *Polistes*. *In*: Ross, K. G. & Matthews, R. W. (eds.) *The Social Biology of Wasps*. Ithaca NY: Comstock Publishing Associates, Cornell University Press, pp. 99–148.

Reeve, H. K. & Gamboa, G. J. (1983) Colony activity integration in primitively eusocial wasps: The role of the queen (*Polistes fuscatus*, Hymenoptera: Vespidae). *Behavioral Ecology and Sociobiology*, **13**, 63–74.

(1987) Queen regulation of worker foraging in paper wasps: A social feedback control mechanism (*Polistes fuscatus*, Hymenoptera: Vespidae). *Behaviour*, **102**, 147–167.

Riabinin, K., Kozhevnikov, M., & Ishay, J. S. (2004) Ventilating activity at the hornet nest entrance. *Journal of Ethology*, **22**, 49–53.

Richard, F.-J. & Hunt, J. H. (2013) Intracolony chemical communication in social insects. *Insectes Sociaux*, **60**, 275–291.

Richards, O. W. (1978) *The Social Wasps of the Americas, Excluding the Vespinae*. London: British Museum (Natural History).

Riggs, A. D. (1975) X inactivation, differentiation and DNA methylation. *Cytogenetics and Cell Genetics*, **14**, 9–25.

Ross, K. G. & Carpenter, J. M. (1991) Population genetic structure, relatedness, and breeding system. *In*: Ross, K. G., & Matthews, R. W. (eds.) *The Social Biology of Wasps*. Ithaca NY: Comstock Publishing Associates, Cornell University Press, pp. 451–479.

Ross, K. G., & Matthews, R. W. (1989a) New evidence for eusociality in the sphecid wasp *Microstigmus comes*. *Animal Behaviour*, **38**, 613–619.

(1989b) Population genetic structure and social evolution in the sphecid wasp *Microstigmus comes*. *The American Naturalist*, **134**, 574–598.

(1991) *The Social Biology of Wasps*. Ithaca NY: Comstock Publishing Associates, Cornell University Press.

Roubaud, E. (1916) Recherches biologiques sur les guêpes solitaires et sociales d'Afrique. La genèse de la vie sociale et l'évolution de l'instinct maternel chez les vespides. *Annales des Sciences Naturelles, 10e série: Zoologie*, **1**, 1–160.

Sakagami, S. F. & Fukushima, K. (1957) *Vespa dybowskii* André as a facultative temporary social parasite. *Insectes Sociaux*, **4**, 1–12.

Santos, B. F., Payne, A., Pickett, K. M., & Carpenter, J. M. (2014) Phylogeny and historical biogeography of the paper wasp genus *Polistes* (Hymenoptera: Vespidae): Implications for the overwintering hypothesis of social evolution. *Cladistics*, **31**, 535–549.

Schmitz, J. & Moritz, R. F. A. (1998) Molecular phylogeny of Vespidae (Hymenoptera) and the evolution of sociality in wasps. *Molecular Phylogenetics and Evolution*, **9**, 183–191.

(2000) Molecular evolution in social wasps. *In*: Austin, A. D., & Dowton, M. (eds.) *Hymenoptera: Evolution, Biodiversity and Biological Control*. Collingwood: CSIRO, pp. 84–89.

Seeley, T. D. & Heinrich, B. (1981) Regulation of temperature in the nests of social insects. *In*: Heinrich, B. (ed.) *Insect Thermoregulation*. New York: Wiley, pp. 159–234.

Seppä, P., Fogelqvist, J., Gyllenstrand, N., & Lorenzi, M. C. (2011) Colony kin structure and breeding patterns in the social wasp, *Polistes biglumis*. *Insectes Sociaux*, **58**, 345–355.

Sheehan, M. J. & Tibbetts, E. A. 2008. Robust long-term social memories in a paper wasp. *Current Biology*, **18**, R851-R852.

Sheehan, M. J., Botero, C. A., Hendry, T. A., *et al.* (2015) Different axes of environmental variation explain the presence vs. extent of cooperative nest founding associations in *Polistes* paper wasps. *Ecology Letters*, **18**, 1057–1067.

Shima, S. N., Noll, F. B., Zucchi, R., & Yamane, S. (1998) Morphological caste differences in the Neotropical swarm-founding polistine wasps IV. *Pseudopolybia vespiceps*, with preliminary considerations on the role of intermediate females in social organization of the Epiponini (Hymenoptera, Vespidae). *Journal of Hymenoptera Research*, **7**, 280–295.

Shreeves, G. & Field, J. (2002) Group size and direct fitness in social queues. *The American Naturalist*, **159**, 81–95.

Sledge, M. F., Fortunato, A., Turillazzi, S., Francescato, E., Hasim, R., *et al.* (2000) Use of Dufour's gland secretion in nest defense and brood nutrition by hover wasps (Hymenoptera, Stenogastrinae). *Journal of Insect Physiology*, **46**, 753–761.

Sledge, M. F., Dani, F. R., Cervo, R., Dapporto, L., & Turillazzi, S. (2001a) Recognition of social parasites as nest-mates: Adoption of colony-specific host cuticular odours by the paper wasp parasite *Polistes sulcifer*. *Proceeding of the Royal Society of London B*, **268**, 2253–2260.

Sledge, M. F., Boscaro, F., & Turillazzi, S. (2001b) Cuticular hydrocarbons and reproductive status in the social wasp *Polistes dominulus*. *Behavioral Ecology and Sociobiology*, **49**, 401–409.

Smith, A. R., O'Donnell, S., & Jeanne, R. L. (2001) Correlated evolution of colony defence and social structure: A comparative analysis in eusocial wasps (Hymenoptera: Vespidae). *Evolutionary Ecology Research*, **3**, 331–344.

Snelling, R. (1952) Notes on nesting and hibernation of *Polistes*. *Pan-Pacific Entomologist*, **29**, 177.

Spradbery, J. P. (1973) *Wasps: An Account of the Biology and Natural History of Solitary and Social Wasps*. Seattle, WA: University of Washington Press.

(1993) Queen brood reared in worker cells by the social wasp, *Vespula germanica* (F.) (Hymenoptera: Vespidae). *Insectes Sociaux*, **40**, 181–190.

Starks, P. T. (1998) A novel 'sit and wait' reproductive strategy in social wasps. *Proceedings of the Royal Society of London B*, **265**, 1407–1410.

Starr, C. K. (1985) Enabling mechanisms in the origin of sociality in the Hymenoptera – the sting's the thing. *Annals of the Entomological Society of America*, **78**, 836–840.

Strassmann, J. E. (1981a) Evolutionary implications of early male and satellite nest production in *Polistes exclamans* colony cycles. *Behavioral Ecology and Sociobiology*, **8**, 55–64.

(1981b) Kin selection and satellite nests in *Polistes exclamans*. *In*: Alexander, R. D. & Tinkle, D. W. (eds.) *Natural Selection and Social Behavior: Resent Research and New Theory*, New York: Chiron Press, pp. 45–58.

(1981c) Wasp reproduction and kin selection: Reproductive competition and dominance hierarchies among *Polistes annularis* foundresses. *Florida Entomologist*, **64**, 74–88.

(2001) The rarity of multiple mating by females in the social Hymenoptera. *Insectes Sociaux*, **48**, 1–13.

Strassmann, J. E. & Meyer, D. C. (1983) Gerontocracy in the social wasp, *Polistes exclamans*. *Animal Behaviour*, **31**, 431–438.

Strassmann, J. E. & Queller, D. C. (1989) Ecological determinants of social evolution. *In*: Breed, M. & Page, R. (eds.) *The Genetics of Social Evolution*. Boulder: Westview Press, pp. 81–101.

Strassmann, J. E., Queller, D. C., & Hughes, C. R. (1988) Predation and the evolution of sociality in the paper wasp *Polistes bellicosus*. *Ecology*, **69**, 1497–1505.

Strassmann, J. E., Queller, D. C., Solís, C. R., & Hughes, C. R. (1991) Relatedness and queen number in the Neotropical wasp, *Parachartergus colobopterus*. *Animal Behavior*, **42**, 461–470.

Strassmann, J. E., Gastreich, K. R., Queller, D. C., & Hughes, C. R. (1992) Demographic and genetic evidence for cyclical changes in queen number in a Neotropical wasp, *Polybia emaciata*. *The American Naturalist*, **140**, 363–372.

Strassmann, J. E., Solís, C. R., Hughes, C. R., Goodnight, K. F., & Queller, D. C. (1997a). Colony life history and demography of a swarm-founding social wasp. *Behavioral Ecology and Sociobiology*, **40**, 71–77.

Strassmann, J. E., Klingler, C. J., Arévalo, E., *et al.* (1997b) Absence of within-colony kin discrimination in behavioural interactions of swarm-founding wasps. *Proceeding of the Royal Society of London B*, **264**, 1565–1570.

Sugden, E. A. & McAllen, R. L. (1994) Observations on foraging, population and nest biology of the Mexican honey wasp, *Brachygastra mellifica* (Say) in Texas [Vespidae: Polybiinae]. *Journal of the Kansas Entomological Society*, **67**, 141–155.

Sumana, A. & Gadagkar, R. (2001) The structure of dominance hierarchies in the primitively eusocial wasp *Ropalidia marginata*. *Ethology Ecology & Evolution*, **13**, 273–281.

Suryanarayanan, S. & Jeanne, R. L. (2008) Antennal drumming, trophallaxis, and colony development in the social wasp *Polistes fuscatus* (Hymenoptera: Vespida). *Ethology*, **114**, 1201–1209.

Suryanarayanan, S., Hantschel, A. E., Torres, C. G., & Jeanne, R. L. (2011a) Changes in the temporal pattern of antennal drumming behavior across the *Polistes fuscatus* colony cycle (Hymenoptera, Vespidae). *Insectes Sociaux*, **58**, 97–106.

Suryanarayanan, S., Hermanson, J. C., & Jeanne, R. L. (2011b) A mechanical signal biases caste development in a social wasp. *Current Biology*, **21**, 231–235.

Suzuki, T. (2003) Queen replacement without gerontocracy in the paper wasp *Parapolybia indica* in temperate Japan. *Ethology Ecology & Evolution*, **15**, 191–196.

Suzuki, T. & Ramesh, M. (1992) Colony founding in the social wasp, *Polistes stigma* (Hymenoptera Vespidae), in India. *Ethology Ecology & Evolution*, **4**, 333–341.

Thornhill, R. & Alcock, J. (1983) *The Evolution of Insect Mating Systems.* Cambridge MA: Harvard University Press.

Tibbetts, E. A. (2002) Visual signals of individual identity in the wasp *Polistes fuscatus. Proceedings of the Royal Society of London B*, **269**, 1423–1428.

(2004) Complex social behaviour can select for variability in visual features: A case study in *Polistes* wasps. *Proceeding of the Royal Society of London B*, **271**, 1955–1960.

(2006) Badges of status in worker and gyne *Polistes dominulus* wasps. *Annales Zoolgici Fennici*, **43**, 575–582.

(2010) The condition dependence and heritability of signaling and nonsignaling color traits in paper wasps. *The American Naturalist*, **175**, 495–503.

Tibbetts, E. A. & Curtis, T. R. (2007) Rearing conditions influence quality signals but not individual identity signals in *Polistes* wasps. *Behavioral Ecology* **18**, 602–607.

Tibbetts, E. A. & Dale, J. (2004) A socially enforced signal of quality in a paper wasp. *Nature*, **432**, 218–222.

Tibbetts, E. A. & Lindsay, R. (2008) Visual signals of status and rival assessment in *Polistes dominulus* paper wasps. *Biology Letters*, **4**, 237–239.

Tibbetts, E. A. & Reeve, H. K. (2003) Benefits of foundress associations in the paper wasp *Polistes dominulus*: Increased productivity and survival, but no assurance of fitness returns. *Behavioral Ecology*, **14**, 510–514.

Tibbetts, E. A. & Sheehan, M. J. (2012) The effect of juvenile hormone on *Polistes* wasp fertility varies with cooperative behavior. *Hormones and Behavior*, **61**, 559–564.

Tibbetts, E. A., Skaldina, O., Zhao, V., *et al.* (2011) Geographic variation in the status signals of *Polistes dominulus* wasps. *PLoS ONE*, **6**, e28173.

Tindo, M., D'Agostino, P., Francescato, E., Dejean, A., & Turillazzi, S. (1997) Associative colony foundation in the tropical wasp *Belonogaster juncea juncea* (Vespidae, Polistinae). *Insectes Sociaux*, **44**, 365–377.

Tindo, M., Kenne, M., & Dejean, A. (2008) Advantages of multiple foundress colonies in *Belonogaster juncea juncea* L.: Greater survival and increased productivity. *Ecological Entomology*, **33**, 1–5.

Toth, A. L., Varala, K., Henshaw, M. T., *et al.* (2010) Brain transcriptomic analysis in paper wasps identifies genes associated with behaviour across social insect lineages. *Proceedings of the Royal Society of London B*, **277**, 2139–2148.

Toth, A. L., Tooker, J. F., Radhakrishnan, S., Henshaw, M. T., & Grozinger, C. M. (2014) Shared genes related to aggression, rather than chemical communication, are associated with reproductive dominance in paper wasps (*Polistes metricus*). *BMC Genomics*, **15**, 75.

Toth, A. L., Sumner, S., & Jeanne, R. L. (2016) Patterns of longevity across a sociality gradient in vespid wasps. *Current Opinion in Insect Science*, **16**, 28–35.

Trostle, G. E. & Torchio, P. F. (1986) Notes on the nesting biology and immature development of *Euparagia scutellaris* Cresson (Hymenoptera, Masaridae). *Journal of the Kansas Entomological Society*, **59**, 641–647.

Turillazzi, S. (1983) Patrolling behaviour in males of *Parischnogaster nigricans serrei* (Du Buysson) and *P. mellyi* (Saussure) (Hymenoptera, Stenogastrinae). *Accademia Nazionale dei Lincei, Rendiconti della Classe di Scienze fisiche, matematiche e naturali, Serie VIII*, **72**, 153–157.

(1985) Colonial cycle of *Parischnogaster nigricans serrei* (Du Buysson) in West Java (Hymenoptera, Stenogastrinae). *Insectes Sociaux*, **32**, 43–60.

(1987) Colony defense in Stenogastrinae wasps (Hymenoptera). *Monitore zoologico italiano – Italian Journal of Zoology*, **21**:205–205.

(1988) Social biology of *Parischnogaster jacobsoni* (Du Buysson) (Hymenoptera, Stenogastrinae). *Insectes Sociaux*, **35**, 133–143.

(1991) The Stenogastrinae. *In*: Ross, K. G., & Matthews, R. W. (eds.) *The Social Biology of Wasps*. Ithaca NY: Comstock Publishing Associates, Cornell University Press, pp. 74–98.

(1996) *Polistes* in perspcetive: Comparative social biology and evolution in *Belonogaster* and Stenogastrinae. *In*: Turillazzi, S., & West-Eberhard, M. J. (eds.) *Natural History and Evolution of Paper-Wasps*. Oxford: Oxford University Press, pp. 235–247.

(2012) *The Biology of Hover Wasps*. Heidelberg: Springer.

Turillazzi, S. & Francescato, E. (1990) Patrolling behavior and related secretory structures in the males of some stenogastrine wasps (Hymenoptera, Vespidae). *Insectes Sociaux*, **37**, 146–157.

Turillazzi, S. & Pardi, L. (1981) Ant guards on nests of *Parischnogaster nigricans serrei* (Buysson) (Stenogastrinae). *Monitore zoologico italiano (Nuova serie)*, **15**, 1–7.

Turillazzi, S. & West-Eberhard, M. J. (1996) *Natural History and Evolution of Paper-Wasps*. Oxford: Oxford University Press.

Turillazzi, S., Francescato, E., Tosi, A. B., & Carpenter, J. M. (1994) A distinct caste difference in *Polybioides tabidus* (Fabricius) (Hymenoptera, Vespidae). *Insectes Sociaux*, **41**, 327–330.

Tuschida, K., & Itô, Y. (1987) Internidal drifting and dominance behaviour in *Polistes jadwigae* Dalla Torre workers (Hymenoptera: Vespidae). *Journal of Ethology*, **5**, 83–85.

Vesy-Fitzgerald, D. (1938) Social wasps (Hym. Vespidae) from Trinidad, with a note on the genus *Trypoxylon* Latreille. *Transactions of the Royal Entomological Society of London A*, **25**, 81–86.

Velthuis, H. H. W., Roeling, A., Imperatriz-Fonseca, V. L. (**2001**) Repartition of reproduction among queens in the polygynous stingless bee Melipona biocolor. *Proceedings of the Section Experimental and Applied Entomology of the Netherlands Entomological Society*. **12**, 45–49.

Vetter, R. S. & Visscher, P. K. (1997) Plasticity of annual cycle in *Vespula pensylvanica* shown by a third year polygynous nest and overwintering of queens inside nests. *Insectes Sociaux*, **44**, 353–364.

Wcislo, W. & Tierney, S. M. (2009) The evolution of communal behavior in bees and wasps: An alternative route to sociality. *In*: Gadau, J. & Fewell, J. (eds.) *Organization of Insect Societies: From Genome to Sociocomplexity*. Cambridge, MA: Harvard University Press, pp. 148–169.

Wcislo, W., West-Eberhard, M. J., & Eberhard, W. G. (1988) Natural history and behavior of a primitively social wasp, *Auplopus semialatus* and its parasite, *Irenangelus eberhardi* (Hymenoptera: Pompilidae) *Journal of Insect Behavior*, **1**, 247–260.

Weiner, S. A., Upton, C. T., Noble, K., Woods, W. A., & Starks, P. T. (2010) Thermoregulation in the primitively eusocial paper wasp, *Polistes dominulus*. *Insectes Sociaux*, **57**, 157–162.

Wenzel, J. W. (1987a) *Ropalidia formosa*, a nearly solitary paper wasp from Madagascar (Hymenoptera: Vespidae). *Journal of the Kansas Entomological Society*, **60**, 679–699.

(1987b). Male reproductive behavior and mandibular glands in *Polistes major* (Hymenoptera: Vespidae). *Insectes Sociaux*, **34**, 44–57.

(1989) Endogenous factors, external cues, and eccentric construction in *Polistes annularis* (Hymenoptera: Vespidae). *Journal of Insect Behavior*, **2**, 679–699.

(1991) Evolution of nest architecture. *In*: Ross, K. G., & Matthews, R. W. (eds.) *The Social Biology of Wasps*. Ithaca NY: Comstock Publishing Associates, Cornell University Press, pp. 480–519.

(1992) Extreme queen-worker dimorphism in *Ropalidia ignobilis*, a small-colony wasp (Hymenoptera: Vespidae). *Insectes Sociaux*, **39**, 31–43.

West-Eberhard, M. J. (1969) The social biology of polistine wasps. *Miscellaneous Publications, Museum of Zoology, University of Michigan*, **140**, 1–101.

West-Eberhard, M. J. (1978) Temporary queens in *Metapolybia* wasps: Nonreproductive helpers without altruism? *Science*, **200**, 441–443.

(1981) Intragroup selection and the evolution of insect societies. *In*: Alexander, R. D., and Tinkle, D. W. (eds.) *Natural Selection and Social Behavior: Recent Research and New Theory*. New York: Chiron Press, pp. 3–17.

(1982) The nature and evolution of swarming in tropical social wasps (Vespidae, Polistinae, Polybiini). *In*: P. Jaisson (ed.) *Social Insects in the Tropics*, Vol. 1. Paris: Université Paris-Nord, pp. 97–128.

(1987a) Flexible strategy and social evolution. *In*: Itô, Y., Brown, J. L., & Kikkawa, J. (eds.) *Animal Societies: Theories and Facts*. Tokyo: Scientific Societies Press Ltd., pp. 35–51.

(1987b) Observations of *Xenorhynchium nitidulum* (Fabricious) (Hymenoptera, Eumeninae), a primitively social wasp. *Psyche*, **94**, 317–324.

(2005) Behavior of the primitively social wasp *Montezumia cortesioides* Willink (Vespidae Eumeninae) and the origins of vespid sociality. *Ethology Ecology & Evolution*, **17**, 201–215.

Wilson, E. O. (1971) *The Insect Societies*. Cambridge MA: Belknap Press of Harvard University Press.

(2008) One giant leap: How insects achieved altruism and colonial life. *BioScience*, **58**, 17–25.

Wilson, E. O. & Hölldobler, B. (2005) Eusociality: Origin and consequences. *Proceedings of the National Academy of Sciences USA*, **102**, 13367–13371.

Yamane, S. (1985) Social relation among females in pre- and post-emergence colonies of a subtropical paper wasp, *Parapolybia varia* (Hymenoptera: Vespidae). *Journal of Ethology*, **2**, 27–38.

(1996) Ecological factors influencing the colony cycle of *Polistes* wasps. *In*: Turillazzi, S. & West-Eberhard, M. J. (eds.) *Natural History and Evolution of Paper Wasps*. Oxford: Oxford University Press.

Yamane, S. & Kawamichi, T. (1975) Bionomic comparison of *Polistes biglumis* (Hymenoptera, Vespidae) at two different localities in Hokkaido, northern Japan, with reference to its probable adaptation to cold climate. *Kontyu*, **43**, 214–232.

Yamane, S., Kojima, J., & Yamane, S. (1983) Queen/worker size dimorphism in an Oriental wasp, *Ropalidia montana* Carl (Hymenoptera: Vespidae). *Insectes Sociaux*, **30**, 416–422.

Yoshikawa, K. (1955) A polistine colony usurped by a foreign queen. Ecological studies of *Polistes* wasps II. *Insectes Sociaux*, **2**, 255–260.

Yoshikawa, K., Ohgushi, R., & Sakagami, S. F. (1969) Preliminary report on entomology of the Osaka City University 5th scientific expedition to southeast Asia 1966 - with descriptions of two new genera of stenogastrine wasps by J. van der Vecht. *Nature and Life in Southeast Asia*, **6**, 153–182.

Zucchi, R., Yamane, S., & Sakagami, S. F. (1976) Preliminary notes on the habits of *Trimeria howardii*, a Neotropical communal masarid wasp, with description of the mature larva (Hymenoptera: Vespoidea). *Insecta Matsumarana, new series*, **8**, 47–57.

Zucchi, R., Sakagami, S. F., Noll, F. B., *et al.* (1995) *Agelaia vicina*, a swarm-founding polistine with the largest colony size among wasps and bees (Hymenoptera: Vespidae). *Journal of the New York Entomological Society*, **103**, 129–137.

5 Sociality in Termites

Judith Korb and Barbara Thorne

Overview

Termites are biologically unique because, across all known living creatures, termites have the most diverse array of distinct body forms encoded by a single genome. These different morphologies characterize separate castes within termite societies, including up to three types of reproductives of each sex, and depending upon the species, several castes of workers and soldiers. No microbe, plant, or animal – including other eusocial taxa – rival the breadth of polyphenism found within termite species. The exceptional behavioral and physiological specializations facilitated by varying internal and external morphologies of these diversified castes, and the resulting flexibility and adaptability afforded an integrated colony, comprise one of the fundamental drivers of termites' ecological success and evolutionary radiation. Another key component of termite abundance and prosperity in many ecosystems is their exclusive niche among eusocial insects (as well as most of other animals): termites feed and often nest in or around dead plant material. Although a cellulose-based diet has its challenges, termites have evolved "clever" innovations in order to thrive as detritivores. Exploiting the vast and relatively uncrowded plant decomposer niche has contributed to termites' success, and has rendered their activities critical to nutrient cycles and other essential dynamics of tropical forests, savannahs, and other ecosystems.

I SOCIAL DIVERSITY

5.1 How Common is Sociality in Termites?

Termites are "eusocial cockroaches," a monophyletic clade of diploid insects (Infraorder Isoptera) within the order Blattaria (Inward, *et al.*, 2007a; Lo, *et al.*, 2007; Engel, *et al.*, 2009; Krishna, *et al.*, 2013). All of the nearly 3,000 species of modern termites (Krishna, *et al.*, 2013) are eusocial, as they have sterile soldiers, a trait which is only

We thank Dustin Rubenstein and Patrick Abbot for the invitation to write this review and for all their and other contributors' efforts invested in this book. J. K. acknowledges NESCent and the German Science Foundation (DFG, KO1895/16-1) for support. Both authors salute our colleagues, and fascinating termites, for exciting adventures, unexpected discoveries, and the delights of integrating field and laboratory work with comparative biology and theory.

secondarily lost in some Termitidae (Noirot & Pasteels, 1987). Termites have castes of reproductives, workers, and soldiers, but differ conspicuously from eusocial Hymenoptera in being hemimetabolous (i.e. gradual development of the same body plan via molts through juvenile stages to adult) and in having morphologically versus allometrically distinct sterile soldiers. Because of hemimetabolous development, termites beyond the first instars are not helpless as are juvenile Hymenoptera larvae. Additionally, like some social vertebrates, but again in contrast to Hymenoptera, termites of both sexes are diploid and serve as helpers: female and male parents (i.e. queen and king) survive after initial mating, help rear the first brood, and continue to mate periodically.

5.2 Forms of Sociality in Termites

Levels of social organization and developmental flexibility differ distinctly across termite taxa, corresponding tightly to nest and foraging habits (so called "life types") (Tables 5.1 and 5.2). Social organization ranges from simple colonies with small colony sizes in which each individual has a high developmental plasticity, to some of the largest and most complex societies among animals in which the caste fate of each member is determined in the egg. Modern termites fall into two fundamental groups, with just a few species showing intermediate characteristics: One Piece Life Type Termites (hereafter OPT; terminology from Abe, 1987), thought to reflect ancestral life history patterns (Noirot, 1985a; Abe, 1987; Noirot & Pasteels, 1987; Noirot & Bordereau, 1988; Grandcolas & D'Haese, 2004; but see also Watson & Sewell, 1981), and Separate Type Termites (hereafter ST; Abe, 1987). This distinction in life type is reflected mainly in the worker caste (reviewed in Roisin & Korb, 2011). Soldiers do not differ as dramatically between OPT and ST species; in most species soldiers have a defensive role within the colony, but they can also serve as scouts and be involved in foraging (Traniello & Leuthold, 2000). A termite soldier is always a final (terminal) instar that develops via two molts through a presoldier intermediate. Soldiers are sterile and hence can gain only indirect fitness.

5.2.1 One Piece Life Type Termites (OPT)

OPT species (also called "single site," Noirot, 1970; "single-site nesters," Shellman-Reeve, 1997; or "wood-dwellers," Korb, 2007a) spend their entire lives nesting and feeding within the same single tree, log, stump, or piece of wood where their founding queen and king first initiated the colony. They are restricted to that one resource, and do not search for or exploit nearby pieces of wood (only rare cases of "house hunting" occur; Rupf & Roisin, 2008). These termites must cope with the nutritional, moisture, thermal, competitive, predator, pathogen and parasite circumstances within their nest wood, including inevitable changes over time. When their sole resource is depleted or otherwise inhospitable, the colony dies. Their only option for relocation is for developmentally agile individuals within the colony to differentiate into winged sexuals (alates) and disperse by flight to attempt to found their own colonies.

Table 5.1 Key ecological and developmental traits of termite families and their sister taxon the *Cryptocercus* woodroaches.

Taxonomy	Number species[1]	Distribution	Food	Build nest[2]	Foraging[3]	True workers[4]	Replacement reproductives[5]
Cryptocercidae		N-America, Oriental	Damp, partially decayed wood	No	No	No	No
Mastotermitidae	1	Australia	Dead wood	Yes	Yes	Yes	Yes; ergatoids
Hodotermitidae	15	Mainly Africa	Grass, leaves, bark	Yes	Yes	Yes	Yes; nymphoids
Archotermopsidae	13	N-America, Oriental, Australia	Damp, partially decayed wood	No	No	No	Yes; all older instars
Stolotermitidae	7	Mainly Australia	Damp, partially decayed wood	No	No	No	
Kalotermitidae	~410	World-wide	Dry, sound dead wood	No	No	No	Yes; all older instars
Stylotermitidae	34	Oriental					Unknown, but likely
***Rhino*termitidae**	~270	World-wide	Dead wood	Yes, except 2 genera	Yes, except 2 genera	Yes, except 2 genera	Yes; ergatoid & nymphoid
Serritermitidae	1	Neotropical	Unknown	Yes	Yes	Yes	Unknown
Termitidae	~1900	World-wide	Diverse: dead wood, grass, leaf litter, humus, soil with decomposed plant material, lichen	Yes	Yes	Yes	Yes, but rare; mainly adultoids or nymphoids

OPT taxa are in *italics and underlined*. The Rhinotermitidae have both OPT and ST species.

[1] approximate number of species

[2] species construct a nest and/or build foraging galleries

[3] species forage away from the nest for separate resources

[4] workers have a reduced developmental flexibility, bifurcated pathway with workers developing from the apterous line

[5] types of replacement reproductives after Myles (1999)

Table 5.2 Major patterns of social organization within Isoptera.

One Piece Life Type Termites (OPT)	Separate Life Type Termites (ST)
Low population size per colony	Larger population size per colony
Nest excavated in one piece of wood which serves as single food source; no foraging or travel away from nest wood except for alate flights; little gallery or other construction; relatively ephemeral	Elaborately constructed, complex, stable nests
Local foraging (nest wood)	Foraging away from the nest, potentially exploiting multiple resources concurrently, and able to search for and colonize new resources
Developmental options more flexible, even in later instars	Development more rigid; largely constrained to apterous and nymphal lines in early instars
When the food source is depleted or degrades, many individuals develop into alates and disperse	When a food source is depleted colony searches and moves to exploit new resource(s)
Monogamy of true (alate-derived) queen and king (assuming no intercolony fusion)	Polygyny and polyandry of alate-derived queens and kings in some derived groups
Single nest 'headquarters' confined to wood in which colony was founded	Polycaly (single colony occupying multiple nest sites) in some derived taxa
Soldier caste present	Secondary loss of soldiers in Apicotermitinae
"Sex-egalitarian" castes – i.e. little morphological or behavioral specialization known between the sexes in helpers	"True worker", taxa with distinct apterous line – sexual dimorphism in castes in some species (e.g. large workers of one sex; small workers the opposite gender) and sex specialization (e.g. all soldiers are of only one sex)
Soldiers usually differentiate from 5th or 6th (or later) instar (then proceed first through a presoldier molt before final molt into soldier)	Soldiers generally develop from earlier instars, even from the 2nd or 3rd instar, although can be later (presoldier instar consistent with more primitive termites)
Soldiers may differentiate from helpers (no wingbuds) or nymphs; in the latter case soldiers have wingbuds	Soldiers differentiate almost exclusively from the apterous (worker) line
Normal soldiers are sterile; unusual fertile neotenics with soldier-like traits (of both sexes, called "reproductive soldiers" or "soldier neotenics") are known from 6 species of Archotermopsidae and Stolotermitidae.	All soldiers are sterile
Soldiers mandibulate; they generally employ physical defense, although chemical defenses are known in *Cryptotermes*	Soldiers use chemical defense in addition to, or in place of, physical defense; mandibles range from strong and elaborate to vestigial
Replacement reproductives (neotenics) can develop from any instar after the third, with and without wingbuds	Replacement neotenics, with and without wingbuds, are also known within Termitidae
Adultoids (alate-derived reproductives that stay within their natal nest, or a satellite from that original nest) unknown	Adultoids of both sexes, sometimes more than one of one or both sexes within a colony, known in some Termitidae
Merged or fused interfamilial colonies occur	Rare fusion of unrelated families into a single cooperating functional colony known in *Reticulitermes*; unknown in Termitidae

Development in OPTs is exceptionally flexible; there is no early decision point in development that determines an individual's caste fate, as there is in most eusocial Hymenoptera (e.g. Noirot, 1985b; Roisin, 2000; Korb & Hartfelder, 2008; Roisin & Korb, 2011). Offspring in these species are often referred to as helpers (analogous to helpers in cooperatively breeding vertebrates and wasps) (Roisin, 2000) or, as pseudergates, which means false workers. The use of pseudergate in the literature has been inconsistent and is therefore a confusing term. Formally (*sensu stricto*) pseudergate means an individual that has gone through a regressive or stationary molt (Grasse & Noirot, 1947; Noirot & Pasteels, 1987; Noirot & Pasteels, 1988; for an expanded discussion of the term *pseudergate*, see Roisin & Korb, 2011). For clarity, pseudergates should refer only to such individuals. Otherwise, the extended terms *pseudergate sensu lato* and *pseudergates sensu stricto* should be used (reviewed in Roisin & Korb, 2011), or to avoid confusion, simply *helper* to indicate the function of the individuals.

In OPT species, even late instar helpers retain the capability to differentiate into reproductives, alates (which disperse and become founding primary reproductives), or soldiers. Such developmental flexibility is achieved through an exceptional diversity of developmental options, including progressive, regressive, stationary, and saltational molts, the latter referring to molts across several instars (e.g. reviewed in Noirot, 1985b; Korb & Hartfelder, 2008). Depending upon circumstances (e.g. death of the queen and king), female and/or male helpers at a variety of developmental stages (including nymphs, i.e. larvae with wing buds) may molt into reproductives, called neotenics. They remain, and inbreed, within their parents' nest. Most commonly, neotenic reproductives are replacements, succeeding a dead or senescent king or queen. In some species neotenics can differentiate as supplementary reproductives, meaning they persist as fertile reproductives in addition to a fecund king or queen of the same sex (Myles, 1999). Neotenics, alates, and soldiers are "terminal" castes, meaning no further molts or developmental changes are possible after reaching one of those castes.

Hence in OPT termites, the caste fate of an individual helper is determined by social context (e.g. presence/abundance of other castes, especially reproductives, colony size) and environmental factors (e.g. resource availability and quality, parasite and predation pressure, Lüscher, 1974; Lenz, 1994; Miura, 2004; Scharf, *et al.*, 2007; Korb & Hartfelder, 2008; Brent, 2009). When food availability declines or colonies reach a certain size, helpers in these developmentally flexible termites predominantly develop into alates and disperse from the nest to attempt to found new colonies (reviewed in Nutting, 1969; Lenz, 1994; Korb & Schmidinger, 2004). The long-term flexibility of development of helpers is widely considered to reflect the ancestral pattern of the earliest termites (Noirot & Pasteels, 1987; Noirot & Pasteels, 1988; Inward, *et al.*, 2007b; Legendre, *et al.*, 2008; but see also Watson & Sewell, 1981). Modern termites fitting this pattern comprise less than 15 percent of extant species in 5 of 8 families: all species of Archotermopsidae, Stolotermitidae, Kalotermitidae, Stylotermitidae, as well as the genus *Prorhinotermes* and *Termitogeton* within the Rhinotermitidae.

5.2.2 Separate Life Type Termites (ST)

ST species (here this comprises the "intermediate-type" where colonies are founded in a piece of wood and individuals only start foraging later in the colony cycle, and true "separate-type" nesting termites where colonies are founded in the soil, Abe, 1987; Noirot, 1970; "central-site" and "multiple site nesters," Shellman-Reeve, 1997; "foraging termites," Korb, 2007a) have a nest that is separate from their multiple food sites. They are central-place foragers that may nest within wood or soil, or build a mound or carton-like nest structure and individuals forage to one or more resources spatially separated from the nest, exploring and exploiting new resources over time. As Noirot (1970) points out, many ST nesters live early colony stages as OPTs, remaining cloistered and hidden within a single piece of wood until colony population size is sufficient to enable foraging away from the original enclave (see also Thorne & Haverty, 2000).

Developmental flexibility in ST species is more constrained due to early instar separation into two developmental pathways (reviewed in Noirot, 1985a; Roisin, 2000; Korb & Hartfelder, 2008; Roisin & Korb, 2011): (1) the wingless (apterous) line from which workers (formal term *workers* in ST species) and soldiers develop; and (2) the nymphal line (wing buds present in later instars) that culminates in alates. As with OPT species, upon death or senescence of the queen and/or king, female and/or male workers or nymphs may differentiate into neotenic reproductives (termed ergatoid neotenics if worker derive and nymphoid neotenics if they have wing buds and molt from the nymphal line, Roisin & Korb, 2011). Both ergatoid and nymphoid neotenics are common in many Rhinotermitidae; they are also known in Termitidae (reviewed in Myles, 1999). The combination of nesting, foraging, and developmental traits in ST termites are generally considered derived, and to have fostered diverse ecological innovations and radiations (Noirot & Pasteels, 1987; Noirot & Pasteels, 1988; Inward, *et al.*, 2007b; Legendre, *et al.*, 2008; but see Watson & Sewell, 1981). More than 85 percent of living species are ST termites, including the single extant species of Mastotermitidae, all Rhinotermitidae except in the genus Prorhinotermes and Termitogeton, and all species of Hodotermitidae, Serritermitidae, and Termitidae (Kambhampati & Eggleton, 2000).

Details of development, number of castes, and task allocation vary widely across ST taxa. The most complex social systems occur among the fungus-growing termites (Macrotermitinae) and nasute termites (Nasutitermitinae). In *Macrotermes* species, two worker and two soldier castes (minor and major) are typical, each differing distinctly in morphology and in task allocation (Gerber, *et al.*, 1988; Lys & Leuthold, 1991, reviewed in Traniello & Leuthold, 2000). In *Macrotermes bellicosus*, major workers forage and transport food back to the colony, while minor workers take care of the brood, the royal pair, fungus, and perform most of the mound building (Gerber, *et al.*, 1988; Lys & Leuthold, 1991). Foraging major workers are protected primarily by accompanying minor soldiers while major soldiers focus on defending the nest. In Macrotermitinae caste fate seems to be predetermined at the egg stage, probably through maternal factors (Okot-Kotber, 1985). In these most socially complex species, the queen can lay up to 20,000 eggs per day (Grasse, 1949; Darlington & Dransfield,

1987; Kaib, *et al.*, 2001) and colony sizes range up to several million individuals (Darlington, 1984, 1900; Darlington & Dransfield, 1987; Darlington, *et al.*, 1992).

The OPT life type is broadly considered to be ancestral in termites because (1) colonies are relatively small and less socially complex, (2) individuals retain life-long flexibility in caste development (except for individuals that reach a terminal caste of reproductive or soldier), and (3) colonies nest and feed within a single piece of wood for their entire life cycle, a trait they share with their sister taxon, the woodroaches. However, Mastotermitidae is the most basal family among modern termites (reviewed in Krishna, *et al.*, 2013), and its single extant species, *Mastotermes darwiniensis*, has an ST life type with large colonies, constrained developmental pathways and true workers, and foraging to multiple resources away from the nest. This paradox has fueled debate regarding whether the ST life type is derived (Noirot, 1985a,b; Grasse, 1986; Noirot & Pasteels 1987, 1988) or ancestral (Watson & Sewell, 1981) in termites. *Mastotermes darwiniensis*, the only living "relict" among many extinct genera and species of Mastotermitidae, has both ancestral and derived morphological and life history traits, featuring a common evolutionary pattern of a living member of an ancient taxon bearing a combination of ancestral and derived anatomical, developmental, and behavioral characteristics (reviewed in Thorne, *et al.*, 2000; Krishna, *et al.*, 2013). It seems unparsimonious that termites first evolved constrained, bifurcated juvenile development and complex foraging away from their nest, characteristics of ST, and only then evolved flexible development with long-term individual plasticity as well as smaller colonies restricted to the single piece of wood in which the colony was founded; i.e., traits of OPT.

5.3 Why Termites Form Social Groups

Termites are a monophyletic clade that evolved eusociality more than 130 million years ago (reviewed in Krishna, *et al.*, 2013). The evolutionary question regarding the selective forces that favored formation of social groups, and especially sterile castes, is difficult to address because all living species are eusocial. Hence, we elaborate here which factors shape social organization in recent species. These factors differ principally between OPT and ST species, but also within each life type considerable variation exists, especially in the latter.

5.3.1 Resource Acquisition and Use

In most animals, local resource competition between parents and offspring selects for offspring dispersal (Hamilton & May, 1977). As in many social vertebrates and social insects, resource competition is mitigated in termites either by colony founding within a "bonanza-type" food source (Wilson, 1971) or by central-place foraging, which can include relocation of the nest (Alexander, 1974; Alexander, *et al.*, 1991; Korb, 2009).

Hence, offspring can stay for prolonged periods with their parents. The established nest also provides a relatively safe haven compared to the high risk of mortality during dispersal. This, together with the low nutritional quality of wood and the necessity of offspring to re-acquire gut symbionts after each molt, are likely factors favoring an extended period of offspring staying in the natal nest and that might have been crucial in facilitating the overlap of generations, opportunities for helping, and the transition to a eusocial life (Cleveland, *et al.*, 1934; Alexander, 1974; LaFage & Nutting, 1978; Alexander, *et al.*, 1991; Thorne, 1997; Korb, 2008). That they might have been necessary but not sufficient prerequisites is reflected in the fact that *Cryptocercus* woodroaches live in a similar type of resource, yet never evolved eusociality (Nalepa, 2015; Korb, 2016).

Termites feed primarily upon plant material (Table 5.1), a diet that has influential consequences on their life history, including the evolution of social life (Cleveland, *et al.,* 1934; Waller & Lafage, 1987; Nalepa, 1994). Termites rely on gut symbionts to assist in digesting cellulose, although many species can also produce endogenous enzymes that break down cellulose directly (Lo, *et al.*, 2011). Most termites harbor a diverse community of microbes (bacteria and/or protists) in their highly structured guts, facilitating termites' exploitation of many microhabitats (reviewed in Brune & Ohkuma, 2011; Ohkuma & Brune, 2011). The highly specialized fungus-growing termites (Macrotermitinae) forage for dead plant material and use it as a pre-digested substrate to culture *Termitomyces* fungi within their nests (Grasse & Noirot, 1951; Wood & Thomas, 1989; Leuthold, 1990; Rouland-Lefèvre, 2000; Nobre, *et al.*, 2011). The fungus garden serves as an essential protein-rich food supplement for the termites. Fungus growers host a lower diversity of gut symbionts. Probably because cultured *Termitomyces* fungi facilitate digestion of cellulose and complex plant compounds (Nobre, *et al.*, 2011; Poulsen, *et al.*, 2014).

Due to both their diet and hemimetabolous development, termites must reacquire obligate gut symbionts following each molt. This is accomplished by proctodeal trophallaxis, or anal feeding, among colony members, a factor proposed to have influenced termites' need for overlap of generations (Cleveland, *et al.*, 1934; Nalepa, 2011). In the sister taxon of termites, the *Cryptocercus* woodroaches, symbiont loss during molting requires that immatures remain with the family through their final molt to an adult (Nalepa, 1994). In termites, siblings rather than parents typically provide proctodeal trophallaxis. However, as in OPT termites, all individuals within a termite colony act as symbiont donors as well as recipients; reinoculation of gut fauna is a cooperative rather than altruistic act (Cleveland, *et al.*, 1934; Korb, 2007b; Korb, *et al.*, 2012). Altruism requires that some individuals invest more than others and that the behavior reduces an individual's direct fitness. This is currently unclear for OPT termites.

Termites' plant-based diet has a relatively high C:N ratio, compelling relatively slow development (Higashi, *et al.*, 1992; Lenz, 1994; Nalepa, 1994). With few exceptions (e.g. fungus growers), termites are considered protein (nitrogen)-limited. N-fixing gut bacteria in some species help moderate this constraint (Brune & Ohkuma, 2011). At present, little is known about the diversity and nutritional/developmental impacts of N-fixing symbionts across termites.

5.3.2 Predator Avoidance

In ST species, once colonies grow to a size at which workers and soldiers forage away from their founding nest, those ST castes experience different mortality risks than those of OPT species. OPT termites may share their single piece of wood with dynamic assemblages of competitors, predators, and/or pathogens, shifting in abundance and composition as the host resource decomposes (Thorne, *et al.*, 2003). In contrast, ST soldiers and workers can be vulnerable to predators when searching and feeding outside the nest. Thus, predation pressure varies markedly among ST termites. Some ST species are highly protected, either underground in cryptic galleries, above ground within foraging tunnels, or as in some soil feeding Apicotermitinae that nest in and feed on carton nests. Such differences in predation risk correlate with soldier morphology and the proportions of soldiers within a colony (Haverty & Howard, 1981). For instance, the highly protected soil-feeding Apicotermitinae have secondarily lost the soldier caste completely (Sands, 1972).

5.3.3 Homeostasis

The wood nest of OPT species, and the fact that these termites never forage outside that resource, insulates them against some environmental fluctuations, probably more so in drywood termites (Kalotermitidae) that nest in sound wood than in dampwood termites (Archotermopsidae) that nest in decaying wood. The degree to which individuals and their colony in ST species are exposed to environmental fluctuations varies markedly among species, depending upon nest type, habitat, and colony size (Grasse & Noirot, 1948; Noirot, 1970; Korb & Linsenmair, 2000; Noirot & Darlington, 2000; Korb, 2011). For example, species nesting deep inside the ground may experience more consistent conditions than those nesting close to the soil surface. Occupied soil depth may be adjusted by termites moving according to environmental conditions (Lepage, 1989; Darlington, 1990).

Most studies on homeostasis focus upon mound-building species (reviewed in Noirot, 1970; Noirot & Darlington, 2000; Korb, 2011). Some of these species adapt mound architecture to local environmental conditions and provide inside-nest temperatures that vary by less than 3°C daily, with a mean annual fluctuation of around 1°C, while outside temperatures can fluctuate by more than 30°C. Mounds also can have efficient ventilation systems that facilitate the input of O_2 and output of CO_2. Such impressive abilities are especially pronounced in fungus-growing termites (reviewed in Korb, 2011). Such self-organized structures are most easily implemented with relatively large colony sizes. Mounds help protect the royal chamber, nursery, and inhabiting individuals against predators and environmental perturbations, facilitating relative homeostasis for the colony. Within a species, homeostasis typically increases with mound size, which is related to colony size (Noirot, 1970; Korb & Linsenmair, 2000). Mounds also provide space for food storage in some species (and in the case of fungus-growing Macrotermitinae, for fungus cultivation) making colonies less dependent on short-term ecological conditions (reviewed in Schmidt, *et al.*, 2014). Hence it is not surprising that mound-building species often have influential ecological impacts.

5.3.4 Mating

The nuptial flight of termites generally occurs during the rainy season after precipitation events, though exceptions exist (Nutting, 1969; Bourguignon, *et al.*, 2009). The common pattern is as follows (Nutting, 1969; Minnick, 1973). Winged sexuals fly from their nest after a rain. Following dispersal, males are attracted to females that "call" by standing in a relatively exposed position while emitting a pheromone. When a male approaches a female, he follows closely behind the female ("tandem-run"), both shed their wings synchronously, and they search for and burrow into a crack in a piece of wood or into soil. As founding primary reproductives (i.e. alate-derived king and queen), they establish a nuptial chamber where mating takes place, eggs are laid, and the first offspring are produced. During this founding period, kings and queens of some species can eat surrounding wood, or they survive by metabolizing stored resources, e.g. their now unnecessary wing muscles. After the first workers develop, these offspring start foraging, with the first broods consuming local resources.

5.3.5 Offspring Care

There are major differences in the importance of offspring brood care between OPT and ST species that correlate with their different resource acquisition types. In central-place foraging ST species, foraging for food is an important part of brood care behavior, and workers in at least some species have a high risk of dying while foraging outside the nest (Korb & Linsenmair, 2002). At the same time – and like social Hymenoptera – foraging workers can increase their indirect fitness by providing resources for their colony. In contrast, entire OPT colonies reside inside their food. Hence, in at least some cases, brood care in OPT colonies may, be less labor intensive than in ST species, as evidence in drywood termites suggests (Korb, 2007b; Korb, *et al.*, 2012). In the OPT genus *Cryptotermes*, for example, helpers are immatures that do not invest in foraging for food for the colony and mainly seem to follow a "sit and wait" tactic to try to reproduce later in life (Korb, 2007b, 2009). Correspondingly, annual growth rates of colonies and colony sizes are small in OPT species compared to those of ST species (Shellmann-Reeve, 1997), and individual body size of helpers and soldiers is often large in OPT species, although there are some exceptionally large body sized ST species such as *Syntermes* or *Macrotermes*.

In OPT species, intensity of brood care might be associated with pathogen load within the nest (Rosengaus, *et al.*, 2003, 2011; Korb, *et al.*, 2012). In *Zootermopsis*, which live in rotten/decaying wood, lab experiments have demonstrated that of all brood care behavior directed towards eggs and larvae, hygienic allogrooming is especially common (Korb, *et al.*, 2012). This strongly contrasts with the drywood termite *C. secundus* that nests in sound wood with extremely low pathogen loads.

Workers of ST species provide extensive care of the nursery, soldiers, and reproductives (e.g. reviewed in Traniello & Leuthold, 2000). They bring food and water to the brood, reproductives, and soldiers, whose mandibles prevent harvesting food resources on their own (as is the case for OPT species). Workers construct and maintain the nest, which in some species involves elaborate and highly functional architecture, with heights reaching several meters (Traniello & Leuthold, 2000; Noirot & Darlington, 2000).

5.4 The Role of Ecology in Shaping Sociality in Termites

Nest and foraging ecology coincide with developmental pathways and not only distinguish OPT from ST species, but are also influential in shaping social patterns and complexity among taxa as outlined earlier. Below, we discuss how most termites share very similar habitat requirements: they are mainly tropical and subtropical detritivores with some taxonomic groups occupying more specialized niches, such as fungus-cultivating, soil-feeding, or living in temperate regions.

5.4.1 Habitat and Environment

In contrast to the differences in nest and foraging ecology, habitat and environmental requirements are similar across termites and often independent of social complexity. Among the OPT termites, dampwood and drywood species represent different termite families (Archotermopsidae and Kalotermitidae) that nest in decaying versus structurally sound wood, respectively. Comparable clear distinctions are lacking for other families. For example, Rhinotermitidae are also called subterranean termites, but they share habitat and nest type with many Termitidae. Termites are predominantly tropical decomposers that prefer temperatures above 15°C and relative high humidity. Indeed, there are only very few species that are resistant to temperatures below zero, a notable exception are the Archotermopsidae, including the only Nearctic temperate endemic genus *Zootermopsis* (Eggleton, 2000; Lacey, *et al.*, 2010). Generally, termites thrive, and reach exceptional diversity, in tropical rainforests. Using exemplar assemblages, Jones & Eggleton (2011) found that tropical rainforests have a higher generic diversity than savanna woodlands and semi-deserts with least genera occurring in temperate woodlands and temperate rainforests.

5.4.2 Biogeography

Termites are mainly tropical and subtropical species, and species richness increases with latitude; some species occur in temperate regions, although with far lower diversity (Eggleton, 1994). The highest known generic richness is found in African tropical rainforests (Eggleton, *et al.*, 1994; Eggleton, 2000; Jones & Eggleton, 2011). Controlling for net primary productivity, the Afrotropical region has more described endemic genera (85) than the Neotropics (63), Australasia (23) or the Oriental region (56) (Eggleton, 2000). Caveats might be that not all regions have been sampled equally for termites and that this pattern applies to generic richness, which is arguably an arbitrary classification. Termites are generally poorly studied at the species level because, among other factors, they often have few external morphological traits that reliably distinguish closely related species (Donovan, *et al.*, 2000; Kambhampati & Eggleton, 2000). Use of molecular markers for species diagnosis (barcoding) suggests cryptic species diversity but also high phenotypic plasticity that risks inaccurate estimates of species richness (Hausberger, *et al.*, 2011). Nevertheless, there is a dramatic drop in species richness in temperate regions, with four and three endemic genera in the Nearctic and Paleoarctic regions, respectively (Eggleton, 2000). Moreover, most of these species occur in subtropical deserts or arid grasslands rather than temperate zones with definitive winters (Eggleton, 2000). The island regions of Madagascar and Papua are also relatively depauperate with four and three endemic genera, respectively (Eggleton, 2000). This distribution pattern illustrates (1) the effect of historical biogeography and (2) that termites thrive in warm climates, from dry deserts to humid rainforests (Abensperg-Traun & Steven, 1997; Eggleton, 2000; Jones & Eggleton, 2011).

5.4.3 Niches

Termites are predominantly detritivores. Compared with other insect orders (or infraorders), Isoptera show relatively low niche differentiation, with most species preferring warm conditions and nutrition from dead, often partially decomposed plant material (e.g. wood, grass, lichen, seeds, herbivore dung; Eggleton & Tayasu, 2001). Hence, termites compete locally for the same foods. Despite niche similarities, several ecologically equivalent species coexist at the local scale in tropical and subtropical habitats (Wood, *et al.*, 1977; Korb & Linsenmair, 2001; Dosso, *et al.*, 2010; Hausberger, *et al.*, 2011). Specialist feeders include fungus-growing termites and "soil feeders" that consume decomposing litter or humus. Fungus-cultivating termites are restricted to the paleotropics, where they occupy a niche resembling that of neotropical fungus-growing ants (Attini) except that the latter generally forage on live plants (Waller, 1988; Nobre, *et al.*, 2011). Specialist humus feeders (i.e. feeding on soil-like substrates containing recognizable plant material) and soil feeding termites (i.e. consuming soil-like material with a high proportion of silica and no recognizable plant material) also belong to the Termitidae (Bignell & Eggleton, 2000; Brauman, *et al.*, 2000; Eggleton & Tayasu, 2001). While soil feeders are common in humid forests in Africa (and perhaps the Neotropics, where the genus *Anoplotermes* has yet to be studied in detail), fungus-growing termites are ecologically dominant in African and Asian grasslands and savannas (Josens, 1983; Deshmukh, 1989; Bignell & Eggleton, 2000; Eggleton, 2000).

5.5 The Role of Evolutionary History in Shaping Sociality in Termites

The stem groups of all known fossil and living termite taxa originated by the early Cretaceous (135–100 MYA) on the Gondwana landmass (Thorne, *et al.*, 2000; Engel, *et al.*, 2009; Krishna, *et al.*, 2013). Although impacted by continental drift, the primary driver of modern biogeographic distributions was an explosive Tertiary radiation followed by rapid dispersal as termites became ecologically dominant detritivores (Eggleton, 2000; Thorne, *et al.*, 2000; Krishna, *et al.*, 2013). The evolution of representatives of the different families was probably fast and the ST species lifestyle radiated rapidly. *Cryptocercus* woodroaches, the sister taxon of the termites, are also one-piece nesters, but they are subsocial rather than eusocial (Cleveland, *et al.*, 1934; Nalepa, 1994; Thorne, 1997; Nalepa & Bandi, 2000), implying that OPT is the ancestral life-type.

II SOCIAL TRAITS

Termites have a long history of meticulous natural history observations and careful taxonomy by early entomologists (see Snyder, 1956; many key contributors are profiled in Volume 1 of Krishna, *et al.,* 2013). Their efforts provided the foundation for the modern study of termites and their diversity. New insights on the social traits of termites will emerge from modern tools, theoretical considerations, careful observations and experiments that address specific question illuminating long-standing questions and generate new avenues of research on termite social biology.

5.6 Traits of Social Species

5.6.1 Cognition and Communication

The communication and cognition system of termites is fundamentally similar to that of their closest relatives, the cockroaches, although more sophisticated due to their increased social complexity. Termites, except for winged sexuals, generally lack well developed eyes. Although phototaxis plays a role during nuptial flights, the main modes of communication in termites are olfactory and tactile, with chemical, vibrational, and behavioral signals playing important roles in species recognition and within colony recognition, as well as in foraging, building, and defensive behaviors (Leuthod, 1979; Bordereau & Pasteels, 2011; Bagnères & Hanus, 2015). Convergent with eusocial Hymenoptera and many other insects, hydrocarbons on the cuticle (cuticular hydrocarbons, CHCs) provide information on identity, both within and among colonies (e.g. Haverty, *et al.,* 1988; Clément & Bagnères, 1998), with long-chained CHCs indicating reproductive status (Liebig, *et al.*, 2009; Weil, *et al.*, 2009; Hoffmann, *et al.*, 2014). Moreover, volatile compounds function for intermediate distance communication within the nest (Lüscher, 1974; Matsuura, *et al.*, 2010) and species-specific trail pheromones allow efficient recruitment of nestmates to food sources (reviewed in Bordereau & Pasteels, 2011). Vibrations are known to be important in alarm communication (e.g. Rohrig, *et al.*, 1999) or, in OPT termites, for measuring food availability and the presence of competing colonies (Evans, *et al.*, 2005, 2009).

5.6.2 Lifespan and Longevity

There are few detailed studies revealing termite life history data (Table 5.3). As is typical for eusocial insects, there are major differences in longevity between castes

Table 5.3 Life history traits for termite families and their sister taxon, the *Cryptocercus* wood roaches. Maximal values or ranges over different species are given. Numbers in (?) are anecdotal or unconfirmed reports.

Taxonomy	Maximum longevity [in years][1]		Fecundity [per year]	Maximum colony size
	Reproductives	Workers/Soldiers		
Cryptocercidae	6	na	1–75	2 parents + offspring
Mastotermitidae	7–17			1.1 Mio
Hodotermitidae	~20			
Archotermopsidae	4–6	>3–5	20–1000	several thousand
Stolotermitidae				
Kalotermitidae	4–5 (14?)	>5–6 (14?)	20 to few hundred	1,500–3,000
Stylotermitidae				
Rhinotermitidae	9–17	>3–5		few thousand–2,750,000
Serritermitidae				
Termitidae	~20 (50?)	<1	10,000–7,300,000	10,000–several million

Given are maximal values or ranges applying to different species. Values with? are anecdotal, unconfirmed reports
[1] reviewed in Shellman-Reeve (1997)

(Wilson, 1971; Keller & Genoud, 1997; Keller, 1998). Because kings and queens typically have longer lifespans than workers or soldiers, the common life history trade-off between longevity and fecundity seems absent in termites. This may be in part due to the sheltered nests, but also because after initial phases of colony founding, eusocial insect reproductives are fed and cared for by other members of the colony (e.g. Wilson, 1971; Keller & Genoud, 1997; Heinze & Schrempf, 2008). This both reduces random extrinsic mortality rates – selecting for increased longevity in reproductives – and alleviates resource allocation trade-offs (e.g. Heinze & Schrempf, 2008).

Data on longevity of termite reproductives in the field are rare (for Termitidae they are mainly restricted to fungus-growers; Table 5.3), but thus far suggest that there are no major differences among taxa. The maximum longevity of Mastotermitidae, Kalotermitidae, Rhinotermitidae and several representatives of Termitidae indicate that they commonly reach an age of between 10 to 20 years with average life spans of 4 to 5 years (reviewed in Shellman-Reeve, 1997; Keller, 1998; for Mastotermitidae, Watson & Abbey, 1989; for Archotermopsidae, Heath, 1907; Thorne, *et al.*, 2002; for Kalotermitidae, Grasse, 1984; Luykx, 1993, J. Korb, *unpublished data*; for Rhinotermitidae, Nutting, 1969; Grasse, 1984; Termitidae: Macrotermitinae, Grasse, 1984; Leuthold, 1979; *personal communication*). Record holders might be some mound building species, such as *Macrotermes* or *Nasutitermes exitiosus*, for which anecdotal notes exist that mounds remained active for up to 50 and 80 years, respectively (Gay & Calaby, 1970; Grasse, 1984). However, it is unknown, and probably unlikely, if those nest structures were continuously inhabited by the same colony with the same reproductives.

For the shorter-lived worker and soldier castes, there may be a trend for a decrease in lifespan from OPT to ST termites (reviewed in Shellman-Reeve, 1997; for Archotermopsidae, Nutting, 1969; for Kalotermitidae, Gay & Calaby, 1970; J. Korb *unpublished data*; for Rhinotermitidae, Pickens, 1934; Gay & Calaby, 1970; for Termitidae: Macrotermitinae, Bouillon, 1970; Collins, 1981; Josens, 1982; Darlington, 1991; Cubitermes, Bouillon, 1970; Grasse, 1984; Table 5.3). This difference in worker longevity is in accord with life history theory (Stearns, 1992; Baudisch, 2005): a high extrinsic mortality in foraging workers in ST species is expected to result in earlier senescence and shorter lifespans compared to OPT workers that are protected within their wooden nest. Accordingly, the intra-specific difference in longevity between reproductives and non-reproductives seems to increase from OPT to ST termites and within the ST species with increasing social complexity. With 4 to 5 years, lifespan is similar for workers, soldiers, and reproductives in *Zootermopsis* (Heath, 1907). On the other hand, the most extreme known example of lifespan occurs in *Macrotermes* where workers only live a few months while reproductives can reach ages of 20 years (Traniello & Leuthold, 2000).

Despite the high potential lifespan of reproductives, there is reasonable probability of dying each year due to ecological circumstances. Especially in OPT species with their totipotent helpers and conspecific competitors within the same limited resource, loss of founding reproductives offers opportunities for offspring to inherit the colony and differentiate into new reproductives (Thorne, *et al.*, 2002, 2003; Korb & Schneider, 2007).

Cryptocercus woodroaches, the extant sister taxon of termites, are semelparous. Both parents help to raise one clutch until that monogamous pair dies. One study on *C. punctulatus* showed that roughly 15 percent of the parents die within the first year after reproducing (Nalepa, 1984). On average, parents seem to stay with the brood/offspring for 3 years (Seelinger & Seelinger, 1983; Nalepa, 1984). As the developmental time until the offspring reach maturity is about six years (Cleveland, *et al.*, 1934), parents are often dead before offspring reach maturity.

5.6.3 Fecundity

Fecundity of primary as well as neotenic reproductives increases with age and increasing physogastry until a plateau is reached. With a maximum of a few hundred eggs per year, the fecundity of *Cryptotermes* species is low relative to most other social insects. *Cryptotermes* colony sizes with a single reproductive queen are generally only a few hundred individuals (Korb & Schmidinger, 2004; Korb, 2008; J. Korb, *unpublished data*). In *Zootermopsis,* OPT dampwood termites, fecundity is higher than in drywood species and colonies reach sizes of a few thousand individuals (Heath, 1903, 1927; B. Thorne, *unpublished data*). However, *Zootermopsis* never attain the fecundity and colony size of some of the most derived Termitidae, which can reach a few million individuals (Heath, 1927; Shellman-Reeve, 1997) (although some Termitidae retain relatively small colonies). Record holders are probably some fungus-growing Macrotermitinae with 20,000 eggs per day (reviewed in Shellman-Reeve, 1997; for Kalotermitidae, Wilkinson, 1962; Lenz, 1987, 1994; Korb & Schneider, 2007; Neoh & Lee, 2011; for Macrotermitinae, Grasse, 1949; Darlington & Dransfield, 1987; Kaib, *et al.*, 2001; Table 5.3). Often fecundity of the female and male founding queen and king is higher than that of neotenics, but this is not the case in all taxa (Myles, 1999). The cumulative fecundity of multiple neotenic reproductives within a single colony may exceed that of the monogamous founding pair.

Comparing termites with *Cryptocercus* woodroaches reveals that OPT species are similar to *Cryptocercus* with respect to the clutch size of first year adults. Unless the first brood is unsuccessful, *Cryptocercus punctulatus* produces one clutch of eggs consisting of about twenty offspring (range 1 to 75, Nalepa, 1984, 1988).

5.6.4 Age at First Reproduction

Termite colony life cycles proceed through the three stages typical of most eusocial insects (Oster & Wilson, 1978): (1) a founding stage; (2) an ergonomic stage during which the colony increases in size through production of workers and soldiers; and (3) a reproductive stage after a colony reaches maturity and produces winged sexuals. At the end of the reproductive stage, which may last over a decade, colony size and the number of sexuals produced decline. The reproductive phase of a colony is difficult to measure in the field and is often not reached before an age of several years (for Archotermopsidae, Thorne, *et al.*, 2002; for Kalotermitidae, 5 years, J. Korb, *unpublished data*; Wilkinson, 1962; Wilkinson, 1963; for Rhinotermitidae and Termitidae,

Grasse, 1982; Shellman-Reeve, 1997) with timing strongly dependent upon environmental conditions (e.g. food quantity and quality, temperature). In the laboratory, *Zootermopsis* can produce the first mature alates after only 18 months following pairing of a queen and king (B. Thorne, *unpublished data*). For some OPT species, it takes at least 3 to 4 years (and often much longer) until an individual can develop from an egg into an alate (for Kalotermitidae, Lüscher, 1952; Wilkinson, 1962; J. Korb, *unpublished data*). As in *Cryptotermes,* alates do not develop until certain colony sizes are reached, and the first sexuals do not occur before an age of five (J. Korb, *unpublished data*). In Termitidae (especially fungus-growers and tropical species), individual development times are considerably faster, and it often takes less than one year for alates to develop from eggs (Johnson, 1981; Grasse, 1982; Noirot, 1985a, 1990). However, when queens actually begin to produce alate-destined eggs seems to depend upon colony size and this minimum size is reached only after several years (Grasse, 1982, 1984; Noirot, 1990; Han & Bordereau, 1992).

5.6.5 Dispersal

In both OPT and ST species, new colonies are founded generally by alates after nuptial flights that occur seasonally. Termite alates are generally regarded as poor flyers (Nutting, 1969). Population genetic studies, however, suggest that dispersal might be less limiting and that local populations are not highly inbred (Vargo & Husseneder, 2011). Major differences likely exist between species (Schmidt, *et al.*, 2013) and depending on local habitat and topography. In some species new colonies can also be founded by budding of colony parts (Nutting, 1969; Thorne 1982, 1984; Adams & Atkinson, 2007; Vargo & Husseneder, 2011). Behavioral details about how such budding occurs, and whether it is started by few individuals with scouts or by colonies breaking apart, are largely unknown (Thorne, 1982, 1984).

5.7 Traits of Social Groups

5.7.1 Genetic Structure

Unlike Hymenoptera that are haplodiploid, termites are diploid. Genetic relatedness varies in termite colonies depending on group structure, which is largely determined by the breeding structure (reviewed by Vargo & Husseneder, 2009, 2011). In almost all termite species, new colonies are founded by a monogamous, unrelated pair of winged sexuals, the future queen and king. Hence, the default genetic relatedness among nuclear family colony members is 0.5 (parent to offspring and vice versa: $r = 0.5$; offspring to offspring: $r = 0.5$). However, within-colony relatedness can decrease in the case of colony fusions (Thorne, *et al.*, 1999; Vargo & Husseneder, 2011). It can also increase due to inbreeding for instance, when an offspring inherits the natal breeding position and mates with the opposite-sex parent or sibling, or in the case of parthenogenetic reproduction, which is documented in several termite species.

5.7.2 Group Structure, Breeding Structure and Sex Ratio

Three family (colony) types are distinguished in termites (Thorne, *et al.*, 1999; Bulmer, *et al.*, 2001; Vargo, 2003). *Simple families* are colonies with genotypes consistent with the progeny of a monogamous pair of reproductives (monogamous colonies; a nuclear family). *Extended families* are headed by reproductives descended from the founding pair, leading to inbreeding and increased relatedness among colony/family members. These colonies can be monogamous or polygamous with several primary reproductives, and/or one or more neotenic reproductives. Finally, *mixed families* contain progeny of multiple unrelated reproductives all functioning as a social unit. Mixed family colonies can be the result of fusion of two or more neighboring colonies, nest foundation by more than one pair of unrelated reproductives (pleometrosis), or – at least in theory, although never unambiguously documented – adoption of unrelated reproductives. These elaborations on colony structure can result in polygynous and sometimes polyandrous mating systems and a decrease in average genetic relatedness among apparently cooperating individuals within a mixed colony.

Colony fusions and inbreeding, that lead to mixed and extended families, respectively, seem to be especially common in OPT species (Myles, 1999; Thorne, *et al.,* 2002, 2003; Korb & Schneider, 2007; Korb & Hartfelder, 2008; Johns, *et al.*, 2009; Korb & Roux, 2012; Luchetti, *et al.*, 2013; Howard, *et al.*, 2013). This is probably due to the common occurrence of several colonies that were founded within the same piece of wood and that meet during "nest" growth and extension while feeding (Thorne, *et al.,* 2002, 2003; Johns, *et al.* 2009; Korb & Roux, 2012; Howard, *et al.*, 2013). High degrees of inbreeding in OPT species likely exist because of the high developmental flexibility of helpers which retain long-term flexibility to develop into neotenic reproductives (reviewed in Noirot, 1969; Korb & Hartfelder, 2008). Among ST species, *Mastotermes* and *Reticulitermes* can have large numbers of neotenics and thus inbreeding can be common too (reviewed in Vargo & Husseneder, 2011). Facultative thelytokous reproduction (i.e. a type of parthenogenesis in which females are produced from unfertilized eggs) can occur in some species from all termite families studied to date (reviewed in Matsuura, 2011).

Termitidae may take advantage of a wide variety of reproductive alternatives. For example, in addition to the typical termite pattern of colonies founded by a single king and queen (monogamy), *Nasutitermes corniger*'s reproductive options include multiple queens, multiple kings, satellite nests, and colony budding (Thorne, 1983; Roisin & Pasteels, 1986; Adams & Atkinson, 2007). This plasticity makes *N. corniger* exceptionally flexible, resilient, and equipped with the reproductive infrastructure for rapid colony growth.

In contrast to social Hymenoptera in which colonies are composed of female workers, male and female workers and soldiers (with caste gender dependent upon species), occur in most termite species with varying sex ratios (Noirot, 1969, 1985a; Roisin, 2001; Bourguignon, *et al.*, 2012). There are exceptions, however, especially among the Termitidae where caste system can align with sexual dimorphism. Notably, there is no consistent phylogenetic pattern between which sex develops into which caste, although some predispositions for soldiers have been more recently proposed (Matsuura, 2006; Muller & Korb, 2008; Bourguignon, *et al.*, 2012).

III SOCIAL SYNTHESIS

5.8 A Summary of Termite Sociality

Termites are the eusocial clade within the cockroaches. They can be grouped into two life types that largely differ in developmental plasticity, social complexity and ecology: the OPT and the ST species. The OPT termites in particular share several similarities with the termite's sister taxon, the *Cryptocercus* woodroaches.

In contrast to woodroaches, and especially relevant in OPT species, termite helpers can develop into replacement reproductives within their natal nest (Table 5.1). Hence, offspring of OPT species can take advantage of a nest and resource inheritance strategy (including the colony population in the case of termites), as do helpers of cooperatively breeding birds and mammals (Wilson, 1971; Thorne, 1997; Thorne, *et al.*, 2003; Korb & Heinze, 2008a, 2008b). The extent to which helper or worker termites are involved in brood care, how much indirect fitness they gain, and their probability of becoming a new reproductive in their parents' colony depends upon taxon and ecological conditions (reviewed in Korb, *et al.*, 2012).

Data for woodroaches and OPT species show that parents may die during the long developmental period from egg to maturity (Nalepa, 1984; Thorne, *et al.*, 2002; Thorne, *et al.*, 2003; Korb & Schneider, 2007; Johns, *et al.* 2009; Korb & Roux, 2012; Howard, *et al.*, 2013; Table 5.4). The possibility of inheritance of the nest and philopatric reproduction in OPT species also has the advantage of providing opportunities for gaining direct and/or indirect fitness benefits without risky dispersal (Myles, 1988; Thorne 1997; Thorne, *et al.*, 2003; Korb & Schneider, 2007; Korb, 2007b, 2008;

Table 5.4 Key life history and morphological differences between modern woodroaches (*Cryptocercus*) and termites.

Characteristic	*Cryptocercus*	Termites
Wings	Wingless; local dispersal	Winged adults; dehiscent wings; dispersal flights
Colony size	Both parents + offspring of single reproductive event	Minimum size ~40 termites; in higher termites can reach hundreds of thousands or even a million or more individuals
Overlap of generations	Parents rarely if ever overlap in time or space with adult offspring	Parents commonly survive past maturity of adult offspring; older offspring may persist indefinitely in parental nest
Castes	Absent	Distinct castes; pronounced division of labor and reproductive skew
Body size (of adults)	Larger	Smaller (adult body size equivalent to juvenile *Cryptocercus*)
Mouthparts	Opistognathus (head facing down; mouthparts directed posteriorly)	Prognathus (head and mouthparts directed anteriorly)

Korb & Hartfelder, 2008; Johns, *et al.*, 2009; Korb & Roux, 2012; Howard, *et al.*, 2013). Associated with a shift to ST life histories is the emergence of true workers (Noirot 1985a; Abe 1987). The occurrence of true workers coincides with morphological differentiation (e.g. increased sclerotization), reduced developmental flexibility, and decreased lifespan in workers. These species resemble ants in many respects, and workers as well as soldiers gain mainly indirect fitness benefits during their lifetimes (Howard & Thorne, 2011). As most living termites belong to the ST category, this striking resemblance inspired the common name of termites as "white ants" in some languages.

5.9 Comparative Perspectives on Termite Sociality

Korb & Heinze (2008b) identified three "sociality syndromes" by comparing ecological and relatedness parameters across a broad range of social vertebrates and invertebrates. These traits included: food acquisition (central-place foraging/nesting inside food), nest type (inside/outside food), main type of helping (allofeeding/defense), chances to inherit the nest (yes/no), inbreeding possible (yes/no), genetic system (diploid/haplodiploid/parthenogenetic) and main altruistic caste (workers/helpers/soldiers). Strikingly, these traits, except the genetic system, grouped well into three syndromes of co-occurring traits, and all social animals could be categorized accordingly. The OPT termites fit with social aphids, thrips, and naked mole-rats in syndrome I. They are characterized by a "bonanza-type" food resource, and fortress defense by altruistic soldiers. Opportunities for offspring to gain direct fitness as winged sexuals and/or through nest inheritance and inbreeding by differentiating into reproductives within their parents' nest are substantial, and systems are characterized by low local competition over food (at least in recently founded resources) but potential conflict over breeding. This syndrome is broadly equivalent with the "fortress defenders" recognized by Crespi (1994) and Queller & Strassmann (1998).

In contrast ST termites, along with ants and honeybees, belong to syndrome II where opportunities for offspring workers to inherit and directly reproduce in a colony are reduced, and allofeeding plays a more important role. Here, feeding typically involves progressive food provisioning, which is costly to reproductives and can be "handed over" to workers. These are the "classically" eusocial insects, largely equivalent with the "life insurers" (Queller & Strassmann, 1998). Overall, for workers there is a shift to a higher importance of indirect fitness compared to direct fitness when comparing ST species to OPT, while soldiers of both groups gain indirect fitness only.

Syndrome III comprises most cooperatively breeding vertebrates and social Hymenoptera with totipotent workers (e.g. wasps and queenless ants). It is intermediate between societies with altruistic, subfertile workers of syndrome II and those consisting of totipotent individuals of syndrome I. Helpers gain indirect fitness through potentially costly alloparental care but they can potentially also gain direct fitness through inheriting the breeding position or, in some cases, by founding their own nest. Depending upon whether brood care is costly, OPT termites might be grouped also in this syndrome. In Queller & Strassmann's (1998) categorization, this group is not described.

Contemporary viewpoints on the fundamental characteristics and dynamics driving the evolution of eusociality in termites center on ecology, behavior, and development (Shellman-Reeve, 1997; Thorne, 1997; Roisin, 2000; Thorne, *et al.*, 2003; Korb, 2008, 2016; Korb & Hartfelder, 2008; Johns, *et al.*, 2009; Korb, *et al.*, 2012; Howard, *et al.*, 2013; see also Nalepa, 2015). Compelling insights have been advanced regarding the selective landscape favoring termite eusocial evolution compared with factors influencing eusocial origins and elaborations in other animals (e.g. Alexander, *et al.*, 1991; Korb & Heinze, 2008a, 2008b; Howard & Thorne, 2011). Comprehensive, truly integrative syntheses, however, will require continued research at molecular through community levels.

5.10 Concluding Remarks

Termites are cellulose- feeding social insects with distinct castes. Soldiers are sterile and are morphologically highly specialized with a fixed developmental pathway and no option for further molts. In contrast, developmental plasticity and function of helpers or workers varies, from species with totipotent immatures to morphologically and behaviorally highly specialized individuals that resemble such patterns in workers of eusocial Hymenoptera. The opportunity to inherit the local breeding position probably played an important role during the transition to eusociality. Hence, depending upon caste, phylogeny, and ecology, termites span a wide range of social systems observed in animals: from systems where workers are similar to helpers of cooperative breeding vertebrates, or soldiers in social thrips and aphids, to the most complex societies which share similarities with ants.

References

Abe, T. (1987) Evolution of life types in termites. *In*: Kawano, S., Connell, J. H., & Hidaka, T. (eds.) *Evolution and Coadaptation in Biotic Communities*. Tokyo: University of Tokyo Press, pp. 125–148.

Abensperg-Traun, M. & Steven, D. (1997) Latitudinal gradients in the species richness of Australian termites (Isoptera). *Australian Journal of Ecology*, **22**, 471–476.

Adams, E. S. & Atkinson, L. (2007) Queen fecundity and reproductive skew in the termite *Nasutitermes corniger*. *Insectes Sociaux*, **55**, 28–36.

Alexander, R. D. (1974) The evolution of social behavior. *Annual Reviews of Ecology and Systematics*, **5**, 325–383.

Alexander, R. D., Noonan, K. M., & Crespi, B. J. (1991) The evolution of eusociality. *In:* Sherman, P. W., Jarvis, J. U. M., & Alexander, R. D. (eds.) *The Biology of the Naked Mole-rat*. Princeton, New Jersey: Princeton University Press, pp. 3–44.

Abensperg Traun, M. & Steven, D. (1997) Latitudinal gradients in the species richness of Australian termites (Isoptera). *Austral Ecology*, 22, 471–476.

Atkinson, L. & Adams, E. S. (1997) The origins and relatedness of multiple reproductives in colonies of the termite *Nasutitermes corniger*. *Proceedings of the Royal Society of London B*, **264,** 1131–1136.

Baudisch, A. (2005) Hamilton's indicators of the force of selection. *Proceedings of the National Academy of Sciences USA*, **102**, 8263–8268.

Bagnères, A.-G. & Hanus, R. (2015) Communication and social regulation in termites. *In:* Aquiloni, L. & Tricarico, E. (eds.) *Social Regulation in Invertebrates.* Heidelberg: Springer, pp. 193–248.

Bignell, D. E. & Eggleton, P. (2000) Termites in ecosystems. *In:* Abe, T., Bignell, D. E. & Higashi, M. (eds.) *Termites: Evolution, Sociality, Symbiosis and Ecology.* Dordrecht, NL: Kluwer Academic Publishers, pp. 363–387.

Bordereau, C. & Pasteels, J. M. (2011) Pheromones and chemical ecology of dispersal and foraging in termites. *In:* Bignell, D. E., Roisin, Y., & Lo, N. (eds.) *Biology of Termites: A Modern Synthesis.* Dordrecht, Heidelberg, London, New York: Springer, pp. 279–320.

Bouillon, A. (1970) Termites of the Ethiopian region. *In:* Krishna, K. & Weesner, F. M. (eds.) *Biology of Termites II.* New York: New York Academic Press.

Bourguignon, T., Leponce, M., & Roisin, Y. (2009) Insights into the termite assemblage of a neotropical rainforest from the spatio-temporal distribution of flying alates. *Insect Conservation and Diversity*, **2**, 153–162.

Bourguignon, T., Hayashi, Y., & Miura, T. (2012) Skewed soldier sex ratio in termites: Testing the size-threshold hypothesis. *Insectes Sociaux*, **59**, 557–563.

Brauman, A., Bignell, D. E., & Tayasu, I. (2000) Soil-feeding termites: Biology, microbial associations and digestive mechanisms. *In*: Abe, T., Bignell, D. E., & Higashi, M. (eds.) *Termites: Evolution, Sociality, Symbioses, Ecology.* Dordrecht: Kluwer Academic Press, pp. 233–259.

Brent, C. S. (2009) Control of termite caste differentiation. *In:* Gadau, J. & Fewell, J. H. (eds.) *Organization of Insect Societies. From Genome to Sociocomplexity.* Cambridge, MA: Harvard University Press, pp. 105–127.

Brune, A. & Ohkuma, M. (2011) Role of the termite gut microbiota in symbiotic digestion. *In:* Bignell, D. E., Roisin, Y., & Lo, N. (eds.) *Biology of Termites: A Modern Synthesis.* Dordrecht, Heidelberg, London, New York: Springer, pp. 439–476.

Bulmer, M. S., Eldridge, A. S., & Traniello, J. F. (2001) Variation in colony structure in the subterranean termite *Reticulitermes flavipes*. *Behavioral Ecology and Sociobiology*, **49,** 236–243.

Clément, J.-L. & Bagnères, A.-G. (1998) Nestmate recognition in termites. *In:* Vander Meer, R. K., Breed, M. D., Winston, M. L., & Espelie, K. (eds.) *Pheromone Communication in Social Insects: Ants, Wasps, Bees and Termites*. Boulder: Westview Press, pp. 125–155.

Cleveland, L. R., Hall, S. K., Sanders, E. P., & Collier, J. (1934) The wood- feeding roach *Cryptocercus*, its protozoa, and the symbiosis between protozoa and roach. *Memoirs of the American Academy of Arts and Sciences*, **17**, 185–382.

Collins, N. M. (1981) Populations, age structure and survivorship of colonies of *Macrotermes bellicosus* (Isoptera: Macrotermitinae). *Journal of Animal Ecology*, **50**, 293–311.

Crespi, B. J. (1994) Three conditions for the evolution of eusociality: Are they sufficient? *Insectes Sociaux*, **41**, 395–400.

Darlington, J. P. E. C. (1984) A method for sampling for populations of large termite nests. *Annals of Applied Biology*, **104**, 427–436.

(1990) Populations in nests of the termite *Macrotermes subhyalinus* in Kenya. *Insectes Sociaux*, **37**, 158–168.

(1991) Turnover in the populations within mature nests of the termite *Macrotermes michaelseni* in Kenya. *Insectes Sociaux*, **38**, 251–262.

Darlington, J. P. E. C. & Dransfield, R. D. (1987) Size relationships in nest populations and mound parameters in the termite *Macrotermes michaelseni* in Kenya. *Insectes Sociaux*, **34**, 165–180.

Darlington, J. P. E. C., Zimmerman, P. R., & Wandiga, S. O. (1992) Populations in nests of the termite *Macrotermes jeanneli* in Kenya. *Journal of Tropical Ecology*, **8**, 73–85.

Deshmukh, I. (1989). How important are termites in the production ecology of African savannas? *Sociobiology*, **15**, 155–168.

Donovan, S. E., Jones, D. T., Sands, W. A., & Al., E. (2000). Morphological phylogenetics of termites (Isoptera). *Biological Journal of the Linnean Society*, **70**, 467–513.

Dosso, K., Konate, S., Aidara, D., & Linsenmair, K. E. (2010) Termite diversity and abundance across fire-induced habitat variability in a tropical moist savanna (Lamto, Central Côte d'Ivoire). *Journal of Tropical Ecology*, **26**, 323–334.

Eggleton, P. (1994) Termites live in a pear-shaped world: A response to platnick. *Journal of Natural History*, **28**, 1209–1212.

(2000) Global patterns of termite diversity. *In:* Abe, T., Bignell, D. E., & Higashi, M. (eds.) *Termites: Evolution, Sociality, Symbiosis and Ecology*. Netherlands: Kluwer Academic Publishers, pp. 25–51.

Eggleton, P. & Tayasu, I. (2001) Feeding groups, lifetypes and the global ecology of termites. *Ecological Research*, **16**, 941–960.

Eggleton, P., Williams, P. H., & Gaston, K. J. (1994) Explaining global termite diversity: Productivity or history? *Biodiversity and Conservation*, **3**, 318–330.

Engel, M. S., Grimaldi, D. A., & Krishna, K. (2009) Termites (Isoptera): Their phylogeny, classification, and rise to ecological dominance. *American Museum Novitates*, **3650**, 1–27.

Evans, T. A., Inta, R., Lai, J. C. S., Prueger, S., Foo, N. W., Fu, E. W., & Lenz, M. (2009) Termites eavesdrop to avoid competitors. *Proceedings of the Royal Society of London B*, **276,** 4035–4041.

Evans, T. A., Lai, J. C. S., Toledano, E., Mcdowall, L., Rakotonarivo, S., & Lenz, M. (2005) Termites assess wood size by using vibration signals. *Proceedings of the National Academy of Science USA*, **102,** 3732–3737.

Gay, F. J. & Calaby, J. H. (1970) Termites of the Australian region. *In:* Krishna, K. & Weesner, F. M. (eds.) *Biology of Termites II*. New York: Academic Press, pp. 393–448.

Gerber, C., Badertscher, S., & Leuthold, R. H. (1988) Polyethism in *Macrotermes bellicosus* (Isoptera). *Insectes Sociaux*, **35,** 226–240.

Grandcolas, P. & D'Haese, C. (2004) The origin of a 'true' worker caste in termites: Phylogenetic evidence is not decisive. *Journal of Evolutionary Biology*, **15,** 885–888.

Grasse, P. P. (1949) Ordre des Isopteres ou termites. *In*: Grasse, P. P. (ed.) *Traite de Zoologie*. Paris: Masson, pp. 408–544

(1982) *Termitologia: Anatomie-Physiologie-Biologie-Systematique des Termites. Tome I*, Paris: Masson.

(1984) *Termitologia. Fondation des Societes-Construction. Tome II*, Paris: Masson.

(1986) *Termitologia. Comportement-Socialité-Écologie-Évolution-Systématique. Tome III*, Paris: Masson.

Grasse, P. P. & Noirot, C. (1947) Le polymorphisme social du termite à cou jaune (*Kalotermes flavicollis*) les faux-ouvriers ou pseudergates et les mues regressives. *Comptes Rendus de l'Academie des Sciences*, **224**, 219–221.

(1948) La “climatisation” de la termitière par ses habitants et le transport de l’eau. *Comptes Rendus de l’Academie des Sciences*, **227**, 869–871.

(1951) Nouvelles recherches sur la biologie de divers termites champignonnistes (Macrotermitinae). *Annales des Sciences Naturelles Zoologie et Biologie Animale*, **11, 13**, 291–342.

Hamilton, W. D. & May, R. M. (1977) Dispersal in stable habitats. *Nature*, **269**, 578–581.

Han, S. H. & Bordereau, C. (1992) From colony foundation to dispersal flight in a higher fungus-growing termite, *Macrotermes subhyalinus*, (Isoptera, Macrotermitinae). *Sociobiology*, **20,** 219–231.

Hausberger, B., Van Neer, A., Kimpel, D., & Korb, J. (2011) Uncovering cryptic species diversity of a termite community in a West African savanna. *Molecular Phylogenetics and Evolution*, **61**, 964–969.

Haverty, M. I. & Howard, R. W. (1981) Production of soldiers and maintenance of soldier proportions by laboratory experimental groups of *Reticulitermes flavipes* (Kollar) and *Reticulitermes virginicus* (Banks) (Isoptera: Rhinotermitidae). *Insectes Sociaux*, **28**, 32–39.

Haverty, M. I., Page, M., Nelson, L. J., & Blomquist, G. J. (1988) Cuticular hydrocarbons of dampwood termites, *Zootermopsis*: Intra- and intercolony variation and potential as taxonomic characters. *Journal of Chemical Ecology*, **14**, 1035–1058.

Heath, H. (1903) The habits of California termites. *Biological Bulletin*, 4, 47–63.

(1907) The longevity of members of different castes of *Termopsis angusticollis*. *Biological Bulletin*, **13**, 161–164.

(1927) Caste formation in the termite genus *Termopsis*. *Journal of Morphology and Physiology*, **43**, 387–423.

Heinze, J. & Schrempf, A. (2008) Aging and reproduction in social insect: A mini-review. *Gerontology*, **54**, 160–167.

Higashi, M., Abe, T., & Burns, T. P. (1992) Carbo-nitrogen balance and termite ecology. *Proceedings of the Royal Society of London B*, **249**, 303–308.

Hoffmann, K., Gowin, J., Hartfelder, K., & Korb, J. (2014) The scent of royalty: A P450 gene signals reproductive status in a social insect. *Molecular Biology and Evolution*, **31**, 2689–2696.

Howard, K. J. & Thorne, B. L. (2011) Eusocial evolution in termites and hymenoptera. *In:* Bignell, D. E., Roisin, Y., & Lo, N. (eds.) *Biology of Termites: A Modern Synthesis*. Dordrecht, Heidelberg, London, New York: Springer, pp. 97–132.

Howard, K. J., Johns, P. M., Breisch, N. L., & Thorne, B. L. (2013) Frequent colony fusions provide opportunities for helpers to become reproductives in the termite *Zootermopsis nevadensis*. *Behavioural Ecology and Sociobiology*, **67**, 1575–1585.

Inward, D., Beccaloni, G., & Eggleton, P. (2007a) Death of an order: A comprehensive molecular phylogenetic study confirms that termites are eusocial cockroaches. *Biology Letters*, **3**, 331–335.

Inward, D. J. G., Vogler, A. P., & Eggleton, P. (2007b) A comprehensive phylogenetic analysis of termites (Isoptera) illuminates key aspects of their evolutionary biology. *Molecular Phylogenetics and Evolution*, **44**, 953–967.

Johns, P. M., Howard, K. J., Breisch, N. L., Rivera, A., & Thorne, B. L. (2009) Nonrelatives inherit colony resources in a primitive termite. *Proceedings of the National Academy of Sciences USA*, **106,** 17452–17456.

Johnson, R. A. (1981) Colony development and establishment of the fungus comb in *Microtermes sp. nr. umbaricus* (Sjöstedt) (Isoptera: Macrotermitinae) from Nigeria. *Insectes Sociaux*, **28,** 3–12.

Jones, D. T. & Eggleton, P. (2011) Global biogeography of termites: A compilation of sources. *In:* Bignell, D. E., Roisin, Y., & Lo, N. (eds.) *Biology of Termites: A Modern Synthesis.* Dordrecht, Heidelberg, London, New York: Springer, pp. 477–498.

Josens, G. (1982) Adaptive strategies in colony foundations of two Termitidae. *In:* Breed, M. D., Michener, C., & Evans, H. E. (eds.) *IUSSI,* 1982 Boulder CO: Westview Press, pp. 66.

(1983) The soil fauna of tropical savannas III: The termites. *In:* Bourliere, F. (ed.) *Tropical Savannas.* Amsterdam: Elsevier, pp. 505–524.

Kaib, M., Hacker, M., & Brandl, R. (2001) Egg laying in monogynous and polygynous colonies of the termite *Macrotermes michaelseni* (Isoptera, Macrotermitinae). *Insectes Sociaux*, **48,** 231–237.

Kambhampati, S. & Eggleton, P. (2000) Taxonomy and phylogeny of termites. *In:* Abe, T., Bignell, D. E., & Higashi, M. (eds.) *Termites: Evolution, Sociality, Symbiosis and Ecology.* Netherlands: Kluwer Academic Publishers, pp. 1–23.

Keller, L. (1998) Queen lifespan and colony characteristics in ants and termites. *Insectes Sociaux*, **45**, 235–246.

Keller, L. & Genoud, M. (1997) Extraordinary lifespans in ants: A test of evolutionary theories of ageing. *Nature*, **389**, 958–960.

Korb, J. (2007a) Termites. *Current Biology*, **17**, R995–999.

(2007b) Workers of a drywood termite do not work. *Frontiers in Zoology*, **4**, e7.

(2008) The ecology of social evolution in termites. *In:* Korb, J. & Heinze, J. (eds.) *Ecology of Social Evolution.* Berlin, Heidelberg: Springer, pp. 151–174.

(2009) Termites: An alternative road to eusociality and the importance of group benefits in social insects. *In:* Gadau, J. & Fewell, J. H. (eds.) *Organization of Insect Societies. From Genome to Sociocomplexity*. Cambridge, MA: Harvard University Press, pp. 128–147.

(2011) Termite mound architecture, from function to construction. *In:* Bignell, D. E., Roisin, Y., & Lo, N. (eds.) *Biology of Termites: A Modern Synthesis.* Dordrecht, Heidelberg, London, New York: Springer, pp. 349–374.

(2016) Towards a more pluralistic view of termite social evolution. *Ecological Entomology*, **41**, 34–36.

Korb, J. & Hartfelder, K. (2008) Life history and development: A framework for understanding the ample developmental plasticity in lower termites. *Biological Reviews*, **83,** 295–313.

Korb, J. & Heinze, J. (2008a) *Ecology of Social Evolution.* Heidelberg: Springer.

(2008b) The ecology of social life: A synthesis. *In:* Korb, J. & Heinze, J. (eds.) *Ecology of Social Evolution.* Heidelberg: Springer, pp. 245–260.

Korb, J. & Linsenmair, K. E. (2000) Thermoregulation of termite mounds: What role does ambient temperature and metabolism of the colony play? *Insectes Sociaux*, **47**, 357–363.

(2001) Resource availability and distribution patterns, indicators of competition between *Macrotermes bellicosus* and other macro-detritivores in the Comoé National Park, Côte d'Ivoire. *African Journal of Ecology*, **39**, 257–265.

(2002) Evaluation of predation risk in the collectively foraging termite *Macrotermes bellicosus. Insectes Sociaux*, 49**,** 264–269.

Korb, J. & Roux, E. A. (2012) Why join a neighbour: Fitness consequences of colony fusions in termites. *Journal of Evolutionary Biology*, **25**, 2161–2170.

Korb, J. & Schmidinger, S. (2004) Help or disperse? Cooperation in termites influenced by food conditions. *Behavioral Ecology and Sociobiology*, **56**, 89–95.

Korb, J. & Schneider, K. (2007) Does kin structure explain the occurrence of workers in a lower termite? *Evolutionary Ecology*, **21**, 817–828.

Korb, J., Buschmann, M., Schafberg, S., Liebig, J., & Bagneres, A. G. (2012) Brood care and social evolution in termites. *Proceedings of the Royal Society of London B*, **279**, 2662–2671.

Krishna, K., Grimaldi, D. A., Krishna, V., & Engel, M. S. (2013) Treatise on the Isoptera of the world. *Bulletin of the American Museum of Natural History*, **377,** 1–2704.

Lacey, M. J., Lenz, M., & Evans, T. A. (2010) Cryoprotection in dampwood termites (Termopsidae, Isoptera). *Journal of Insect Physiology*, **56**, 1–7.

LaFage, J. P. & Nutting, W. L. (1978) Nutrient dynamics of termites. *In*: Brian, M. V. (ed.) *Production Ecology of Ants and Termites*. Cambridge: Cambridge University Press.

Legendre, F., Whiting, M. F., Bordereau, C., Cancello, E. M., Evans, T. A., & Grandcolas, P. (2008) The phylogeny of termites (Dictyoptera: Isoptera) based on mitochondrial and nuclear markers: Implications for the evolution of the worker and pseudergate castes, and foraging behaviors. *Molecular Phylogenetics and Evolution*, **48**, 615–627.

Lenz, M. (1987) Brood production by imaginal and neotenic pairs of *Cryptotermes brevis* (Walker): The significance of helpers (Isoptera: Kalotermitidae). *Sociobiology*, **13,** 59–66.

(1994) Food resources, colony growth and caste development in wood-feeding termites. *In*: Hunt, J. & Nalepa, C. A. (eds.) *Nourishment and Evolution in Insect Societies.* New Delhi: Oxford and IBH Publishing Co. Prt. Ltd, pp. 159–209.

Lepage, M. (1989) Ecologie et adaptations des sociétés de termites en Afrique tropicale aride. *Bulletin d'Ecologie*, **20,** 59–63.

Leuthold, R. H. (1979) Chemische Kommunikation als Grundlage des Soziallebens bei Termiten. *In:* Lüscher, M. (ed.) *Insektenstaaten. Neuere Erkenntnisse.* Bern: Naturhistorisches Museum.

(1990) L'organisation sociale chez des termites championnistes du genre. *Macrotermes. Actes de Coloques Insectes Sociaux*, **6**, 9–20.

Liebig, J., Eliyahu, D., & Brent, C. S. (2009) Cuticular hydrocarbon profiles indicate reproductive status in the termite *Zootermopsis nevadensis*. *Behavioral Ecology and Sociobiology*, **63**, 1799–1807.

Lo, N., Engel, M. S., Cameron, S., Nalepa, C. A., *et al.* (2007) Save Isoptera: A comment on Inward, *et al. Biology Letters*, **3,** 564–565.

Lo, N., Tokuda, G., & Watanabe, H. (2011) Evolution and function of endogenous termite cellulases. *In:* Bignell, D. E., Roisin, Y., & Lo, N. (eds.) *Biology of Termites: A Modern Synthesis.* Dordrecht, Heidelberg, London, New York: Springer, pp. 51–68.

Long, C. E. & Thorne, B. L. (2006) Resource fidelity, brood distribution and foraging dynamics in complete laboratory colonies of *Reticulitermes flavipes* (Isoptera, Rhinotermitidae). *Ethology, Ecology & Evolution*, **18,** 113–125.

Luchetti, A., Dedeine, F., Velona, A., & Mantovani, B. (2013) Extreme genetic mixing within colonies of the wood-dwelling termite *Kalotermes flavicollis* (Isoptera, Kalotermitidae). *Molecular Ecology*, **22**, 3391–3402.

Lüscher, M. (1952) Untersuchungen über das individuelle Wachstum bei der Termite *Kalotermes flavicollis* Fabr. (Ein Beitrag zum Kastenbildungsproblem). *Biologisches Zentralblatt*, **71**, 529–543.

(1974) Kasten und Kastendifferenzierung bei Niederen Termiten. *In:* Schmidt, G. H. (ed.) *Sozialpolymorphismus bei Insekten.* Stuttgart: Wissenschaftliche Verlagsgesellschaft, pp. 694–739.

Luykx, P. (1993) Turnover in termite colonies: A genetic study of colonies of *Incisitermes schwarzi* headed by replacement reproductives. *Insectes Sociaux*, **40**, 191–205.

Lys, J. A. & Leuthold, R. H. (1991) Task-specific distribution of the worker castes in extranidal activities in *Macrotermes bellicosus* (Smeathman): Observations of behaviour during food acquisition. *Insectes Sociaux*, **38**, 161–170.

Matsuura, K. (2006) A novel hypothesis for the origin of the sexual division of labor in termites: Which sex should be soldiers? *Evolutionary Ecology*, **20**, 565–574.

(2011) Sexual and asexual reproduction in termites. *In:* Bignell, D. E., Roisin, Y., & Lo, N. (eds.) *Biology of Termites: A Modern Synthesis.* Dordrecht, Heidelberg, London, New York: Springer, pp. 255–278.

Matsuura, K., Himuro, C., Yokoi, T., Yamamoto, Y., Vargo, E. L., & Keller, L. (2010) Identification of a pheromone regulating caste differentiation in termites. *Proceedings of the National Academy of Science USA*, **107**, 12963–12968.

Minnick, D. R. (1973) The flight and courtship behavior of the drywood termite, *Cryptotermes brevis*. *Environmental Entomology*, **2**, 587–591.

Miura, T. (2004) Proximate mechanisms and evolution of caste polyphenism in social insects: From sociality to genes. *Ecological Research*, **19**, 141–148.

Muller, H. & Korb, J. (2008) Male or female soldiers? An evaluation of several factors on soldier sex ratio in lower termites. *Insectes Sociaux*, **55**, 213–219.

Myles, T. (1988) Resource inheritance in social evolution from termite to man. *In:* Slobodchikoff, C. N. (ed.) *The Ecology of Social Behavior.* New York: Academic Press, pp. 379–423.

Myles, T. G. (1999) Review of secondary reproduction in termites (Insecta: Isoptera) with comments on its role in termite ecology and social evolution. *Sociobiology*, **33,** 1–91.

Nalepa, C. & Bandi, C. (2000) Characterizing the ancestors: Paedomorphosis and termite evolution. *In*: T. Abe, D. E. B. a. M. H. (eds.) *Termites: Evolution, Sociality, Symbioses, Ecology.* Dordrecht: Kluwer Academic Press, pp. 53–76.

Nalepa, C. A. (1984) Colony composition, protozoan transfer and some life history characteristics of the woodroach *Cryptocercus punctulatus* Scudder (Dictyoptera: Cryptocercidae). *Behavioural Ecology and Sociobiology*, **14**, 273–279.

(1988) Cost of parental care in the woodroach *Cryptocercus punctulatus* Scudder (Dictyoptera: Cryptocercidae). *Behavioural Ecology and Sociobiology*, **23**, 135–140.

(1994) Nourishment and the origin of termite eusociality. *In:* Hunt, J. H. & Nalepa, C. A. (eds.) *Nourishment and Evolution in Insect Societies.* Westview Press, Inc, pp. 57–104.

(2011) Altricial development in wood-feeding cockroaches: The key antecedent to termite eusociality. *In:* Bignell, D. E., Roisin, Y., & Lo, N. (eds.) *Biology of Termites: A Modern Synthesis.* Dordrecht, Heidelberg, London, New York: Springer, pp. 69–96.

(2015) Origin of termite eusociality: Trophallaxis integrates the social, nutritional, and microbial environments. *Ecological Entomology*, **40**, 323–335.

Neoh, K. B. & Lee, C. Y. (2011) Developmental stages and caste composition of a mature and incipient colony of the drywood termite, *Cryptotermes dudleyi* (Isoptera: Kalotermitidae). *Journal of Economic Entomology*, **104,** 622–628.

Nobre, T., Rouland-Lefevre, C., & Aanen, D. K. (2011) Comparative biology of fungus cultivation in termites and ants. *In:* Bignell, D. E., Roisin, Y., & Lo, N. (eds.) *Biology of Termites: A Modern Synthesis.* Dordrecht, Heidelberg, London, New York: Springer, pp. 193–210.

Noirot, C. (1969) Formation of castes in the higher termites. *In:* Krishna, K. & Weesner, F. M. (eds.) *Biology of Termites Vol. 1.* New York: Academic Press.

(1970) The nests of termites. *In:* Krishna, K. & Weesner, F. M. (eds.) *Biology of Termites Vol. 2.* New York: Academic Press.

(1985a) The caste system in higher termites. *In:* Watson, J. A. L., Okot-Kotber, B. M., & Noirot, C. (eds.) *Caste Differentiation in Social Insects.* Oxford: Pergamon Press, pp. 75–86.

(1985b) Pathways of caste development in the lower termites. *In:* Watson, J. A. L., Okot-Kotber, B. M. & Noirot, C. (eds.) *Caste Differentiation in Social Insects.* Oxford: Pergamon Press, pp. 41–58.

(1990) Sexual castes and reproductive strategies in termites. *In:* Engels, W. (ed.) *An Evolutionary Approach to Castes and Reproduction.* Berlin: Springer Verlag, pp. 3–35.

Noirot, C. & Bordereau, C. (1988) Termite polymorphism and morphogenetic hormones. *In:* Gupta, A. P. (ed.) *Morphogenetic Hormones of Arthropods.* New Brunswick: Rutgers University Press, pp. 293–324.

Noirot, C. & Darlington, J. P. E. C. (2000) Termite nests: Architecture, regulation and defence. *In:* Abe, T., Bignell, D. E., & Higashi, M. (eds.) *Termites: Evolution, Sociality, Symbiosis and Ecology.* Netherlands: Kluwer Academic Publishers, pp. 121–139.

Noirot, C. & Pasteels, J. M. (1987) Ontogenic development and evolution of the worker caste in termites. *Experientia*, **43**, 851–860.

(1988) The worker caste is polyphyletic in termites. *Sociobiology*, **14**, 15–20.

Noirot, C. & Thorne, B. L. (1988) Ergatoid reproductives in *Nasutitermes columbicus* (Isoptera, Termitidae). *Journal of Morphology*, **195**, 83–93.

Nutting, W. L. (1969) Flight and colony foundation. *In:* Krishna, K. & Weesner, F. M. (eds.) *Biology of Termites Vol. 1.* New York: Academic Press.

Ohkuma, M. & Brune, A. (2011) Diversity, structure, and evolution of the termite gut microbial community. *In:* Bignell, D. E., Roisin, Y., & Lo, N. (eds.) *Biology of Termites: A Modern Synthesis.* Dordrecht, Heidelberg, London, New York: Springer, pp. 413–438.

Okot-Kotber, B. M. (1985) Mechanisms of caste determination in a higher termite, *Macrotermes michaelseni* (Isoptera, Macrotermitidae). *In*: Watson, J. A. L., Okot-Kotber, B. M., & Noirot, C. (eds.) *Caste Differentiation in Social Insects.* Oxford: Pergamon Press, pp. 267–306.

Oster, G. F. & Wilson, E. O. (1978) *Caste and Ecology of Social Insects.* Princeton, Princeton University Press.

Pickens, A. L. (1934) The biology and economic significance of the Western subterranean termite, *Reticulitermes hesperus. In:* Kofoid, C. A. (ed.) *Termites and Termite Control.* Berkeley: University of California Press, pp. 157–183.

Poulsen, M., Hu, H., Li, C., Chen, Z., Nygaard, S., *et al.* (2014) Holobiomic division of labor in fungus-farming termites. *Proceedings of the National Academy of Sciences USA*, **111**, 14500–14505.

Queller, D. C. & Strassmann, J. E. (1998) Kin selection and social insects. *Bioscience*, **48,** 165–178.

Rohrig, A., Kirchner, W. H., & Leuthold, R. H. (1999) Vibrational alarm communication in the African fungus-growing termite genus Macrotermes (Isoptera, Termitidae). *Insectes Sociaux*, **46**, 71–77.

Roisin, Y. (2000) Diversity and evolution of caste patterns. *In:* Abe, T., Bignell, D. E., & Higashi, M. (eds.) *Termites: Evolution, Sociality, Symbioses, Ecology.* Dordrecht, Netherlands: Kluwer Academic Publishers, pp. 95–119.

(2001) Caste sex ratios, sex linkage, and reproductive strategies in termites. *Insectes Sociaux*, **48**, 224–230.

Roisin, Y. & Korb, J. (2011) Social organisation and the status of workers in termites. *In:* Bignell, D. E., Roisin, Y., & Lo, N. (eds.) *Biology of Termites: A Modern Synthesis.* Dordrecht, Heidelberg, London, New York: Springer, pp. 133–164.

Roisin, Y. & Pasteels, J. M. (1987) Caste developmental potentialities in the termite *Nasutitermes novarumhebridarum. Entomologia Experimentalis et Applicata*, **44**, 277–287.

Rosengaus, R. B., Moustakas, J. E., Calleri, D. V., & Traniello, J. F. A. (2003) Nesting ecology and cuticular microbial loads in dampwood (*Zootermopsis angusticollis*) and drywood termites (*Incisitermes minor*, Schwarzi, *Cryptotermes cavifrons*). *Journal of Insect Science*, **3,** e31.

Rosengaus, R. B., Traniello, J. F. A., & Bulmer, M. S. (2011) Ecology, behavior and evolution of disease resistance in termites. *In*: Bignell, D. E., Roisin, Y., & Lo, N. (eds.) *Biology of Termites: A Modern Synthesis*. Dordrecht, Heidelberg, London, New York: Springer, pp. 165–192.

Rouland-Lefèvre, C. (2000) Symbiosis with fungi. *In:* Abe T, B. D., Higashi M (ed.) *Termites: Evolution, Sociality, Symbioses, Ecology*. Dordrecht: Kluwer Academic Publishers, pp. 289–306.

Rupf, T. & Roisin, Y. (2008) Coming out of the woods: Do termites need a specialized worker caste to search for new food sources? *Naturwissenschaften*, **95**, 811–819.

Sands, W. A. (1961) Foraging behavior and feeding habits in five species of Trinervitermes in West Africa. *Entomologia Experimentalis et Applicata*, **4,** 277–288.

(1972) The soldierless termites of Africa (Isoptera: Termitidae). *Bulletin of the British Museum (Natural History) Entomology*, **S18,** 1–244.

Scharf, M. E., Buckspan, C. E., Grzymala, T. L., & Zhou, X. (2007) Regulation of polyphenic caste differentiation in the termite *Reticulitermes flavipes* by interaction of intrinsic and extrinsic factors. *Journal of Experimental Biology*, **210,** 4390–4398.

Schmidt, A. M., Jacklyn, P., & Korb, J. (2013) Isolated in an ocean of grass: Low levels of gene flow between termite subpopulations. *Molecular Ecology*, **22,** 2096–2105.

(2014) 'Magnetic' termite mounds: Is their unique shape an adaptation to facilitate gas exchange and improve food storage? *Insectes Sociaux*, **41**, 61–69.

Seelinger, G. & Seelinger, U. (1983) On the social organisation, alarm and fighting in the primitive cockroach *Cryptocercus punctulatus* Scudder. *Zeitschrift für Tierpsychologie*, **61,** 315–333.

Shellman-Reeve, J. S. (1997) The spectrum of eusociality in termites. *In*: Choe, J. C. & Crespi, B. J. (eds.) *The Evolution of Social Behavior in Insects and Arachnids*. Cambridge: Cambridge University Press, pp. 52–93.

Snyder, T. E. (1956) *Annotated, Subject-Heading Bibliography of Termites 1350 B.C. to A.D. 1954*. Washington D.C.: Smithsonian Institution.

Stearns, S. C. (1992) *The Evolution of Life Histories*. Oxford, Oxford University Press.

Thorne, B. & Traniello, J. (2003) Comparative social biology of basal taxa of ants and termites. *Annual Reviews of Entomology*, **48,** 283–306.

Thorne, B., Breisch, N., & Haverty, M. I. (2002) Longevity of kings and queens and first time of reproduction of fertile progeny in dampwood termite (Isoptera; Termopsidae; *Zootermopsis*) colonies with different reproductive structures. *Journal of Animal Ecology*, **71,** 1030–1041.

Thorne, B. L. (1982) Polygyny in termites: Multiple primary queens in colonies of *Nasutitermes corniger* (Motschulsky) (Isoptera: Termitidae). *Insectes Sociaux*, **29**, 102–107.

(1983) Alate production and sex ratio in colonies of the Neotropical termite *Nasutitermes corniger* (Isoptera; Termitidae). *Oecologia*, 58, 103–109.

(1984) Polygyny in the Neotropical termite *Nasutitermes corniger*: Life history consequences of queen mutualism. *Behavioral Ecology and Sociobiology*, **14**, 117–136.

(1997) Evolution of eusociality in termites. *Annual Review of Ecology and Systematics*, **28,** 27–54.

Thorne, B.L. & Haverty, M. I. (2000) Nest growth and survivorship in three species of Neotropical *Nasutitermes* (Isoptera: Termitidae). *Environmental Entomology*, **29**, 256–264.

Thorne, B. L., Traniello, J. F. A., Adams, E. S., & Bulmer, M. (1999) Reproductive dynamics and colony structure of subterranean termites of the genus Reticulitermes (Isoptera, Rhinotermitidae): A review of the evidence from behavioral, ecological, and genetic studies. *Ethology Ecology Evolution*, **11,** 149–169.

Thorne, B. L., Breisch, N., & Muscedere, M. (2003) Evolution of eusociality and the soldier caste in termites: Influence of intraspecific competition and accelerated inheritance. *Proceedings of the National Academy of Sciences USA*, **100,** 12808–12813.

Traniello, J. F. & Leuthold, R. H. (2000) Behavior and ecology of foraging in termites. *In:* Abe, T., Bignell, D. E. & Higashi, M. (eds.) *Termites: Evolution, Sociality, Symbiosis and Ecology.* Netherlands: Kluwer Academic Publishers, pp. 141–168.

Vargo, E. (2003) Hierarchical analysis of colony and population genetic structure of the eastern subterranean termite, *Reticulitermes flavipes*, using two classes of molecular markers. *Evolution*, **57**, 2805–2818.

Vargo, E. L. & Husseneder, C. (2009) Biology of subterranean termites: Insights from molecular studies of Reticulitermes and Coptotermes. *Annual Review of Entomology*, **54**, 379–403.

(2011) Genetic structure of termite colonies and populations. *In:* Bignell, D. E., Roisin, Y. & Lo, N. (eds.) *Biology of Termites: A Modern Synthesis.* Dordrecht, Heidelberg, London, New York: Springer, pp. 321–348.

Waller, D. A. (1988) Ecological similarities of fungus-growing ants (Attini) and termites (Macrotermitinae). *In:* Troger, J. C. (ed.) *Advances in Myrmecology.* New York: E.J. Bill, pp. 337–345.

Waller, D. A. & La Fage, J. P. (1987) Nutritional ecology of termites. *In*: Slansky, F. & Rodriguez, J. G. (eds.) *Nutritional Ecology of Insects, Mites, and Spiders.* Chichester, New York: John Wiley, pp. 487–532.

Watson, J. A. L. & Abbey, H. M. (1989) A 17-year old primary reproductive of *Mastotermes darwiniensis* (Isoptera). *Sociobiology*, **15**, 279–284.

Watson, J. A. L. & Sewell, J. J. (1981) The origin and evolution of caste systems in termites. *Sociobiology*, **6**, 101–118.

Weil, T., Hoffmann, K., Kroiss, J., Strohm, E., & Korb, J. (2009) Scent of a queen-cuticular hydrocarbons specific for female reproductives in lower termites. *Naturwissenschaften*, **96,** 315–319.

Wilkinson, W. (1962) Dispersal of alates and establishment of new colonies in *Cryptotermes havilandi* (Sjöstedt) (Isoptera, Kalotermitidae). *Bulletin of Entomological Research*, **53**, 265–286.

(1963) The alate flight and colony foundation of *Cryptotermes havilandi* (Sjöstedt) (Isoptera, Kalotermitidae). *Symposium of Genetics and Biologie Italy*, **11**, 269–275.

Wilson, E. (1971) *Insect Societies.* Cambridge, MA: Belknap Press of Harvard University Press.

Wood, T. G. & Thomas, R. J. (1989) The mutualistic association between macrotermitinae and termitomyces. *In:* Wilding, N., Collins, N. M., Hammond, P. M., & Webber, J. F. (eds.) *Insect-fungus Interactions.* New York: Academic Press, pp. 69–92.

Wood, T. G., Johnson, R. A., & Ohiagu, C. E. (1977) Populations of termites (Isoptera) in natural and agricultural ecosystems in Southern Guinea savanna near Mokwa, Nigeria. *GeoEco-Trop*, **1,** 139–148.

6 Sociality in Aphids and Thrips

Patrick Abbot and Tom Chapman

Overview

After insect sociobiology emerged as a discipline in its own right (Hamilton, 1964; Wilson, 1971), a number of social insect species were discovered that had been previously unknown to science. Aoki (1976) and Crespi (1992a,b) described unusual patterns of social behavior in the plant-feeding aphids and thrips, respectively. Among their many species, a small number express patterns of social organization resembling those found in some other eusocial animals. These are the eusocial aphids and thrips, and while they are not closely related (i.e. they are in separate orders in an ancient insect group within the superorder Paraneoptera, Grimaldi & Engels, 2005), they share a number of convergent social traits, the most important being that they form defense-based social groups composed reproductives and defenders, or "soldiers". The discoveries of this and the research that followed accompanied a renewed enthusiasm for broadly comparative approaches in social evolution (Choe & Crespi, 1997), a legacy that this and other recent volumes share.

Because social aphids and thrips share a number of features, considering them jointly provides insights into the evolution of defense-based societies in insects (Table 6.1, Figure 6.1). Unlike eusocial insects like the ants, which are both ecologically and—judging by their persistence in the fossil record, evolutionarily successful—the rarity of sociality in aphids and thrips implies that the conditions for the emergence of complex social traits are not common in either group. Like the cooperatively breeding birds or mammals, however, it is the uniqueness of sociality that invites comparative analyses of traits that vary between social and non-social taxa. Crespi, *et al.* (2004) reviewed the ecology and evolution of Australian *Acacia* thrips, including sociality in this group, and pointed to several convergent aspects of sociality in thrips and aphids. In his recent book *The Other Social Insects*, Costa (2006) provided broad overviews of the biology of social aphids and thrips, as well as a diversity of other minor social insects. Our goal for this chapter is to focus mainly on the evolutionary transitions both have made to eusociality, leaving aside the variety of cooperative behaviors that non-eusocial species express. In contrast to most other groups reviewed in this volume, the body of research on aphids and thrips remains relatively young and uneven. Trends have emerged

The authors gratefully acknowledge funding from US NSF IOS-1147033 and NSERC, and helpful comments from three anonymous reviewers and Dustin Rubenstein.

nonetheless. Research on social evolution in aphids and thrips has largely followed two routes: one, from more evolutionary-minded researchers, particularly those working on thrips, has generally asked questions such as: what traits do eusocial aphids or thrips have that converge on those expressed by social invertebrates? The other, from researchers with interests in behavioral biology or natural history, has asked questions such as: what are the ecological and behavioral properties of individual species? The literature on social aphids and thrips is thus divided along axes defined by natural history and evolution. Uniting the two themes across aphids and thrips is the gall-forming habit, which shapes much of their natural history and governs many of the features of their societies.

I SOCIAL DIVERSITY

6.1 How Common is Sociality in Aphids and Thrips?

As highlighted by Costa (2006), thrips and aphids are two of the most "studied of the understudied" social insect groups. Perhaps the most important starting point for understanding their social behavior is that, even under the most generous descriptions, it is extremely rare. To date, advanced sociality (i.e. eusociality) has only been formally described in about 1 percent of aphids, and an even smaller fraction of thrips.

True aphids in the family Aphididae are small, soft-bodied herbivorous insects (Blackman & Eastop, 1994). Aphids feed on plant sap and are hemimetabolous insects, meaning that like grasshoppers or termites, they do not undergo complete metamorphosis. There are about 5,000 species of aphids, and many have complex life-cycles, including generations that alternate between distinct host plant species, and asexual and sexual reproduction. Because of their prodigious abilities for rapid reproduction and the various strategies that they have for dispersal and persistence in seasonal environments, many are agricultural pests in temperate regions, particularly in the Northern Hemisphere, where they are more diverse than in lower or southern latitudes. Currently, twenty-five subfamilies within the Aphididae are recognized. The majority of species are found in one subfamily, the Aphidinae. However, eusociality has only been described from two small subfamilies, the Hormaphidinae and Eriosomatinae (formerly Pemphiginae). In these two subfamilies, there have been repeated origins of larval soldiers in sixty or so species in multiple genera. Both subfamilies include three tribes each, and social species are more common in some tribes than others (Stern & Foster, 1996).

Like aphids, thrips are small insects that feed primarily on plants or fungi. Unlike aphids, thrips can tap into individual cells of a plant using their maxillary laciniae and left mandible (the right mandible ceases development early in this order), like a straw inserted into a juice box. Not all thrips are plant feeders; a minority of thrips species have co-opted their plant piercing mouthparts for predation (Minaei, 2014) and perhaps even for occasional feeding on vertebrates (Childers, *et al.*, 2005). The order Thysanoptera includes two suborders with nine extant families, the largely plant-feeding

Table 6.1 A comparison of key similarities and differences between eusocial aphids and thrips.

Trait	Eusocial aphids	Eusocial thrips
Number of social species (%)	~60 (1.3%)	~7 (0.1%)
Eusocial definition	Presence of nymphal soldiers	Presence of non-dispersing, adult soldiers
Life cycle	May include social or non-social phase on alternate host plant with no gall	Social & gall-dwelling throughout life cycle
Biogeography	Largely northern temperate, seasonal	Australia, arid
Reproduction	Viviparous; clonal within social group	Oviparous; fertilized = females; unfertilized = males
Colony founding	Female only	Female typically, may include males in some species
Sex ratio	Female only	Female biased to even
Skew	Most species have only 1 reproducing female	Variable
Inbreeding	Common	Common
Sources of mortality that favor soldiers	Predation by various (mostly) larval insects	Exploitation by specialist thrips
Microbes and pathogens	Largely unknown. Aphids produce copious wax with antimicrobial properties.	Likely important
Demographic drivers	Sociality associated with longer-lived, often smaller galls	Sociality associated with longer-lived, elongated and smaller galls
Functional studies of caste regulation?	More extensive	Less extensive
Ecological and behavioral studies of sociality	More extensive	Less extensive
Phylogenetic and evolutionary studies of sociality?	Less extensive	More extensive
Composite scenario for evolution of sociality?	Possibly multiple routes because of primary and secondary host soldiers. High relatedness + slow growth rate + natural enemies likely important	Valuable resource + slow growth rate + high relatedness + sex ratio manipulation + high skew + natural enemies

Terebrantia, and the Tubulifera, which is composed of a single family (Phlaeothripidae) of about 3,500 species and 450 genera, many of which are fungus feeders (Mound & Morris, 2007). The Tubulifera includes all of the eusocial species. Unlike social aphids, which are widely distributed both taxonomically and geographically, and are found on diverse host plants, eusocial thrips are restricted to a single genus (*Kladothrips*) of

Figure 6.1. Galls and aphid soldiers attacking a *Drosophila* larva (left) and galls and a thrips soldier (right). Photo credit P. Abbot and H. Caravan.

twenty-four described species on Australian *Acacia*. The genus radiated perhaps in the last 10 million years. The seven described eusocial species are monophyletic, and include two species where the soldier caste was lost (Crespi, *et al.*, 2004; McLeish, *et al.*, 2007). The described diversity of *Kladothrips* is an underestimate, as cryptic species complexes have been identified in the gall-inducer clade, suggesting that there are many independently evolving populations of thrips beyond those that have been formally described (McLeish, *et al.*, 2006).

6.1.1 Defining Sociality in Aphids and Thrips

Defining sociality in both groups is not straightforward, as they do not fit neatly into the definitions for eusociality outlined by Michener (1969), Wilson (1975) and others. The hallmark of aphid and thrips social behavior is self-sacrificial defense of relatives, which is a behavioral phenotype that may or may not be accompanied by morphological and life history specialization. Thus, in both aphids and thrips, social species have been described by a varying combination of metrics. Practically all described social species cause swellings (known as galls) or manipulate plant tissues in various ways. But the simplicity of this observation obscures important complications. For example, all of the gall-forming aphid and thrips species live in multi-generational family groups, but only a minority of these express unique adaptations for defense. Even in the absence of galls, aphids and thrips may aggregate in kin groups. Moreover, many aphids have complex life cycles that encompass two host plants. On one, (the *primary* host), a sexual generation occurs, where a gall is formed. Confusingly, in many aphid species, social behavior is sometimes expressed on the *secondary* host, even in the absence of a gall.

Adding to the difficulty in defining sociality in both lineages, what constitutes a social aphid or thrips may depend upon whom you ask. The current lists of social aphids and thrips have been defined by different combinations of experimental and diagnostic criteria, consisting of various descriptions of morphology and of behavior. For example, in what is recognized as the first description of a social aphid, Aoki's (1976) detailed dimorphic first instar larva in *Colophina clematis* that differed in the degree of robustness in their legs and length of the feeding mouth parts. Aoki recognized these as "soldiers" and soon after, described juvenile nymphs of various species that did not molt, and were therefore sterile. Due in large part to Aoki's efforts, new discoveries of social aphids are almost always accompanied by an initial species description, including both morphological keys and notes on behavior.

In thrips, the first descriptions of social behavior derived from a more behavioral tradition. For example, the initial descriptions of eusociality in thrips by Crespi (1992a) involved puncturing the galls of two polymorphic Australian thrips species, *Kladothrips intermedius* and *K. habrus* (which were referred to as *Oncothrips tepperi* and *Oncothrips habrus*, respectively, before they were synonymized with *Kladothrips*). By inserting another species of thrips into the breach with forceps, Crespi observed that "micropterous" or wing-reduced individuals within galls would grasp the invaders that he had inserted. These individuals had more robust forelimbs compared to winged or "macropterous" individuals of this species. Crespi interpreted these observations as evidence for a division of labor in these insect colonies. Analogous to Aoki's description of aphid soldiers, he used the term "soldier" to describe these defense-minded, morphologically specialized castes.

6.2 Forms of Sociality in Aphids and Thrips

Behaviorally or morphologically specialized soldiers defend relatives in the eusocial aphids and thrips. However, in both, there is increasing evidence of complexity in social organization (the composition and cohesion of their groups). In thrips, and to a lesser extent in aphids, various forms of communal, cooperative, and helping behaviors are recognized. For example, there are group living species in each that are categorized as "non-social", because they lack the defender-based hallmarks of sociality. Most species have received little scrutiny, however, and in both lineages, there is still a large number of species for which the categorization of social status is based on anecdotal observations. Only recently have social categories been defined in aphids and thrips, and organizational schemes have been proposed that are analogous to r/K life history syndromes. Pike & Foster (2008) suggested that gall-forming aphids exhibit three basic strategies: "hide and run," involving a life cycle with only a short duration within galls and no soldiers; "stay a little, fights a little, but then run," involving a somewhat longer duration within galls, and weakly aggressive nymphs without morphological specialization; and "stay and fight," involving long-lived galls and eusociality. For thrips, Crespi, *et al.* (2004) proposed four categories of gall-forming *Acacia* thrips: either a "flier" strategy involving large galls, fast-developing brood, and little or no investment in defense; a "hider" strategy that develops large, tightly sealed galls in the arid interior of Australia and no soldiers; a "fighter" strategy involving small volume and more elongate or hemispherical galls, smaller dispersing brood numbers, and the production of soldiers (Crespi & Worobey, 1998; Wills, *et al.*, 2001); or finally an "exploiter strategy" that does not induce galls, but instead invades and usurps the galls of other thrips species. We use three headings (non-social group living, communal, and eusocial, Chapters 7 and 8) that approximate the hider/flier-to-fighter life history continuum described by Pike & Foster (2008) and Crespi, *et al.* (2004).

6.2.1 Non-social Group Living Species

Many species of herbivorous, non-gall-forming aphids and thrips can be found commonly in groups on plants, but they do not express coordinated social behaviors (Dixon, 1998). Such groups are sometimes described as social, but "colonial breeders" would be a more accurate description. Aggregations of aphids and thrips ultimately derive from the fact that they feed on plants. Other reasons include the fact that only certain plant tissues may be suitable for feeding, the tendency of both to join existing groups, low vagility, and reasons related to reproduction.

Most aphids are plant specialists and can regularly be observed in feeding aggregations on their host plants (Prokopy & Roitberg, 2001). Feeding aggregations may be composed of unrelated individuals or of highly related individuals from only a small number of clones (Abbot, 2011). Aphid aggregations provide various ecological advantages, including greater per capita resource acquisition from host plants, mutual sinks of plant assimilates, or protection from predators (Cappuccino, 1987; Prokopy & Roitberg, 2001). Aphids are agricultural pests, and consequently there is a large applied literature on how and why they aggregate and disperse.

While some are important horticultural pests, there is not a comparably broad literature on non-social thrips aggregations. The genus *Frankliniella*, which includes the western flower thrips and other species that transmit tospoviruses, has received the most study by far. But like aphids, the plants and fungi on which they feed are patchy resources that give rise to aggregations. Many thrips have claustral habits, unlike sap-feeding aphids, which aggregate in open colonies on or near growing plant tissues such as leaves and stems. The claustral tendencies of thrips probably favor dense groups in many thrips species (Lewis, 1973). Many species produce aggregation pheromones, for many of the same reasons that aphids aggregate: food, protection, shelter, and particularly in the case of thrips, for mating (Lewis, 1973).

6.2.2 Communal Species

There are various definitions of communal nesting and group breeding, but the formation of casteless groups of cooperative females is common in bees and wasps, as well as in various groups of mammals and birds (Clutton-Brock, 2009; Wcislo & Tierney, 2009). In some aphid and thrips species, unrelated females either obligately or facultatively nest and reproduce together. While they may exhibit various cooperative behaviors, such as coordinated responses to threat, they do not exhibit defensive behaviors that define the eusocial species.

In aphids, there have been a handful of descriptions of communal groups of asexual females called foundresses co-inhabiting nest-like galls or domiciles. Communally nesting aphids produce multi-generational, mixed-family groups, but otherwise express little in the way of cooperative behavior. In *Tamalia cowenii*, for example, multiple foundresses occupy galls on their host plants, *Arctostaphylos* (Miller 1998, 2005). Such groups form under conditions of high density and limitations on suitable sites to form galls on their host plants. Even mixed-species groups of females can sometimes occur. For example, galls of *T. cowenii* are often invaded by females from a related species that do not form their own galls. A similar pattern occurs in species in the genus *Eriosoma*, where female *Eriosoma* foundresses on *Ulmus* invade each other's galls, resulting in a mixed species group (Akimoto, 1989). Although not communal in the strict sense, mixed groups of offspring can occur in some species when juveniles move between galls (Aoki, 1979; Abbot, 2009).

Colonial or communal groups are more common in thrips. Because thrips have claustral habits, females of some species may mate, oviposit and "bivouac" in common, protected sites (Kiester & Strates, 1984). Such female groups attract males, which fight in order to establish dominance (Crespi, 1986, 1988). Pleometrotic group formation has been studied most thoroughly in a single genus, *Dunatothrips*. Adult females characteristically form domiciles that they cooperatively construct on their host plants by gluing the leaf-like phyllodes (modified petioles) of acacia trees together with silky threads (Morris & Schwarz, 2002; Bono & Crespi, 2006, 2008; Gilbert & Simpson, 2013). The benefits of communal nesting includes increased foundress survival, protection of brood against exploiters and adverse abiotic conditions, and mutualistic advantages of collective construction of domiciles when costs of construction are high (Crespi, *et al.*, 2004; Gilbert, 2014). Like *Tamalia* aphids (Miller, 2005), however, there are costs in the form of density-dependent effects of group size on per capita productivity (Morris, *et al.*, 2002; Bono & Crespi, 2006). It may well be that species like the *Dunatothrips* have more complex patterns of social organization than a "communal" label implies.

6.2.3 Eusocial Species

Eusocial aphids and thrips have given rise to some debate about the definition of eusociality (Costa & Fitzgerald, 2005). While there is not currently a universal definition that incorporates "actual phylogenetic patterns" that would usefully facilitate comparative studies, as Wcislo (1997) recommends, eusocial aphids and thrips have behaviorally or morphologically specialized group members that express costly, self-sacrificial traits. In particular, they are among the species that express fortress defense sociality, in which highly related groups of individuals occur in valuable and/or defensible resources and express in traits for defense by some group members (Alexander, *et al.*, 1991; Crespi, 1994; Queller & Strassmann, 1998).

Most would agree that aphids have a primitive form of eusociality. This involves family groups composed of a single foundress and one or more generations, some or all of the juvenile nymphs of which act as soldiers at some stage during development (typically first or second larval instars). In most Hormaphidinae species, there are clear castes of morphologically specialized, non-dispersing sterile soldiers. In others, particularly those in the Eriosomatinae, all juvenile nymphs aggressively defend the group, but they are monomorphic and totipotent. (Some authors refer to these as defenders rather than soldiers.) A typical aphid society within a gall generally develops in the following way. A foundress forms a gall on her host plant, and asexually produces a generation of daughters, some or all of which are morphologically and/or behaviorally specialized for defense. Either this or subsequent generations produced by the foundress disperse, and the entombed foundress dies within the gall. Species that express sociality on the secondary host have roughly the same features: group formation by a reproductive female, defense by some specialized juvenile soldiers, and dispersal by winged adults.

Seven of the seven species of gall-forming thrips on Australian *Acacia*, all in the single genus *Kladothrips* Froggatt, have a nondispersing soldier caste, and are considered to be eusocial (Crespi, *et al.*, 2004; Wills, *et al.*, 2004). As with aphids, eusocial thrips species differ in important aspects of their life histories, but generally speaking, a typical eusocial thrips society develops in the following way. A foundress forms a gall on actively growing *Acacia* phyllodes inside of which she lays eggs. The foundress-produced cohort is morphologically and behaviorally specialized for defense (Crespi, *et al.*, 1997; Crespi, *et al.*, 2004; Mound, 2005). The soldiers mate with siblings and, together with the foundress, produce another cohort destined to be dispersers. The foundress typically dies soon after the soldiers eclose. There is both intra- and interspecific variation in the relative contributions of the foundress and soldiers to the dispersing cohort. Soldiers tend to become progressively less fecund in the socially more derived species (Chapman & Perry, 2006). Note, however, that unlike some social aphid species, soldiers are never obligately sterile in eusocial thrips.

6.3 Why Aphids and Thrips Form Social Groups

Not unlike shrimp (Chapter 8) and naked mole-rats (Alexander, *et al.*, 1991), there was an expectation of advanced social organization in aphids and thrips prior to their discoveries. Hamilton (1964) discussed the discrepancy between the expectation of altruism in free-living clonal organisms such as aphids and the observation that it was not common. Later, addressing both aphids and thrips, he wrote "*We have to explain why other male-haploid groups have not evolved social life, and also why clonal aggregations like those of aphids have not done so.*" (Hamilton, 1972). That social life would subsequently be described in both-in conditions that favor the evolution of defense traits by non-dispersing castes of juvenile soldiers-would later bring Hamilton to reflect with some amusement at the explanations he came up with to explain the discrepancy (Hamilton, 1996). But what are those conditions? Certainly they include the advantages of aggregating for food or protection (within galls, for example), combined with some combination of ecological opportunities and constraints that favor delayed or non-dispersal by juveniles, the high relatedness that can act as a cause or a consequence of group formation (Chapter 7), and high pressure from predators and other forms of exploitation that favor defense. Why they are *able to* form social groups may be a separate, but equally important question. Answers to this may include things having to do with their flexible, termite-like development (Chapter 5), or in the case of thrips, haplodiploidy which permits complex patterns of inbreeding and sex ratio manipulation that promote the evolution of social behavior (Crespi, *et al.*, 2004; Chapman, *et al.*, 2008). The challenge for understanding social evolution in both aphids and thrips is to better define the differences between species that bear the signposts of sociality (i.e. soldier and other forms of self-sacrificial traits by juveniles) and those that do not. In both groups, there are tremendous opportunities for evolutionary, functional, and ecological studies of closely related social and non-social species.

6.3.1 Resource Acquisition and Use

Social aphids and thrips are herbivores. To a large degree, however, neither lineage has been studied in the broader context of plant–insect interactions, and little is known about how host plants shape social evolution. Since all eusocial aphid and thrips occur in gall-forming lineages, a discussion of the role of resource acquisition in aphid and thrips sociality starts naturally with some background on galls (Crespi, 1992b; Foster & Northcott, 1994).

About 6 to 7 percent of all insect species form galls on plants, and the gall-forming habit has evolved multiple times independently in a wide variety of species. The functions of galls overlap ecologically with that of sociality itself (Gilbert, 2014). Galls are composed of the living, reactive tissues of plants, and are thought to provide both nutrition and protection (Stone & Schönrogge, 2003). Certain gall traits are correlated with aphid and thrips life history traits. In both lineages, species with soldiers are thought to inhabit longer-lived galls than species without soldiers, and investing in soldiers tends to be associated with a decline in fecundity and growth (Stern & Foster, 1996; Chapman, *et al.*, 2006), hence the the likening of sociality to a k-selected life history strategy (Crespi, *et al.*, 2004; Pike, *et al.*, 2007).

In aphids, descriptions of life history correlates of social evolution remain somewhat anecdotal. Perhaps as a consequence, robust phylogenetic comparative tests of the correlates of life history and social traits between social and non-social aphids remain rare. In the scheme presented by Pike, *et al.* (2007) scheme, the non-social, "hide and run" species produces galls that are the largest by volume but shortest-lived. Because of the short duration, these species have the lowest reproductive output from galls. The eusocial, "stay and fight" species produce a smaller, but longer-lived gall, and have higher reproductive success. In addition to these interspecific patterns, investment in defense in aphids can vary within species as well, particularly over the course of a season as a colony grows. The general observation is that, for species with true castes, the largest colonies tend to have the greatest proportion of soldiers (Aoki & Kurosu, 2004). Some species may vary the proportion of soldiers they produce over a season, while in others, soldiers themselves may vary in behavior, morphology or life span, presumably in a manner that optimizes trade-offs between defense and reproduction (see Pike & Foster, 2008).

In thrips, patterns of life history and social evolution are better resolved (Chapman, *et al.*, 2008). The transition from a solitary to social life history appears to include a reduction in gall volume and a consequent reduction in disperser numbers, as well as an elongated period for the development of dispersing brood (Crespi & Worobey, 1998; Wills, *et al.*, 2001; Wills, *et al.*, 2004). Among the species with soldiers, the species with the lowest fecundity and smallest galls (*K. waterhousei*, *K. habrus*, and *K. intermedius*) contain the most lethal soldiers (Perry, *et al.*, 2003, 2004). The evolution of smaller galls in the most derived social species seems paradoxical. One might expect that better-protected galls should evolve to be larger. Chapman, *et al.* (2006) hypothesized that smaller galls are the result of selection on the rate at which soldiers are produced. Foundresses are not defended until their soldier brood develops into adults. Because insect galls require growing plant tissue, there is a tendency for the development of galls to be synchronized on a plant, resulting in density-dependent mortality by natural enemies. Consequently, faster developing galls are favored, which results in the reduction in the size of the mature galls, and a reduction in the productivity of social thrips groups relative to solitary species.

6.3.2 Predator (and Parasite) Avoidance

The primary function of soldiers in both aphids and thrips is defense against natural enemies. Increasingly, evidence indicates that microbial pathogens may have favored certain adaptations that soldiers exhibit in both groups (Pike, *et al.*, 2002; Turnbull, *et al.*, 2011). In aphids, some soldiers are "pseudo-scorpion-like," with enlarged forelimbs for grasping their enemies, and/or sharp pronotal horns that they use to butt their enemies. The soldiers of many social aphids pierce their enemies with the needle-like mouthparts called stylets, and in some species, the stylets themselves can be polymorphic. It has long been suspected that aphid and thrips soldiers express venom secretions, and this appears to be the case in at least one aphid species (Kutsukake, *et al.*, 2004). Venoms have not been described in thrips, although as described further later, social thrips appear to secrete antimicrobial compounds.

A long list of various insect predators commonly prey upon aphids (see Dixon, 1998), but there have been few comprehensive surveys of the natural enemies of a social aphid species, and virtually no surveys of the microbial entomopathogens. The predators of social aphids include anthocorid bugs, syrphid and chamaemiid fly larvae, lacewing larvae, various predaceous or opportunistic lepidopteran larvae, beetles, ants, and earwigs (Alton, 1999; Wilch, 1999). A number of studies have experimentally introduced predators that demonstrated the effectiveness of soldiers in both the lab and field (e.g. Foster, 1990; Foster & Rhoden, 1998; Shibao, *et al.*, 2010; Hattori, *et al.*, 2012; Lawson, *et al.*, 2014a).

Microbes may be important selective agents shaping many features of sociality in aphids. Lawson, *et al.* (2014b) showed that the fungal endophytes in the gall tissue differs between various non-social and social species, suggesting that foundresses select sites depending on endophyte composition, or that different species have different effects on endophyte growth. Active "housecleaning" behaviors have been described in a number of aphid species, involving removal of wastes from galls by larval aphids (Benton & Foster, 1992). However, while the cornicle secretions and waxy exudates that many aphids produce have been studied in other, non-social contexts (e.g. Pope, 1983; Pike, *et al.*, 2002), there have been no chemo-ecological studies of derived antimicrobial adaptations that social aphids may possess.

In thrips, communal and eusocial groups are invaded by various opportunists, including lepidopteran and hymenopteran larvae (Crespi & Abbot, 1999; Bono, 2007; Gilbert & Mound, 2012). In general, thrips soldiers protect against a lethal thrips that is specialized to invade their galls. Thrips in the genus *Koptothrips*, which invade and exploit the galls of both social and non-social thrips, are thought to be the major selective force underlying the evolution of soldiers, as well as impacting the behavior and life history of solitary gall-inducing species (Crespi, 1996; Crespi, *et al.*, 1997, 2004; Crespi & Abbot, 1999; Chapman, *et al.*, 2006, 2008). There are only four described species of *Koptothrips*, but a lack of morphological distinction is thought to be hiding significant species diversity related to host specialization (Crespi & Abbot, 1999). Gonsalves (2010) studied invasion behavior of two gall-invading species in both solitary and social thrips, and observed notable differences in the invasion behavior between the two species. However, gall surveys and genetic sequence data confirmed that there was no specialization on social type. It remains to be determined whether *Koptothrips* species is the primary driver of social evolution in thrips, and further investigation is warranted.

Microbial pathogens may have important effects on social evolution in thrips, but only recently have studies emerged demonstrating the scope of these effects. Turnbull, *et al.* (2011, 2012) found evidence that soldiers possess adaptations for defense against entomopathogens. Turnbull, *et al.* (2011) analyzed the antimicrobial capacity of individuals across eight thrips species that differ in sociality and group size, and showed that the non-social species lacked antimicrobial capacity, and the strength of the antimicrobials increased across species with larger group size. More recently, Turnbull, *et al.* (2012) showed that it is the soldiers in eusocial thrips colonies that are secreting compounds that can suppress entomopathogenic fungi that grow within thrips galls. These soldiers appear to have a dual role in defense of the colony.

6.3.3 Homeostasis

How the abiotic factors shape sociality, and how features of sociality alter the abiotic environment, are important "homeostatic" aspects of social evolution. A general feature of insect-induced plant galls is that they provide protected, stable microenvironments that buffer their inhabitants from abiotic variation, principally temperature and humidity (Fernandes & Price, 1988; Stone & Schönrogge, 2003). Given that many aphids and thrips species form galls but are not eusocial, there is little evidence of unique homeostatic adaptations associated with social life *per se*, such as those expressed by honeybees (Chapter 3). Three points are worth noting, however. First, gall traits themselves, such as size, shape, and longevity, undoubtedly affect the microenvironment in which aphid and thrips social groups occur (Crespi & Worobey, 1998). Second, recall that some eusocial aphid groups occur in the absence of galls. If soldier production is correlated with such traits as longer-lived groups in these species, then there may possess unique homeostatic adaptations, such as the ability to resist desiccation or mobilize energy reserves (Jedličková, *et al.*, 2015). Finally, while aphids and thrips are hemimetabolous insects, thrips also express traits such as pupation like holometabolous insects (Lewis, 1973; Gilbert, 2014). Whether social aphids and thrips possess cryptic homeostatic adaptations that accompany or even facilitate the various life histories they express (such as long-lived colonies) is an open question.

6.3.4 Mating

Strictly speaking, for most social aphids, mating itself has little to do with ecological and behavioral aspects of sociality, since sociality is akin to a "vegetative phase" in their life cycles that occurs after and distinct from bouts of sexual reproduction. Only in a few species do the sexuals and soldiers ever co-occur (e.g. *Astegopteryx spinocephala*, Kurosu, *et al.*, 2006; Aoki, *et al.*, 2007). Nearly all aphid species express cyclical apomictic parthenogenesis with viviparous reproduction during the asexual phase. Aphids are diplodiploids, with a XX/X0 sex determination mechanism that, for reasons having to do with parthenogenesis, differs from other XX/X0 systems found in some invertebrates (Jaquiery, *et al.*, 2012). Aphid life cycles are complex, but despite many exceptions, three basic types are recognized (Blackman & Eastop, 1994). For most species, the entire life cycle, involving both sexual and asexual generations, occurs on one host plant species, spanning a single year. As described earlier, others alternate between host plants, usually over the span of a year but sometimes longer (Kurosu & Aoki, 2009), and the sexual and asexual generations are spatially segregated. In other aphid species, the sexual stages seem to have been dropped altogether, or at least have not yet been discovered.

Although there is much variation, spatial structure and local mate competition favor female-biased investment ratios and inbreeding in most social aphids (Foster & Benton, 1992; Foster, 2009). Some of this variation has to do with how the sexual morphs are formed in aphids. In the social aphid subfamilies, the sexual males and oviparous females are produced parthenogenetically by specialized winged morphs called the sexuparae. The behavior (or even the existence) of sexuparae may seem arcane, but the consequences are important for understanding mating systems in social aphids, as well as patterns of inbreeding, outbreeding, and sex ratio variation (Foster, 2009), and indeed, to read some of the literature on social aphids (e.g. Kurosu, *et al.*, 2006).

Briefly, in host-alternating species, the sexuparae are produced on the secondary host in the summer or autumn, and fly back to the primary host, where they deposit the wingless sexuals. In some non-host-alternating species, sexuparae are formed on the primary host, which in most cases, disperse to other conspecific plants. In the eriosomatine aphids, the sexuparae deposit the sexuals in crevices in the bark, whereas in the hormaphidine aphids, she deposits them on the leaves (Blackmon & Eastop, 1994).

Whereas aphid groups are formed by asexual reproduction, thrips are haplodiploid, sexually reproducing insects with life cycles that promote inbreeding. Like the Hymenoptera, fertilized eggs develop into diploid females, while unfertilized eggs develop into haploid males. Reproduction by soldiers can occur in eusocial thrips, and the dispersing brood is produced by a combination of reproduction by soldiers and the foundress (Chapman, *et al.*, 2002). Soldier reproduction ranges across species. In some, reproduction is almost equivalent to that of their foundress mother, while in others, soldiers produce only a few parthenogenetically produced male offspring (Chapman, *et al.*, 2002). The next generation of gall-formers may be composed of both foundress and soldier-derived-dispersers. In some species, foundresses produce females first (protogyny), and in others, males are produced first (protandry). Protogyny may be generally favored because females may be better fighters than males, or males may be short-lived, and thus require females to be immediately available upon male eclosion (Kranz, *et al.*, 2001a,b). Subsequent work in a social species with protogyny found no difference in the fighting ability between sexes (*K. hamiltoni, unpublished data*). Alternatively, protandry may be favored for reasons having to do with sex ratio variation and inbreeding. Protandry promotes sib-mating and ensures that female soldiers can mate soon after eclosion and produce daughters (Kranz, *et al.*, 1999). Moreover, protandry may be favored by kin selection because, when coupled with sib-mating, it results in the unusual situation in which foundresses are more related to their granddaughters than to their daughters

6.3.5 Offspring Care

A defining feature of the hemimetabolous eusocial insects is that direct brood care is precluded by their mode of development. Offspring care, as it is typically conceived, is thus not an important feature of sociality in aphids or thrips. However, in aphids in which three or more generations occur within groups, from the clone's perspective, soldiers protect the offspring of clonal siblings. For example, in one aphid species with an unusual life cycle, soldiers guard eggs produced by a sexual generation that occur within galls (Kurosu, *et al.*, 2006). In addition to the housecleaning behaviors described earlier, some social aphids have adaptations for self-sacrificial repair of galls that indicate traits not directly related to defense that are analogous to brood care (Kutsukake, *et al.*, 2009).

Similarly, the function of the housecleaning adaptations expressed by thrips is to promote the successful development of offspring, beyond defense of the colony from macro invaders. Coates (*unpublished data*) has shown that foundresses of two species of gall thrips (without soldiers) can produce a substance that can suppress the germination of fungal spores, like that of the soldiers in *Kladothrips intermedius* (Turnbull, *et al.*, 2012).

The foundress' brood and the conspecifics of the soldiers (siblings and offspring) presumably benefit from this retarding of fungal growth, which would suggest that soldier thrips are involved in brood care beyond defense of the colony from macro invaders. Evidence from *K. intermedius* suggests that the enlarged forelimbs of soldiers does not predict fighting ability (Turnbull, *et al.*, 2012), casting doubt on the importance of changes in the forelimb morphology as specialization for fighting. As well, foundresses of two species (*K. intermedius* and *K. habrus*) are able to kill invaders, despite not having the robust forelimbs of the soldiers (Turnbull, *et al.*, 2012). The implication is that non-functional explanations for soldier morphology (e.g. morphometric changes induced by the evolution of wing reduction, or differences in the age of group members) need further investigation in eusocial thrips. If the soldier morphology is not entirely explained by specialization for defense, and foundresses themselves are successful fighters, then parallels emerge between soldiers and foundresses that suggests that traits related to maternal behavior and offspring care may have served as important pre-adaptations for the evolution of soldier castes (Toth, *et al.*, 2007).

6.4 The Role of Ecology in Shaping Sociality in Aphids and Thrips

While it is impossible to generalize about the ecological factors that have collectively shaped sociality in aphids and thrips, the fact that they are herbivores, and the ecological consequences of life on plants, is the central organizing principle of their social lives. Plants provide resources, attract predators, facilitate the growth of pathogens, facilitate and constrain reproduction, development, population growth and dispersal (Crespi, *et al.*, 2004; Abbot, 2015).

6.4.1 Habitat and Environment

Gall forming in insects is associated with various latitudinal and ecological trends, such as the tendency to be rare in wet, tropical climates, and more common in warmer, xeric climates (Fernandes & Price, 1988). As far as host plants themselves are concerned, many gall-forming insects tend to be on plants that occur in environmentally stressful conditions (Fernandes & Price, 1988), which may be better defended against generalist herbivores that threaten gall-formers, have long-lived leaves, or occur in habitats where the adverse conditions themselves protect gall-formers from herbivores, predators and parasites.

In aphids, it is difficult to generalize about the traits of host plants because they are diverse and many species alternate and/or form social groups on unrelated hosts. Of nearly sixty social aphid species, about 65 percent form galls on either *Styrax* (Ericales: Styracaceae; sometimes known as snowbells) or *Populus* (Malpighiales: Salicaceae; commonly known as poplars or cottonwoods). Generally, *Styrax* tend to occur in the open woodland understory or forest edges, much like dogwood, whereas poplars are shade-intolerant colonizers of disturbed habitats that often occur in stands. The secondary hosts include species of bamboo, ginger, climbing vines, broad-leaf grasses, and various crucifers and composites. There has been little systematic treatment of the role of host plants in aphid sociality. Stern (1998) proposed that shifts to host plants that provide inadequate nutrition for rapid growth might coincide with the origins of soldiers in aphids, because slow growth would favor investment in defense. This hypothesis has not been fully tested.

Eusocial thrips, by contrast, have been described exclusively from Australian *Acacia* (Fabales: Fabaceae). Unlike *Styrax* or *Populus*, *Acacia* are drought-hardy and are often associated with semi-arid or arid regions, which is where the social species are found (Crespi, *et al.*, 2004). Crespi, *et al.* (2004) suggested that plant traits themselves in arid regions may favor sociality. The slower-growing plants typical of these regions may be better protected from desiccation or herbivory, and the insects on them more vulnerable as a consequence, setting the stage for various demographic and ecological drivers of sociality, such as long-lived galls and increased predation. Kranz, *et al.* (2002) point out that one species that has secondarily lost sociality occurs in the more temperate regions of Australia, implying a role of ecology in determining the expression of social traits. Overall, the ecology of sociality in aphids and thrips clearly needs greater study.

6.4.2 Biogeography

Although aphids are concentrated in the northern temperate zones, gall-forming species have been described from every continent except Antarctica, with perhaps the greatest diversity concentrated in central and southeast Asia. Social aphids, on the other hand, have been studied almost entirely in Japan, Taiwan, Europe, and North America. There is no evidence that the distribution of social aphids differs from that of non-social or communal aphids. Different aphid species can be rare or common, but the reasons are elusive. To date, the only large scale biogeographic study of social aphids is that of Stern (1998), who characterized biogeographic patterns in southeast Asia and Japan of social and non-social species in the hormaphidine tribe Cerataphidini, and identified a likely origin of social species in the Southeast Asian tropics and subtropics, and subsequent colonization of Japan by dispersal.

Like aphids, gall-forming thrips are distributed globally, and are found in tropical, temperate, and arid regions (Mound, 2005). Most records of thrips galls are from Asia and Australia, and there has been little study of thrips galls in the Neotropics. As noted earlier, social thrips on *Acacia* are found in the arid and semi-arid parts of Australia, and seasonal patterns of rainfall tightly constrains the window for gall formation, favoring survival strategies such as usurpation of conspecific or heterospecific galls, and gall defense. The non-social gall-forming *K. rodwayi*, for example, occurs in temperate regions of Australia with less seasonal patterns of rainfall, enabling more opportunities for gall induction and, consequently, reduced rates of gall usurpation by *Koptothrips* species (Crespi & Abbot, 1999).

6.4.3 Niches

Cooperative species are ecologically successful, often dominating the terrestrial environments in which they are found or thriving in harsh, seasonal environments unavailable to solitary species (Wilson, 1971; Rubenstein & Lovette, 2007). There have been relatively few studies of how or whether niche relationships differ between social aphids and thrips relative to non-social species. It is not uncommon for both eusocial aphids and thrips to co-occur on their host plants with other non-social species. Potentially, gall-forming aphids may compete indirectly through the host plant for resources (Inbar, *et al.*, 1995). However, it is unlikely that interference competition is common between social and non-social aphids and thrips, and directly related to the ecological or selective advantages or disadvantages of sociality in either lineage (Inbar, 1998). There is some evidence in aphids that social species can colonize and persist on more long-lasting leaves than non-social species (Abbot & Withgott, 2004; Abbot, 2015), a trend that is analogous to a trade-off between persistence and dispersal in social and solitary thrips (Crespi, *et al.*, 2004). As well, there is evidence in both aphids and thrips that the tendency of gall foundresses to aggressively compete for gall initiation sites may have served as important pre-adaptations for the evolution of aggressive soldiers (Whitham, 1979; Crespi, 1986). However, in general, the question in aphids and thrips is how sociality extends the gall-forming and plant-feeding niches themselves, relative to the many related gall-forming and plant-feeding species that are not social. At present, there is no study that explicitly addresses this question.

6.5 The Role of Evolutionary History in Shaping Sociality in Aphids and Thrips

While varying degrees of coordinated group and parental behaviors can be found among the aphids and thrips, the most advanced forms of sociality are exceedingly rare relative to the overall diversity in both lineages. There has been no comprehensive phylogenetic analysis of social evolution in aphids, but anecdotal evidence suggests that the evolution of altruistic soldiers in gall-forming species is associated with life history factors or plant traits that prolong the galling phases of aphid life cycles and/or slow population growth, favoring investment in defense over reproduction (Pike, *et al.*, 2007). Stern (1994, 1998) conducted the most thorough phylogenetic analysis of soldier evolution in aphids. Two results from these studies are worth highlighting. The origins of primary and secondary host soldiers has represented a mystery in aphids. Are they independent, or do they represent shared phenotypes expressed at alternative points in aphid life cycles? Stern (1994, 1998) showed that primary and secondary host soldiers likely represent independent origins of sociality in aphids. Another question is the role of host plant shifts in the origins of sociality. Do some host plants favor slow growth or heightened vulnerability that favors the evolution of defense in aphids?

The development of a molecular phylogeny for the gall-inducers in thrips (Crespi, *et al.*, 1998; Morris, *et al.*, 2002) has been instrumental in revealing the likely trajectory that sociality has taken for this group of animals. Mapping of various life history and genetic traits onto the gall-inducers' phylogeny suggests that altruistic soldiers in the gall-inducers appear to have evolved from a non-dispersing, sib-mating, reproductive morph (Chapman, *et al.*, 2000, 2002). Soldiers evolved following a host shift within *Acacia* group, accompanied by subsequent host shifts and changes in gall duration, shape and size, and the mode of colony-founding, which progressively favored high skew groups with increasingly efficient soldiers (Crespi, *et al.*, 2004; Chapman, *et al.*, 2008).

Thus, in both aphids and thrips, host shifts may have played important roles in shaping the demographic and life history traits associated with the replacement of the "hide and run" syndrome and emergence of "stay and fight" sociality. Additionally, a number of pleisiomorphic traits are thought to have been important in social evolution in both groups. These include plasticity in development, inbreeding, clonal or haplodiploid genetics, fighting for gall initiation sites by foundresses (Whitham, 1979; Foster, 1996; Crespi, *et al.*, 2004), or in the case of thrips, male fighting for mates or defense of brood by foundresses. Given the suitable ecological drivers, the transition to sociality may simply have involved the redeployment of ancestral behavioral and developmental modules in non-dispersing aphids or thrips.

II SOCIAL TRAITS

A feature that aphids and thrips share with many other species discussed in this volume is that far more is known about many non-social species than about social species. Because aphids and thrips are crop pests and important vectors of plant viruses, entomologists have intensively studied them for decades. Much of what we know about social aphids and thrips, therefore, can be placed against the backdrop of this entomological literature. One of the promising avenues for research on social aphids and thrips is a better integration with the large body of biological knowledge and tools available for the study of aphids and thrips (Crespi, *et al.*, 2004; Abbot, 2015).

6.6 Traits of Social Species

6.6.1 Cognition and Communication

Unlike most other eusocial insects, which must coordinate resource acquisition and colony development in extended family groups that may persist for years, most social aphids and thrips form comparatively short-lived groups often comprised of only two or three generations, and they live in or on their food and do not forage outside of the nest. Consequently, there is less scope for the development of complex patterns of communication. However, there has been little study of this in aphids or thrips, despite a wealth of knowledge about the chemical ecology of species in both groups. The issue is important because of the demographic consequences of soldier investment in both groups, discussed further below.

Aphids produce sex pheromones during their sexual generation, but as noted earlier, sexual reproduction is not immediately relevant to aphid sociality. Many species emit warning pheromones when threatened, and the typical response is to drop off of a plant or move away from the perceived threat. How or whether social aphids use warning pheromones in a different manner has been little investigated. It is possible that social aphids coordinate their aggressive responses to predators via pheromones. Schütze & Maschwitz (1991) showed that the horned soldiers of *Pseudoregma sundanica* (Aphididae: Hormaphidinae) respond to specific contact semiochemicals in the hemolymph of other insects, rather than to non-specific or volatile stimuli, indicating that soldiers are tuned to respond to the wounds of predators. Rhoden & Foster (2002) documented a similar pattern in *Pemphigus* aphids. In the North American species, *Pemphigus obesinymphae* (Aphididae: Eriosomatinae), soldiers respond developmentally to the presence or absence of the foundress, suggesting that the foundress supplies a developmentally suppressive cue (Withgott, *et al.*, 1997). And in the same species, Abbot, *et al.* (2001) demonstrated that soldier phenotypes can be conditionally expressed when soldiers move between galls. However, there is no evidence that aphids can discriminate between kin and non-kin (Aoki, *et al.*, 1991; Miller, 1998; Shibao, 1999). This has come as some surprise, especially given in the secondary host aphid societies, kin recognition would facilitate group cohesion in the absence of galls.

Sophisticated coordination of group activity has been described in non-social thrips, but as with aphids, far more is known about the chemical ecology of mating and host plant location than of coordination of group activity, caste polyphenism, or gall defense (Lewis, 1973; Kutsukake, *et al.*, 2004; Shibao, *et al.*, 2010). However, many species of thrips express a defensive anal secretion that deters predators. De Facci, *et al.* (2013, 2014) hypothesized that eusocial thrips use anal secretions in defense or as alarm pheromones. They described various carboxylic acids in the anal secretions, and a difference in the effect of these secretions in the response of soldiers and dispersers. In one species, dispersers and soldiers both produced this droplet similarly, but it was only the soldier-produced droplet that induced behavioral changes in both castes. The soldiers' signal led to increases in the length of their movements while the dispersers decrease their movements, potentially enhancing the soldiers' ability to defend the colony. It is not clear whether *Acacia* thrips express kin recognition. Recently, de Bruijn & Egas (2014) found evidence suggesting that non-social Western flower thrips larvae express a form of environmentally cued kin recognition. Possibly, *Acacia* thrips express similar environmentally cued recognition mechanisms. However, for both aphids and thrips, the habit of living in galls means that group cohesion would normally be maintained in the absence of recognition mechanisms.

6.6.2 Lifespan and Longevity

In eusocial insects, discussion of longevity and reproduction are complicated by the issue of whether the colony or individual is the unit of interest. Although there are exceptions, most social aphids and thrips exhibit annual or biennial life cycles, and individual lifespans are relatively short. In both aphids and thrips, the first generation may die before the dispersing generation reaches maturity (Kranz, *et al.*, 2001a,b). Extrinsic sources of mortality likely shorten the lifespan of both thrips and aphid soldiers, and gall-living may be associated with various ecological constraints that favor staying and helping rather than dispersing and reproducing independently

(Emlen, 1982). As mentioned, an important demographic correlate of sociality in both social aphids and thrips is that soldier production tends to be associated with species that have longer, slower-growing galls or colonies, and thus longer generations, particularly in soldiers (Moran, 1993a; Crespi, *et al.*, 2004; Pike, *et al.*, 2004; 2007). The functional basis of prolonged soldier development in aphids is not fully understood (Shibao, *et al.*, 2010). Demographic patterns associated with galls have been more thoroughly described in thrips. The transition from a solitary to social life history appears to include a reduction in gall volume and a consequent reduction in disperser numbers, as well as an elongated period for the development of dispersing brood (Crespi & Worobey 1998; Wills, *et al.*, 2001, 2004); that is, gall-inducing thrips lineages have evolved to adopt either of two strategies under selective pressure from *Koptothrips* invasions: (1) produce large galls and fast-developing brood in order to ameliorate losses due to *Koptothrips* exploitation; or (2) produce small galls and smaller dispersing brood numbers, but increase the survival of these galls by producing soldiers. Chapman, *et al.* (2006) proposed a mechanistic hypothesis for how *Koptothrips* select for smaller galls in social species, based on the idea that there is trade-off between soldier production and the development of galls. Social thrips are particularly vulnerable before soldiers, and there may have been selection to develop soldiers quickly at the expense of larger, longer developing galls.

6.6.3 Fecundity

Aphids are viviparous parthenogens during most of their life cycle. Fecundity of aphids is positively related to body size, and is negatively affected by density, plant quality, and other habitat-specific attributes (Dixon, 1998). Aphids have prodigious capacities for reproduction, with fast development times and telescopic generations in which offspring develop within embryonic females. Gall-forming aphids represent an exception, since the gall-dwelling females are potentially limited by within-gall density and gall size, as well as between-gall competition for plant resources. Whitham (1986) found that the foundress in gall-forming non-social species *Pemphigus betae* produces about 142 offspring over her lifetime in the absence of competition with other females. This is about two offspring per day over her lifetime, less than the daily fecundity of most of a number of non-gall-forming species (Lamb, *et al.*, 2012). The issue of fecundity is complicated in social aphids, however, since sterile soldiers represent an investment in defense, not reproduction, and in many species, soldiers are not sterile (as in thrips). Generally, investment in soldiers is thought to increase with the value of dispersing offspring (Stern & Foster, 1996).

Thrips exhibit a range of reproductive modes, including oviparity, ovoviparity, and viviparity, and exhibit wide variation in lifetime fecundity (Lewis, 1973). In gall-forming thrips, adult females lay eggs within the galls, which is rare for gall-forming insects. Some thrips can have many generations per year, but in gall-forming species, two to three generations occur within galls. Soldier thrips are sexually mature adults. However, female soldier ovaries are typically smaller with fewer, but larger eggs with chorionated oocytes compared to foundresses (Crespi, 1992b). The larger eggs of soldiers may be due to competitive, density-dependent effects associated with reduced fecundity (Kranz, 2005). The per capita fecundity of foundress females tends to be higher than that of the soldier females, and the extent of soldier reproduction ranges across species from equivalent to that of their foundress mother to the production of only a few parthenogenetically produced male offspring (Chapman, *et al.*, 2002). Because soldier defensive competence appears to vary directly with reproductive sacrifice (Perry, *et al.*, 2003, 2004), there is a trend of increased egg size relative to body size, and reduction in fecundity and brood size in the evolutionarily more derived social species relative to solitary species (Kranz, *et al.*, 2001, 2002).

6.6.4 Age at First Reproduction

As with longevity, in social aphids and thrips, the age at first reproduction is most applicable to the colony life cycle, rather than the individual life span. For both groups, seasonal events and plant phenologies determine general features of life cycle timing. For aphids, for example, reproduction by adults other than the foundress typically occurs after juveniles complete a final molt into winged adults and depart the galls for alternative host plants. However, in species with multiple generations within galls, or those that do not host alternate, reproduction can occur within the galls themselves. Regardless, reproduction is always asexual in aphid galls (there are rare exceptions in which sexual males and females are produced within galls). Likewise, in thrips, eusocial species share life cycles that are similar to those of non-social species. Two generations occur within galls, and the second generation either pupates in the soil after departing the gall, or within the gall itself (Crespi, *et al.*, 2004). From the colony's point of view, the function of sociality in aphids and thrips is the production of winged adults. Winged adults can be produced by the gall foundress, or by subsequent generations of within galls.

6.6.5 Dispersal

Herbivory requires dispersal between plants, and the life cycles of aphids and thrips are organized seasonally around colonizing their host plants. In both aphids and thrips, colonies are initiated by single reproductives. There is no analog of the swarm foundation of new colonies common in bees. To a large extent, general features of dispersal do not differ between social and non-social species in aphids and thrips. As in termites, dispersal by various winged morphs is an obligate feature of their life cycles, thereby placing some practical limits on the explanatory scope of ecological constraints in the evolution of their sociality (Roisen, 2006). Nevertheless, as described earlier, there is evidence in both of some temporal delay in dispersal from groups of social relative non-social species. In aphids, is likely due to the benefits of accumulating a large number of soldiers during the early phases of colony development (Stern & Foster 1996). In thrips, ecological constraints that favors staying and helping over dispersal may play a more important role (Crespi, *et al.*, 2004).

At a proximate level, both are weak flyers. But because of their importance as crop pests, we know a good bit about dispersal mechanisms, host plant location, and resource tracking. In host-alternating gall-forming aphids, dispersal biology can be exceedingly complex (Moran, 1992; Dixon, 1998; note that some species do not alternate between hosts, and therefore their dispersal biology differs). In host-alternating species, winged parthenogenetic adult females emerge from galls and fly to secondary hosts, where they initiate several rounds of reproduction. In most species, dispersal distances probably range from hundreds of meters to several kilometers (Miller, *et al.*, 2005).

As weak fliers, colonizing thrips are easily buffeted by the wind, and exhibit uncontrolled flight except near their host plants or ground (Lewis, 1973). Weak flying in thrips results in virgin females sometimes inducing galls, in which case a foundress will lay unfertilized, male eggs first. Because the foundress dies before the last generation of dispersers are produced, there is a change over the course of gall development in the relative production of soldier and disperser classes. Initially, the foundress produces all soldiers and the first round of dispersers. Soldiers then reproduce and produce both soldiers and dispersers. In some of the more derived social species, there is an increase in the disparity between foundresses and other soldier females in both per capita fecundity and total contribution to the dispersing class (i.e. higher reproductive skew, Kranz, *et al.*, 1999, 2001c). McLeish, *et al.* (2003) studied dispersal patterns in the social thrips *K. intermedius* (formerly called *Oncothrips tepperi*) and found dispersal distances in excess of one kilometer. This suggests that there is little evidence of local population structure in social thrips, and consistent with thrips in general.

6.6.6 Other Traits: Body Size

Body size is a strong predictor of various ecological and life history traits in aphids, including population growth rate, developmental rate, fecundity, survival and dispersal distances (relatively smaller individuals being better dispersers; Dixon, 1998). Both aphids and thrips are minute by nearly any measure. Kranz, *et al.* (2000) point out that small body size makes for ephemeral adult lives, in which case clonal reproduction, haplodiploidy and inbreeding are advantageous. Thus, the genetics favorable to social evolution in both lineages may have something to do with their extremely small body size.

In some social aphids, soldiers tend to be larger than non-soldiers (Hattori & Itino, 2008), but outside of taxonomic descriptions, there are no comparative studies of body size variation across social and non-social species. In gall-forming aphids generally, larger first instar foundresses are more successful in competition for galling sites than smaller foundresses (Whitham, 1986; Inbar, 1998). Competitive success and fecundity of gall-forming aphids and their various reproductive morphs are likely related to body size (Foster & Benton 1992; Akimoto & Yamaguchi, 1997). Gall size is a strong predictor of clone size and the number of emerging alates. In *Pemphigus* aphids living on poplars, larger leaves are associated with larger galls. In the social aphid *Pemphigus spyrothecae*, Alton (1999) showed that greater leaf area predicts larger galls, larger adult size of gall foundresses, and larger groups. Body size of *P. spyrothecae* foundresses declines over the season, likely as a result of reproductive effort and an overall decline in resource availability as the leaf and gall senesce. The number of galls on a petiole or leaf can negatively affect gall size and thus various life history traits, yet Alton (1999) found no effect of such competition on foundress body size. However, in the communal species *Tamalia coweni* (Aphididae: Aphidinae), Miller (1998) found that foundress size decreases with the number of co-occupying foundresses within galls.

In thrips, there is a more nuanced relationship between body size and fecundity, but little is known about body size patterns in eusocial, gall-forming species. In the western flower thrips, which is a non-social species in the Terebrantia, there is no clear relationship between adult female body size and fecundity (de Kogel, *et al.*, 1999). However, in *Elaphrothrips tuberculatus*, a tubuliferan thrips, female fecundity is positively correlated with adult body size. Males tend to be larger than females in the Tubulifera (Lewis, 1973). In species in which males compete for access to females or guard eggs, larger male body size leads to greater success in bouts, and may promote the evolution of alternative reproductive tactics among males (Crespi, 1988).

6.7 Traits of Social Groups

6.7.1 Genetic Structure

The genetic systems of aphids and thrips are favorable to the evolution of cooperation: aphids reproduce clonally during most of the year, thrips are haplodiploid, like ants and other eusocial Hymenoptera, and inbreeding is common in the gall-formers (Chapman & Crespi, 1998; Chapman, *et al.*, 2000). Thus, in both lineages, when family groups do form, relatedness among group members is not only high, but higher than in most other lineages of social animals. However, more is known about genetic structure of thrips societies than of aphids. Moreover, the haplodiploid genetic system of thrips generates more potential for complexity and conflict over reproduction than does the clonal system of the aphids.

In aphids, parthenogenesis and all-female societies preclude conflict over reproduction. Rather, the only source of relatedness asymmetries within aphid is if individual move between conspecific or heterospecific groups. By and large, relatedness has not been formally estimated in most species. However, it is almost certainly the case that the groups of gall-forming aphids are either entirely clonal or characterized by high relatedness. The galls of many aphids are completely sealed for most of the season, precluding admixture between groups. For species with galls that are open to the outside, and for species that are social on the secondary host, admixture can occur between groups (Akimoto, 1989; Abbot, *et al.*, 2001). Abbot (2009) used molecular markers to estimate clonal variation in the galls of various social and non-social North American species, and found that multiclonal groups can be common. In social aphids, if invading individuals do not contribute to defense, then they are unrelated cheaters, which are known to occur in at least one species (Abbot, *et al.*, 2001; Abbot & Chhatre, 2007).

In thrips, singly mated foundresses tend to be highly inbred. Chapman, *et al.* (2000, 2002) showed that high levels of inbreeding were common at the origins of *Kladothrips*, likely setting the stage for the evolution of sociality in the genus. Moreover, sib-mating by soldiers creates unusual patterns of relatedness that can favor helping over reproduction. However, not all species of eusocial thrips are inbred. In some species, foundresses are multiply mated and outbred, and more derived social species exhibit lower rates of inbreeding and within-gall relatedness. Possibly, reduction in inbreeding coefficients were necessary for the evolution of the more extreme cases of altruism in more derived species since high inbreeding reduces relatedness asymmetries that govern the trade-off between helping and dispersal (Chapman, *et al.*, 2008).

6.7.2 Group Structure, Breeding Structure and Sex Ratio

Group size can vary enormously in social aphids. The galls of *Asteropteryx styracicola* (Hormaphidinae) may contain as many as 200,000 aphids (Aoki, *et al.*, 1977), although tens or hundreds is more common in most species. As described earlier, all social aphid groups are entirely female, and the production of males and females during the sexual phase tends to be female-biased. While reproductive skew, breeding structure, and sex ratio are not directly relevant to the single foundress, all-female, clonal groups in social aphids, these factors do determine patterns of genetic variation in social aphid populations. Aphids characteristically produce extreme sex ratios (Moran, 1993b; Foster, 2009). Males and females are produced parthenogenetically, and sex is not determined during meiosis. Rather, males are XX/XO. The extent to which inbreeding occurs depends heavily upon whether males and females are winged. In the two subfamilies in which eusocial aphids occur, males and females are wingless, sib-mating is common, and sex ratios are strongly female-biased (Moran, 1993b).

The regulation of the different morphs within aphid social groups is not well understood, but apparently can be influenced by a number of environmental cues, including crowding or colony size, tactile and possibly non-volatile stimulation, season or colony age, host plant characteristics, the presence of predators, and the presence of tending ants (Stern, 1994; Shingleton & Foster, 2000, 2001; Shibao, *et al.*, 2003, 2004a,b, 2010; Ijichi, *et al.*, 2005). Thus, across species, group structure is likely sensitive to local environmental and seasonal conditions, and may vary considerably within species. Larger patterns in life cycle variation across species are important as well, such as whether or not aphids alternate between host plants, the structure of galls, and whether the social group occurs on the primary or secondary host (Aoki & Kurosu, 2010).

Sexual reproduction and haplodiploidy facilitate complex breeding structures and sex ratio variation in eusocial thrips. As noted previously, the species with the most efficacious soldiers appear also tend to be the least reproductive. Most eusocial thrips have female-biased sex ratios in the dispersing morph, but there is considerable variation across species in soldier sex ratios (Crespi, *et al.*, 2004). Species such as *K. hamiltoni* produce female-biased disperser morphs, but unbiased soldier morphs (Kranz, *et al.*, 1999, 2001a). *K. hamiltoni* also produces small galls, and resource competition between soldiers may underlie even sex ratios in this species. The female-biased disperser sex ratios are probably associated with inbreeding and local mate competition (Chapman & Crespi, 1998; Chapman, *et al.*, 2000). Indeed, where outbreeding occurs, disperser sex ratios are unbiased (Chapman, *et al.*, 2002). Females eclose first in most eusocial thrips, but in some species, the opposite is true (Kranz, *et al.*, 1999; Wills, *et al.*, 2004). Protandry may be an adaptation associated with ensuring reproduction by short-lived female soldiers. However, sib-mating of this sort has the unusual effect of elevating the relatedness of the foundress to her granddaughters, such that she is more related to them than to her own daughters (Kranz, *et al.*, 1999). This combination of reproduction assurance and relatedness may favor a protandrous strategy in some species. Finally, there is a trend in thrips for an increase in reproductive skew as gall volume decreases (Chapman, *et al.*, 2002). It is possible that smaller galls are associated with smaller soldier numbers and higher soldier efficiency, both of which may favor foundress reproduction over that of soldiers (Perry, *et al.*, 2004).

III SOCIAL SYNTHESIS

6.8 A Summary of Aphid and Thrips Sociality

We have described general features of the eusocial aphids and thrips, and although there are aspects of their social biology that are remarkably similar, their differences are important as well (Table 6.1). Some of the evolutionary, ecological, and social traits that eusocial aphids and thrips share include the following: (1) Both are in tremendously diverse orders of hemimetabolous insects, within which they are taxonomically rare and closely related to non-social species and which they broadly overlap ecologically and biogeographically. The significance of this point is that both lineages are unusually promising for comparative studies (Chapman, *et al.*, 2008), but also that, from an ecological perspective, the distinction between eusocial species and other solitary or group living species can be subtle; (2) In both lineages, social species are plant-feeding gall-formers; (3) Sociality is often associated with long gall durations, longer lived colonies, and trade-offs between investment in defense and reproduction; (4) Their diagnostic eusocial feature is the presence of soldier morphs; (5) Natural enemies are the primary selective agent driving social evolution; (6) They do not provide direct brood care; (7) By virtue of their gall-forming habit, they occur in highly related family groups; (8) Traits derived from life on plants (e.g. intraspecific aggression over gall initiation sites) may have served as important pre-adaptations for eusocial traits; (9) Eusociality has been lost in some lineages.

The similarities eusocial aphids and thrips share are counterbalanced by a number of important differences between the two taxonomic groups. These include: (1) Aphid groups are always initiated by a single, asexually reproducing foundress, whereas thrips groups are founded by a sexual haplodiploid female, sometimes accompanied by a male. Consequently, questions regarding the role of reproductive skew are not relevant to aphids, whereas there is evidence of increased skew in the more derived eusocial thrips; (2) Aphid soldiers are always juvenile, and in some species may be sterile, whereas thrips soldiers are fertile adult females and males, a difference that has implications for differences in their social complexity and the scope for within-group conflict between the two lineages; (3) Eusocial aphids are found on diverse plant genera worldwide, whereas eusocial thrips are more biogeographically restricted, occurring only on Australian *Acacia*; (4) Eusocial aphid life cycles often include extensive "non-galling" phases on secondary host plants on which they do not form galls, and where they may or may not form soldiers, whereas the life cycles of eusocial thrips, encompassing rounds of growth within galls followed by dispersal and reproduction, are more analogous to those of other social insects and cooperatively breeding vertebrates. This has implications for the ability to develop single, comprehensive explanations for eusociality that encompass both lineages: the added complexity of aphid life cycles, and especially the presence of soldiers on plants in the absence of galls has no parallel in thrips or most other social insects and vertebrates; (5) There have been multiple origins of soldiers in aphids, but only a single origin in thrips; (6) Although natural enemies have favored defense in both lineages, in eusocial thrips, exploitation by specialized thrips has been a prime agent favoring the evolution of soldiers. There is nothing as broadly comparable in aphids; (7) Finally, although not directly related to their biology, there are distinct differences in how the research programs on eusocial aphids and thrips have developed since their initial discoveries. These differences have implications for the extent to which inferences can confidently be drawn when comparing the two. Phylogenetic and comparative studies of social evolution in thrips have been extensive, and to a far greater degree than in aphids. By contrast, most descriptions of eusociality in aphids – and less so in thrips – have been accompanied by detailed behavioral descriptions, and in some species, sophisticated molecular methods have been brought to bear on mechanistic and developmental aspects of their biology.

Thus, we know more about the evolution of sociality in thrips than in aphids, and more about the functional and behavioral biology of sociality in aphids than in thrips. And in both groups, we know comparatively little about the ecological milieu out of which eusociality has evolved; there is need for detailed comparative ecological studies of species with and without soldiers. Overall, the shared features of eusocial aphids and thrips invite broad comparisons that reveal how independent transitions to sociality converged on a set of shared features. And while the differences between the two are informative, they also place cautionary limits on the confidence that can be derived when comparing social evolution in aphids and thrips.

Finally, a general issue is what exactly is sociality in aphids and thrips. Both aphids and thrips have morphologically distinct castes that fight invertebrate and microbial invaders. Even though this is self-sacrificial, soldiers in most species of aphids and thrips have the potential to reproduce. This observation has led some to categorize thrips colonies as subsocial or facultative eusocial (Boomsma, *et al.*, 2014). However, reproductive division of labor occurs in both lineages, whether or not soldiers are sterile. Furthermore, the potential of thrips soldiers for reproduction is a characteristic shared with many worker castes within the Hymenoptera, where there is little or no debate regarding their inclusion as examples of eusociality (Crespi & Yanega, 1995). Thus, we think that the fortress defense species like aphids and thrips represent, if not classically defined eusocial organisms, forms of eusociality that broaden the scope of the field.

6.9 Comparative Perspectives on Aphid and Thrips Sociality

As fortress defenders, eusocial aphids and thrips share with other insects and some other invertebrates and vertebrates a mode of social evolution in which a form of sociality evolves involving: (1) reproduction within a valuable, limiting, and defensible resource; (2) the absence of direct brood care or foraging for food beyond the nest; and (3) the production of a soldier caste early in colony development. Features of eusociality in aphids and thrips thus bear remarkable similarities to the likes of wood-dwelling termites (Chapter 5), snapping shrimps (Chapter 8), and naked mole-rats (Chapter 10) (Queller & Strassmann, 1998; Chapman, *et al.*, 2008; Chapters 5 & 8). Broad empirical evidence of a distinct fortress defense mode of social evolution represents one of the successes in the modern study of social evolution, and a validation of the predictive power of social evolutionary theory (Hamilton, 1964, 1972; Wilson, 1971; Alexander, *et al.*, 1991; Strassmann & Queller, 2010). It is an encouraging fact for researchers new to social evolution that, given that this mode of social evolution occurs in species typically dwelling in confined and difficult to observe spaces, it is likely that the full diversity of fortress defense social species has not yet been described.

That said, the fortress defense model of social evolution remains in need of further conceptual development. Ecological constraints and limitations on dispersal and independent reproduction form core conceptual motifs in our understanding of social evolution in both primitively eusocial insects (i.e. wasps) and cooperatively breeding birds and mammals (Emlen, 1982). At the moment, how and whether ecological constraints shape social evolution in fortress defense social taxa are not clear. In the herbivorous aphids and thrips, whether constraints are markedly different between social and non-social taxa has not yet been explored. Because social and non-social species often share similar life cycles, or even share host plants, careful study will be needed to tease apart how ecological factors determine transitions to sociality. Moreover, compared to thrips, eusocial aphids may sit more comfortably in the special category of fortress defense sociality. The soldier caste is expressed in the juveniles alone, implying they evolved purely in the context of defense, whereas in thrips, the possibility that soldiers evolved in the context of redirecting caring for offspring to caring for conspecifics suggests a commonality with Hymenoptera not shared by aphids (but see Inbar, 1998). It may well be that in other fortress defense species, parental and alloparental traits have contributed to the ground plan out of which traits for defense have evolved. If so, this may be an unappreciated commonality in the social evolution of diverse species, and the categories that have been usefully erected to organize this diversity may not be as distinct as they may appear, as others in this volume make plain.

6.10 Concluding Remarks

Research on aphids or thrips shares the goal of sociobiology broadly: to understand the unifying principles that guide the development of altruistic behavior itself (Strassmann & Queller, 2010). Years of study suggest that Alexander, *et al.* (1991) and Emlen (1982) captured the general features of the ecological drivers of social behavior. Constraints and vulnerability of sedentary family groups on plants, coupled with perhaps precursors that favored the elaboration of social phenotypes. Aphids are poised for broader, cladistic scale studies of species that differ in solitary and social traits, similar to what has been done in the thrips (Chapman, *et al.*, 2008), and thrips are poised for broader behavioral and functional work. There is cause for renewed enthusiasm for comparative evolutionary studies of social behavior in these diverse but relatively poorly studied groups.

References

Abbot, P. (2009) On the evolution of dispersal and altruism in aphids. *Evolution*, **63**, 2687–2696.

(2011) A closer look at the spatial architecture of aphid clones. *Molecular Ecology*, **20**, 4587–4589.

(2015) The physiology and genomics of social transitions in aphids. In: Zayed, A. & Kent, C.F. (eds.) *Genomics, Physiology and Behavior of Social Insects. Advances in Insect Physiology*, Volume 48. New York: Academic Press, pp. 163–188.

Abbot, P. & Chhatre, V. (2007) Kin structure provides no explanation for intruders in social aphids. *Molecular Ecology*, **16**, 3659–3670.

Abbot, P. & Withgott, J. H. (2004) Phylogenetic and molecular evidence for allochronic speciation in gall-forming aphids (*Pemphigus*). *Evolution*, **58**, 539–553.

Abbot, P., Withgott, J. H., & Moran, N. A. (2001) Genetic conflict and conditional altruism in social aphid colonies. *Proceedings of the National Academy of Sciences of the United States of America*, **98**, 12068–12071.

Akimoto, S. (1989) Gall-invading behavior of *Eriosoma* aphids (Homoptera, Pemphigidae) and its significance. *Japanese Journal of Entomology*, **57**, 210–220.

(1996) Ecological factors promoting the evolution of colony defense in aphids: Computer simulations. *Insectes Sociaux*, **43**, 1–15.

Akimoto, S., & Yamaguchi, Y. (1997) Gall usurpation by the gall-forming aphid, *Tetraneura sorini* (Insecta Homoptera). *Ethology Ecology and Evolution*, **9**, 159–168.

Alexander, R. D., Noonan, K. M., & Crespi, B. J. (1991) The evolution of eusociality. In: Sherman, P.W., Jarvis, J. U. M., & Alexander, R.D. (eds.) *The Biology of the Naked Mole Rat*. Princeton: Princeton University Press, pp. 3–44.

Alton, K. (1999) *The biology of Pemphigus spyrothecae galls on poplar leaves*. PhD thesis, Nottingham: University of Nottingham.

Aoki, S. (1976) Occurrence of dimorphism in the first instar larva of *Colophina clematis* (Homoptera, Aphidoidea) *Kontyû*, **44**, 130–137.

(1979) Dimorphic first instar larvae produced by the fundatrix of *Pachypappa marsupialis* (Homoptera: Aphidoidea) *Kontyû*, **47**, 390–398.

Aoki, S. & Kurosu, U. (2004) How many soldiers are optimal for an aphid colony? *Journal of Theoretical Biology*, **230**, 313–317.

Aoki, S. & Kurosu, U. (2010) A review of the biology of Cerataphidini (Hemiptera, Aphididae, Hormaphidinae), focusing mainly on their life cycles, gall formation, and soldiers. *Psyche: A Journal of Entomology*, **2010**, 1–34.

Aoki, S., Yamane, S., & Kiuchi, M. (1977) On the biters of *Astegopteryx styracicola* (Homoptera, Aphidoidea). *Kontyû*, **45**, 563–570.

Aoki, S., Kurosu, U., & Stern, D. L. (1991) Aphid soldiers discriminate between soldiers and non-soldiers, rather than between kin and non-kin, in *Ceratoglyphina bambusae*. *Animal Behaviour*, **42**, 865–866.

Aoki, S., Kurosu, U., & Buranapanichpan, S. (2007) Female production within the gall and male production on leaves by individual alates of a social aphid. *Insectes Sociaux*, **54**, 356–362.

Benton, T. G. & Foster, W. A. (1992) Altruistic housekeeping in a social aphid. *Proceedings of the Royal Society of London. Series B: Biological Sciences*, **247**, 199–202.

Blackman, R. L. & Eastop, V. F. (1994) *Aphids on the World's Trees: An Identification and Information Guide*. London: University Press.

Boomsma, J. J., Huszár, D. B., & Pedersen, J. S. (2014) The evolution of multiqueen breeding in eusocial lineages with permanent physically differentiated castes. *Animal Behaviour*, **92**, 241–252.

Bono, J. M. (2007) Patterns of kleptoparasitism and inquilinism in social and non-social *Dunatothrips* on Australian Acacia. *Ecological Entomology*, **32**, 411–418.

Bono, J. M. & Crespi, B. J. (2006) Costs and benefits of joint colony founding in Australian *Acacia* thrips. *Insectes Sociaux*, **53**, 489–495.

(2008) Cofoundress relatedness and group productivity in colonies of social *Dunatothrips* (Insecta: Thysanoptera) on Australian Acacia. *Behavioral Ecology and Sociobiology*, **62**, 1489–1498.

Cappuccino, N. (1987) Comparative population dynamics of two goldenrod aphids: Spatial patterns and temporal constancy. *Ecology*, **68**, 1634–1646.

Chapman, T. W. & Crespi, B. J. (1998) High relatedness and inbreeding in two species of haplodiploid eusocial thrips (Insecta: Thysanoptera) revealed by microsatellite analysis. *Behavioral Ecology and Sociobiology*, **43**, 301–306.

Chapman, T. W. & Perry, S. P. (2006) Evolution of fighting ability in soldiers of Australian gall thrips. *In:* Kipyatkov, V. (ed.) *Life Cycles in Social Insects: Behaviour, Ecology and Evolution*. St. Petersburg: St. Petersburg University Press, pp. 113–120.

Chapman, T. W., Crespi, B. J., Kranz, B. D., & Schwarz, M. P. (2000) High relatedness and inbreeding at the origin of eusociality in gall-inducing thrips. *Proceedings of the National Academy of Sciences of the United States of America*, **97**, 1648–1650.

Chapman, T. W., Kranz, B. D., Bejah, K. L., & Crespi, B. J. (2002) The evolution of soldier reproduction in social thrips. *Behavioral Ecology*, **13**, 519–525.

Chapman, T. W., Geyer, K. F., & Schwarz, M. P. (2006) The impact of kleptoparasitic invasions on the evolution of gall-size in social and solitary Australian *Acacia* thrips. *Insect Science*, **13**, 391–400.

Chapman, T. W., Crespi, B. J., & Perry, S. P. (2008) The evolutionary ecology of eusociality in Australian gall thrips: A "model clades" approach. *In:* Korb, J. & Heinze, J. (eds.) *Ecology of Social Evolution*. Berlin: Springer-Verlag, pp. 57–83.

Childers, C. C., Beshear, R. J., Frantz, G., & Nelms, M. (2005) A review of thrips species biting man including records in Florida and Georgia between 1986–1997. *Florida Entomologist*, **88**, 447–451.

Choe, J. C. & Crespi, B. J. (1997) *The Evolution of Social Behavior in Insects and Arachnids*. Cambridge: Cambridge University Press.

Clutton-Brock, T. (2009) Cooperation between non-kin in animal societies. *Nature*, **462**, 51–57.

Costa, J. T. (2006) *The Other Social Insect Societies*. Cambridge, MA: Harvard University Press.

Costa, J. T. & Fitzgerald, T. D. (2005) Social terminology revisited: Where are we ten years later? *Annales Zoologici Fennici*, **2**, 559–564.

Crespi, B. J. (1986) Territoriality & fighting in a colonial thrips, *Hoplothrips pedicularius*, and sexual dimorphism in Thysanoptera. *Ecological Entomology*, **11**, 119–130.

(1988) Alternative male mating tactics in a thrips: Effects of sex ratio variation and body size. *American Midland Naturalist*, **119**, 83–92.

(1992a) Eusociality in Australian gall thrips. *Nature*, **359**, 724–726.

(1992b) Behavioral ecology of Australian gall thrips (Insecta, Thysanoptera). *Journal of Natural History*, **26**, 769–809.

(1994) Three conditions for the evolution of eusociality: are they sufficient? *Insectes Sociaux*, **41**, 395–400.

(1996) Comparative analysis of the origins and losses of eusociality: Causal mosaics and historical uniqueness. In: Martins, E. (ed.) *Phylogenies and the Comparative Method in Animal Behavior*. Oxford: Oxford University Press, pp. 253–287.

Crespi, B. J. & Abbot, P. (1999) The behavioral ecology and evolution of kleptoparasitism in Australian gall thrips. *The Florida Entomologist*, **82**, 147.

Crespi, B. J. & Worobey, M. (1998) Comparative analysis of gall morphology in Australian gall thrips: The evolution of extended phenotypes. *Evolution*, **52**, 1686.

Crespi, B. J. & Yanega, D. (1995) The definition of eusociality. *Behavioral Ecology*, **6**: 109–115.

Crespi, B. J., Carmean, D. A., & Chapman, T. W. (1997) Ecology and evolution of galling thrips and their allies. *Annual Review Of Entomology*, **42**, 51–71.

Crespi, B. J., Carmean, D. A., Mound, L. A., & Worobey, M. (1998) Phylogenetics of social behavior in Australian gall-forming thrips: Evidence from mitochondrial DNA sequence, adult morphology and behavior, and gall morphology. *Molecular Phylogenetics and Evolution*, **9**, 163–180.

Crespi, B. J., Morris, D. C., & Mound, L. A. (2004) *Evolution of ecological and behavioral diversity: Australian Acacia thrips as model organisms*. CSIRO, Canberra: Australian Biological Resources Study & Australian National Insect Collection.

de Bruijn, P. J. A. & Egas, M. (2014) Effects of kinship or familiarity? Small thrips larvae experience lower predation risk only in groups of mixed-size siblings. *Behavioral Ecology and Sociobiology*, **68**, 1029–1035.

De Facci, M., Svensson, G. P., Chapman, T. W., & Anderbrant, O. (2013) Evidence for caste differences in anal droplet alarm pheromone production and responses in the eusocial thrips Kladothrips intermedius. *Ethology*, **119**, 1118–1125.

De Facci, M., Wang, H-L., Yuvaraj, J. K., et al. (2014) Chemical composition of anal droplets of the eusocial gall-inducing thrips *Kladothrips intermedius*. *Chemoecology*, **24**, 85–94.

de Kogel, W. J., Bosco, D., Van der Hoek, M., & Mollema, C. (1999) Effect of host plant on body size of *Frankliniella occidentalis* (Thysanoptera: Thripidae) and its correlation with reproductive capacity. *European Journal of Entomology*, **96**, 365–368.

Dixon, A. F. G. (1998) *Aphid Ecology*. 2nd Edn. London: Chapman & Hall.

Emlen, S. T. (1982) The evolution of helping. I. An ecological constraints model. *American Naturalist*, **119**, 29–39.

Fernandes, G. W. & Price, P. W. (1988) Biogeographical gradients in galling species richness. *Oecologia*, **76**, 161–167.

Foster, W. A. (1990) Experimental evidence for effective and altruistic colony defense against natural predators by soldiers of the gall-forming aphid *Pemphigus spyrothecae* (Hemiptera: Pemphigidae). *Behavioral Ecology and Sociobiology* **27**, 421–430.

(1996) Duelling aphids: Intraspecific fighting *in Astegopteryx minuta* (Homoptera: Hormaphididae). *Animal Behavior*, **51**, 645–655.

(2009) Aphid sex ratios. In: Hardy, I.C.W. (ed.) *Sex Ratios Concepts and Research Methods*. Cambridge: Cambridge University Press, pp. 254–265.

Foster, W. A. & Benton, T. G. (1992) Sex ratio, local mate competition and mating behavior in the aphid *Pemphigus spyrothecae*. *Behavioral Ecology and Sociobiology*, **30**, 297–307.

Foster, W. A., & Northcott, N. A. (1994) Galls and the evolution of social behavior in aphids. In: Williams, M.A.J. (ed.) *Plant Galls: Organisms, Interactions, Populations*. Oxford: Oxford University Press, pp. 161–182.

Foster, W. A. & Rhoden, P. (1998) Soldiers effectively defend aphid colonies against predators in the field. *Animal Behaviour*, **55**, 761–765.

Gilbert, J. D. J. (2014) Thrips domiciles protect larvae from desiccation in an arid environment. *Behavioral Ecology*, **25**, 1338–1346.

Gilbert, J. D. J. & Mound, L. A. (2012) Biology of a new species of socially parasitic thrips (Thysanoptera: Phlaeothripidae) inside *Dunatothrips* nests, with evolutionary implications for inquilinism in thrips. *Biological Journal of the Linnean Society*, **107**, 112–122.

Gilbert, J. D. J. & Simpson, S. J. (2013) Natural history and behavior of *Dunatothrips aneurae* Mound (Thysanoptera: Phlaeothripidae), a phyllode-gluing thrips with facultative pleometrosis. *Biological Journal of the Linnean Society*, **109**, 802–816.

Gonsalves, G. (2010) *Host exploitation and fidelity in* Acacia *gall-invading parasites*. Masters thesis, Memorial University of Newfoundland. Newfoundland: St. Johns.

Grimaldi, D. & Engels, M. S. (2005) *Evolution of the Insects*. New York: Cambridge University Press, pp. 755.

Hamilton, W. D. (1964) The genetical evolution of social behavior. *II. Journal of Theoretical Biology* **7**, 1–52.

(1972) Altruism and related phenomena, mainly in social insects. *Annual Review of Ecology and Evolution*, **3**, 193–232.

(1996) *Narrow Roads of Gene Land*, Vol. 1. Oxford: Oxford University Press.

Hattori, M. & Itino, T. (2008) Soldiers' armature changes seasonally and locally in an eusocial aphid (Homoptera: Aphididae). *Sociobiology*, **52**, 429–436.

Hattori, M., Kishida, O., & Itino, T. (2012) Buying time for colony mates: The anti-predatory function of soldiers in the eusocial aphid *Ceratovacuna japonica* (Homoptera, Hormaphidinae). *Insectes Sociaux*, **60**, 15–21.

Ijichi, N., Shibao, H., Miura, T., Matsumoto. T., & Fukatsu, T. (2005) Analysis of natural colonies of a social aphid *Colophina arma*: Population dynamics, reproductive schedule, and survey for ecological correlates with soldier production. *Applied Entomology and Zoology*, **40**, 239–245.

Inbar, M. (1998) Competition, territoriality & maternal defense in a gall-forming aphid. *Ethology Ecology and Evolution*, **10**, 159–170.

Inbar, M., Eshel, A., & Wool, D. (1995) Interspecific competition among phloem-feeding insects mediated by induced host-plant sinks. *Ecology*, **76**, 1506–1515.

Jaquiery, J., Stoeckel, S., Rispe, C., Mieuzet, L., Legeai, F., & Simon, J. C. (2012) Accelerated evolution of sex chromosomes in aphids, an X0 system. *Molecular Biology and Evolution*, **29**, 837–847.

Jedličková, V., Jedlička, P., & Lee, H.-J. (2015) Characterization and expression analysis of adipokinetic hormone and its receptor in eusocial aphid *Pseudoregma bambucicola*. *General and Comparative Endocrinology*, **223**, 1–9.

Kiester, A. R., & Strates, E. (1984) Social behavior in a thrips from Panama. *Journal of Natural History*, **18**, 303–314.

Kranz, B. D. (2005) Egg size and reproductive allocation in eusocial thrips. *Behavioral Ecology*, **16**, 779–787.

Kranz, B. D., Schwarz, M. P., Mound, L. A., & Crespi, B. J. (1999) Social biology and sex ratios of the eusocial gall-inducing thrips *Kladothrips hamiltoni*. *Ecological Entomology*, **24**, 432–442.

Kranz, B. D., Schwarz, M. P., Giles, L. C., & Crespi, B. J. (2000) Split sex ratios and virginity in a gall-inducing thrips. *Journal of Evolutionary Biology*, **13**, 700–706.

Kranz, B. D., Schwarz, M. P., Mound, L. A., & Crespi, B. J. (2001a) Social biology and sex ratios of the eusocial gall-inducing thrips *Kladothrips hamiltoni*. *Ecological Entomology*, **24**, 432–442.

Kranz, B. D., Chapman, T. W., Crespi, B. J., & Schwarz, M. P. (2001b) Social biology and sex ratios in the gall-inducing thrips, *Oncothrips waterhousei* and *Oncothrips habrus*. *Insectes Sociaux*, **48**, 315–323.

Kranz, B. D., Schwarz, M. P., Wills, T., Chapman, T. W., Morris, D. C., & Crespi, B. J. (2001c) A fully reproductive fighting morph in a soldier clade of gall-inducing thrips (*Oncothrips morrisi*). *Behavioral Ecology and Sociobiology*, **50**, 151–161.

Kranz, B. D., Schwarz, M. P., Morris, D. C., & Crespi, B. J. (2002) Life history of *Kladothrips ellobus* and *Oncothrips rodwayi*: Insight into the origin and loss of soldiers in gall-inducing thrips. *Ecological Entomology*, **27**, 49–57.

Kurosu, U. & Aoki, S. (2009) Extremely long-closed galls of a social aphid. *Psyche: A Journal of Entomology*, 2009, 1–9.

Kurosu, U., Buranapanichpan, S., & Aoki, S. (2006) *Astegopteryx spinocephala* (Hemiptera: Aphididae), a new aphid species producing sterile soldiers that guard eggs laid in their gall. *Entomological Science*, **9**, 181–190.

Kutsukake, M., Shibao, H., Nikoh, N., et al. (2004) Venomous protease of aphid soldier for colony defense. *Proceedings of the National Academy of Sciences of the United States of America*, **101**, 11338–11343.

Kutsukake, M., Shibao, H., Uematsu, K., & Fukatsu, T. (2009) Scab formation and wound healing of plant tissue by soldier aphid. *Proceedings of the Royal Society of London. Series B: Biological Sciences*, **276**, 1555–1563.

Lamb, R. J., MacKay, P. A., & Migui, S. M. (2012) Measuring the performance of aphids: Fecundity versus biomass. *The Canadian Entomologist*, **141**, 401–405.

Lawson, S. P., Legan, A. W., Graham, C., & Abbot, P. (2014a) Comparative phenotyping across a social transition in aphids. *Animal Behaviour*, **96**, 117–125.

Lawson, S. P., Christian, N., and Abbot, P. (2014b) Comparative analysis of the biodiversity of fungal endophytes in insect-induced galls and surrounding foliar tissue. *Fungal Diversity*, **66**, 89–97.

Lewis, T. (1973) *Thrips, Their Biology, Ecology and Economic Importance*. New York: Academic Press.

McLeish, M. J., Perry, S. P., Gruber, D., & Chapman, T. W. (2003) Dispersal patterns of an Australian gall-forming thrips and its host tree (*Oncothrips tepperi* and *Acacia oswaldii*). *Ecological Entomology*, **28**, 243–246.

McLeish, M. J., Chapman, T. W., & Crespi, B. J. (2006) Inbreeding ancestors: The role of sibmating in the social evolution of gall thrips. *Journal of Heredity*, **97**, 31–38.

McLeish, M. J., Crespi, B. J., Chapman, T. W., & Schwarz, M. P. (2007) Parallel diversification of Australian gall-thrips on Acacia. *Molecular Phylogenetics and Evolution*, **43**, 714–725.

Michener, C. D. (1969) Comparative social behavior of bees. *Annual Review Of Entomology*, **14**, 299–342.

Miller, D. G. III (1998) Consequences of communal gall occupation and a test for kin discrimination in the aphid *Tamalia coweni* (Cockerell) (Homoptera: Aphididae). *Behavioral Ecology and Sociobiology*, **43**, 95–103.

(2005) Ecology and radiation of galling aphids (*Tamalia*; Hemiptera: Aphididae) on their host plants (Ericaceae). *Basic and Applied Ecology*, **6**, 463–469.

Miller, N. J., Kift, N. B., & Tatchell, G. M. (2005) Host-associated populations in the lettuce root aphid, *Pemphigus bursarius* (L.). *Heredity*, **94**, 556–564.

Minaei, K. (2014) New record of predatory thrips, *Aeolothrips melaleucus* (Thysanoptera, Aeolothripidae). *Linzer Biologische Beitraege*, 46, 637–642.

Moran, N. A. (1992) The evolution of aphid life cycles. *Annual Review of Entomology*, **37**, 321–348.

(1993a) Defenders in the North American aphid *Pemphigus obesinymphae*. *Insectes Sociaux*, **40**, 391–402.

(1993b) Evolution of sex ratio variation in aphids. In: D. L. Wrensch & M. A. Ebbert (eds.) *Evolution and Diversity of Sex Ratio in Insects and Mites*. New York: Chapman and Hall, pp. 346–368.

Morris, D. C. & Schwarz, M. P. (2002) Pleometrosis in phyllode-glueing thrips (Thysanoptera: Phlaeothripidae) on Australian *Acacia*. *Biological Journal of the Linnean Society*, **75**, 467–474.

Morris, D. C., Schwarz, M. P., Crespi, B. J., & Cooper, S. J. B. (2001) Phylogenetics of gall-inducing thrips on Australian *Acacia*. *Biological Journal of the Linnean Society*, **74**, 73–86.

Morris, D. C., Schwarz, M. P., Cooper, S. J. B., & Mound, L. A. (2002) Phylogenetics of Australian *Acacia* thrips: The evolution of behavior and ecology. *Molecular Phylogenetics and Evolution*, **25**, 278–292.

Mound, L. A. (2005) Thysanoptera: Diversity and interactions. *Annual Review of Entomology*, **50**, 247–269.

Mound, L. A. & Morris, D. C. (2007) The insect order Thysanoptera: Classification versus systematics. *Zootaxa*, **1668**, 395–411.

Perry, S. P., McLeish, M. J., Schwarz, M. P., & Boyette, A. H. (2003) Variation in propensity to defend by reproductive gall morphs in two species of gall-forming thrips. *Insectes Sociaux*, **50**, 54–58.

Perry, S. P., Chapman, T. W., Schwarz, M. P., & Crespi, B. J. (2004) Proclivity and effectiveness in gall defense by soldiers in five species of gall-inducing thrips: Benefits of morphological caste dimorphism in two species (*Kladothrips intermedius* and *K. habrus*). *Behavioral Ecology and Sociobiology*, **56**, 602–610.

Pike, N. & Foster, W. A. (2008) The ecology of altruism in a clonal insect. In: Korb, J. & Heinze, J. (eds.) *Ecology of Social Evolution*. Berlin: Springer-Verlag, pp. 37–56.

Pike, N., Richard, D., Foster, W., & Mahadevan, L. (2002) How aphids lose their marbles. *Proceedings of the Royal Society of London. Series B: Biological Sciences*, **269**, 1211–1215.

Pike, N., Braendle, C., & Foster, W. A. (2004) Seasonal extension of the soldier instar as a route to increased defense investment in the social aphid *Pemphigus spyrothecae*. *Ecological Entomology*, **29**, 89–95.

Pike, N., Whitfield, J.A., & Foster, W. A. (2007) Ecological correlates of sociality in *Pemphigus* aphids, with a partial phylogeny of the genus. *BMC Evolutionary Biology*, **7**, 185.

Pope, R. D. (1983) Some aphid waxes, their form and function (Homoptera: Aphididae). *Journal of Natural History*, **17**, 489–506.

Prokopy, R. J. & Roitberg, B. D. (2001) Joining and avoidance behavior in nonsocial insects. *Annual Review Of Entomology*, **46**, 631–665.

Queller, D. C. & Strassmann, J. E. (1998) Kin selection and social Insects. *Bioscience*, **48**, 165–175.

Rhoden, P. K. & Foster, W. A. (2002) Soldier behavior and division of labour in the aphid genus *Pemphigus* (Hemiptera, Aphididae). *Insectes Sociaux*, **49**, 257–263.

Roisen, Y. (2006) Life history, life types and caste evolution in termites. *In:* Kipyatkov, V. (ed.) *Life Cycles in Social Insects: Behaviour, Ecology and Evolution*. St. Petersburg: St. Petersburg University Press, pp. 85–95.

Rubenstein, D. R. & Lovette, I. J. (2007) Temporal environmental variability drives the evolution of cooperative breeding in birds. *Current Biology*, **17**, 1414–1419.

Schütze, M. & Maschwitz, U. (1991) Enemy recognition and defense within trophobiotic associations with ants by the soldier caste of *Pseudoregma sundanica* (Homoptera: Aphidoidea). *Entomologia Generalis*, **16**, 1–12.

Shibao, H. (1999) Lack of kin discrimination in the eusocial aphid *Pseudoregma bambucicola* (Homoptera: Aphididae). *Journal of Ethology*, **17**, 17–24.

Shibao, H., Lee, J.-M., Kutsukake, M., & Fukatsu, T. (2003) Aphid soldier differentiation: Density acts on both embryos and newborn nymphs. *Naturwissenschaften*, 90, 501–504.

Shibao, H., Kutsukake, M., & Fukatsu, T. (2004a) The proximate cue of density-dependent soldier production in a social aphid. *Journal of Insect Physiology*, **50**, 143–147.

(2004b) Density triggers soldier production in a social aphid. *Proceedings of the Royal Society of London. Series B: Biological Sciences*, **271**, *Suppl*, **3**, S71–S74.

Shibao, H., Kutsukake, M., Matsuyama, S., Fukatsu, T., & Shimada, M. (2010) Mechanisms regulating caste differentiation in an aphid social system. *Communicative and Integrative Biology*, **3**, 1–5.

Shingleton, A. W. & Foster, W. A. (2000) Ant tending influences soldier production in a social aphid. *Proceedings of the Royal Society of London. Series B: Biological Sciences*, **267**, 1863–1868.

(2001) Behavior, morphology and the division of labour in two soldier-producing aphids. *Animal Behaviour*, **62**, 671–679.

Stern, D. L. (1994) A phylogenetic analysis of soldier evolution in the aphid family Hormaphididae. *Proceedings of the Royal Society of London. Series B: Biological Sciences*, **256**, 203–209.

(1998) Phylogeny of the tribe Cerataphidini (Homoptera) and the evolution of the horned soldier aphids. *Evolution*, **52**, 155.

Stern, D. L. & Foster, W. A. (1996) The evolution of soldiers in aphids, *Biological Reviews of the Cambridge Philosophical Society*, **71**, 27–79.

Strassmann, J. E. & Queller, D. C. (2010) The social organism: Congresses, parties, and committees. *Evolution*, **64**, 605–616.

Stone, G. N. & Schönrogge, K. (2003) The adaptive significance of insect gall morphology. *Trends in Ecology and Evolution*, **18**, 512–522.

Toth, A. L., Varala, K., Newman, T. C., et al. (2007) Wasp gene expression supports an evolutionary link between maternal behavior and eusociality. *Science*, **318**, 441–444.

Turnbull, C., Hoggard, S., Gillings, M., et al. (2011) Antimicrobial strength increases with group size: Implications for social evolution. *Biology Letters*, **7**, 249–252.

Turnbull, C., Caravan, H., Chapman, T., et al. (2012) Antifungal activity in thrips soldiers suggests a dual role for this caste. *Biology Letters*, **8**, 526–529.

Wcislo, W. T. (1997) Are behavioral classifications blinders to studying natural variation? In: Choe, J. C. & Crespi, B. J. (eds.) *The Evolution of Social Behavior in Insects and Arachnids*. London: Cambridge University Press, pp. 8–13.

Wcislo, W. T. & Tierney, S. M. (2009) The evolution of communal behavior in bees and wasps: An alternative to eusociality. In: Gadau, J. & Fewell, J. (eds.) *Organization of Insect Societies*. Cambridge: Harvard University Press, pp. 148–169.

Whitham, T. G. (1979) Territorial behavior of *Pemphigus* gall aphids. *Nature*, **279**, 324–325.

(1986) Cost of benefits of territoriality: Behavioral and reproductive release by competing aphids. *Ecology*, **67**, 139–147.

Wilch, M. H. (1999) *Predation and prey response in the galls of* Pemphigus populi-ramulorum. Masters thesis, Tucson: University of Arizona.

Wills, T. E., Chapman, T. W., Kranz, B. D., & Schwarz, M. P. (2001) Reproductive division of labour coevolves with gall size in Australian thrips with soldiers. *Naturwissenschaften*, **88**, 526–529.

Wills, T. E., Chapman, T. W., & Mound, L. A. (2004) Natural history and description of *Oncothrips kinchega*, a new species of gall-inducing thrips with soldiers (Thysanoptera: Phlaeothripidae). *Australian Journal of Entomology*, **43**, 169–176.

Wilson, E. O. (1971) *The Insect Societies*. Cambridge: Harvard University Press.

(1975) *Sociobiology: The New Synthesis*. Cambridge, MA: Harvard University Press, Belknap Press.

Withgott, J. H., Abbot, D. K., &. Moran, N. A (1997) Maternal death relaxes developmental inhibition in nymphal aphid defenders. *Proceedings of the Royal Society B: Biological Sciences*, **264**, 1197–1202.

7 Sociality in Spiders

Leticia Avilés and Jennifer Guevara

Overview

Three factors need to be considered when addressing why some species form social groups and others do not: (1) the set of environmental challenges and opportunities that group living and cooperation allow particular species to meet; (2) the life history and other intrinsic features of those organisms relevant to those challenges and opportunities; and (3) the phylogenetic and historical contexts that determine whether particular organisms can be molded into social lifestyles. The interplay between the first two factors should explain why social and nonsocial species in a given taxon occupy particular environments, and why, within a particular habitat, closely related species differ in their social systems. History and phylogeny should determine the availability of taxa of appropriate characteristics to move into a social niche. Here, we consider how these three factors play out in explaining the diversity of spider social systems and their phylogenetic and geographic distributions. We argue that the architecture of the webs built is a key determinant of the social system that develops – species with irregular, shareable webs can develop systems involving cooperation in web building, foraging, and brood care (social species), whereas those with orbicular webs are limited to aggregating individually used webs (colonial species). These intrinsic traits, in interaction with particular environmental conditions, would then determine the fitness of individuals as a function of group size and thus whether and where colonies form and the range of sizes they develop. Social spiders would have evolved to occupy environments where factors such as strong rains or high predation make living solitarily or in very small groups untenable for spiders with costly webs and strict dependence on maternal care. For colonies to grow, however, increasingly large insects must be captured to make up for a decline in web prey capture area per spider as the spiders' colonies and tri-dimensional webs grow in size. This would explain why social spiders are

Jennifer Guevara was supported by a NSERC Canada Graduate Fellowship, CGS-D, and by post-doctoral funding from the Avilés laboratory. The Avilés laboratory was funded by the US National Science Foundation (DEB-707474 and DEB-9815938), the James S. McDonnell Foundation (2010 to 2014), and NSERC Canada (Discovery and Accelerator awards). Research in Ecuador was sponsored by the Pontificia Universidad Católica del Ecuador, the Corporación Sociedad para la Investigación y Monitoreo de la Biodiversidad Ecuatoriana SIMBIOE, and the Museo Ecuatoriano de Ciencias Naturales, and in Brazil by the University of Campinas and the field station at Serra do Japi. We thank members of the Avilés lab (A. Gonzalez, R. Sharpe, C Hoffmann, P. Fernández and M. Robertson), two anonymous reviewers, and the editors, for comments on the manuscript.

concentrated at lower latitudes and elevations where temperatures are warm and insects large. Colonial spiders, on the other hand, would have evolved to take advantage of an ecological opportunity at the microhabitat level – the availability of open spaces with high insect abundance in habitats, ranging from deserts to rainforests, where species of like characteristics could nonetheless live solitarily. We review evidence suggesting that fitness is a humped function of colony size in both types of spiders and propose that the hump is likely to be taller and the minimum viable group size larger in the social species. Conditions giving rise to social species may thus result in obligate group living, whereas those giving rise to colonial spiders would allow for facultative coloniality. Such a difference would then explain the different breeding systems of social and colonial spiders – spiders bound to their colonies and mating with colony mates from generation to generation, in the former, and dispersal and outbreeding, in the latter. We suggest that consideration of how environmental conditions interact with intrinsic features of the organisms to shape fitness as a function of group size can help us understand not only the spiders', but animal's social systems, in general.

I SOCIAL DIVERSITY

7.1 How Common is Sociality in Spiders?

Although spiders are well known for their predatory and solitary habits, out of approximately 45,400 or more described species to date, a few dozen, belonging to a variety of genera and families, have evolved some form of social behavior. Sociality in these species ranges from temporary aggregations of individual webs to tightly knit social groups where multiple females and their offspring share a single communal nest and cooperate in prey capture, feeding, and brood care (Avilés, 1997; Uetz & Hieber, 1997; Bilde & Lubin, 2011; Yip & Rayor, 2014).

7.2 Forms of Sociality in Spiders

Two criteria have been used to classify sociality in spiders: (1) whether individuals occupy a common space or maintain individual territories within nests (non-territorial vs. territorial); and (2) whether individuals aggregate for part of or for their entire life cycle (periodic- vs. permanent-social) (D'Andrea, 1987; Avilés, 1997). According to these two criteria, most spider species displaying some form of social behavior can be assigned to one of four categories: (1) the non-territorial permanent-social, or simply, social spiders (also known as quasisocial, Wilson, 1971; or cooperative spiders, Avilés, 1997; Bilde & Lubin, 2011) form colonies typically containing multiple adult females and their offspring and lasting for several generations; (2) the non-territorial periodic-social, or subsocial species, form colonies that normally contain a single adult female and offspring that disperse prior to producing offspring of their own (Yip & Rayor, 2014); (3) the territorial permanent-social, or colonial spiders, aggregate individual webs within a web complex that may last for several generations (Uetz & Hieber, 1997); or (4) the territorial periodic-social spiders, also labeled colonial, are single-family groups where offspring maintain individual territories within a web complex and disperse prior to reproduction. As we elaborate below, whether colonies are permanent versus temporary may depend upon environmental conditions, whereas whether individual territories are maintained may depend upon the architecture of the webs. Specifically, orb-weaving species would be required to maintain separate territories since orbicular webs are designed to be built and used by single individuals (Lubin, 1974; Uetz & Hieber, 1997), whereas irregular webs can be built, enlarged, and occupied simultaneously by multiple individuals. As we will see, however, even species with irregular webs may maintain individual territories if environmental conditions do not allow groups of closely spaced individuals to form. Before considering what those conditions may be, we will review the characteristics of social, subsocial, and colonial spiders.

7.2.1 Non-Territorial Permanent-Social or "Social" Species

With the reclassification of species in the genera *Delena* (Sparassidae) and *Diaea* (Thomisidae) as subsocial (Lubin & Bilde, 2007; Yip & Rayor, 2014), social spiders are now represented by 19 species in 8 genera and 5 families (Figure 7.1), all of which build irregular webs – two dimensional meshes, in the case of the neotropical *Aebutina binotata* (Dictynidae), three-dimensional ones in the remaining cases (Figure 7.2A) (Avilés, 1997). All activities of the colony take place in these communal webs, including cooperative web-building and maintenance, prey capture, feeding, and brood care. Rather than dispersing, offspring remain in the natal nest to mate and reproduce. Depending upon the species, colonies may grow over multiple generations to contain dozens to thousands or tens of thousands of individuals. Colonies that have reached relatively large sizes propagate by budding or fission (e.g. Lubin & Robinson, 1982; Seibt & Wickler, 1988; Avilés, 2000) or by inseminated females dispersing either singly or in groups (e.g. Vollrath, 1982; reviewed in Avilés, 1997). With one known exception (Avilés, 1994), there is apparently no or very little gene flow among the colony lineages (e.g. Johannesen, *et al.*, 2009; Smith, *et al.*, 2009; Agnarsson, *et al.*, 2010; Settepani, *et al.*, 2014), causing spiders and their colonies to be highly inbred (reviewed in Avilés, 1997; Lubin & Bilde, 2007; Avilés & Purcell, 2012). Such strongly subdivided population structure is thought to have facilitated the evolution of the highly female-biased sex ratios characteristic of the inbred social species (Avilés, 1993a). There are no reproductive castes within the colonies (e.g. Avilés, 1993b; Lubin, 1995), although in some species the proportion of females that reproduce may decrease with increasing colony size (Avilés & Tufiño, 1998). The tendency of individuals to participate in communal activities may also vary with age (Rypstra, 1990; Lubin, 1995; Yip & Rayor, 2011; Settepani, *et al.*, 2013), condition (Ebert, 1998; Keiser, *et al.*, 2014), or behavioral type (Pruitt, *et al.*, 2012; Grinsted, *et al.*, 2013).

7.2.2 Non-Territorial Periodic-Social or "Subsocial" Species

According to Yip & Rayor (2014), at least 70 species in 16 families and representing 18 independent origins can be classified as subsocial. Many of these are congeners of the

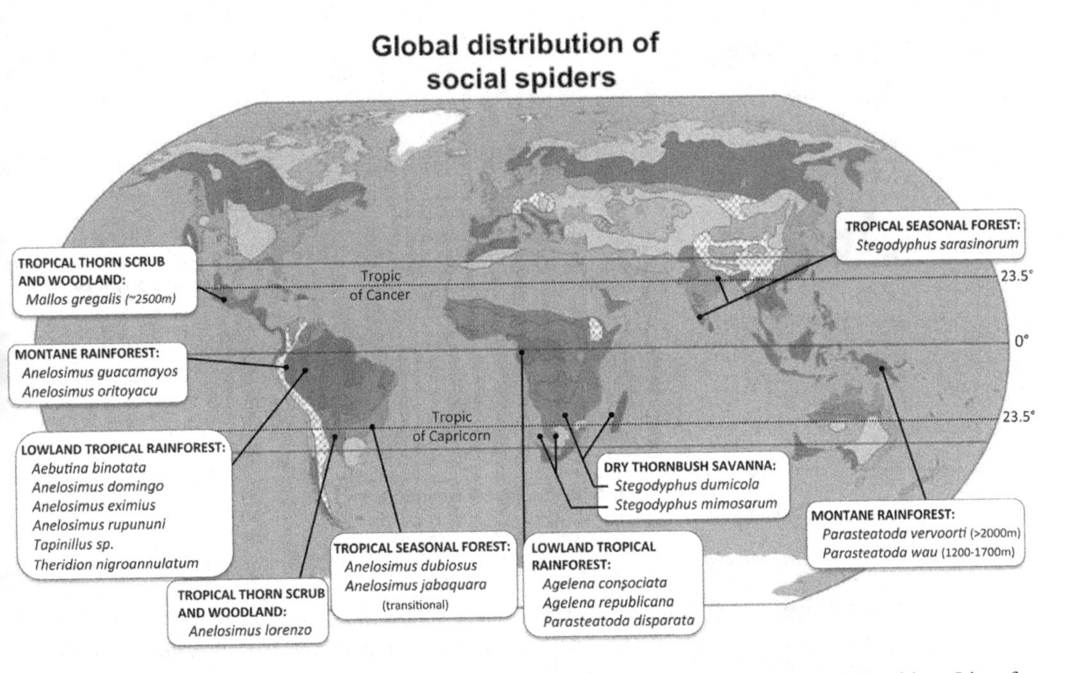

Figure 7.1 Global distribution of the non-territorial permanent social or "social" spiders. List of species revised from Avilés (1997) and Lubin & Bilde (2007) by removing species in the Australian genera *Diaea* (Thomisidae) and *Delena* (Sparassidae), which have been reclassified as subsocial (Lubin & Bilde, 2007; Yip & Rayor, 2014). *Anelosimus puravida* from Costa Rica (Agnarsson, 2006) is not included, as there are no formal studies confirming its status as a social species. *Anelosimus jabaquara*, on the other hand, is included, but noted as being "transitional" between subsocial and social (Gonzaga & Vasconcellos-Neto, 2001). The genera listed belong to five spider families: Agelenidae or funnel web spiders (*Agelena*), Theridiidae or cobweb spiders (*Anelosimus*, *Parasteatoda* = *Achaearaneae*, and *Theridion*), Oxyopidae or lynx spiders (*Tapinillus*), Dictynidae (*Mallos* and *Aebutina*), and Eresidae or velvet spiders (*Stegodyphus*). Dictynidae and Eresidae are families of cribellate spiders, whereas the remaining three are ecribellate. *Tapinillus* is the only web-building genus in the Oxyopidae, a family of web-less wandering spiders. All species listed build relatively dense tri-dimensional webs, with the exception of *Aebutina binotata*, whose webs are two-dimensional meshes covering the surfaces of leaves. For illustrations of the nests of these genera, see Avilés (1997). For a list and geographical distribution of subsocial species, see Yip & Rayor (2014).

social species and inferred to resemble the ancestral system from which the social species originated (Agnarsson, *et al.*, 2006; Johannesen, *et al.*, 2007). With some exceptions, in subsocial spiders offspring disperse prior to reaching reproductive maturity, sex ratios are generally unbiased, and nest re-use is not the rule (Yip & Rayor, 2014). In two Australian species in the genus *Diaea* (Avilés, 1997), however, offspring mature and mate in the natal nest prior to dispersing and sex ratios are female biased. Nest reuse over multiple generations, on the other hand, is common in the Australian *Delena cancerides*, whose colonies typically contain multiple cohorts of offspring (Yip & Rayor, 2011). Some subsocial species in the genera *Anelosimus* (Theridiidae) and

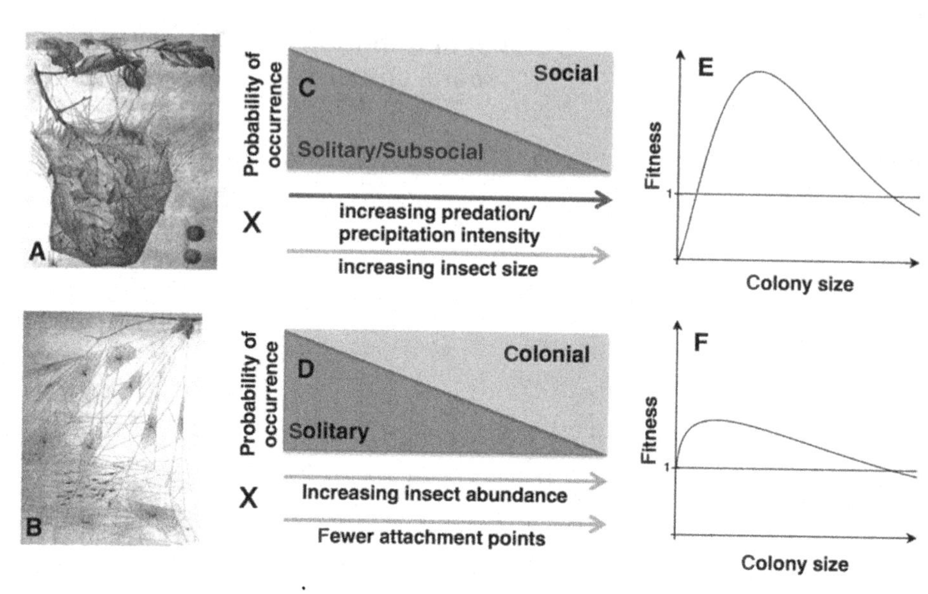

Figure 7.2 Nests of (A) a social and (B) a colonial spider, *Anelosimus eximius* and *Philoponella republicana*, respectively, as illustrated in Simon (1891). (C) and (D) Models that predict the occurrence of the various types of spider sociality as a function of gradients of abiotic and biotic factors at the habitat (social/subsocial/solitary) or microhabitat (colonial/solitary) level. (C) According to the precipitation intensity and *predation rate* hypotheses (Avilés, *et al.*, 2007), species that depend on costly tri-dimensional webs and extended maternal care will be increasingly less likely to occur as solitary individuals as the intensity of these factors increases, whereas the likelihood that groups of increasing size will form will increase along with these factors, so long as there is a parallel increase in the size of insects available for prey capture (the prey size hypothesis). (D) Colonial species will be more likely to occur, and solitary orb weaving species less likely to occur, in microhabitats where there are few attachment points for individual webs, but insect abundance is high. (E) and (F) Postulated shape of the function representing the fitness of individuals (females) in colonies of different size for (E) social and (F) colonial spiders. Note the much steeper shape of the social spider fitness function and the fact that in these species individuals in colonies of very small size are predicted to be unable to replace themselves in the environments in which they occur, i.e. they would exhibit an Allee effect.

Stegodyphus (Eresidae) may also vary across their geographic range, having in some areas colonies that contain multiple adult females (Riechert & Jones, 2008; Avilés & Harwood, 2012; Lubin & Bilde, 2007). Another *Anelosimus* species, *A. jabaquara*, appears to be transitional between subsocial and social (Marques, *et al.*, 1998; Avilés & Harwood, 2012).

7.2.3 Territorial Periodic- and Permanent-Social or "Colonial" Species

The most common and best-studied colonial spiders build individual orbicular webs embedded within a scaffolding of communally built frame lines (Figure 7.2B) (Uetz & Hieber, 1997). One of these species, the South American *Parawixia* (= *Eriophora*) *bistriata* (Araneidae) (e.g. Fowler & Diehl, 1978; Fernández-Campón, 2007), is periodic-social, as its single-family units disperse prior to mating and reproduction. The remaining species, in the genera *Cyclosa* (Araneidae), *Cyrtophora* (Araneidae), *Metepeira* (Araneidae), *Metabus* (Tetragnatidae), *Leucage* (Tetragnatidae), and *Philoponella* (Uloboridae) (Figure 7.3), are considered permanent-social as their colonies contain multiple families and may last several generations. Depending upon the species, colony sizes range from few to thousands (Uetz & Hieber, 1997). With few exceptions (Breitwisch, 1989; Binford & Rypstra, 1992; Masumoto, 1998; Fernández-Campón, 2007), group foraging is absent and orb webs are utilized as individual territories, likely due to constraints of their architectural design. Allomaternal care is also absent, even though nest complexes of species in the genera *Metabus*, *Parawixia*, and *Philoponella* have common areas where females guard their egg sacs (Uetz & Hieber, 1997). New colonies may be established either by single females, by the aggregation of individuals that have dispersed singly, or by groups of juveniles dispersing together from their natal nest (Smith, 1983; Uetz & Hieber, 1997; Johannesen, *et al.*, 2012). Males may also move between colonies at the time of mating (Johannesen, *et al.*, 2012). The mating system is thus outbred, with no biased sex ratios reported.

Species that build irregular webs in the families Dictynidae and Pholcidae have also been found in aggregations sufficiently close to be considered colonial. In the dictynid genera *Dictyna*, *Mallos*, and *Mexitlia*, singly-occupied irregular webs may occur in close proximity to one another (Uetz & Hieber, 1997; Whitehouse & Lubin, 2005), whereas colonial pholcid webs may house multiple individuals (Jakob, 2004; Avilés, *et al.*, 2001). Finally, non-web building cursorial spiders in the families Salticidae (jumping spiders) and Scytodidae (spitting spiders) have in a few cases been reported to form complexes of individual nest units interconnected by silk (e.g. Bowden, 1991; Miller, 2006; Jackson, *et al.*, 2008; Yip & Rayor, 2014). We will concentrate our discussion on the colonial orb weavers, however, as there is limited information on other types of colonial spiders.

7.3 Why Spiders Form Social Groups

A leading hypothesis to explain why organisms live in groups is that grouping causes some components of fitness to increase as group size increases (Avilés, 1999; Krause & Ruxton, 2002). With costs of group living being unavoidable, however, other components of fitness are expected to decrease with increasing group size, causing average fitness to peak at intermediate group sizes (Avilés, 1999). An alternative hypothesis is that there are no intrinsic benefits to grouping, but that individuals are bound to their natal site or patchy resource by limited availability of suitable habitat or by high costs of dispersal (Emlen, 1982). These hypotheses have been formally tested in at least two of the social spiders and partially in a colonial and a fourth social species (Figure 7.4). In all four cases, a clear benefit to grouping was found. In the social *Anelosimus eximius* and in one of the populations of *Stegodyphus dumicola*, fitness was greatest at intermediate colony sizes (Figures 7.4A & B), owing mostly to the fact that colonies with few individuals failed (Avilés & Tufiño, 1998; Bilde, *et al.*, 2007). In the colonial

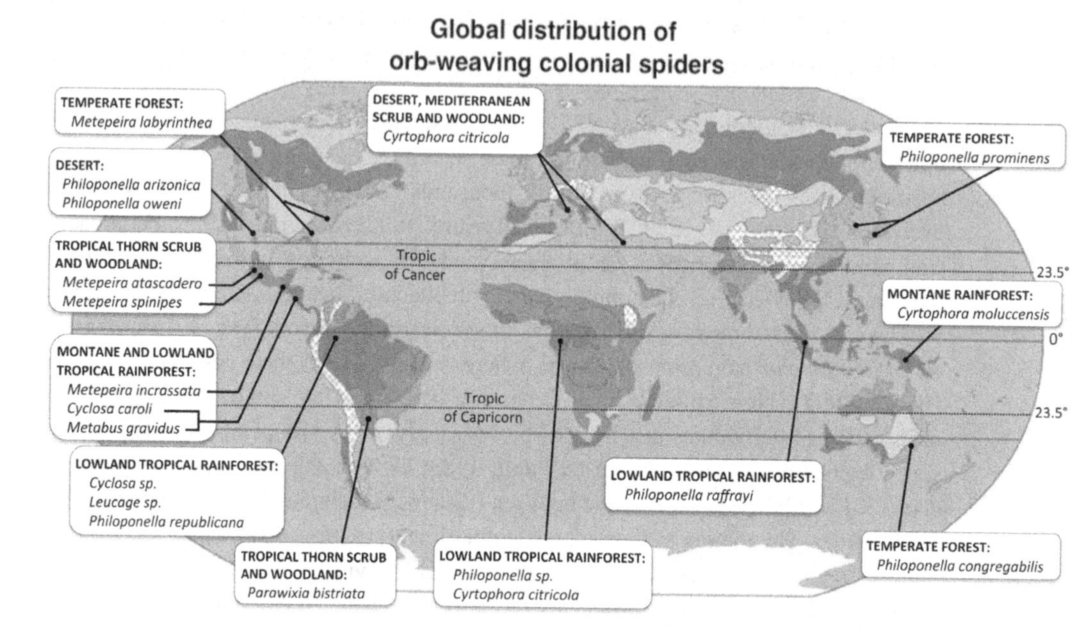

Figure 7.3 Global distribution of the orb-weaving territorial periodic- (*Parawixia bistriata*) and permanent-social (the remaining species) or "colonial" species. List of species modified from Uetz & Hieber (1997) and Whitehouse & Lubin (2005), by removing *Cyrtophora* species (*C. cicatrosa, C. monulfi*) described by Lubin (1974) as being solitary, and the addition of two new colonial spiders from Ecuador (*Cyclosa* sp. and *Leucage* sp., Avilés, *et al.*, 2001; Salomon, *et al.*, 2010) and one from Korea (*Philoponella prominens*, Park, 1999). The genera listed belong to three orb-weaving familes: Araneidae (*Cyrtophora, Cyclosa, Metepeira, Parawixia* [formerly *Eriophora*]); Tetragnatidae (*Metabus* and *Leucage*), and the Uloboridae (*Philoponella*).

Metepeira incrassata, Uetz & Hieber (1997) showed that females in colonies of intermediate size produced the greatest number of hatchlings (Figure 7.4D) and that net energy gain per capita, after accounting for metabolic and material costs of web building, increased with colony size. Finally, in the social *Aebutina binotata*, pilot data suggested that females in colonies of intermediate size produced the greatest number of offspring (Figure 7.4C) (Avilés, 2000). Data on subsocial spiders, on the other hand, suggest that offspring benefit from the protection offered by the natal nest and the presence of the mother, but that foraging competition within groups may cause fitness to be either a flat or a decreasing function of the number of offspring in the group (Schneider, 1995; Jones & Parker, 2000; Avilés & Bukowski, 2006).

In the sections that follow, we elaborate on how benefits of group living associated with food acquisition, predator avoidance, nest maintenance, mate finding, and off-spring care may shape the fitness of social, subsocial, and colonial spiders. We argue that the relative importance of these components of fitness in explaining spider social systems likely depends on the architecture – and presumed cost – of their webs. Thus, for social and subsocial spiders with their (presumably) expensive irregular webs, the

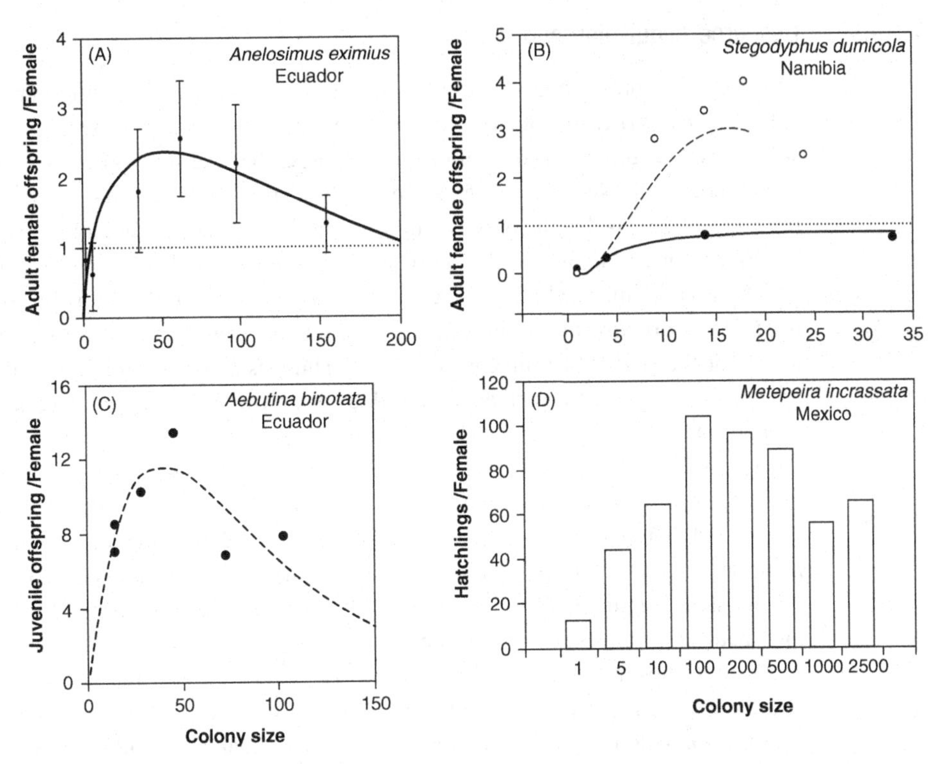

Figure 7.4 Data on three social ((A) *Anelosimus eximius*, (B) *Stegodyphus dumicola*, (C) *Aebutina binotata*) and one colonial ((D) *Metepeira incrassata*) species showing that the number of offspring per capita at various stages of development (hatchlings, in *M. incrassata*; juveniles, in *A. binotata*; and adult female offspring, in *A. eximius* and *S. dumicola*) peaks at some intermediate colony size, demonstrating intrinsic benefits of group living in all four species. The data on *A. eximius* and *S. dumicola*, which correspond to average overall female lifetime reproductive success, show that the females living solitarily or in very small groups are not able to replace themselves (i.e. their fitness is below 1) in the environments in which they occur. (Note that one of the populations of *S. dumicola* was entirely below replacement value at the time the data was collected.) Data reproduced with permission from Avilés & Tufiño (1998) ©1998 by The University of Chicago Press, Bilde, *et al.* (2007) ©2007 by John Wiley & Sons, Avilés (2000) © 2000 by Elsevier, and Uetz & Hieber (1997).

most immediate benefit of group living might have been retaining the protection afforded by the natal nest and sharing the costs of its maintenance, followed by a need for additional caregivers in adverse environments. We suggest that the foraging advantage of accessing large prey (Nentwig, 1985; Ward, 1986; Rypstra, 1990) is likely secondary in social species and even less important in subsocial ones. For the orb-weaving colonial spiders with their (presumably) relatively inexpensive webs, in contrast, the primary driver of their social system may have been accessing, through the aggregation of individual webs, open spaces where prey are abundant (a foraging advantage), with secondary benefits in predator avoidance or defense (see also Whitehouse & Lubin, 2005).

7.3.1 Resource Acquisition and Use

The capture of prey too large for solitary individuals to access is an important, albeit as we suggested earlier, secondary benefit of group living for the social spiders. That social spiders can capture prey orders of magnitude larger than the spiders themselves has long been recognized (Nentwig, 1985; Ward, 1986; Rypstra, 1990). What has only recently been appreciated is that the presence of large insects is essential for large social colonies to form, a concept known as the "prey size" hypothesis (Avilés, *et al.*, 2007; Powers & Avilés, 2007; Yip, *et al.*, 2008). As shown by Yip, *et al.* (2008) for the Neotropical *Anelosimus eximius*, availability of large insects is necessary to address scaling challenges of the spiders' tri-dimensional nests. Thus, as colonies grow large, the surface area of the web, which intercepts prey, does not grow as fast as the volume, which is proportional to the number of spiders in the nests. As a result, number of prey per capita declines with colony size (Figure 7.5A). If large insects are available in the environment, however, the spiders can compensate for this decline by cooperatively capturing increasingly large insects as the size of their webs, and the number of potential hunters, increase (Figure 7.5B). Prey biomass per capita is then maximized at intermediate colony sizes (Figure 7.5C) (Yip, *et al.*, 2008). That biomass per capita likely peaks at intermediate colony sizes would then explain both why social spiders are better off living in groups and why they disperse at large colony sizes. Studies on *Stegodyphus* species are consistent with these observations. Thus, Ward (1986) showed that larger *S. mimosarum* colonies captured larger prey, but contained smaller spiders, as prey availability did not keep pace with colony size (see also Crouch & Lubin, 2000).

In subsocial species, resource acquisition does not appear to be a particularly important driver of their social systems. Instead, within group competition for resources appears to predominate (Schneider, 1995; Jones & Parker, 2000; Powers & Avilés, 2003; Avilés & Bukowski, 2006). Thus, even though groups may obtain more food per capita than singletons, within groups per capita food intake (Jones & Parker, 2000) and survival (Avilés & Bukowski, 2006) may decline with group size. Fitness in subsocial spiders is thus likely a flat or declining function of the number of siblings in the group, a finding that, based on the prey size hypothesis, we attribute to an absence of large insects in their respective environments. Consistent with this suggestion, Powers & Avilés (2007) showed that young of subsocial *Anelosimus* spiders at high latitudes and elevations (in Arizona and Ecuador) dispersed from the natal nest once they grew to a size that matched the average size of insects intercepting their webs.

A foraging benefit, on the other hand, may be the primary advantage of grouping in the colonial orb weaving spiders, since by aggregating individual webs they are able to access spaces that are prey rich, yet too vast for solitary spiders to utilize (Uetz & Hieber, 1997), such as gaps between trees or other solid structures (Lubin, 1974; Rypstra, 1979) or open spaces over bodies of water (Buskirk, 1975; Salomon, *et al.*, 2010). Even in the absence of cooperative foraging, colonial orb weaving species may further benefit by catching insects that rebound off adjacent webs, a phenomenon known as the "ricochet effect" (Rypstra, 1979; Uetz, 1989). As an exception, the periodic-colonial *Parawixia bistriata* is able to engage in joint prey capture thanks to

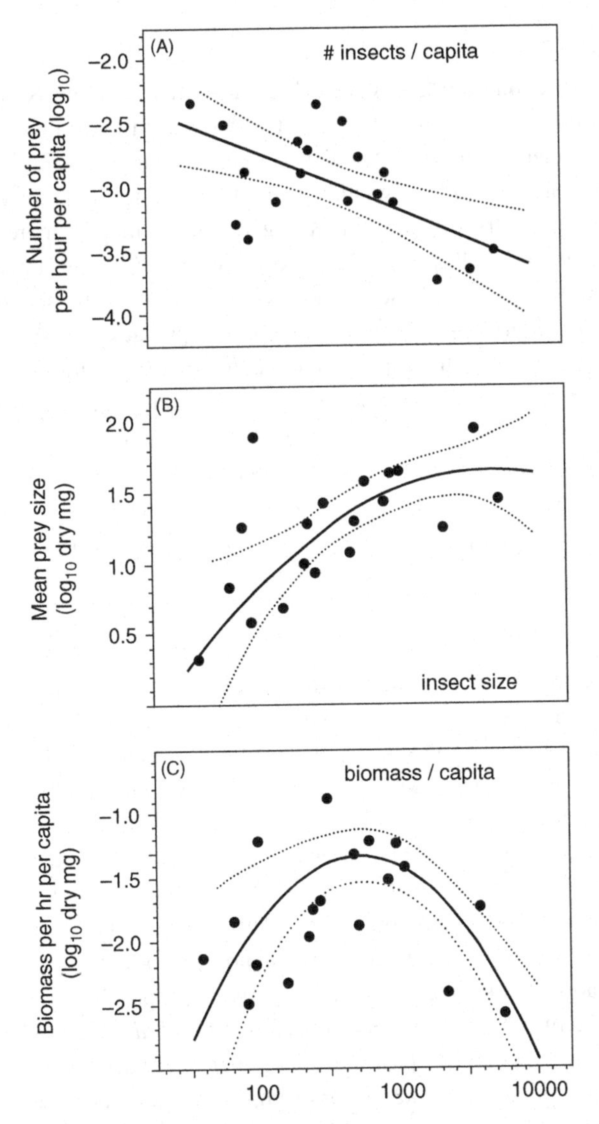

Figure 7.5 Data from the Neotropical social spider *Anelosimus eximius* in eastern Ecuador showing the relationships between colony size and (A) the number of insects captured per hour per spider, (B) the size of insects captured (dry mg), and (C) the prey mass (dry mg) captured per hour per spider (C). Dotted lines indicate 95 percent confidence intervals for the fits. Each dot in the graphs represents a colony's average based on a median of 27 prey capture events per colony (range 1 to 62) on 21 forest interior and 7 river edge colonies. Reproduced with permission from Yip, *et al.* (2008) ©2008 by National Academy of Sciences, U.S.A.

the placement of its orbs along a single plane (Uetz & Hieber, 1997; Fernández-Campón (2007)). Rather than accessing open habitats, an abundance of small insects may be behind coloniality in species with singly-occupied irregular webs, such as the colonial *Dictyna* and *Mallos* (Jackson, 1977).

7.3.2 Predator Avoidance

Given the physical protection afforded by their dense tri-dimensional nests, we suspect that protection from predation and the elements may have been the primary driver of subsocial and social spider systems (i.e. "predation rate" hypothesis). Additional benefits, relevant also to colonial spiders, would result from dilution and early warning effects. After accounting for the possibility that larger colonies may be more attractive to predators (e.g. Lubin, 1974; Rypstra, 1979; Keiser, *et al.*, 2015), all three benefits should increase as the size of the colonies increases. Uetz, *et al.* (2002) showed that in the colonial *Metepeira incrassata*, successful attacks by predatory wasps decreased significantly with colony size and the spiders benefited from "early warning" effects. In the social *Stegodyphus dumicola*, Henschel (1998) showed that large groups were less vulnerable to predation as they could prevent predators from entering the core of the colonies. In subsocial spiders, delayed offspring dispersal well past the mother's death may be due to the protection afforded by the maternal nest, as shown for species in the Australian genera *Diaea* (Evans, 1998) and *Delena* (Yip & Rayor, 2011).

7.3.3 Homeostasis

To the extent that the web is an extension of the spiders' body and essential for their livelihood, its maintenance can be considered a homeostatic function of the colony. We suggest that in social and subsocial spiders, the relative permanency and protection afforded by their dense tri-dimensional webs (Figure 7.2) may have initially selected for offspring to remain in the maternal nest. Likely savings in the amount of resources per capita needed to maintain these relatively expensive webs would then have reinforced this initial benefit, with additional benefits arising from avoiding dispersal costs. The benefits of remaining in the natal nest would also apply for species in the genera *Diaea* and *Delena*, which do not build webs for prey capture, but occupy nests that offer protection from predation and the elements and may also be expensive to produce and maintain (*Diaea*) or be of limited availability (*Delena*, Yip, *et al.*, 2012). Being more ephemeral, the webs of the colonial spiders, on the other hand, may be less likely to promote site fidelity. Colonial *Philoponella*, for example, tend to relocate as a group in response to prey scarcity (e.g. Smith, 1985). The webs of colonial spiders in the genera *Cyrtophora* and *Metepeira*, however, do have some degree of permanency (Lubin, 1974; Uetz & Hieber, 1997). Site fidelity in colonial spiders may also arise as a result of advantages of the site itself.

7.3.4 Mating

A mating benefit is unlikely to be a primary driver of group living in either social, subsocial, or colonial spiders. Nonetheless, once groups have formed, baring potential costs of inbreeding depression, males should benefit by finding mates among colony mates rather than dispersing. Following Waser, *et al.* (1986), Avilés & Purcell (2012) argue that the extent of this benefit would depend upon the number of potential mates in the local nest relative to the distance to other nests, the mortality risk while dispersing, the dispersal ability of the spiders, and any benefits of group living potentially lost by abandoning the natal group. The main deterrent against local mating, at least initially, would have been potential costs of inbreeding depression, which are expected when closely related individuals of an initially outbred species mate with each other. Clearly, social and colonial spiders have had different outcomes of the balance between dispersal and philopatry, as the former have evolved strongly inbred social systems with female biased sex ratios, whereas the latter have not. Avilés & Purcell (2012) calculated from various studies on social and subsocial *Anelosimus* spiders that the benefits of philopatry for both males and females likely far exceeded potential costs of inbreeding depression in the transition from an outbred to an inbred system, noting that a similar pattern likely applies to species in other genera (e.g. Lubin & Bilde, 2007). As colonies of the colonial spiders appear of mixed genetic origin or subject to gene flow (see below), in these species males can take advantage of the concentration of females without paying the costs of inbreeding depression. So, in both social and colonial spiders a secondary mating benefit from grouping may accrue to males. However, the required data to appraise the magnitude of this benefit, genetic and otherwise, are not yet available. In contrast, since in most subsocial spiders individuals disperse from the natal nest prior to reaching reproductive maturity (Yip & Rayor, 2014), neither males nor females derive a clear mating benefit from the social system. The only exceptions are two *Diaea* species, where mating takes place in the natal nest prior to dispersal and, as a result, sex ratios are highly female biased (Rowell & Main, 1992; Avilés, 1997).

7.3.5 Offspring Care

There is substantial evidence that maternal or allomaternal care confers significant benefits to the offspring of subsocial and social spiders. In the subsocial *Anelosimus arizona*, for example, offspring size and survival correlated positively with the lifespan of the mother (Avilés & Bukowski, 2006) and in *Stegodyphus lineatus* young dispersed at a later age when the mother lived longer (Schneider, 1995). In social spiders, and in subsocial or transitional species where multiple adult females share a common nest, the availability of surrogate caregivers in the event of the mother's death may be essential for the survival of the offspring. According to the "maternal death" hypothesis (Jones, *et al.*, 2007), surrogate caregivers would allow a species to occupy environments where offspring development is slow (Riechert & Jones, 2008) or maternal mortality is high (Avilés, *et al.*, 2007; Purcell & Avilés, 2008).

In addition to egg sac guarding, prey sharing, and retention of the offspring within the maternal nest, maternal or allomaternal care behaviors may involve in some species regurgitation feeding (e.g. Schneider, 2002; Viera, *et al.*, 2006) and/or matriphagy, where mothers offer themselves as food source to their offspring (e.g. Evans, *et al.*, 1995; Kim, *et al.*, 2000; Viera, *et al.*, 2007). There is confirmation that these maternal care behaviors do indeed involve alloparenting (e.g. Salomon & Lubin, 2007; Samuk & Avilés, 2013), albeit with some variability among species. Thus, Samuk & Avilés (2013) documented frequent switching of females guarding egg sacs in the highly social *A. eximius*, but significantly less so in the less social *A. guacamayos*. Likewise, Marques, *et al.* (1998) found that in the transitional *A. jabaquara*, egg sacs and young belonging to different mothers were kept in more or less separate areas of the common nest up to their fourth instar, whereas offspring of the more highly social *A. dubiosus* already mixed with other broods in their second instar.

It is also of interest how allomaternal care behaviors originated and the consequences they may have on the evolution of a species. Observations on *Anelosimus* suggest that alloparenting may simply reflect retention of an ancestral lack of discrimination against foreign offspring, as subsocial *Anelosimus* females experimentally given a foreign egg sac readily cared for it (Samuk & Avilés, 2013). The reproductive state of the female may be important, however. In the subsocial *S. lineatus*, for example, only mated females cared for other's young (Schneider, 2002). Finally, Samuk, *et al.* (2012) suggested that alloparenting may have relaxed selection on individual maternal care effort. For example, social *Anelosimus* females were more likely to leave egg sacs unattended, guard them from greater distances, or relinquish them, when compared to females of phylogenetically close subsocial species (Samuk, *et al.*, 2012).

Limited to egg sac guarding and tolerance of the young within the complex, offspring care is an individual activity in the colonial spiders. Here females guard egg sacs in their individual webs (e.g. *Cyrtophora spp*., Lubin, 1974; *Philoponella arizonica*, Smith, 1997) or in common retreats, but attached to their own bodies (e.g. *P. republicana* and *P. oweni*, Smith, 1985, 1997). Eclosing offspring are then tolerated within the web complex (Smith, 1982, 1997; Mestre & Lubin, 2011) where they may place their orbs in centrally located, and presumably better protected, areas of the complex, as in the creek spider *Leucage* sp. (Salomon, *et al.*, 2010).

7.4 The Role of Ecology in Shaping Sociality in Spiders

As is clear from the discussion earlier, along with characteristics of the species themselves, ecology is an essential component of the equation explaining why social groups form. Below we summarize the main ecological factors that we believe are responsible for the development of the various types of spider social systems and their geographical distribution (Figures 7.1 & 7.3). We then consider how the social systems, in turn, may influence the ecological niche (space and resources) that social, subsocial, and colonial species occupy.

7.4.1 Habitat and Environment

Regarding social and subsocial spider systems, we suggest that three main environmental factors would explain whether and where species with dense tri-dimensional webs and strict dependence on maternal care can live as single females or in groups (Figure 7.2C): precipitation intensity, predation risk, and insect size, corresponding to the hypotheses described above of the same name (Avilés, *et al.*, 2007; Purcell & Avilés, 2008; Yip, *et al.*, 2008; Guevara & Avilés, 2015). Precipitation intensity and predation rate, alone or in combination (Figure 7.2C), may explain why subsocial species are excluded from certain areas, whereas insect size would address where large social colonies can form. Thus, subsocial species, with colonies that contain a single mother and her offspring, may not be able to cope with frequent disturbance by strong rains or high predation rates threatening the nest or the mother (Riechert, *et al.*, 1986; Purcell & Avilés, 2008). Social spiders, being able to cope with these conditions, on the other hand, would only be able to occupy environments with a sufficient supply of large insects to address the scaling challenge of their tri-dimensional webs (Yip, *et al.*, 2008) (Figure 7.2C). Predation risk, in particular by ants (Henschel, 1998; Purcell & Avilés, 2008; Keiser, *et al.*, 2015), is a factor likely common to species in all social spider genera, whereas strong rains are only relevant to species occupying rainforest environments (all but species in *Stegodyphus* and *Mallos*) (Figure 7.1). That these factors may exclude subsocial species from certain habitats is suggested by the fact that solitary-living females of social *Anelosimus* and *Stegodyphus* species are not able to replace themselves in the core habitats in which they live (Figures 7.4A & B). Single-female nests of the social *A. eximius* in the lowland tropical rainforest, for example, have a 90 percent failure rate (Avilés & Tufiño, 1998). Seasonality, on the other hand, is not a factor in our models (Figure 7.2C), as we suspect that absence of large insects is the more significant impediment for the formation of large social colonies at high latitudes, as some social species do occupy relatively seasonal environments (Avilés, 1997; Majer, *et al.*, 2013; Guevara & Avilés, 2015) (Figure 7.1).

Based upon our observations on *Anelosimus* species, which may be generally applicable to species with three-dimensional webs, insect size, rather than simply abundance, seems to be key for the presence of social species in a given habitat (Powers & Avilés, 2007; Yip, *et al.*, 2008). For example, upper elevation cloud forest habitats in Ecuador, where only subsocial *Anelosimus* occur, harbor as much or even greater total insect biomass than the lowland tropical rainforest, albeit encapsulated in mostly small insects (Powers & Avilés, 2007; Guevara & Avilés, 2007). The same situation likely applies to areas in North America where subsocial *Anelosimus* occasionally form colonies with multiple females, but do not grow to larger sizes (Riechert & Jones, 2008), and where colonial *Dictyna* and *Mallos* live singly in their irregular webs (Jackson, 1977): without large insects to capture cooperatively, the spiders are better off capturing prey individually or in very small groups (Yip, *et al.*, 2008). In general, and consistent with the prey size hypothesis, therefore, an abundance of small insects does not appear sufficient to support large social colonies.

For the orb-weaving colonial species, factors at the micro-habitat level appear key for their occurrence: namely the presence of open spaces with few web attachment points, but high prey abundance (Figure 7.2D) (Lubin, 1974; Smith, 1982). Because colonial orb-weavers, for the most part, capture prey individually, insect abundance, rather than size, may be the main requirement for colony formation. What is likely key in these cases is that the density of potential prey be sufficiently high to sustain spiders that are closely spaced together (Burkirk, 1975; Rypstra, 1979). For example, Smith (1985) found that individuals in *Philoponella republicana* adjust their spacing within colonies based on insect abundance. Coloniality in orb-weaving spiders would thus be a strategy to access areas of high prey density that, because of insufficient web attachment points, are inaccessible to solitary spiders (Figure 7.2D).

7.4.2 Biogeography

Social spiders have long been recognized as having a primarily tropical distribution, with congeneric subsocial species extending into higher latitudes and elevations (Avilés, 1997) (Figure 7.1). Our studies on the genus *Anelosimus* provide a working hypothesis to explain this distribution in this and perhaps other spider genera. We argue that the geographic distribution of social and subsocial *Anelosimus* species likely reflects the distribution of insect sizes, precipitation intensity, and predation risk along latitudinal and elevational gradients (Avilés, *et al.*, 2007; Powers & Avilés, 2007; Purcell & Avilés, 2008; Guevara & Avilés, 2015). Social *Anelosimus* species in America and Ecuador, for example, are restricted to low to mid-elevation wet tropical habitats, whereas subsocial species are excluded from the lowland tropical rainforest, but occur at higher elevations and latitudes and in dry environments (Agnarsson, 2006; Avilés, *et al.*, 2007; Purcell, 2011; Guevara & Avilés, 2015). Consistent with the prey size hypothesis, and the expectation that the maximum size insects can reach is directly proportional to ambient temperature (Makarieva, *et al.*, 2005), insects in warm lowland tropical rainforests are considerably larger, on average, than insects at the cooler higher elevations and higher latitudes (Guevara & Avilés, 2007; Powers & Avilés, 2007; Yip, *et al.*, 2008). Such macroecological patterns of insect body size would explain the decline in colony size and level of sociality of *Anelosimus* species both among and within species along elevation gradients in Ecuador (Avilés, *et al.*, 2007; Purcell & Avilés, 2007) and the absence of social *Anelosimus* in the temperate zones. This would also explain why even though colonies of the subsocial *A. studiosus* in North America may contain multiple females at higher latitudes (Riechert & Jones, 2008), they do not grow to larger sizes over multiple generations. Rain intensity and predation risk, in particular by ants, also increase with proximity to the lowland tropical rainforest (Purcell & Avilés, 2008; Hoffman & Avilés, 2017; see also Kaspari, *et al.*, 2000), which, according to the rain intensity and predation risk hypotheses, would explain the absence of subsocial *Anelosimus* from that habitat (Purcell & Avilés, 2008; Purcell, 2011). At intermediate elevations and latitudes, where insects, precipitation, and predation are intermediate in size or intensity, both social systems are viable. Here, differences in level of sociality among the various congeneric species may reflect differences in their life history traits.

Social spider species in other genera also have a mostly tropical distribution (Figure 7.1), but with a key distinction between species found in wet tropical environments (species in the genera *Parasteatoda* – formerly *Achaearaneae* – *Agelena*, *Aebutina*, *Theridion*, and *Tapinillus*) and those that occur under drier conditions (species in the genera *Stegodyphus*, *Mallos*, and one *Anelosimus*) (Figure 7.1). Wet tropical habitats likely present similar challenges and opportunities for species with dense tri-dimensional webs and a dependence on maternal care, namely strong rains and high predation rates, requiring the spiders to remain in their natal nest (e.g. Lubin & Robinson, 1982; Riechert, *et al.*, 1986; Avilés, 1997), and availability of large insects, allowing colonies to grow to large sizes. Species in the genus *Stegodyphus*, on the other hand, extend into the subtropics and occupy dry scrub savanna habitats (Majer, *et al.*, 2013) (Figure 7.1) where rain is likely not a problem for web maintenance. In drier savanna habitats, large insects still appear important (Ward, 1986; Crouch & Lubin, 2000), whereas predation, in particular by ants, may be a critical factor requiring group living (Henschel, 1998; Keiser, *et al.*, 2015). When analyzing the distribution of group- and solitary-living *Stegodyphus* across continents, Majer, *et al.* (2013) showed that group-living species are associated with high vegetation productivity, which in turn should be associated with high insect biomass.

Rather than being associated with particular habitats, local micro-habitat differences appear key for the colonial orb-weavers, as they occur in a variety of habitat types (Figure 7.3). Although most are either tropical (*Philoponella, Metabus, Leucage, Metepeira*) or subtropical (*Philoponella, Cyrtophora, Metepeira*), some extend into temperate regions (e.g. *C. citricola*, Leborge, *et al.*, 1998; *P. prominens*, Park, *et al.*, 1999) and occupy habitats ranging from tropical rainforests (*Philoponella, Cyrtophora, Metepeira,* and *Metabus*, Lubin, 1974; Buskirk, 1975; Smith, 1985; Breitwisch, 1989; Uetz, *et al.*, 2002) to grassland and desert environments (*Philoponella, Metepeira, Cyrtophora*, Smith, 1997; Uetz, *et al.*, 1982). Even single species may inhabit a variety of habitats, as is the case of *Parawixia bistriata*, which is found from semi-arid scrub to wet forests in southeastern South America (Fernández-Campón, 2010). *Cyrtophora citricola* is also broadly distributed in Asia, Africa, and around the Mediterranean (Johannesen, *et al.*, 2012). All evidence thus suggests that spatial patterning for the colonial spiders occurs at the microhabitat level, with colonies occupying open spaces where prey are abundant at least part of the year (Lubin, 1974; Smith, 1983; Uetz & Hieber, 1997).

7.4.3 Niches

By living in groups, social and colonial spiders are able to occupy ecological niches not accessible to solitary individuals: in the case of colonial spiders, open spaces (above creeks or between tree tops) that are too vast for the span of a single web; in the case of social ones, regions of frequent disturbance by strong rains or predators, but with an abundance of large prey (Figure 7.2). Recent studies in the genus *Anelosimus* suggest that further ecological niche differentiation among congeneric species may occur as a result of finer differences in body and colony size and degree of cooperation within colonies. For example, five *Anelosimus* species, ranging from nearly solitary to fully social, co-occur at Serra do Japi, Brazil. Here, species with larger colonies and more developed cooperation captured larger insects than less social ones (Guevara, *et al.*, 2011; Harwood & Avilés, 2013). More social species with larger colonies also tended to live inside the forest, where sturdier and longer-lived vegetation was available, whereas less social species were more common at the forest edge (Purcell, *et al.*, 2012). More recently, Agnarsson, *et al.* (2015) studied a local community of ten co-occurring subsocial *Anelosimus* species in Madagascar, where the species were found to overlap considerably in body size, web size, and microhabitat use, but to be overdispersed in their phenology.

Although not yet formally explored, spatial niche segregation based on degree of sociality and/or body size may characterize assemblages involving species in other genera. For example, subsocial or solitary *Stegodyphus* in southern Africa tended to occupy vegetation close to the ground, whereas social species occurred higher up on trees: *S. mimosarum*, at a variety of heights; *S. dumicola*, closer to the ground (Seibt & Wickler, 1988). Likewise, the microhabitats of the two social *Agelena* in Gabon differed – the larger *A. consociata* occurred in the forest understory, whereas the smaller *A. republicana* occurred along the forest edge, higher up from the ground, and on the distal part of branches (Darchen, 1967).

Potential niche differentiation based on body size and the tendency to live in groups has also been reported for colonial species. Lubin (1974) reported that solitary species with small body sizes (and small webs) in an assemblage of five *Cyrtophora* species in Papua New Guinea tended to occupy grassland habitats or forest edges. In contrast, larger species tended to be either colonial, thus taking advantage of spaces between treetops where environmental conditions were harsher (e.g. strong rains and winds) but resources more abundant, or else to live solitarily in the more stable (but rarer) forest interior microhabitats. Likewise, two sympatric *Philoponella* in southeastern Arizona differed in their microhabitat use: *P. arizonica*, which forms aggregations of only a few individuals, was found to build webs on shrubs and grasses, whereas *P. oweni*, which forms aggregations of up to a few dozen individuals, made use of rigid substrates such as fallen logs and niches in hollow trees and under rocks (Smith, 1997). These apparent microhabitat differences, which are associated with differences in colony size, point to a potentially key role of group living in niche segregation and species coexistence in a variety of spider systems (Guevara, *et al.*, 2011; Purcell, *et al.*, 2012).

7.5 The Role of Evolutionary History in Shaping Sociality in Spiders

When considering the phylogenetic distribution of spider sociality, it is clear that the presence and type of web play a critical role in determining whether sociality develops and the type of social system that evolves (Uetz & Hieber, 1997). Among species that do not rely on a web for prey capture, few have evolved social behavior that goes beyond early offspring care. These include the Australian subsocial spiders in the genera *Delena* and *Diaea* (Yip & Rayor, 2014), a few colonial ant-mimicking jumping spiders (Salticidae) (e.g. Jackson, *et al.*, 2008), and a handful of species in the family Scytodidae, which capture prey by spitting a sticky venomous substance, rather than using a web. Among the latter, subsocial (Yip & Rayor, 2014), colonial (Bowden, 1991), and social (Miller, 2006) species have been reported.

As discussed, among species that use webs for prey capture, there is a clear distinction between species that build orbicular webs and those that build irregular tridimensional webs (Figure 7.2) – the former giving rise to systems where individual territories are maintained, the latter being associated with non-territorial permanent sociality. By requiring that the spiders maintain individual prey capture webs, the orb web constrains the development of cooperative behaviors other than in the construction of a communal scaffolding for the web complex. As the orb-weaving colonial spiders belong to different genera in three families that contain largely solitary species, it is evident that they represent at least seven independent origins of colonial behavior, one for each of the genera to which the species belong (Figure 7.3).

Likewise, our updated list of 19 social species (not including the transitional *Anelosimus jabaquara*) (Figure 7.1) represents at least 16 independent origins of social behavior: 7 in *Anelosimus* (Agnarsson, *et al.*, 2006), 3 in *Stegodyphus* (Johannesen, *et al.*, 2007), 4 for the singleton species in *Aebutina*, *Mallos*, *Theridion*, and *Tapinillus*, and at least 1 each for *Parasteatoda* and *Agelena*, for which phylogenetic relationships are still unknown (Avilés, 1997). Notably, all these are genera that build some sort of irregular web (illustrations in Avilés, 1997), including, interestingly, *Tapinillus*, which is the only web-building genus among the lynx spiders (Oxyopidae), known otherwise for their non-web building cursorial habits (Avilés, 1994). With the exception of the Dictynidae, subsocial behavior is common in all these families and, in the genera *Anelosimus* and *Stegodyphus*, inferred to be ancestral to the social species (Agnarsson, *et al.*, 2006; Johannesen, *et al.*, 2007; Yip & Rayor, 2014).

II SOCIAL TRAITS

7.6 Traits of Social Species

The influence of environmental factors in animal sociality largely depends upon their interaction with relevant intrinsic features of the organisms. Thus, allomaternal care in the context of group living may allow species with low fecundity or long development time to occupy environments where solitary individuals would not be able to replace themselves (Stern & Foster, 1996; Jones, *et al.*, 2007). Likewise, cooperative hunting or information exchange may evolve when foragers of a given size or ability have the opportunity to exploit resources that are too large or too difficult for single individuals to capture or locate successfully. In the sections that follow, we consider the role of intrinsic organismal traits in shaping spider social systems.

7.6.1 Cognition and Communication

The web is a key medium for inter-individual attraction and communication in social, subsocial, and colonial spiders (Seibt & Wickler, 1988; Evans & Main, 1993; Bernard & Krafft, 2002; Trabalon & Assi-Bessekon, 2008). Web vibrations allow spiders to detect prey and to coordinate prey capture efforts (Krafft & Pasquet, 1991), as well as to detect and deter intruders (Henschel, 1998). The web may also mediate chemical communication. In the subsocial *Coelotes terrestris*, for instance, lipids deposited on the web change from a contact attractant pheromone during the juveniles' gregarious phase to an airborne repellent pheromone during dispersal (Trabalon & Assi-Bessekon, 2008). Likewise, juveniles and non-gravid females of the subsocial *Diaea socialis* are attracted to a chemical deposited onto the silk, but gravid females are not (Evans &

Main, 1993). Finally, in the social *Anelosimus eximius* hungry spiders are less attracted to silk and thus move to the periphery of the nest, thus possibly initiating its expansion (Bernard & Krafft, 2002). Cuticular hydrocarbons may also play a role. In the subsocial *Stegodyphus lineatus*, young exhibit group-specific cuticular hydrocarbons that may aid in discrimination against foreign spiders (Grinsted, *et al.*, 2011). In the social *Anelosimus eximius*, however, no discrimination against non-colony members has been detected, even though colonies exhibit quantitative differences in their cuticular chemical profiles (Pasquet, *et al.*, 1997).

7.6.2 Lifespan and Longevity

The relationship between sociality and lifespan in spiders has not yet been formally investigated, although these traits likely coevolve. Selection for extended maternal care, for instance, may explain the substantially longer life spans of females relative to males in subsocial and social spiders (e.g. Smith, 1983; Schneider, 1995; Avilés & Gelsey, 1998; Evans, 1998; Avilés & Salazar, 1999; Avilés, 2000). Likewise, nest sharing by multiple females may have been selected for in areas where slow offspring development (Jones, *et al.*, 2007; Riechert & Jones, 2008) or high risk of maternal mortality (Avilés, *et al.*, 2007) require that surrogate caregivers care for orphaned offspring. The protection conferred by a communal nest and the presence of surrogate caregivers may, in turn, have allowed the evolution of lower fecundity and/or longer development times in some social species relative to congeneric subsocial or solitary relatives.

7.6.3 Fecundity

Smith (1982) noted that social and colonial spiders appear to exhibit opposite trends on how group living affects fecundity: in social species, the number of eggs per sac appears to decrease as level of sociality increases, whereas in colonial species the opposite trend is apparent. Smith (1982) attributes this difference, which needs confirmation through studies that control for body size and phylogeny, to the fact that extended maternal care characterizes the social, but not the colonial species. With extended maternal care, greater levels of sociality would presumably involve greater investment in the offspring, which, in turn, would require a reduction in the number of offspring produced (Kullman, 1972). In contrast, in the absence of extended maternal care, additional resources gained by colonial females can be allocated to the production of more eggs. Thus, in the colonial *Philoponella oweni*, colonial females produced larger clutches than solitary ones because they occupied areas with greater prey abundance (Smith, 1983) and the colonial *Cyrtophora citricola* produced larger colonies and exhibited greater reproductive investment in prey-rich than in prey-poor areas (Mestre & Lubin, 2011). Fecundity also appears associated with colony size within social species, as females in larger colonies of social *Stegodyphus* species produced smaller clutches (Seibt & Wickler, 1988; Bilde, *et al.*, 2007), whereas the opposite was the case in *Anelosimus* (Avilés & Tufiño, 1998; Gonzaga & Vasconcellos-Neto, 2001). In these genera, eggs appear also to be larger in the social species compared to subsocial ones (Grinsted, *et al.* 2014).

7.6.4 Age at First Reproduction

There is some evidence that retention through adulthood of traits characteristic of early developmental stages, such as tolerance or philopatric tendencies, may have been the proximate mechanisms behind the origin of sociality in certain spider genera (e.g. Kullmann, 1972; Kraus & Kraus, 1990; Wickler & Seibt, 1993). That social *Stegodyphus* exhibit juvenile anatomical characteristic as adults, such as a small body size and low leg to the body ratio (Kraus & Kraus, 1988), is consistent with this suggestion (Kullmann, 1972; Kraus & Kraus, 1990). The process may be different in the genus *Anelosimus*, where the number of molts to reach maturity appears similar for species with very different body sizes (L. Avilés, *personal observations*). Unfortunately, similar data from colonial species are lacking.

7.6.5 Dispersal

New nests and colonies in social, subsocial, and colonial spiders are initiated when individuals disperse some distance away from the natal nest, either singly or in groups. In subsocial spiders, both males and females disperse singly at a juvenile to young adult instar (reviewed in Yip & Rayor, 2014). In social species, adult inseminated females may disperse either alone or in groups, or colonies may fission or bud-off daughter colonies with a mixed age composition (e.g. Vollrath, 1982; Lubin & Robinson, 1982; Seibt & Wickler, 1988; Avilés, 2000; reviewed in Avilés, 1997). Less is known of the dispersal patterns of the colonial spiders, although data from *Philoponella oweni* (Smith, 1983) and *Cyrtophora citricola* (Johannesen, *et al.*, 2012) suggest that juveniles of both sexes may disperse as a group or move between colonies, whereas adult females stay put (see also Uetz & Hieber, 1997). As previously discussed, these modes of dispersal are likely driven by competition for resources within the growing colonies (e.g. Gunderman, *et al.*, 1993; Schneider, 1995; Berger-Tal, et al., 2016). Additionally, dispersal, or colony relocation, may occur when colonies are heavily burdened by kleptoparasitic or other types of nest inquilines (e.g. Lubin & Robinson, 1982; Avilés, *et al.*, 2006; Pruitt, 2012). In all these cases, relatively short dispersal distances are involved (Vollrath, 1982; Lubin & Robinson, 1982; Seibt & Wickler, 1988; Avilés & Gelsey, 1998; Powers & Avilés, 2003; Bilde, *et al.*, 2005). These models of dispersal may lead to colony mixing in subsocial and colonial spiders, but not in social ones. Further mixing occurs when males undergo a mating dispersal phase in colonial and subsocial spiders (Smith, 1983; Schneider & Lubin, 1996; Avilés & Gelsey, 1998; Johanessen, *et al.*, 2012), a phase that is absent in the social species (e.g. Lubin, *et al.*, 2009; reviewed in Avilés & Purcell, 2012). Corcobado, *et al.* (2012) showed that the different dispersal tendencies across *Anelosimus* species are reflected in a reduced leg to body size ratio – a correlate of dispersal ability – in the more social species, in particular in males.

Two modes of longer distance dispersal have been reported for the social species. Schneider, *et al.* (2001) found that adult females of the African *Stegodyphus dumicola* may disperse by ballooning, and Avilés (2000) found that colonies of the neotropical

Aebutina binotata underwent a nomadic phase prior to egg laying during which nest mates migrated as a group for several weeks and potentially hundreds of meters.

7.6.6 Other Traits: Body Size

There is no systematic analysis of the relationship between spider sociality and body size, but some trends seem apparent. In *Stegodyphus*, for example, social species have been shown to exhibit smaller body sizes than their subsocial or solitary sister species, a pattern consistent with the suggested origin of their sociality via paedogenesis (Kraus & Kraus, 1988). By increasing offspring survival, sociality may also have released a constraint on body size by making small, low fecundity species viable, as some of the smallest species in the genus *Anelosimus* (*A. rupununi, A. lorenzo*, and *A. oritoyacu*, Agnarsson, 2006) are social.

Within some social species, individuals in larger colonies have been found to be smaller (Seibt & Wickler, 1988; Bilde, *et al.*, 2007) or to exhibit greater body size variability (Rypstra, 1993; Ebert, 1998; Amir, *et al.*, 2000). In the Neotropical *Theridion nigroannulatum*, females may occur in two discrete size classes (Avilés, *et al.*, 2006). Such patterns may result from resource competition within the colonies combined, in some cases, with unequal distribution of resources. In addition to fecundity (e.g. Seibt & Wickler, 1988; Gonzaga & Vasconcellos-Neto, 2001), adult female body size may also affect competitive and foraging ability, as larger females may have an advantage ingesting food (Riechert, *et al.*, 1986; Amir, *et al.*, 2000). Larger females in some species also appear more likely to disperse (Lubin & Robinson, 1982; Gonzaga & Vasconcellos-Neto, 2001).

7.7 Traits of Social Groups

As we discuss earlier, not only the size, but also the structure and dynamics of social groups should reflect how the interaction between organisms and their environments shape the fitness of individuals living solitarily and in groups.

7.7.1 Genetic Relatedness

As would be expected from their highly subdivided population structure, social spiders exhibit levels of intracolony relatedness exceeding those of full sibs and approaching levels found in clonally reproducing organisms (e.g. Smith & Engel, 1994; Johannesen, *et al.*, 2009; Agnarsson, *et al.*, 2010; Settepani, *et al.*, 2014). Subsocial species, on the other hand, which are thought to resemble the ancestral system from which the social spiders originated, exhibit unremarkable levels of relatedness, ranging from half to full sibs, depending upon whether more than one father has sired the progeny (reviewed in Yip & Rayor, 2014). Within-colony relatedness may be further diluted in these species by dispersing juveniles joining neighboring colonies (Evans & Goodisman, 2002; Yip, *et al.*, 2012). Where subsocial colonies with multi-females occur, relatedness among co-nesting females has been found to be only slightly greater than that between pairs of neighboring solitary individuals (e.g. $r = 0.25$ vs. 0.18, Duncan, *et al.*, 2010). Given dispersal patterns that lead to colony mixing (see above), within-colony relatedness in the colonial species is expected to be low. In fact, Johannesen, *et al.* (2012) found that in *Cyrtophora citricola* most colonies had levels of relatedness lower than that of half sibs (range of $r = 0.0$ to 0.67) and that males tended to be less related to nestmates than females.

In general, we argue that in all three types of spider social organization, levels of relatedness are a result, rather than a cause, of the social system. Thus, it may be tempting to suggest that the colonial spiders exhibit low levels of cooperation because of the low relatedness within their colonies. We argue, however, that the direction of causality is the reverse: due to the constraints imposed by the orbicular web, colonial spiders are limited in the extent to which they are able to cooperate, resulting in benefits of group living not being sufficiently high to select for individuals remaining in the natal group beyond a certain age. As a result, juveniles disperse, males venture outside their colonies to find mates, and relatedness within colonies is low. In the social spiders, on the other hand, benefits of group living and costs of giving up philopatry (i.e. dispersing and forfeiting benefits of group living) would be substantial enough for both males and females to remain in the natal group, thus causing levels of relatedness to be high.

7.7.2 Group Composition, Size, Breeding Structure and Sex Ratio

Social and subsocial species exhibit significant variability in colony size, largely reflecting the size at which colonies become established and when dispersal happens. In subsocial species colonies are typically initiated by single females and contain the offspring of a single clutch. Exceptions include the Australian *Delena cancerides*, where colonies may contain the offspring of up to five clutches (Yip & Rayor, 2011) and species where in certain areas colonies may contain multiple mothers and their offspring (e.g. Lubin & Bilde, 2007; Riechert & Jones, 2008; Avilés & Harwood, 2012).

In contrast, social spider colonies grow through internal recruitment to contain dozens, hundreds, thousands, or tens of thousands of individuals (reviewed in Avilés, 1997). They are thus not only social groups but also self-sustaining populations. In addition to giving rise to daughter colonies after reaching relatively large sizes, social spider colonies may also be subject to high rates of extinction (Avilés, 1997; see also Crouch & Lubin, 2001). Avilés (1997) suggested that such high colony extinction rates may reflect a boom and bust pattern of colony growth arising from a combination of fast rates of growth, discrete generations, and negative density dependence (Avilés, 1997, 1999; Hart & Avilés, 2014). The suddenness of such extinction events (L. Avilés, *personal observations*) suggests that demographic processes, such as overproduction of offspring followed by scramble competition, may be involved (Sharpe & Avilés, 2016), although disease (Crouch & Lubin, 2001) or excessive inquiline or parasite burden (Riechert, *et al.*, 1986) have also been implicated. There is evidence that social spider colonies also exhibit Allee effects, i.e. a minimum colony size below which individuals are unable to replace themselves (Figure 7.4A-C) (Avilés & Tufiño, 1998; Bilde, *et al.*, 2007; Hart & Avilés, 2014).

Much less is known about the population structure and dynamics of the orb-weaving colonial species. Their colonies also range in size from few to thousands (reviewed by Uetz & Hieber, 1997), with broad variability within species (e.g. Smith, 1982; Leborge, *et al.*, 1998; Whitehouse & Lubin, 2005). With the exception of *Parawixia bistriata*, whose colonies constitute single-family groups (Fowler & Diehl, 1978), colonies of other species contain multiple females and their offspring and tend to be of mixed genetic origin. Whitehouse & Lubin (2005) suggest that colonial spiders would lack a boom and bust pattern of growth because of their ability to respond to current conditions by dispersing, potentially resulting in extremely long lived colonies in some species (e.g. Lubin, 1974).

In addition to their strong population subdivision and inbreeding, social spiders are unusual among social organisms in having highly female-biased sex ratios (reviewed in Avilés, 1997; Lubin & Bilde, 2007; Avilés & Purcell, 2012). The mechanisms for the biased sex ratios – 8 to 28 percent males, already at the embryo stage – are still unknown; spiders are diplodiploid with chromosomal sex determination and thus males, the heterogametic sex, are expected to produce equal male and female sperm as a product of meiosis (reviewed in Avilés & Maddison, 1991). According to Avilés (1993a), the biased sex ratios may have evolved through intercolony selection for faster rates of colony growth and proliferation, made possible by the isolation of the colony lineages and their fast rate of turnover. Molecular studies have shown that, in fact, colonies are fixed to different alleles, even when close in space, and that individuals and colonies exhibit extremely low (to nil) genetic variability (Lubin & Crozier, 1985; Roeloffs & Riechert, 1988; Smith & Engel, 1994; Smith & Hagen, 1996; Smith, *et al.*, 2009; Agnarsson, *et al.*, 2010; Settepani, *et al.*, 2014). The only social species that is apparently outbred and lacks female biased sex ratios, the social lynx spider *Tapinillus* sp. (Avilés, 1994), belongs to a family of fast moving cursorial spiders whose males are likely capable of efficient dispersal.

The more mixed population structure of subsocial and colonial spiders, on the other hand, is expected to result in 1:1 sex ratios according to the usual Fisherian argument (Fisher, 1930). Primary sex ratios have indeed been shown to be unbiased in two subsocial *Anelosimus* species (Avilés & Maddison, 1991; Avilés & Harwood, 2012) and are suspected to be so in other species (reviewed in Yip & Rayor, 2014). The only clear exceptions are two subsocial *Diaea* where individuals apparently mate before dispersing and sex ratios are female biased (Rowel & Main, 1992; Avilés, 1997). It cannot be ruled out, however, that in some subsocial or transitional species, such as *A. elegans*, *A. jabaquara*, and in northernmost populations of *A. studiosus*, a fraction of individuals may mate before dispersing, as somewhat biased subadult sex ratios have been reported (Marques, *et al.*, 1998; Avilés & Harwood, 2012; J. Pruitt, *unpublished data*). No biased sex ratios have been reported for any of the colonial spiders. Johannesen, *et al.* (2012) confirmed that *Cyrtophora citricola* colonies are highly polymorphic, with little genetic differentiation among them. It does appear, therefore, that, despite short dispersal distances leading to population viscosity and mild inbreeding in the subsocial spiders (Johannesen & Lubin, 2001; Evans & Goodisman, 2002; Bilde, *et al.*, 2005; Ruch, *et al.*, 2009), by and large the conditions are not present in most subsocial or colonial spiders for biased sex ratios to evolve.

The reasons for the origin of the highly inbred systems of the social spiders and their short- and long-term consequences are of interest. The extreme philopatry of members of both sexes is likely the result of benefits of philopatry far exceeding the costs of inbreeding depression for both males and females (Avilés & Purcell, 2012). Furthermore, the transition to inbreeding could have been facilitated by mild levels of inbreeding in the ancestral subsocial species (Bilde, *et al.*, 2005), and by maternal and social effects buffering against extreme inbreeding depression (Avilés & Bukowski, 2006). In the long term, however, it has been argued that the inbred social spiders would represent evolutionary dead ends due to low genetic variability of individuals, colonies, and, due to the dynamics of extinction and recolonization of the colony lineages, entire metapopulations (Johanessen, *et al.*, 2002, 2009; Agnarsson, *et al.*, 2006, 2013). As a result, inbred social spiders are likely to be more susceptible to disease and parasites and less able to respond to changing environments than their outbred less social counterparts (Avilés, 1997; Lubin & Bilde, 2007; Avilés & Purcell, 2012). The result would be species that are short-lived and/or unable to speciate, consistent with the fact that the vast majority of inbred social spider species represent isolated tips on the phylogenetic tree of spiders (Agnarsson, *et al.*, 2006, 2013; Johannesen, *et al.*, 2007; Settepani, *et al.*, 2014).

III SOCIAL SYNTHESIS

7.8 A Summary of Spider Sociality

Social, subsocial, and colonial spiders provide a fascinating example of the interplay between extrinsic and intrinsic factors in the evolution of animal social systems, as well as one of the finest examples of how abiotic and biotic factors combine to produce broad-scale patterns of sociality. Among the intrinsic factors, the type of web spiders build appears to either enable or hinder the development of certain types of cooperative behaviors, which, in turn, will determine whether the spiders can take advantage of opportunities afforded by particular environments (Figure 7.2). Thus, by making cooperative prey capture possible, irregular webs allow access to insects that are many times larger than the spiders themselves, whereas colonial orb weavers are limited by their own size in the size of insects they can capture. By providing a common space for co-habitation, irregular webs also make communal brood care possible, whereas care is limited to egg sac guarding and done individually in the colonial species.

The interaction between intrinsic and extrinsic factors should then yield a characteristic fitness for individuals living solitarily and in groups (Figure 7.2). The shape and magnitude of this function would then indicate whether or not groups should form and when dispersal should happen. All evidence suggests that fitness is likely a humped function of group size in permanent-social and colonial spiders (Figure 7.4), but a flat or declining function of the number of offspring in the group, in subsocial ones. Subsocial young likely remain in the natal nest to avoid costs of dispersing at an early age and to take advantage of benefits derived from the nest itself and the presence of the mother. As they grow, however, their energetic needs would eventually exceed what their environments can provide, shifting the balance in favor of dispersal. In social and colonial spiders, on the other hand, fitness likely peaks at group sizes that exceed the number of offspring in a clutch, favoring colonies that contain dozens, hundreds, or thousands of spiders, depending on the species. Given their limited repertoire of cooperative behaviors, the fitness hump is likely small in colonial spiders, whereas a much taller hump is expected in the more cooperative social species (Figures 7.2E & F). Dispersal would be predicted when the fitness of individuals in increasing large groups drops below that of individuals living alone or in small groups. The shape and magnitude of the fitness function should further indicate whether solitary living is possible and the minimum viable group size. With costly webs and their offspring's dependence upon maternal care, solitary living is likely untenable for social species and their congeners in the core habitats where the social species occur. With webs that are presumably considerably less expensive, larger clutch sizes, and little reliance on maternal care, colonial spiders, in contrast, should be able to live solitarily in the same environments where congeneric colonial species occur. Finally, the differences in the shape of the fitness function between social and colonial spiders may explain their vastly different breeding systems: the combination of high benefits of group living and inability to persist as solitary individuals would bind social spiders to their colonies, whereas colonial spiders would maintain their ability to move in and out of colonies and also live solitarily.

A further significant difference between social and colonial spiders is on the scale at which their presence or absence is determined – at the microhabitat level in the colonial spiders, at the biome or habitat level in the social ones. Colonial spiders can occupy a variety of habitats (Figure 7.3), from deserts to rainforests, as long as within them prey rich microhabitats exist that can be accessed when webs are aggregated. In contrast, social spiders live in environments where factors such as strong rains or high predation rates would render them inaccessible to solitary or subsocial species with costly webs and strict dependence on maternal care (i.e. the rain intensity and predation rate hypotheses). Likewise, scaling properties of tri-dimensional webs make the availability of large insects necessary for the development of large social colonies (the insect size hypothesis). As the maximum size insects can reach is positively correlated with ambient temperature (Makarieva, *et al.*, 2005), this largely would explain why the social spiders are concentrated at low elevation tropical areas of the world (Figure 7.1). Lack of seasonality in these areas, on the other hand, appears secondary, as a few of the social species (e.g. *Stegodyphus* spp, *Mallos gregalis, Anelosimus dubiosus*) do occur in relatively seasonal environments (Figure 7.1). The question of how life history traits, other than web type, interact with these environmental factors and with level of sociality is yet to be systematically explored. We believe that therein lays the next frontier in the study of spider sociality.

7.9 Comparative Perspectives on Spider Sociality

Relative to other social species, vertebrate in particular, subsocial spiders can be said to resemble species with helpers at the nest, where offspring are retained within the natal group for an extended period of time during which they help each other and their parents with communal activities. They differ from these vertebrate systems, however, in producing only one cohort of offspring within the nests (with the exception of *Delena cancerides*), in having offspring disperse prior to reaching reproductive maturity, and in having only mothers (not fathers) initiate the groups and provide parental care. Eusocial insect systems, such as termites and ants, are also extended family groups, but with reproductive division of labor and multiple cohorts of offspring. In these cases, and in some of the vertebrate cooperative breeding systems, non-reproductive roles are typically socially enforced, either through a morphologically determined caste system or by behavioral or hormonal suppression of reproduction (e.g. Oster & Wilson, 1978; Jarvis, *et al.*, 1994). In the spiders, on the other hand, there is no evidence for socially-enforced caste differentiation. Instead, it appears that competition for resources within the growing colonies may be responsible for a fraction of females in certain species failing to reproduce as colony size increases. Nonetheless, since non-reproducing females do participate in communal activities, it can be argued that their continuing presence and activity is reinforced by the indirect fitness benefits individuals would gain from the increased survival and productivity of their colonies.

The social system of colonial spiders, on the other hand, is more akin to that of colonial birds or mammals (e.g. colonial marine birds, prairie dogs) where single individuals or basic family units aggregate to take advantage of information sharing opportunities or predator protection (e.g. Hoogland, 1981; Rolland, *et al.*, 1998). Thus, colonial spiders also appear to aggregate to take advantage of locally abundant resources and also benefit from predator protection. Subsocial and colonial spiders are, thus, not unlike a variety of other social organisms. It is the social spiders, however, that are quite a unique social system.

Social spiders have at least three characteristics that make them unlike the vast majority of social organisms: (1) their colonies are not only social groups, but also self-sustaining populations that grow through internal recruitment over multiple generations; (2) they are highly inbred; and (3) they exhibit highly female 0 biased sex ratios. Only two arthropod systems (spider mites and psocids in the F. Archipsocidae), and no vertebrate ones, exhibit a comparable combination of features (reviewed in Avilés & Purcell, 2012). Among vertebrates, the naked mole rats are highly inbred because their colonies are initiated by single females who have been inseminated in their natal group prior to dispersal (Jarvis, *et al.*, 1994). With the founding female being the sole breeder, however, naked mole rat colonies are extended family groups rather than self-sustaining populations. Sex ratios are also not female biased (Brett, 1991). Species in two arthropod groups – psocids belonging to the Family Archipsocidae and spider mites – exhibit remarkably similar social systems to those of the social spiders. In both of these groups, colonies are housed within dense, cooperatively-spun webs that surround a piece of vegetation and are thought to serve an antipredator function (Mockford, 1957; New, 1973; Mori & Saito, 2005). Colonies of both of these organisms are initiated by females who have apparently been inseminated in their natal nest prior to dispersal and who produce offspring with highly female biased sex ratios (Mockford, 1957; Mitchell, 1973; Norton, *et al.*, 1993). As in the spiders, there are no reproductive castes and individuals remain in the natal nest throughout their lives, likely mating with each other to produce successive generations. In spider mites, Saito & Mori (2005) note that females disperse only to establish new nests and males disperse rarely, whereas New (1973) notes that psocid colonies may grow to contain many hundreds of individuals and may also fragment or coalesce with neighboring colonies. Notably, in addition to producing dense webs that are likely costly to build and to maintain, in both psocids and mites, as in the spiders, individuals feed and carry all their activities without leaving the shelter of the web. Unlike the spiders, neither group is predatory (both feed upon vegetable matter) and both have a haplodiploid sex determination system, whereas spiders are diplodiploid with chromosomal sex determination (reviewed in Avilés & Purcell, 2012). Finally, psocids have distinct dispersing and sedentary morphs (New, 1973), whereas the spiders are monomorphic.

According to Avilés & Purcell (2012), three other arthropod groups exhibit an association between sociality, inbreeding, and biased sex ratios: socially parasitic ants with intra-nest mating (*Myrmica* and *Myrmoxenus*), Australian gall-inducing thrips, and ambrosia beetles. In all three cases, the groups – established within host ant colonies, self-induced galls, and tree galleries, respectively – are initiated by single females who have been inseminated by a nest mate prior to dispersal and who will produce a female-biased progeny (Kirkendall, 1983; Buschinger, 1989; Chapman, *et al.*, 2000). Rather than being self-sustaining populations, as in the social spiders, these are family groups that dissolve once offspring complete development. In this sense, these systems resemble those of the inbred subsocial *Diaea*, rather than those of the social spiders. Unlike subsocial spiders, the thrips, ants, and beetles may produce multiple cohorts of offspring within the host resource; the eusocial thrips also exhibit reproductive castes (sterile soldiers and winged reproductive females) (Chapman, *et al.*, 2000; Chapter 6), whereas the socially parasitic ants with intra-nest mating have lost the worker caste and produce queens and reproductive males only (Buschinger, 1989).

7.10 Concluding Remarks

We have seen how in spiders, diverse social systems have arisen as a result of an interaction between intrinsic features of the organisms and the environments in which they live. Group living is predicted when the product of this interaction causes one or more components of fitness to increase as group size increases, so that overall fitness is maximum at some intermediate group size. For the social spiders, we have suggested that group living and cooperation may be strategies to colonize environments that, given particular characteristic of the organisms, are inaccessible to solitary individuals (e.g. species with dense, expensive webs in environments with strong torrential rains). In the colonial spiders, on the other hand, group living may be a strategy to take advantage of an ecological opportunity – open spaces with abundant prey – in environments where solitary life is nonetheless possible. The former condition will result in obligate group living, whereas the latter would allow for facultative coloniality. We suggest that similar principles should apply across the animal kingdom as an explanation for the diversity of animal social systems. As the research described in this chapter suggests, social, subsocial, and colonial spiders are ideal study systems on which to test the assumptions and predictions of these models.

References

Agnarsson, I. (2006) A revision of the New World *eximius* lineage of *Anelosimus* (Araneae, Theridiidae) and a phylogenetic analysis using worldwide exemplars. *Zoological Journal of the Linnean Society*, **146**, 453–593.

Agnarsson, I., Avilés, L., Coddington, J., & Maddison, W. (2006) Sociality in Theridiid spiders: Repeated origins of an evolutionary dead end. *Evolution*, **60**, 2342–2351.

Agnarsson, I., Maddison, W. P., & Avilés, L. (2010) Complete separation along matrilines in a social spider metapopulation inferred from hypervariable mitochondrial DNA region. *Molecular Ecology*, **19**, 3052–3063.

Agnarsson, I., Avilés, L., & Maddison, W. P. (2013) Loss of genetic variability in social spiders: Genetic and phylogenetic consequences of population subdivision and inbreeding. *Journal of Evolutionary Biology*, **26**, 27–37.

Agnarsson, I., Gotelli, N. J., Agostini, D., & Kuntner, M. (2015) Limited role of character displacement in the coexistence of congeneric *Anelosimus* spiders in a Madagascan montane forest. *Ecography*, **38**, 001–011.

Amir, N., Whitehouse, M. E. A., & Lubin, Y. (2000) Food consumption rates and competition in a communally feeding social spider, *Stegodyphus dumicola (Eresidae) Journal of Arachnology*, **28**, 195–200.

Avilés, L. (1993a) Interdemic selection and the sex-ratio - a social spider perspective. *American Naturalist*, **142**, 320–345.

(1993b) Newly-discovered sociality in the neotropical spider *Aebutina binotata* Simon (Dictynidae) *Journal of Arachnology*, **21**, 184–193.

(1994) Social behavior in a web-building lynx spider, *Tapinillus* sp (Araneae, Oxyopidae). *Biological Journal of the Linnean Society*, **52**, 163–176.

(1997) Causes and consequences of cooperation and permanent-sociality in spiders. *In:* Choe, J. C. & Crespi, B. J. (eds.) *Evolution of Social Behavior in Insects and Arachnids*. Cambridge, MA: Cambridge University Press, pp. 476–498.

(1999) Cooperation & non-linear dynamics: An ecological perspective on the evolution of sociality. *Evolutionary Ecology Research*, **1**, 459–477.

(2000) Nomadic behaviour and colony fission in a cooperative spider: Life history evolution at the level of the colony? *Biological Journal of the Linnean Society*, **70**, 325–339.

Avilés, L. & Bukowski, T. C. (2006) Group living and inbreeding depression in a subsocial spider. *Proceedings of the Royal Society of London B*, **273**, 157–163.

Avilés, L. & Gelsey, G. (1998) Natal dispersal and demography of a subsocial *Anelosimus* species and its implications for the evolution of sociality in spiders. *Canadian Journal of Zoology*, **76**, 2137–2147.

Avilés, L. & Harwood, G. (2012) A quantitative index of sociality and its application to group-living spiders and other social organisms. *Ethology*, **118**, 1219–1229.

Avilés, L. & Maddison, W. (1991) When is the sex ratio biased in social spiders?: Embryo and male meiosis chromosome studies in *Anelosimus* spp. *Journal of Arachnology*, **19**, 126–135.

Avilés, L. & Purcell, J. (2012) The evolution of inbred social systems in spiders and other organisms: From short-term gains to long-term evolutionary dead ends?. *Advances in the Study of Behavior*, **44**, 99–133.

Avilés, L. & Salazar, P. (1999) Notes on the social structure, life cycle, and behavior of *Anelosimus rupununi*. *Journal of Arachnology*, **27**, 497–502.

Avilés, L. & Tufiño, P. (1998) Colony size and individual fitness in the social spider *Anelosimus eximius*. *The American Naturalist*, **152**, 403–418.

Avilés, L., Maddison, W. P., Salazar, P. A., Estevez, G., Tufino, P., & Cañas, G. (2001) Social spiders of the Ecuadorian Amazonia, with notes on six previously undescribed social species, *Revista Chilena De Historia Natural*, **74**, 619–638.

Avilés, L., Maddison, W., & Agnarsson, I. (2006) A new independently derived social spider with explosive colony proliferation and a female size dimorphism. *Biotropica*, **38**, 743–753.

Avilés, L., Agnarsson, I., Salazar, P. A., Purcell, J., Iturralde, G., *et al.* (2007) Natural history miscellany - Altitudinal patterns of spider sociality and the biology of a new midelevation social *Anelosimus* species in Ecuador. *The American Naturalist*, **170**, 783–792.

Berger-Tal, R., Berner-Aharon, N., Aharon, S., Cristina Tuni, C., & Lubin, Y. (2016). Good reasons to leave home: proximate dispersal cues in a social spider. *Journal of Animal Ecology*, **85**, 1035–1042.

Bernard, A. & Krafft, B. (2002) Silk attraction: Base of group cohesion and collective behaviours in social spiders. *Comptes Rendus Biologies*, **325**, 1153–1157.

Bilde, T., Coates, K. S., Birkhofer, K., *et al.* (2007) Survival benefits select for group living in a social spider despite reproductive costs. *Journal of Evolutionary Biology*, **20**, 2412–2426.

Bilde, T., Lubin, Y., Smith, D., Schneider, J. M., Maklakov, A. A. (2005) The transition to social inbred mating systems in spiders: Role of inbreeding tolerance in a subsocial predecessor. *Evolution*, **59**, 160–174.

Bilde, T. & Lubin, Y. (2011) Group living in spiders: Cooperative breeding and coloniality. *In:* Herberstein, M.E. (ed.) *Spider Behavior, Flexibility and Versatility*. New York: Cambridge University, pp. 275–307.

Bilde, T., Lubin, Y., Smith, D., Schneider, J., & Maklakov, A. (2005) The transition to social inbred mating systems in spiders: Role of inbreeding tolerance in a subsocial predecessor. *Evolution*, **59**, 160–174.

Binford, G. J. & Rypstra, A. L. (1992) Foraging behavior of the communal spider, *Philoponella-republicana* (Araneae, Uloboridae). *Journal of Insect Behavior*, **5**, 321–335.

Bowden, K. (1991) The evolution of sociality in the spitting spider, *Scytodes-fusca* (Araneae, Scytodidae): Evidence from observations of intraspecific interactions. *Journal of Zoology*, **223**, 161–172.

Breitwisch, R. (1989) Prey capture by a West-African social spider (Uloboridae, *Philoponella* sp). *Biotropica*, **21**, 359–363.

Brett, R. A. (1991) The population structure of naked mole-rat colonies. *In:* Sherman, P.W., Jarvis, J. U. M., & Alexander, R.D. (eds.) *The Biology of the Naked Mole-Rat*. Princeton: Princeton University Press, pp. 97–136.

Buschinger, A. (1989) Evolution, speciation, and inbreeding in the parasitic ant genus *Epimyrma* (Hymenoptera, Formicidae). *Journal of Evolutionary Biology*, **2**, 265–283.

Buskirk, R. E. (1975) Coloniality, activity patterns and feeding in a tropical orb-weaving spider. *Ecology*, **56**, 1314–1328.

Chapman, T. W., Crespi, B. J., Kranz, B. D., & Schwarz, M. P. (2000) High relatedness and inbreeding at the origin of eusociality in gall-inducing thrips. *Proceedings of the National Academy of Sciences USA*, **97**, 1648–1650.

Corcobado, G., Rodriguez-Girones, M. A., Moya-Larano, J., & Avilés, L. (2012) Sociality level correlates with dispersal ability in spiders. *Functional Ecology*, **26**, 794–803.

Crouch, T. & Lubin, Y. (2000) Effects of climate and prey availability on foraging in a social spider, *Stegodyphus mimosarum* (Araneae, Eresidae). *Journal of Arachnology*, **28**, 158–168.

(2001) Population stability and extinction in a social spider *Stegodyphus mimosarum* (Araneae: Eresidae). *Biological Journal of the Linnean Society*, **72**, 409–417.

Darchen, R. (1967) Une nouvelle araignée sociale du Gabon *Agelena republicana* Darchen (Aranéide labidognathe). *Biologia Gabonica*, **3**, 31–42.

Duncan, S. I., Riechert, S. E., Fitzpatrick, B. M., & Fordyce, J. A. (2010) Relatedness and genetic structure in a socially polymorphic population of the spider *Anelosimus studiosus. Molecular Ecology*, **19**, 810–818.

D'Andrea, M. (1987) Social behavior in spiders (Arachnida, Araneae). *Italian Journal of Zoology, Monograph,* **3**, 1–156.

Ebert, D. (1998) Behavioral asymmetry in relation to body weight and hunger in the tropical social spider *Anelosimus eximius* (Araneae, Theridiidae). *Journal of Arachnology*, **26**, 70–80.

Emlen, S. T. (1982) The evolution of helping. I. An ecological constraints model. *The American Naturalist*, **119,** 29–39.

Evans, T. (1998) Factors influencing the evolution of social behaviour in Australian crab spiders (Araneae: Thomisidae). *Biological Journal of the Linnean Society*, **63**, 205–219.

Evans, T. & Goodisman, M. (2002) Nestmate relatedness and population genetic structure of the Australian social crab spider *Diaea ergandros* (Araneae: Thomisidae). *Molecular Ecology*, **11**, 2307–2316.

Evans, T. A. & Main, B. Y. (1993) Attraction between social crab spiders - silk pheromones in *Diaea-socialis. Behavioral Ecology*, **4**, 99–105.

Evans, T. A., Wallis, E. J., & Elgar, M. A. (1995) Making a meal of mother. *Nature*, **376**, 299–299.

Fernández-Campón, F. (2007) Group foraging in the colonial spider *Parawixia bistriata* (Araneidae): Effect of resource levels and prey size. *Animal Behaviour*, **74**, 1551–1562.

(2010) Cross-habitat variation in the phenology of a colonial spider: Insights from a reciprocal transplant study. *Naturwissenschaften*, **97**, 279–289.

Fisher, R. A. (1930) *The Genetical Theory of Natural Selection*. Dover, New York.

Fowler, H. G. & Diehl, J. (1978) Biology of a Paraguayan colonial orb-weaver, *Eriophora bistriata* (Rengger) (Araneae, Araneidae). *Bulletin of the British Arachnological Society*, **4**, 241–250.

Gonzaga, M. O. and Vasconcellos-Neto, J. (2001) Female body size, fecundity parameters and foundation of new colonies in *Anelosimus jabaquara* (Araneae, Theridiidae). *Insectes Sociaux*, **48**, 94–100.

Grinsted, L., Bilde, T., & d'Ettorre, P. (2011) Cuticular hydrocarbons as potential kin recognition cues in a subsocial spider. *Behavioral Ecology*, **22**, 1187–1194.

Grinsted, L., Pruitt, J. N., Settepani, V., & Bilde, T. (2013) Individual personalities shape task differentiation in a social spider. *Proceedings of the Royal Society of London B*, 280, 20131407.

Grinsted, L., Breuker, C. J., & Bilde, T. (2014) Cooperative breeding favors maternal investment in size over number of eggs in spiders. *Evolution*, **68**, 1961–1973.

Guevara, J. & Avilés, L. (2007) Multiple techniques confirm elevational differences in insect size that may influence spider sociality. *Ecology*, **88**, 2015–2023.

(2011) Sociality and resource use: Insights from a community of social spiders in Brazil. *Behavioral Ecology*, **22**, 630–638.

(2015) Ecological predictors of spider sociality in the Americas. *Global Ecology and Biogeography*, **24**, 1181–1191.

Gundermann, J. L., Horel, A., & Krafft, B. (1993) Experimental manipulations of social tendencies in the subsocial spider *Coelotes-terrestris*. *Insectes Sociaux*, **40**, 219–229.

Hart, E. M. & Avilés, L. (2014) Reconstructing local population dynamics in noisy metapopulations: The role of random catastrophes and Allee effects. *Plos ONE*, **9,** e110049.

(2013) Differences in group size and the extent of individual participation in group hunting may contribute to differential prey-size use among social spiders. *Biology Letters*, **9**, 20130621.

Henschel, J. R. (1998) Predation on social and solitary individuals of the spider *Stegodyphus dumicola* (Araneae, Eresidae). *Journal of Arachnology*, **26**, 61–69.

Hoffman, C. R. & Avilés, L. (2017). Rain, predators, and spider sociality: a manipulative experiment. *Behavioral Ecology*, in press.

Hoogland, J. L. (1981) The evolution of coloniality in white-tailed and black-tailed prairie dogs (Sciuridae, *Cynomyus leucurus* and *Cynomus ludovicianus*). *Ecology*, **62,** 252–272.

Jackson, R. R. (1977) Comparative studies of *Dictyna* and *Mallos* (Araneae:Dictynidae): III. *Prey and feeding behavior. Psyche*, **83**, 267–280.

Jackson, R. R., Nelson, X. J., & Salm, K. (2008) The natural history of *Myrmarachne melanotarsa*, a social ant-mimicking jumping spider. *New Zealand Journal of Zoology*, **35**, 225–235.

Jakob, E. M. (2004) Individual decisions and group dynamics: Why pholcid spiders join and leave groups. *Animal Behavior*, **68**, 9–20.

Jarvis, J. U. M., Oriain, M. J., Bennet, N. C., & Sherman, P. W. (1994) Mammalian eusociality - a family affair. *Trends in Ecology and Evolution*, **9**, 47–51.

Johannesen, J. & Lubin, Y. (2001) Evidence for kin-structured group founding and limited juvenile dispersal in the sub-social spider *Stegodyphus lineatus* (Araneae, Eresidae). *Journal of Arachnology*, **29**, 413–422.

Johannesen, J., Hennig, A., Dommermuth, B., & Schneider, J. M. (2002) Mitochondrial DNA distributions indicate colony propagation by single matri-lineages in the social spider *Stegodyphus dumicola* (Eresidae). *Biological Journal of the Linnean Society*, **76**, 591–600.

Johannesen, J., Lubin, Y., Smith, D., Bilde, T., & Schneider, J. (2007) The age and evolution of sociality in *Stegodyphus* spiders: A molecular phylogenetic perspective. *Proceedings of the Royal Society of London B*, **274**, 231–237.

Johannesen, J., Wickler, W., Seibt, U., & Moritz, R. F. A. (2009) Population history in social spiders repeated: Colony structure and lineage evolution in *Stegodyphus mimosarum* (Eresidae). *Molecular Ecology*, **18**, 2812–2818.

Johannesen, J., Wennmann, J. T., & Lubin, Y. (2012) Dispersal behaviour and colony structure in a colonial spider. *Behavioral Ecology and Sociobiology*, **66**, 1387–1398.

Jones, T. C. & Parker, P. G. (2000) Costs and benefits of foraging associated with delayed dispersal in the spider *Anelosimus studiosus* (Araneae, Theridiidae). *Journal of Arachnology*, **28**, 61–69.

Jones, T. C., Riechert, S. E., Dalrymple, S. E., & Parker, P. G. (2007) Fostering model explains variation in levels of sociality in a spider system. *Animal Behaviour*, **73**, 195–204.

Kaspari, M., Alonso, L., & O'Donnell, S. (2000) Three energy variables predict ant abundance at a geographical scale. *Proceedings of the Royal Society of London B*, **267**, 485–489.

Keiser, C. N., Jones, D. K., Modlmeier, A. P., & Pruitt, J. N. (2014) Exploring the effects of individual traits and within-colony variation on task differentiation and collective behavior in a desert social spider. *Behavioral Ecology*, **68**, 839–850.

Keiser, C. N., Wright, C.M., & Pruitt, J. N. (2015) Warring arthropod societies: Social spider colonies can delay annihilation by predatory ants via reduced apparency and increased group size. *Behavioral Processes*, **119**, 14–21.

Kim, K., Roland, C., & Horel, A. (2000) Functional value of matriphagy in the spider *Amaurobius ferox*. *Ethology*, **106**, 729–742.

Kirkendall, L. R. (1983) The evolution of mating systems in bark and ambrosia beetles (Coleoptera: Scolytidae and Platypodidae). *Zoological Journal of the Linnean Society*, 77, 293–352.

(1993) Ecology and evolution of biased sex ratios in bark and ambrosia beetles. *In:* Wrench, D.L. & Ebbert, M.A. (eds.) *Evolution and Diversity of Sex Ratio in Insects and Mites*. New York: Chapman and Hall, pp. 235–345.

Krafft, B. & Pasquet, A. (1991) Synchronized and rhythmic activity during the prey capture in the social spider *Anelosimus-eximius* (Araneae, Theridiidae). *Insectes Sociaux*, **38**, 83–90.

Kraus, O. & Kraus, M. (1988) The genus *Stegodyphus* (Arachnida, Araneae). Sibling species, species groups, and parallel origin of social living. *Verhandlungen des Naturwissenschaftlichen Vereins in Hamburg*, **30**, 151–254.

(1990) The genus *Stegodyphus*: Systematics, biogeography and sociality (Araneidae, Eresidae). *Acta Zoologica Fennica*, **190**, 223–228.

Krause, J. & Ruxton, G. (2002) *Living in Groups*. Oxford: Oxford University Press.

Kullman, E. (1972) Evolution of social behavior in spiders. *American Zoologist*, **12**, 419–426.

Leborgne, R., Cantarella, T. and Pasquet, A. (1998) Colonial life versus solitary life in *Cyrtophora citricola* (Araneae, Araneidae). *Insectes Sociaux*, **45**, 125–134.

Lubin, Y. (1995) Is there division-of-labor in the social spider *Achaearanea wau* (Theridiidae). *Animal Behaviour*, **49**, 1315–1323.

Lubin, Y. & Bilde, T. (2007) The evolution of sociality in spiders. *Advances in the Study of Behavior*, **37**, 83–145.

Lubin, Y. D. (1974) Adaptive advantages and evolution of colony formation in *Cyrtophora* (Araneae-Araneidae). *Zoological Journal of the Linnean Society*, **54**, 321.

Lubin, Y. D. & Crozier, R. H. (1985) Electrophoretic evidence for population differentiation in a social spider *Achaearanea-wau* (Theridiidae). *Insectes Sociaux*, **32**, 297–304.

Lubin, Y. D. & Robinson, M. H. (1982) Dispersal by swarming in a social spider. *Science*, **216**, 319–321.

Lubin, Y. D., Birkhofer, K., Berger-Tal, R., & Bilde, T. (2009) Limited male dispersal in a social spider with extreme inbreeding. *Biological Journal of the Linnean Society*, **97**, 227–234.

Majer, M., Svenning, J. C., & Bilde, T. (2013) Habitat productivity constrains the distribution of social spiders across continents: Case study of the genus *Stegodyphus*. *Frontiers in Zoology*, **10**, 9.

Makarieva, A. M., Gorshkov, V. G., & Li, B. L. (2005) Temperature-associated upper limits to body size in terrestrial poikilotherms. *Oikos*, **111**, 425–436.

Marques, E., Vasconcelos-Netto, J., & de Mello, M. (1998) Life history and social behavior of *Anelosimus jabaquara* and *Anelosimus dubiosus* (Araneae, Theridiidae). *Journal of Arachnology*, **26**, 227–237.

Masumoto, T. (1998) Cooperative prey capture in the communal web spider, *Philoponella raffrayi* (Araneae, Uloboridae). *Journal of Arachnology*, **26**, 392–396.

Mestre, L. & Lubin, Y. (2011) Settling where the food is: Prey abundance promotes colony formation and increases group size in a web-building spider. *Animal Behaviour*, **81**, 741–748.

Miller, J. (2006) Web-sharing sociality and cooperative prey capture in a Malagasy spitting spider (Araneae: Scytodidae). *Proceedings of the California Academy of Sciences*, **57**, 25–38.

Mitchell, R. (1973) Growth and population dynamics of a spider mite (*Tetranychus urticae* K., Acarina: Tetranychidae). *Ecology*, **54**, 1349–1355.

Mockford, E.L. (1957) Life history studies on some Florida insects of the genus *Archipsocus* (Psocoptera). *Bulletin of the Florida State Museum, Biological Sciences*, **1**, 253–274.

Mori, K. & Saito, Y. (2005) Variation in social behavior within a spider mite genus, *Stigmaeopsis* (Acari: Tetranychidae). *Behavioral Ecology*, **16**, 232–238.

Nentwig, W. (1985) Social spiders catch larger prey: A study of *Anelosimus-eximius* (Araneae, Theridiidae). *Behavioral Ecology and Sociobiology*, **17**, 79–85.

New, T.R. (1973) The Archipsocidae of South America (Psocoptera). *Transactions of the Royal Entomological Society of London*, **125**, 57–105.

Norton, R. A., Kethley, J. B., Johnston, D. E., & O'Connor, B. M. (1993) Phylogenetic perspectives on genetic systems and reproductive modes of mites. *In:* Wrench, D.L. & Ebbert, M.A. (eds.). *Evolution and Diversity of Sex Ratio in Insects and Mites*. New York: Chapman and Hall, pp. 8–99.

Oster, G. F. & Wilson, E. O. (1978) *Caste and Ecology in the Social Insects*. Princeton University Press, Princeton.

Park, T. S., Namkung, J., & Choe, J. C. (1999) Life history of a colonial spider *Philoponella prominens* (Araneae: Uloboridae) in Korea. *Korean Journal of Biological Sciences*, **3**, 167–172.

Pasquet, A., Trabalon, M., Bagneres, A. G., & Leborgne, R. (1997) Does group closure exist in the social spider *Anelosimus eximius*? Behavioural and chemical approach. *Insectes Sociaux*, **44**, 159–169.

Powers, K. S. & Avilés, L. (2003) Natal dispersal patterns of a subsocial spider *Anelosimus* cf. *jucundus* (Theridiidae). *Ethology*, **109**, 725–737.

(2007) The role of prey size and abundance in the geographical distribution of spider sociality. *Journal of Animal Ecology*, **76**, 995–1003.

Pruitt, J. N. (2012) Behavioural traits of colony founders affect the life history of their colonies. *Ecology Letters*, **15**, 1026–1032.

Pruitt, J. N., Oufiero, C. E., Avilés, L., & Riechert, S. E. (2012) Iterative evolution of increased behavioral variation characterizes the transition to sociality in spiders and proves advantageous. *The American Naturalist*, **180**, 496–510.

Purcell, J. (2011) Geographic patterns in the distribution of social systems in terrestrial arthropods. *Biological Reviews*, **86**, 475–491.

Purcell, J. & Avilés, L. (2007) Smaller colonies and more solitary living mark higher elevation populations of a social spider. *Journal of Animal Ecology*, **76**, 590–597.

(2008) Gradients of precipitation and ant abundance may contribute to the altitudinal range limit of subsocial spiders: Insights from a transplant experiment. *Proceedings of the Royal Society of London B*, **275**, 2617–2625.

Purcell, J., Vasconcellos-Neto, J., Gonzaga, M. O., Fletcher, J. A., & Avilés, L. (2012) Spatio-temporal differentiation and sociality in spiders. *PLoS ONE*, **7**, e34592.

Riechert, S. E. & Jones, T. C. (2008) Phenotypic variation in the social behaviour of the spider *Anelosimus studiosus* along a latitudinal gradient. *Animal Behaviour*, **75**, 1893–1902.

Riechert, S. E., Roeloffs, R., & Echternacht, A. C. (1986) The ecology of the cooperative spider *Agelena-consociata* in equatorial africa (Araneae, Agelenidae). *Journal of Arachnology*, **14**, 175–191.

Roeloffs, R. & Riechert, S. E. (1988) Dispersal and population-genetic structure of the cooperative spider, *Agelena-consociata*, in west-african rainforest. *Evolution*, **42**, 173–183.

Rolland, C., Danchin, E., & de Fraipont, M (1998) The evolution of coloniality in birds in relation to food, habitat, predation, and life-history traits: A comparative analysis. *The American Naturalist*, **151**, 514–529.

Rowell, D. M. & Main, B. Y. (1992) Sex-ratio in the social spider *Diaea-socialis* (Araneae, Thomisidae). *Journal of Arachnology*, **20**, 200–206.

Ruch, J., Heinrich, L., Bilde, T., & Schneider, J. M. (2009) The evolution of social inbreeding mating systems in spiders: Limited male mating dispersal and lack of pre-copulatory inbreeding avoidance in a subsocial predecessor. *Biological Journal of the Linnean Society*, **98**, 851–859.

Rypstra, A. L. (1979) Foraging flocks of spiders: Study of aggregate behavior in *Cyrtophora-citricola* forskal (Araneae, Araneidae) in West-Africa. *Behavioral Ecology and Sociobiology*, **5**, 291–300.

(1990) Prey capture and feeding efficiency of social and solitary spiders: A comparison. *Acta Zoologica Fennica*, **190**, 339–343.

(1993) Prey size, social competition, and the development of reproductive division-of-labor in social spider groups. *The American Naturalist*, **142**, 868–880.

Salomon, M. & Lubin, Y. (2007) Cooperative breeding increases reproductive success in the social spider *Stegodyphus dumicola* (Araneae, Eresidae). *Behavioural Ecology and Sociobiology*, **61**, 1743–1750.

Salomon, M., Sponarski, C., Larocque, A., & Avilés, L. (2010) Social organization of the colonial spider *Leucauge sp* in the Neotropics: Vertical stratification within colonies. *Journal of Arachnology*, **38**, 446–451.

Samuk, K. & Avilés, L. (2013) Indiscriminate care of offspring predates the evolution of sociality in alloparenting social spiders. *Behavioral Ecology and Sociobiology*, **67**, 1275–1284.

Samuk, K. M., LeDue, E. E., & Avilés, L. (2012) Sister clade comparisons reveal reduced maternal care behavior in social cobweb spiders. *Behavioral Ecology*, **23**, 35–43.

Schneider, J. M. (1995) Survival and growth in groups of a subsocial spider (*Stegodyphus lineatus*). *Insectes Sociaux*, **42**, 237–248.

Schneider, J. (2002) Reproductive state and care giving in *Stegodyphus* (Araneae: Eresidae) and the implications for the evolution of sociality. *Animal Behaviour*, **63**, 649–658.

Schneider, J. M. & Lubin, Y. (1996) Infanticidal male eresid spiders. *Nature*, **381,** 655–656.

Schneider, J. M., Roos, J., Lubin, Y., & Henschel, J. R. (2001) Dispersal of *Stegodyphus dumicola* (Araneae, Eresidae): They do balloon after all! *Journal of Arachnology*, **29**, 114–116.

Seibt, U. & Wickler, W. (1988) Bionomics and social structure of 'Family spiders' of the genus *Stegodyphus*, with special reference to the African species *S. dumicola* and *S. mimosarum* (Araneida, Eresidae). Verh. naturwiss. Ver. *Hamburg*, **30**, 255–303.

Settepani, V., Grinsted, L., Granfeldt, J., Jensen, J. L., & Bilde, T. (2013) Task specialization in two social spiders, *Stegodyphus sarasinorum* (Eresidae) and *Anelosimus eximius* (Theridiidae), *Journal of Evolutionary Biology*, **26**, 51–62.

Settepani, V., Bechsgaard, J., & Bilde, T. (2014) Low genetic diversity and strong but shallow population differentiation suggests genetic homogenization by metapopulation dynamics in a social spider. *Journal of Evolutionary Biology*, **27**, 2850–2855.

Sharpe, R. V. & Avilés, L. (2016) Prey size and scramble vs. contest competition in a social spider: Implications for population dynamics. *Journal of Animal Ecology*, 85, 1401–1410.

Simon, E. (1891) Observations biologiques sur les arachnides. *Annales de la Societé Entomologique Française*, **60**, 5–14.

Smith, D., van Rijn, S., Henschel, J., Bilde, T., & Lubin, Y. (2009) Amplified fragment length polymorphism fingerprints support limited gene flow among social spider populations. *Biological Journal of the Linnean Society*, **97**, 235–246.

Smith, D. R. (1982) Reproductive success of solitary and communal *Philoponella-oweni* (Araneae, Uloboridae). *Behavioral Ecology and Sociobiology*, **11**, 149–154.

(1997) Notes on the reproductive biology and social behavior of two sympatric species of *Philoponella* (Araneae, Uloboridae). *Journal of Arachnology*, **25**, 11–19.

Smith, D. R. & Engel, M. S. (1994) Population-structure in an Indian cooperative spider, *Stegodyphus-sarasinorum* karsch (Eresidae). *Journal of Arachnology*, **22,** 108–113.

Smith, D. R. & Hagen, R. H. (1996) Population structure and interdemic selection in the cooperative spider *Anelosimus eximius. Journal of Evolutionary Biology*, **9**, 589–608.

Smith, D. R. R. (1983) Ecological costs and benefits of communal behavior in a presocial spider. *Behavioral Ecology and Sociobiology*, **13**, 107–114.

(1985) Habitat use by colonies of *Philoponella-republicana* (Araneae, Uloboridae). *Journal of Arachnology*, **13**, 363–373.

Stern, D. L. & Foster, W. A. (1996) The evolution of soldiers in aphids. *Biological Reviews of the Cambridge Philosophical Society*, **71**, 27–79.

Trabalon, M. & Assi-Bessekon, D. (2008) Effects of web chemical signatures on intraspecific recognition in a subsocial spider, *Coelotes terrestris* (Araneae). *Animal Behaviour*, **76**, 1571–1578.

Uetz, G. W. (1989) The ricochet effect and prey capture in colonial spiders. *Oecologia*, **81**, 154–159.

Uetz, G. W. & Hieber, C. S. (1997) Colonial web-building spiders: Balancing the costs and benefits of group living. *In:* Choe, J. C. & Crespi, B. J. (eds.) *The Evolution of Social Behavior in Insects and Arachnids*. Cambridge: Cambridge University Press, pp. 458–475.

Uetz, G. W., Kane, T. C., & Stratton, G. E. (1982) Variation in the social grouping tendency of a communal web-building spider. *Science*, **217**, 547–549.

Uetz, G. W., Boyle, J., Hieber, C. S., & Wilcox, R. S. (2002) Antipredator benefits of group living in colonial web-building spiders: The "early warning effect". *Animal Behaviour*, **63**, 445–452.

Viera, C., Ghione, S., & Costa, F. G. (2006) Regurgitation among penultimate juveniles in the subsocial spider *Anelosimus cf. studiosus* (Theridiidae): Are males favored? *Journal of Arachnology*, **34**, 258–260.

Viera, C., Costa, F. G., Ghione, S., & Benamu-Pino, M. A. (2007) Progeny, development and phenology of the sub-social spider *Anelosimus cf. studiosus* (Araneae, Theridiidae) from Uruguay. *Studies on Neotropical Fauna and Environment*, **42**, 145–153.

Vollrath, F. (1982) Colony formation in a social spider. *Zietschrift für Tierpsychologie*, **60**, 313–324.

Ward, P. I. (1986) Prey availability increases less quickly than nest size in the social spider *stegodyphus-mimosarum*. *Behaviour*, **97**, 3–4.

Waser, P. M., Austad, S. N., & Keane, B. (1986) When should animals tolerate inbreeding. *The American Naturalist*, **128**, 529–537.

Whitehouse, M. E. A. & Lubin, Y. (2005) The functions of societies and the evolution of group living: Spider societies as a test case. *Biological Reviews*, **80**, 347–361.

Wickler, W. & Seibt, U. (1993) Pedogenetic sociogenesis via the sibling-route and some consequences for *Stegodyphus* spiders. *Ethology*, **95**, 1–18.

Wilson, E. O. (1971) *The Insect Societies*. Cambridge, MA: Belknap Press.

Yip, E. C. & Rayor, L. S. (2011) Do social spiders cooperate in predator defense and foraging without a web? *Behavioral Ecology and Sociobiology*, **65**, 1935–1947.

(2014) Maternal care and subsocial behaviour in spiders. *Biological Reviews*, **89**, 427–449.

Yip, E. C., Powers, K. S., & Avilés, L. (2008) Cooperative capture of large prey solves scaling challenge faced by spider societies. *Proceedings of the National Academy of Sciences USA*, **105**, 11818–11822.

Yip, E. C., Rowell, D. M., & Rayor, L. S. (2012) Behavioural and molecular evidence for selective immigration and group regulation in the social huntsman spider, *Delena cancerides*. *Biological Journal of the Linnean Society*, **106**, 749–762.

8 Sociality in Shrimps

Kristin Hultgren, J. Emmett Duffy, and Dustin R. Rubenstein

Overview

The genus *Synalpheus* is a species-rich group of snapping shrimps (Alpheidae) common to coral-reef habitats worldwide. The informal "gambarelloides group" (Coutière, 1909; Dardeau, 1984) is a monophyletic clade (Morrison, *et al.*, 2004; Hultgren, *et al.*, 2014) of approximately forty-five currently known species of *Synalpheus* that live symbiotically within sponges and are mostly restricted to the tropical West Atlantic. Sponge-dwelling *Synalpheus* species exhibit a range of social systems, from the family's ancestral condition of pair-living, to social groups with varying numbers of queens and workers (Duffy, 1996a; Duffy & Macdonald, 1999; Duffy, *et al.*, 2000; Duffy, 2003; Duffy, 2007). This social diversity is evident in the distribution of social structures and patterns of reproductive skew among species of *Synalpheus*, which are qualitatively similar to those observed across the entire range of social vertebrate and invertebrate taxa. Eusociality has evolved independently multiple times within *Synalpheus* (Duffy, *et al.*, 2000; Morrison, *et al.*, 2004). Thus, this socially diverse group – including the only known eusocial species from the marine realm – offers a unique opportunity to study the evolution of sociality in the sea.

I SOCIAL DIVERSITY

8.1 How Common is Sociality in Shrimps?

The Crustacea is one of the most phylogenetically, morphologically, and ecologically diverse groups of organisms in the marine realm, with over 50,000 species living in nearly every conceivable ocean habitat (Martin & Davis, 2001). Crustaceans also exhibit a wealth of interesting behavioral variation, including a range of social systems

Kristin Hultgren was supported by the National Geographic Society, the Smithsonian Institution, and the Murdock Charitable Trust. Emmett Duffy was funded by the US National Science Foundation (DEB-9201566, DEB-9815785, IBN-0131931, IOS-1121716), the National Geographic Society, and the Smithsonian Institution's Caribbean Coral Reef Ecosystem Program. Dustin Rubenstein was supported by the US National Science Foundation (IOS-1121435 and IOS-1257530), the American Museum of Natural History, and the Miller Society for Basic Research at the University of California, Berkeley. Early drafts of this chapter were greatly improved by the comments of Patrick Abbot, Solomon Chak, Martin Thiel, and Nancy Knowlton.

as diverse as in their terrestrial relatives (Duffy & Thiel, 2007). Although crustacean social behavior has been less studied than that of insects or vertebrates, group living has been documented in a wide range of terrestrial, freshwater, and marine species (Linsenmair, 1987; Shuster & Wade, 1991; Diesel, 1997; Duffy, 2010). Crustacean social behavior has reached its apex in the diverse shrimp genus *Synalpheus*. Shrimp in the genus *Synalpheus* belong to the snapping shrimp family Alpheidae, whose common name derives from an enlarged claw – used primarily for communication, aggression, and defense against predators, conspecifics, and heterospecifics – that "snaps" upon closing to produce a powerful jet of water (Nolan & Salmon, 1970; Versluis, *et al.*, 2000). The vast majority of *Synalpheus* species and other alpheid shrimp live in pairs that are apparently monogamous, within a burrow or host (typically a sponge or a crinoid echinoderm in the case of species in the genus *Synalpheus*) that they defend vigorously against intruders (Duffy, 2007; Hughes, *et al.*, 2014). Within the single clade of approximately 45 West Atlantic *Synalpheus* species in the *Synalpheus gambarelloides* species group, eusociality has evolved multiple times (Duffy, *et al.*, 2000; Morrison, *et al.*, 2004; Duffy & Macdonald, 2010). Species within this highly social gambarelloides group dwell exclusively in the interior canals of sponges, which they depend upon as a long-lived, predator-free host and food source. Thus, all sponge-dwelling *Synalpheus* species – social or otherwise – meet the criteria of the "fortress defender" social insects that live inside their food sources (Queller & Strassmann, 1998). Specifically, the shrimps are engaged in a symbiotic relationship with their sponge host – a living and continually growing food source – much like some gall-dwelling insect species (Crespi & Mound, 1997; Stern & Foster, 1997; Chapter 6).

8.1.1 Instances of Social Behavior in Snapping Shrimps

Eusociality was first reported in the species *S. regalis*, which exhibits extreme reproductive skew (i.e. colonies with a single breeding female or "queen") and lives in large kin-based colonies of tens to a few hundred individuals, apparently the full-sib offspring of the queen and a single male (Duffy, 1996a). The original discovery of eusociality was based on demographic data showing only a single ovigerous queen (i.e. female with ovaries or eggs) within a sponge, allozyme evidence of close relatedness among colony members, and behavioral experiments demonstrating size-based division of defensive labor (Duffy, 1996a). Eusociality has since been reported from eight other species in this group, and comparative analyses suggest eusociality has arisen independently at least four times (Duffy, *et al.*, 2000; Morrison, *et al.*, 2004; Duffy & Macdonald, 2010) (Figure 8.1). However, group living is not confined to the eusocial species, but rather varies along an apparent continuum in the gambarelloides group, where it ranges from eusocial colonies with single or multiple queens, to communal groups with approximately equal sex ratios (i.e. mated pairs), to pair-living species.

Synalpheus is a globally-distributed lineage, and eusociality is not confined to the West Atlantic gambarelloides group. For example, large colonies of the sponge-dwelling species *S. neptunus neptunus* with a single ovigerous queen have been reported from Indonesia (Didderen, *et al.*, 2006), and colonies with two queens and more than

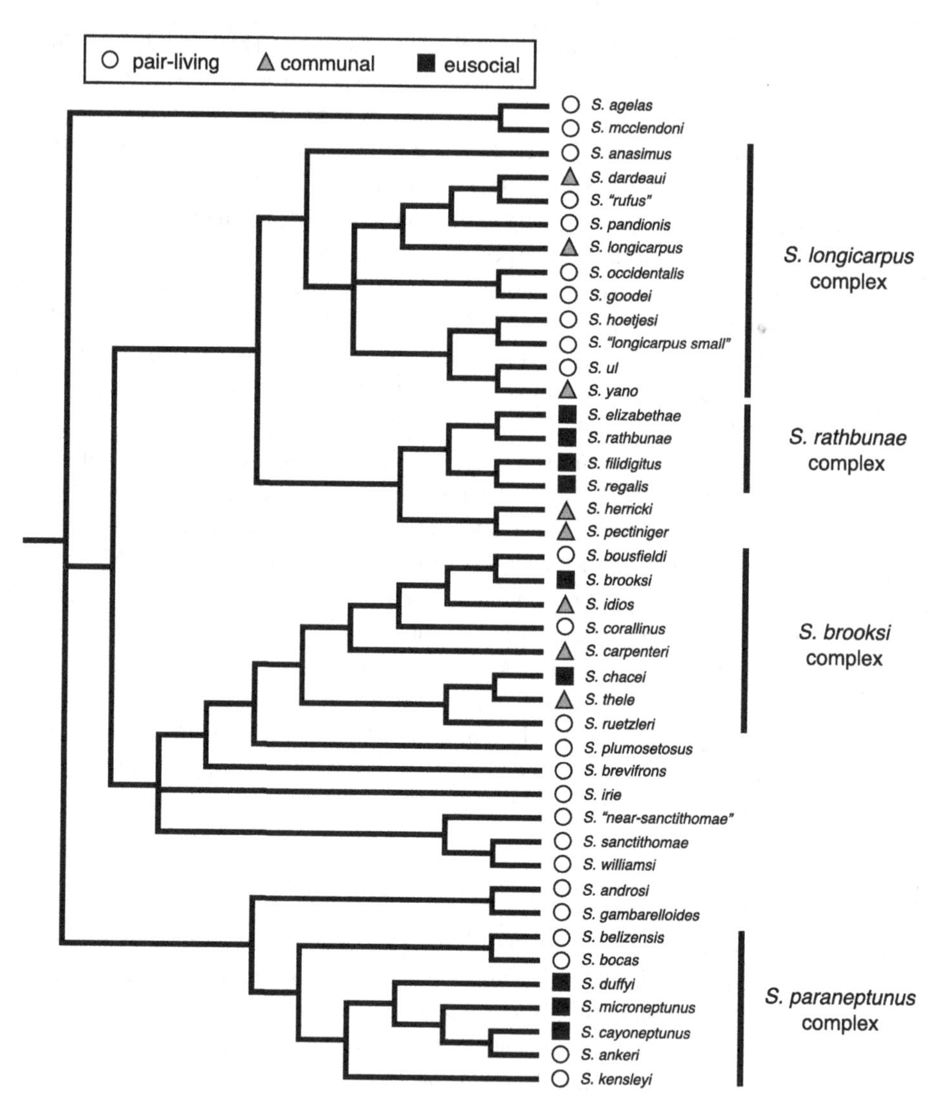

Figure 8.1. Bayesian phylogenetic tree of West Atlantic *Synalpheus* (after Hultgren & Duffy, 2011). The original tree was built from thirty-three morphological characters and three genetic markers: (1) the mitochondrial cytochrome oxidase I gene (COI, ~600 bp of the 5' coding region); (2) the mitochondrial large-subunit ribosomal gene (16S, ~510 bp); and (3) the nuclear gene elongation factor 2 (EF2, ~700 bp). The social system of each of the 42 species depicted in the tree is indicated with symbols defined in the legend. Importantly, some normally pair-living species occasionally occur as communal groups. Identified but undescribed species are noted in quotations.

100 non-ovigerous individuals have been reported in the sponge-dwelling species *S. paradoxus* from the Red Sea (Banner & Banner, 1981). Similarly, large colonies with high reproductive skew have been found in several other species from Indonesia (*S. fossor, S. hastilicrassus*, and *S.* aff. *neomeris*: Didderen, *et al.*, 2006) and East Africa (*S. crosnieri*: Banner & Banner, 1983). In at least two other sponge-dwelling *Synalpheus*

species, large colonies of non-ovigerous individuals with no queens have been reported, including *S. dorae* (Bruce, 1988) and *S. neptunus germanus* (Banner & Banner, 1975). Virtually nothing else is known about these putatively eusocial Indo-Pacific taxa, but interestingly, all reports of potential eusociality outside of the gambarelloides group also come from obligate sponge-dwelling species, suggesting that inhabitation of sponge hosts is a crucial component of sociality in these crustaceans.

8.1.2 Dimensions of Shrimp Sociality

Sociality in snapping shrimp varies in several dimensions. Three of these components – reproductive division of labor (i.e. reproductive skew), overlapping generations, and cooperative social behavior – comprise the classic criteria for eusociality (Wilson, 1971; Sherman, *et al.*, 1995). A fourth component, group or colony size, is also a crucial component of social diversity in snapping shrimp, as well as in other taxonomic groups (Bourke, 1999).

(1) Reproductive Skew. This describes the degree of asymmetry in distribution of direct reproduction among same-sex individuals within a group (Vehrencamp, 1983; Rubenstein, 2012). Reproductive skew among *Synalpheus* species varies from colonies in which nearly all females are ovigerous and breeding, to those in which only one is. Although we do not know how reproduction is shared among co-breeding females within a colony, they often appear to produce similarly sized clutches of eggs. Mature breeding females in *Synalpheus* can be easily distinguished by colored ovaries or eggs (and occasionally other morphological characteristics such as rounded pleura, i.e. the flaps surrounding the abdominal segments), but mature males and non-breeding females are morphologically indistinguishable under ordinary light microscopy (Duffy, 2007). However, identification of gonopores using scanning electron microscope studies showed that most non-breeders or "workers" of adult size class across several eusocial species consist of equal ratios of males and non-breeding females, although hermaphroditic (i.e. intersex) individuals occasionally occur in some species (Tóth & Bauer, 2007, 2008; Chak, *et al.*, 2015a).

(2) Overlapping Generations. This refers to the cohabitation of genetically related adults of different ages or cohorts (i.e. kin groups or family units larger than the mated pair) within a host sponge. Kin structure was first documented with allozymes in *S. regalis;* colonies in this species are composed primarily of full-sib offspring of a mated pair, the queen and an otherwise undifferentiated male (Duffy, 1996a). Similarly, microsatellite analysis in *S. brooksi* suggested cohabitation of family groups within a single host sponge (Rubenstein, *et al.*, 2008). Indirect evidence of overlapping generations in various *Synalpheus* species comes from co-occurrence of different size classes of a single species (i.e. juveniles, non-breeding adults, and breeding females) – often comprising visibly distinct cohorts – within a single sponge, and by behavioral responses of colony members to intruders. For example, whereas pair-living species generally do not tolerate individuals other than their mate, eusocial *Synalpheus* cohabit with large numbers of conspecifics (generally kin) and in some cases distinguish these colony members from heterospecific shrimp and sometimes from foreign (presumably non-kin) conspecifics (Duffy, *et al.*, 2002).

(3) Cooperative Social Behavior. This refers to coordinated behaviors, engaged in by group members, that benefit others in the colony. Early studies of social insects focused upon cooperative care and feeding of young as a criterion of eusociality (Wilson, 1971), but since the underlying premise of cooperative social behavior is altruism and therefore encompasses more than just offspring care, such cooperation can also include coordinated defenses against predators (see below) or other altruistic behaviors. Direct care or feeding of juveniles has not been observed in *Synalpheus*, but we have frequently collected small juveniles in close proximity within a sponge to an ovigerous female and a large male in *S. brooksi*, suggesting some sort of parental care. Genetic analysis of these associations confirmed that the cohabiting adults were indeed the parents of both the eggs and juveniles, suggesting that in at least this species, families associate together for an extended period of time (D. Rubenstein & J. Duffy, *unpublished data*).

(4) Colony or Group Size. This refers to the number of individuals of a single species living together in an individual host sponge. Shrimp abundance within an individual sponge is strongly correlated to sponge volume (Hultgren & Duffy, 2010), and very small sponge fragments generally only host a single pair of shrimp. Many *Synalpheus* species inhabit small encrusting sponges that live in the spaces found in dead coral rubble, whereas others live in much larger, free-living sponges that grow on the reef. However, several different *Synalpheus* species can co-occur in a single individual sponge, with different, but species-specific, group sizes. In general, eusocial species are found in larger groups within an individual sponge, while pair-living species occur as one or occasionally a few heterosexual pairs, even in the largest sponges. It is possible that some large sponges might contain multiple eusocial colonies of the same species living together in distinct portions of the sponges, but current behavioral or molecular data are insufficient to test this hypothesis.

8.2 Forms of Sociality in Shrimps

Sociality in *Synalpheus* takes a number of forms that may be considered to span a continuum, with different species varying in patterns of reproductive skew, group size, coexistence of overlapping generations, and cooperative social behavior. Since sociality has been best studied in the gambarelloides group, our discussion focuses on this clade of sponge-dwelling, West Atlantic *Synalpheus* species. The approximately 45 species in this group exhibit a range of social systems from the family's ancestral condition of pair-living, to communal societies with a variable number of paired breeding males and females (low skew), to eusocial societies with one or, rarely, a few breeding queens and up to hundreds of non-breeders or workers (high skew) (Duffy, 1996a, 2003, 2007 Duffy & Macdonald, 1999; Duffy, *et al.*, 2000). Below we describe in detail these forms of social living in *Synalpheus* shrimps.

8.2.1 Pair-Living Species

Formation of socially monogamous, heterosexual pairs is the ancestral form of sociality in the alpheid snapping shrimp (Knowlton, 1980; Mathews, 2002). More than half of

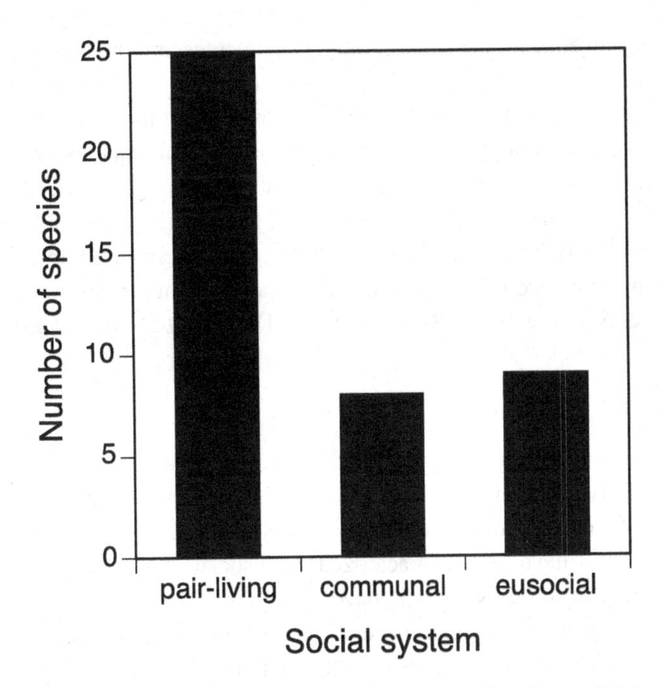

Figure 8.2. The number of pair-living, communal, and eusocial *Synalpheus* species in the gambarelloides group (N = 42 species). Some normally pair-living species occasionally occur as communal groups.

the sponge-dwelling *Synalpheus* in the gambarelloides group exhibit this pair-living lifestyle (Figure 8.2). In most alpheid shrimp, males and females form monogamous pairs, and cooperative behavior is limited to joint defense of their territory, burrowing, or hosting against other conspecific or heterospecific individuals (Mathews, 2002; Duffy, 2007). Although some pair-living *Synalpheus* species inhabit sponge species too small to house more than a pair of shrimp, other species occur occasionally in larger sponges (often with one or several other species of *Synalpheus*), but nevertheless are reliably found in small assemblages of one or a few pairs, even in the largest sponges. Like other alpheids, females in most pair-living *Synalpheus* species typically brood a clutch of tens to a few hundred small eggs, which hatch directly into free-swimming larvae that leave the natal sponge to spend time in the plankton as is typical of decapod crustaceans (Dobkin, 1965; Duffy & Macdonald, 2010). This life history trait (i.e. free-swimming larvae) thereby precludes the opportunity for overlapping generations in the natal sponge or for any type of extended parental care in these pair-living species.

8.2.2 Communal Species

Approximately 20 percent of the species in the gambarelloides group can be classified as communal, meaning that they typically live in groups with most adults breeding in equal sex ratios and low reproductive skew (Figure 8.2). Thus, communal species fall somewhere on the social continuum between the highly eusocial species and the pair-living species. In these communal species, field collections have often yielded an adult male and female living together in the same sponge canal, suggesting that they are a mated pair. Most communal species produce free-swimming larvae that presumably disperse away from the natal sponge, suggesting that these groups of adults are unlikely to consist of kin. This hypothesis is supported by observations of communal species exhibiting aggressive snapping towards other conspecifics in the same sponge – typically tolerating only their mate – that form a striking contrast with the generally peaceful interactions among members of closely related eusocial colonies. These behavioral observations are also supported by evidence of low genetic relatedness among group members in the communal species *S. dardeaui* (D. Rubenstein & J. Duffy, *unpublished data*).

8.2.3 Eusocial Species

Eusocial *Synalpheus* species are characterized by high reproductive skew, overlapping generations, typically large colonies, and by cooperative defense of the host sponge in the few species where behavioral observations have been made. At least nine described species of *Synalpheus* have been characterized as eusocial, most with extreme reproductive skew (i.e. typically only a single queen) (Figure 8.2). However, the magnitude of reproductive skew varies among eusocial *Synalpheus*, ranging from species with invariably only a single breeding queen and often hundreds of workers (e.g. *S. regalis*), to colonies with typically only a few queens (e.g. *S. elizabethae*), to large colonies with occasionally more than a dozen queens (e.g. *S. brooksi*). Group size in eusocial species ranges from tens to hundreds of individuals, but can vary widely among and within species, and is likely limited by both age of the colony and the maximum size of the host sponge. For example, *S. cayoneptunus* is typically found in small colonies of 8 to 30 individuals, living in encrusting sponges within coral rubble (Hultgren & Brandt, 2015), while *S. regalis* forms colonies of a few hundred and sometimes more than 350 individuals in sponges more than 1000 ml in volume (Macdonald, *et al.*, 2009). In addition to exhibiting large colony sizes and high reproductive skew, all eusocial species of *Synalpheus* in which newly hatched juveniles have been observed undergo direct development, with eggs hatching directly into non-swimming, crawling larvae that remain in the natal sponge (Dobkin, 1965, 1969; Duffy & Macdonald, 2010); limited observations suggest that direct development is occasionally seen in one communal species (*S. idios*). In contrast, most communal and all pair-living species of *Synalpheus* release swimming larvae, which live in the plankton for days to weeks and have a much greater potential for long-distance dispersal. Thus, differences in larval development – specifically, direct development in eusocial species – appear to be the primary mechanism allowing for natal philopatry and the accumulation of close relatives and overlapping generations within a single host sponge, and thus a key prerequisite for the evolution of eusociality in *Synalpheus* (Duffy & Macdonald, 2010).

8.3 Why Shrimp Form Social Groups

Three years before the discovery of eusocial *Synalpheus* (Duffy, 1996a), Spanier & colleagues (1993) wrote a paper entitled "Why are there no reports of eusocial marine crustaceans?" The authors predicted that eusocial crustaceans, if they were to be discovered, would have a suite of traits predisposing them to living in family groups – including some form of parental care and non-dispersing juveniles – and that they would use a long-lived, predator-free domicile. With the exception of parental care, eusocial *Synalpheus* fulfill these conditions. However, it is not yet clear why some *Synalpheus* species form social groups and others do not. Indeed, eusocial species are often sister to pair-living and communal species, and eusocial species have evolved independently from pair-living ancestors at least four times (Duffy, *et al.*, 2000; Morrison, *et al.*, 2004; Duffy & Macdonald, 2010). Moreover, pair-living and eusocial species can be found inhabiting the same species of long-lived, predator-free host sponges, often even the same individual sponge. Thus, *Synalpheus* offers a rich opportunity to explore the traits that are most important in facilitating the evolution of eusociality using a phylogenetic comparative approach (e.g. Duffy & Macdonald, 2010). Of the numerous factors known to influence social behavior in other group living organisms, two have been particularly well-studied in *Synalpheus*: (1) the use of sponge resources (which is closely tied to predator avoidance, population size, and persistence); and (2) the variation in reproductive mode, specifically larval development. A variety of other factors could also influence sociality in snapping shrimp, and we briefly discuss them below.

8.3.1 Resource Acquisition and Use

As far as is known, all *Synalpheus* species in the gambarelloides group live essentially their entire lives within the internal spaces of a single host sponge, which serves as both a stable, long-lived predator-free habitat (Duffy & Macdonald, 1999; Duffy, 2003; Hultgren, 2014) and a lifelong food source (Duffy, 1996b). Therefore, the importance of the sponge resource in *Synalpheus* ecology cannot be underestimated. The relationship between *Synalpheus* shrimp and their sponge appears to represent a mutualism. Although *Synalpheus* consume their sponge hosts (or at least the bacteria growing upon the sponge surface), experiments indicate that some species of *Synalpheus* also actively protect their hosts against sea star predators, enlarge sponge canals, and facilitate increased sponge growth under some conditions (Hultgren, 2014). Furthermore, surveys across multiple regions of the Caribbean suggest that the sponge habitat is saturated and available hosts are limiting; more than 95 percent of appropriate sponge habitat in Belize (i.e. the 20 species of sponges most commonly inhabited by *Synalpheus)* is typically occupied by shrimp (Macdonald, *et al.*, 2006). Thus, sponge hosts are essential to *Synalpheus* survival, but they are also in short supply.

Ecological constraints and the lack of available habitat are known to drive the evolution of sociality in vertebrates (Emlen, 1982; Koenig, *et al.*, 1992). Sponge use by *Synalpheus* appears to be consistent with this ecological constraints hypothesis because unoccupied sponges are rare (Macdonald, *et al.*, 2006). Furthermore, the few species of *Synalpheus* from outside of the gambarelloides group that have been characterized as eusocial are also reported as living in sponge hosts (Didderen, *et al.*, 2006). However, obligate sponge dwelling is a synapomorphy uniting all members of the gambarelloides group, from pair-living to communal to eusocial species. Thus, while sponge host use appears to be an important prerequisite for eusociality, it is not sufficient to explain the evolution of complex social behavior in this group (i.e. the "fortress defender" hypothesis alone does not explain sociality in shrimp; Queller & Strassmann, 1998).

8.3.2 Predator Avoidance

Predation avoidance in *Synalpheus* is tightly linked to sponge host use. Although formal studies have not been done on the effects of host sponge on predation risk, shrimp removed from their host sponge in the field are almost immediately consumed by fish (K. Hultgren & J. Duffy, *unpublished data*), and the majority of the sponge species inhabited by *Synalpheus* are chemically defended against fish (Pawlik, *et al.*, 1995). Although cooperative defense of the sponge habitat has been documented in some eusocial species (Tóth & Duffy, 2005), there has been no experimental work on relative rates of predation, colony failure, or colony turnover between host sponges dominated by pair-living, communal, or eusocial species of *Synalpheus*. In addition to the potential vulnerability of *Synalpheus* to predation, some species of sponge hosts are themselves susceptible to predation by fish or invertebrate enemies. While *Synalpheus* shrimps actively defend their sponge host against sea star predators (Hultgren, 2014), the magnitude of defense by species with different social systems has not been investigated.

8.3.3 Homeostasis

As all sponge-dwelling *Synalpheus* live underwater in tropical environments, group living should not have an appreciably large effect on regulation of temperature and other abiotic factors (i.e. physiological components of homeostasis). However, sponge hosts may provide a stable environment for shrimp. *Synalpheus* can be found inhabiting sponges from the intertidal zone to 30 m (or deeper) on some reefs (Macdonald, *et al.*, 2006, 2009, Hultgren, *et al.*, 2011). Although not yet tested, sponges could buffer shrimp from daily and annual changes in salinity, water temperature, and/or dissolved oxygen concentrations at shallower depths.

8.3.4 Mating

Little is known about mating in alpheid shrimp. In most alpheids, mating takes place after the female has molted, when her carapace is soft (Duffy & Thiel, 2007). Experimental work has shown that alpheids are not able to store sperm, suggesting that a female must mate every time she ovulates (summarized in Duffy, 2003). Laboratory studies of captive *S. brooksi* confirm that *Synalpheus* shrimp become receptive around molt, which occurs with the lunar cycle (D. Rubenstein, *unpublished data*). Moreover, limited observations of captive *S. regalis* colonies revealed rapid transfer of a spermatophore from male to female shortly after the female molts (E. Tóth & J. Duffy, *unpublished data*).

Unlike the Hymenoptera, genetic and experimental data suggest that *Synalpheus* are not haplodiploid (Duffy, 1993; Duffy, 1996c). In fact, many *Synalpheus* in the gambarelloides have large genomes and were hypothesized to either be polyploid or have large genome duplications (Rubenstein, *et al.*, 2008). However, subsequent examination suggested that differences in genome size among *Synalpheus* species are related to differences in chromosome size rather than chromosome number (Jeffrey, *et al.*, 2016). It is not clear how sex is determined in *Synalpheus*; the several species studied in detail tend to show equal sex ratios (Chak, *et al.*, 2015a), but hermaphroditic (intersex) individuals have been identified by external morphology in some species (Tóth & Bauer, 2007; Tóth & Bauer, 2008). Preliminary examination of gonadal development suggests that most species with intersex individuals are sequentially hermaphroditic (Chak, *et al.*, 2015a), though it is not clear whether they are protandrous (i.e. male to female) or protogynous (i.e. female to male). It also remains to be determined if hermaphroditic species of *Synalpheus* can – or at least have at some time in the past – reproduce via selfing, as has been observed in malacostracan – but not decapod – crustaceans (Kakui & Hiruta, 2013).

8.3.5 Offspring Care

Extended parental care has evolved repeatedly in several other species of crustaceans (summarized in Duffy & Thiel, 2007), but direct parental care of juveniles in *Synalpheus* has never been documented. Newly-hatched crawling juveniles are evidently able to feed themselves and have been observed feeding from the surface of sponge canals in captive colonies (Duffy, 2007). The formation of family groups with multiple overlapping generations of offspring in *S. brooksi* (D. Rubenstein & J. Duffy, *unpublished data*) suggests that rudimentary parental care could exist in this group, but numerous observations of captive colonies have yet to produce any evidence of direct care. Further work is needed to determine whether cooperation extends to offspring care in *Synalpheus*.

8.4 The Role of Ecology in Shaping Sociality in Shrimp

Life in tropical coral reefs, though physically benign, can be biologically challenging for small invertebrates such as *Synalpheus*: predation rates on crustaceans in the tropics are extremely high relative to temperate regions (Bertness, *et al.*, 1981; Freestone, *et al.*, 2011; Ory, *et al.*, 2014). As suitable host sponges are a limiting resource, *Synalpheus* shrimps face challenges during dispersal to and colonization of sponge hosts, such as competition for space within the sponge and long-term defense of the sponge resource. The challenges associated with founding colonies, surviving predation, and interacting with heterospecific competitors (such as sponge-dwelling polychaete worms and brittle stars) likely played a strong role in selecting for reduction of these risks via direct development, natal philopatry, and perhaps the evolution of eusociality (Duffy, *et al.*, 2002; Tóth & Duffy, 2005; Macdonald, *et al.*, 2006).

8.4.1 Habitat and Environment

Although *Synalpheus* shrimp are restricted to sponges in tropical marine habitats, the type of sponge habitats can vary greatly. Some shrimp species inhabit small cryptic sponges embedded in coral rubble or growing under rocks (e.g. the sponge *Hymeniacidon caerulea*), whereas others inhabit free-living sponges growing on more exposed reef surfaces (e.g. the sponge *Agelas clathrodes*). These host sponges can be found in reefs in shallow wave-exposed habitats (0 to 1 m depth), but also at much deeper depths (30 m). Cryptic host sponges occur in the interstices of live or recently dead coral rubble (predominantly *Madracis* spp.). Appropriate host sponges (and *Synalpheus*) are often uncommon in degraded reef environments (less than 10 percent coral cover) or extremely pristine reef environments (more than 80 percent coral cover) (Hultgren, *et al.*, 2010). Many *Synalpheus* species live in free-living sponges in protected seagrass beds (*Thalassia testudinum*) bordering coral reefs or mangroves; these areas tend to be more buffered than reefs from wave exposure (Macdonald, *et al.*, 2006). Although some shrimp species (e.g. *S. brooksi*) live in sponge hosts from a range of different habitats (e.g. exposed coral reefs and protected seagrass beds), little comparative work has been done on *Synalpheus* communities inhabiting different tropical marine habitats.

8.4.2 Biogeography

Synalpheus shrimp are distributed widely across the globe. Species in the gambarelloides group are largely restricted to the tropical West Atlantic, except for *S. occidentalis* and *S. gambarelloides*, which are endemic to the Mediterranean. Over the past two decades, sampling for sponge-dwelling *Synalpheus* has largely been restricted to fewer than a dozen sites or islands in the Caribbean, and largely at relatively shallow depths. Despite these limitations, these efforts have yielded a database of more than 60,000 specimens and relatively complete species lists for several regions. Although we have not rigorously quantified the effects of sociality on species' ranges, in general, eusocial species do not appear to have wider or narrower geographic distributions than pair-living or communal species. Rather, some shrimp species are distributed widely across the Caribbean, while others are endemic to certain regions, and geographic distribution in some cases is related to sponge host use (J. E. Duffy & K. Hultgren, *unpublished data*). For example, several widespread shrimp species (e.g. the pair-living species *S. agelas*) are specialists in common, cosmopolitan sponge species (*Agelas* spp.), whereas others (e.g. the pair-living species *S. bousfieldi*) live in a range of different sponge hosts in different regions. Some endemic species of shrimp live in what appear to be rare or endemic sponge hosts (e.g. *S. irie* has been found only in a white tube-like sponge observed only in Jamaica, Macdonald, *et al.*, 2009), while other endemic shrimp species live in cosmopolitan sponges (e.g. the eusocial *S. microneptunus* has only been found in *Xestospongia* spp. in Barbados, Hultgren, *et al.*, 2011). Rigorous sampling of additional Caribbean locations will be necessary to more comprehensively examine how sociality, sponge host use, and larval dispersal mode interact to affect *Synalpheus* biogeography.

The biogeographic distribution of species and genetic connectivity of populations is likely to be affected by larval developmental mode (i.e. direct development versus swimming larvae). Allozyme studies comparing a eusocial species with direct development (*S. brooksi)* to a communal species with swimming larvae (*S. pectiniger*) demonstrated significantly higher genetic structuring within and between regions in the eusocial species with direct development, suggesting population structure is correlated with the potential for larval dispersal (Duffy, 1993). Because eusocial species have direct-developing larvae that do not disperse from the natal sponge, we might expect eusocial species to have a lower colonization potential, and consequently a smaller geographic range, than pair-living or communal species. Although comparative studies on biogeography and host ranges have not been conducted, exhaustive surveys in some regions of the Caribbean (e.g. Curaçao) indicated a complete absence of eusocial species, despite the presence of appropriate host sponges and a wide diversity of pair-living and communal species (Hultgren, *et al.*, 2010). Curaçao is upstream to most other Caribbean regions via prevailing surface currents and larval exchange is generally low in Curaçao for many taxa (Roberts, 1997; Vollmer & Palumbi, 2007; Kough, *et al.*, 2013). It is possible that direct development in eusocial *Synalpheus* impedes dispersal to this region, suggesting that ecological constraints beyond simply host limitation might influence the evolution of social diversity in this group.

8.4.3 Niches

Social species or populations have been suggested to occupy a wider niche breadth than non-social species (Sun, *et al.*, 2014). In *Synalpheus*, a wider niche breadth would be a greater range of host species. Comparative work, based on decades of field surveys in Belize, has demonstrated that eusocial *Synalpheus* species are far more abundant in terms of frequency of occurrence and abundance, and occupy a greater range of sponge hosts than pair-living and communal species (Macdonald, *et al.*, 2006; Duffy & Macdonald, 2010; but see Duffy, *et al.*, 2013). Together, these data suggest that eusocial species may be able to competitively exclude pair-living and communal species from host sponges. The large colony sizes of eusocial species, paired with their cooperative defense behaviors (Tóth & Duffy, 2005), likely allow them to successfully dominate and defend large sponges that would be difficult for a single pair or a small group to hold on their own. Furthermore, with sponge habitat saturated (i.e. nearly all sponges are occupied by *Synalpheus*) and unoccupied hosts in short supply, it may be a less risky strategy for juveniles to remain in the natal sponge, as they do in eusocial species, than to disperse and colonize a new sponge. Thus, while sponge host use appears to be necessary for eusociality to evolve in snapping shrimp, and eusocial species are generally more ecologically successful at defending (and possibly acquiring) sponge resources in the field, sponge use alone is insufficient to explain the evolution of sociality in *Synalpheus*. However, by being able to exploit a wider niche, eusocial species appear to maintain a significant competitive advantage over pair-living and communal species.

8.5 The Role of Evolutionary History in Shaping Sociality in Shrimps

Evolutionary history likely plays an important role in explaining social diversity in *Synalpheus* and interacts with other life history and ecological factors. For example, direct larval development (in which newly hatched juveniles remain in the natal sponge) is almost perfectly phylogenetically correlated with eusociality (Duffy & Macdonald, 2010). In previous phylogenetic studies on this group, eusociality has been quantified using a modified version of the eusociality (E) index (Keller & Perrin, 1995) that takes into account both reproductive skew and group size (Duffy, *et al.*, 2000; Duffy & Macdonald, 2010). These data suggest that eusociality has evolved independently at least four times in the gambarelloides group (Figure 8.1), although the phylogenetic clustering of eusocial species clearly suggests strong phylogenetic signal. For example, four of the nine eusocial species come from the *S. rathbunae* complex, a group of four morphologically and ecologically similar species that together account for nearly half of the overall shrimp abundance recorded from three decades of surveys across the Caribbean (Duffy, *et al.*, 2000; Morrison, *et al.*, 2004; Hultgren & Duffy, 2012). Three other eusocial species – the recently described *S. microneptunus*, and *S. duffyi*, and a newly discovered eusocial species, *S. cayoneputunus*, from Florida (Hultgren & Brandt, 2015) – occur in the morphologically homogeneous *S. paraneptunus* complex, while the remaining two (*S. chacei* and *S. brooksi*) are closely related to several other communal (*S. idios*, *S. carpenteri*) and pair-living species (*S. bousfieldi*) in the *S. brooksi* complex (Figure 8.1).

II SOCIAL TRAITS

Vertebrate and social insect biologists often take for granted the wealth of information about the basic biology and life history that is known from their study organisms. Studies of social shrimps do not enjoy this advantage. Early workers on the genus *Synalpheus* (Coutière, 1909; Chace, 1972; Dardeau, 1984) established an invaluable foundation for the taxonomy of this large and difficult group over the last century, but it has only been in the last two decades that research has progressed beyond taxonomy and general distribution to begin revealing the remarkable behavior and life history of *Synalpheus*. *Synalpheus* snapping shrimp have been little studied in the wild, largely because they spend nearly their entire lives inside particular species of sponges, which are themselves very challenging taxonomically, and often at deep depths where doing direct observations is difficult. Therefore, what we know about *Synalpheus* life history traits is limited, and much of it comes from observations in the lab.

8.6 Traits of Social Species

8.6.1 Cognition and Communication

Alpheid shrimp are known to use both visual and chemical signals in conspecific interactions, including mate and competitor recognition (Nolan & Salmon, 1970; Hughes,

1996a; Hughes, 1996b; Obermeier & Schmitz, 2003; Bauer, 2011; Chak, *et al.*, 2015c). Moreover, alpheid shrimp that live symbiotically with gobies use a complex series of signals to communicate with their goby partners (Karplus & Thompson, 2011). Work on cognition and communication in the gambarelloides group has primarily focused upon group recognition and communication in the eusocial species. For example, experimental work in *S. regalis* and observations of other eusocial species demonstrate that individuals can discriminate colony members from conspecific and heterospecific intruders (Duffy, *et al.*, 2002). Individual contacts typically are initiated by antennal palpations, which can then be followed by bouts of snapping. Snapping bouts are typically higher in response to heterospecific than conspecific intruders, whereas colony members are often accepted into the colony without snapping (Duffy, *et al.*, 2002). In addition to individual communication, snapping also facilitates an important form of group communication and cooperative defense in eusocial *Synalpheus.* This so-called coordinated snapping typically occurs when initial warning snaps fail to repel intruders, and consists of multiple colony members snapping in unison for several seconds, causing a distinctive, escalating crackling sound that serves as an intense warning signal towards intruders (Tóth & Duffy, 2005).

8.6.2 Lifespan and Longevity

We know very little about individual lifespan or colony longevity in *Synalpheus* since individuals have not been kept for extended period of times in the lab and long-term field studies have not yet been initiated, largely because many sponge hosts are embedded in coral rubble, making it difficult to monitor colonies repeatedly in the field. Limited experimental data suggest that shrimp colonies may be able to grow along with the sponge host. During short-term experiments in the field (3 weeks), *Synalpheus*-inhabited sponges grew more slowly than empty sponges, but most sponges experienced positive net growth (Hultgren, 2014). However, sponges are typically slow-growing organisms, and it is unknown whether slow sponge growth limits colony growth or keeps pace with colony expansion.

In terms of individual lifespan, *S. brooksi* individuals have been housed successfully in self-contained aquaria in the lab for up to a year (D. Rubenstein, *unpublished data*), and the safe, long-lasting nature of their host sponges (some of which can live for decades or more) suggests by analogy with social insects (Keller & Genoud, 1997) that the lifespans of some social shrimp may be much longer. Colony longevity will be much longer than individual lifespan, since sponges are extremely slow-growing, long-lived organisms; the largest specimens of the sponge *Xestospongia muta* – a close relative of the sponge genus *Neopetrosia* that hosts *Synalpheus* – have been estimated from growth rates to have been alive for more than 2,300 years on reefs in the Florida Keys, where *Synalpheus* are common (McMurry, *et al.*, 2008).

8.6.3 Fecundity

Fecundity varies greatly within and among *Synalpheus* species, with average clutch size ranging from a few to hundreds of eggs (Duffy, 2007; Ríos & Duffy, 2007;

Hernáez, *et al.*, 2010). Some eusocial species with crawling larvae have very low fecundity (e.g. *S. filidigitus*, median clutch size is 4.5), suggesting high offspring survival (Duffy, 2007). However, correlations among fecundity, egg size, and sociality have not been explored across different species. Within a species, the number of eggs a female can produce is related to body size. In the eusocial species *S. regalis*, queen clutch size, as well as queen body size, are positively correlated with colony size (Duffy, 1996a; Duffy, *et al.*, 2002). These data suggest either that non-breeding colony members may enhance queen fecundity, or that fecundity simply increases steadily with size and age of the queen and colony (see above). Females in many species are often externally parasitized by bopyrid isopod parasites, which occur in either the branchial cavity or the abdominal area, and abdominal parasites have been shown to significantly suppress clutch size in infected females (Hernáez, *et al.*, 2010).

8.6.4 Age at First Reproduction

Due to the challenges of rearing sponges and breeding shrimp in the laboratory, little is known about when reproduction commences in *Synalpheus*. Experiments in which empty sponge fragments were colonized by the eusocial species *S. rathbunae* indicate that after 45 days, sponges were inhabited by a male-female pair of shrimp, with the female showing some signs of ovarian development but no embryos (Tóth & Bauer, 2007). Moreover, after queen removal in lab colonies of *S. elizabethae*, female workers developed mature ovaries within 33 days (Chak, *et al.*, 2015b). Although little is known about age of first reproduction, we do know that like most caridean shrimp, *Synalpheus* show continuous (indeterminate) growth throughout their lives, with no terminal molt (Hartnoll, 2001). Species with swimming larvae exhibit distinct larval stages that metamorphose into adults, whereas those with larvae that exhibit gradual development do not (Dobkin, 1965, 1969).

8.6.5 Dispersal

Differences in larval development and consequent mode of dispersal appear to be the primary mechanism underlying the formation of kin-based colonies and thus the evolution of eusociality in *Synalpheus*. As discussed earlier, eggs of all pair-living and communal species studied (with a single possible exception) hatch directly into free-swimming larvae and are released into the water column (Dobkin, 1965; Duffy & Macdonald, 2010), reducing the opportunity for overlapping generations in the natal sponge, any type of extended parental care, or association of kin. In contrast, in all eusocial species that have been studied, eggs undergo direct development, hatching into non-swimming, crawling larvae that remain in the natal sponge (Dobkin, 1965, 1969; Duffy & Macdonald, 2010).

Despite the role of dispersal mode in *Synalpheus* social evolution, few direct experimental studies of colonization and dispersal have been conducted, and basic questions (e.g. how eusocial species establish new colonies) have yet to be fully investigated. Tóth & Bauer (2007) used scanning electron microscopy to determine the sex of individuals of the eusocial species *S. rathbunae* that had colonized unoccupied sponge fragments after 45 days in the field. In most cases, colonists of these small fragments consisted of a single heterosexual pair, often with a sexually immature female, suggesting that eusocial species may colonize available sponges as juveniles or subadults. In another set of field experiments in Panama examining how *Synalpheus* impact sponge growth, unoccupied sponge fragments were rapidly colonized by multiple *Synalpheus* species (50 percent recolonization within approximately 17 days) (Hultgren, 2014). Colonists in empty sponges consisted primarily of newly settled postlarval juveniles of two communal species with swimming larvae, *S. dardeaui* and *S. yano*, and occasionally juveniles of the eusocial species *S. elizabethae* (K. Hultgren, *unpublished data*). Finally, limited evidence from the eusocial species *S. brooksi* demonstrated that males are more genetically related to each other than are ovigerous females, suggesting that dispersal in this eusocial species could be sex-biased (D. Rubenstein, *unpublished data*). That is, *S. brooksi* may exhibit male-biased philopatry and female-biased dispersal, as in most cooperatively breeding birds (Greenwood, 1980). Together, these studies suggest that pair-living and communal species have the ability to rapidly recolonize empty sponges via swimming larvae, but that eusocial species may colonize sponges on a longer time scale as sexually immature juveniles or subadults.

8.6.6 Other Traits: Body Size

Body size is a master trait of sorts among organisms, with pervasive effects on ecological interactions, life history, and distribution (Woodward, *et al.*, 2005). Eusocial *Synalpheus* species tend to be smaller in size on average than pair-living species (Duffy & Macdonald, 2010). Since size is a strongly phylogenetically conserved trait in *Synalpheus* (Hultgren & Duffy, 2012), it is unclear whether small body size is directly related to evolution of eusociality, or a byproduct of the close phylogenetic relationships of many eusocial species. Regardless, given the pervasive inverse relationship between size and abundance in many animal communities, smaller body size could partially explain many of the ecological traits more common in eusocial *Synalpheus*, such as increased abundance. However, comparative analysis examining the dual effects of eusociality (E index) and body size on these ecological traits suggested that eusociality is a stronger correlate of sponge host range and percentage of sponges occupied than body size; body size alone was significantly correlated only to relative abundance (Duffy & Macdonald, 2010). Thus, increased abundance and sponge host breadth in eusocial species appear to be a direct result of social life, rather than of small body size.

8.7 Traits of Social Groups

8.7.1 Genetic Structure

Despite ongoing genetic work, currently there exist few quantitative data on genetic structure in *Synalpheus* colonies. Allozyme studies of eusocial *S. regalis* showed that colonies exhibited high relatedness and consisted primarily of full-sib offspring of a mated pair (Duffy, 1996a). Similarly, microsatellite analysis in eusocial *S. brooksi* suggested cohabitation of family groups within a single host sponge (Rubenstein, *et al.*, 2008; D. Rubenstein & J. Duffy, *unpublished data*). In contrast, microsatellite analyses of communal *S. dardeaui* colonies indicate very low genetic relatedness within colonies (D. Rubenstein, *unpublished data*). These patterns accord with the presence of crawling juveniles in *S. brooksi* versus swimming larvae in *S. dardeaui*. Thus, high genetic structure is expected in other eusocial species with crawling larvae, and similarly low genetic structure in communal species with swimming larvae.

Genetic relatedness among colony members can depend not only upon dispersal mode but also upon mating patterns. Lifetime monogamy has been hypothesized to underlie the evolution of eusociality in social insects because it produces full-sib offspring (Boomsma, 2007, 2009, 2013), as suggested by allozyme data for *S. regalis* (Duffy, 1996a). Preliminary microsatellite-based molecular work in five *Synalpheus* species shows no evidence of multiple paternity (i.e. polyandry) in broods of eggs (D. Rubenstein & J. Duffy, *unpublished data*). This is likely to be the case for all *Synalpheus* species because of the absence of sperm storage in alpheids (Knowlton, 1980) and the constraint that females only mate immediately after they molt and males generally guard them aggressively during this short period of receptivity (Duffy & Thiel, 2007). However, while pairs are genetically monogamous in each reproductive event, females are unlikely to exhibit lifetime monogamy. For example, parentage analysis of family groups and ovigerous females in *S. brooksi* suggest that females can mate sequentially with different males (D. Rubenstein & J. Duffy, *unpublished data*).

8.7.2 Group Structure, Breeding Structure and Sex Ratio

Sociality in *Synalpheus* is defined by the structure and sex ratio of the group. Group living species, within which most adult-sized females breed, are defined as communal. In most of these communal species, a female and male are found together in the same sponge canal, and are likely to be a mated pair. In contrast, eusocial species are typically defined by the highly unequal numbers of ovigerous (i.e. queens) and non-ovigerous individuals (i.e. workers). Determining the exact sex ratios of *Synalpheus* colonies is difficult because non-ovigerous *Synalpheus* lack obvious external sexual characteristics. Based upon external morphology under a scanning electron microscope, Tóth & Bauer (2007) found that worker sex ratios in four eusocial species (*S. regalis, S. rathbunae, S. chacei*, and *S. filidigitus*) generally conform to a 50:50 sex ratio. However, preliminary evidence from histological analysis of gonads in five species of *Synalpheus* suggests that sex ratios in some colonies can vary considerably in both directions from 50:50 among eusocial species (Chak, *et al.*, 2015a). Nonetheless, the ratio of ovigerous to non-ovigerous members of a colony provides a reasonable approximation of the degree of reproductive skew within a group and can be used to differentiate pair-living from communal from eusocial species.

Eusocial species vary considerably in their group structure and degree of reproductive skew. Most social species in the *S. rathbunae* and *S. paraneptunus* complexes are characterized by a single queen. Indeed, rarely is more than a single ovigerous female found in these colonies, despite a range of group sizes. Species like *S. microneptunus* are always found in small colonies, whereas its sister species, *S. duffyi*, can be found in much larger groups. This may be partially due to the size of the host sponges that these species use. In contrast to these obligately eusocial species, a variety of species in the *S. brooksi* complex have multiple queens and much lower reproductive skew. For example, *S. chacei* typically has a single queen, but can be found in colonies with multiple queens. In many parts of its range where it inhabits the large sponge *Spheciospongia vesparium*, *S. brooksi* is almost always found in large colonies with multiple queens (D. Rubenstein & J. Duffy, *unpublished data*). However, in other parts of its range where it lives in the small sponge *Hymeniacidon caerulea, S. brooksi* is found in small groups or even heterosexual pairs (J. Duffy, K. Hultgren & D. Rubenstein, *unpublished data*). It is not yet clear whether this social plasticity occurs because *S. brooksi* social structure varies in different hosts, since this species inhabits a range of sponge species.

8.7.3 Other Traits: Competitive Ability

The primary cooperative benefit that non-breeding colony members provide to the colony (including offspring in species with crawling larvae) is defense against other shrimp. As discussed earlier, evidence for this cooperative "coordinated snapping" in eusocial *Synalpheus* comes from behavioral and morphological data (Tóth & Duffy, 2005). Several eusocial species exhibit size-based and likely individual-level variation in behavior or morphological defense, constituting a kind of division of labor (Duffy, *et al.*, 2002; Tóth & Duffy, 2008). Although the division of labor seen in *Synalpheus* is not as extreme as that seen in social insects, it is a topic that warrants further study. In *S. regalis*, larger, non-breeding colony members spend more time defending the colony than do the queen and smaller juveniles (Duffy, *et al.*, 2002), and larger individuals also are more likely to occupy the peripheral parts of the sponge where intruders are first met (Duffy, 2003). In *S. chacei* and *S. regalis*, queens have smaller major chela (relative to body size) than non-reproductive workers (Tóth & Duffy, 2008). Allometric studies indicate that in larger eusocial colonies, the largest non-breeding individuals have disproportionately large major (fighting) chela, suggesting the formation of a "fighting" caste in some species (Tóth & Duffy, 2008). The most extreme example of this defensive division of labor is seen in *S. filidigitus*, in which the queen typically loses her primary weapon – the major chela – and instead bears two minor-form chelae (Duffy & Macdonald, 1999). Interestingly, this morphological pattern has been noted in the eusocial species *S. rathbunae* (Chace, 1972), as well as a putatively eusocial species from outside of the gambarelloides group, *S. crosnieri* (Banner & Banner, 1983). Coordinated social behavior has not been well-studied in communal species, but data suggest that allometry of fighting claw size (relative to body size) in communal and pair-living species is significantly less steep than in eusocial species, suggesting little morphological (and presumably behavioral) division of labor in larger individuals within a communal group (Tóth & Duffy, 2008).

The competitive advantage of eusociality that we discussed earlier is also consistent with distributional and ecological evidence. First, comparative analyses demonstrate that eusocial *Synalpheus* species constitute the majority of sponge-dwelling *Synalpheus* abundance in localities where quantitative sampling has occurred (Macdonald, *et al.*, 2006). Moreover, eusocial species use a significantly higher number of host sponge species than pair-living or communal species (Macdonald, *et al.*, 2006). Stronger competition in eusocial species has also been supported by studies of phylogenetic community ecology, or the phylogenetic relatedness of co-occurring species in a community. Hultgren & Duffy (2012) examined phylogenetic relatedness of hundreds of *Synalpheus* communities (defined as the community of different species inhabiting a single sponge host) and found striking differences between sponge communities containing eusocial species and those lacking them. Specifically, shrimp communities containing only pair-living or communal species tended to be phylogenetically closely related and similar in body size, consistent with a strong effect of habitat filtering on community assembly. However, shrimp communities containing eusocial species showed a contrasting pattern: communities were less phylogenetically related and more dissimilar in size, suggesting that competitive exclusion is an important determinant of community structure in *Synalpheus* communities, but only when eusocial species were present (Hultgren & Duffy, 2012). Strikingly, survey data collected over nearly three decades across six Caribbean regions failed to find a single instance of two different eusocial species co-occurring in a sponge (Hultgren & Duffy, 2012). Thus, the stronger competitive abilities of eusocial species, paired with data indicating lower dispersal potential of direct-developing eusocial species, suggest that competition-colonization trade-offs may shape *Synalpheus* community assembly within and between regions.

Finally, despite being more competitive, eusocial species may be more susceptible population collapse than communal and pair-living species. Evidence from long-term field surveys throughout the Caribbean suggest a drastic and recent decline in eusocial species along with an associated increase in the relative abundance of pair-living species, due in part to changes in the coral assemblages and associated sponge community (Duffy, *et al.*, 2013). These changes, which could be environmentally-driven, human-induced, or represent natural cycles of population collapse similar to those seen in other non-classically eusocial species (reviewed in Aviles & Purcell, 2012), have led to the local extinction of some (e.g. Panama) or all (e.g. Belize) eusocial species in some regions of the Caribbean, and hint towards furthered extinction in other areas (e.g. Jamaica) (Duffy, *et al.*, 2013).

III SOCIAL SYNTHESIS

8.8 A Summary of Shrimp Sociality

Despite having fewer species than nearly any other taxonomic lineage with highly social representatives (e.g. Hymenoptera), the *Synalpheus gambarelloides* species group exhibits a wide range of social behavior and numerous evolutionary transitions between social states. Nearly half of the species in this clade live in groups of more than two individuals, some forming eusocial colonies with extreme reproductive skew and the beginning of behavioral caste formation. However, snapping shrimp lack sterile castes (Chak, *et al.*, 2015a, 2015b) and therefore have not reached the degree of reproductive specialization as many social insects (Boomsma, 2013). Nonetheless, *Synalpheus* in the gambarelloides group represent the pinnacle of social evolution not only in crustaceans, but also in the sea.

Although we have divided *Synalpheus* into three discrete social categories (pair-living, communal, and eusocial), social structure varies widely and continuously in this group. Nearly all species that typically live as heterosexual pairs are occasionally found in small groups. Several species that we characterize as eusocial because colonies typically have a single queen and high reproductive skew, nevertheless occasionally have colonies with several queens. Moreover, some of these multi-queen colonies are demographically similar to those of the communal species. However, fundamental differences in larval dispersal mode (i.e. swimming larvae that disperse in the water column versus crawling larvae that remain in the host sponge) underlie key differences in kin structure between eusocial and communal species. Although we do not yet know the colony genetic structure of all species, for those that have been studied, eusocial species tend to live in kin groups and communal species do not. All species are likely to be monogamous in a single breeding event – like alpheid shrimp generally (Nolan & Salmon, 1970; Knowlton, 1980; Mathews, 2002) – so variation in genetic structure must be the result of either mate-switching (i.e. sequential polyandry), queen replacement, or reproduction by multiple females within a single colony (i.e. polygyny). Together, these results strongly suggest that kin selection plays an important role in the evolution and maintenance of sociality in this group. The life history differences that mediate kin structure may also have consequences for biogeography and even diversification patterns within this group. We might expect eusocial species to have a lower colonization potential, and consequently smaller geographic ranges, than pair-living and communal species. Interestingly, the same pattern has been observed in cooperatively breeding birds. Non-cooperative breeders tend to have greater capacity for colonization than cooperative breeders, which results in broader ranges and more species-rich clades in the non-cooperative lineages (Cockburn, 2003).

All *Synalpheus* shrimp live symbiotically with other organisms, and species in the gambarelloides group associate only with sponges. Since all species in this group are obligate sponge users, ecological differences are unlikely to fully explain the evolution of eusociality (Duffy, 2007). This does not mean, however, that ecology is unimportant in the discussion of social evolution in snapping shrimp. Several lines of evidence show that eusocial species appear to have a competitive advantage over communal and pair-living species; eusocial species use a significantly higher number of host sponges (Macdonald, *et al.*, 2006), they tend to exclude ecologically similar species from co-occurring in the same sponges (Hultgren & Duffy, 2012), and they cooperatively defend their host sponges (Tóth & Duffy, 2005). Although eusocial species occur at a higher abundance than communal and pair-living species on some reefs (Macdonald, *et al.*, 2006), recent evidence suggests that they may be more susceptible to factors that drive population decline, having gone locally extinct in a number of regions of the Caribbean (Duffy, *et al.*, 2013). This could be due in part to the slower colonization potential of eusocial species with non-dispersing larvae than pair-living or communal species with free-swimming larvae.

8.9 Comparative Perspectives on Shrimp Sociality

Synalpheus species share key life history traits with both social vertebrates and insects, and hence could serve as a model system to bridge the gap between them. Like vertebrates and termites, all eusocial *Synalpheus* species exhibit gradual development (i.e. not discrete larval and adult stages), though eusocial species bear young that look like miniature adults and grow by molting, whereas pair-living species have distinct, swimming larval stages that do not resemble miniature adults and live and mature in a distinct environment (i.e. open water vs. sponge). At least one species of eusocial *Synalpheus* appears to have female-biased dispersal, like many birds (Greenwood, 1980). Conversely, these small-bodied arthropods most resemble social insects in the way they form large colonies and inhabit and defend valuable protective host "fortresses" (Queller & Strassmann, 1998). Like vertebrates, there appears to be minimal morphological and physiological caste specialization in eusocial *Synalpheus*, and the most likely specialization for non-breeding workers in *Synalpheus* involves defense of host sponges against competitors (Tóth & Duffy, 2008), almost like a specialized fighting or defender class such as that seen in termites (Shellman-Reeve, 1997; Korb, 2008), aphids (Stern & Foster, 1997), and thrips (Crespi & Mound, 1997) (see also Chapters 5 and 6). Also like termites, some thrips, and most vertebrates – but unlike most Hymenoptera – the queen in all social *Synalpheus* species must cohabit with her mate. Thus, eusocial *Synalpheus* species are perhaps most similar to the wood-dwelling termites that also live inside their food sources (Chapter 5), as well as the gall-dwelling aphids and thrips (Chapter 6).

8.10 Concluding Remarks

Of the more than 50,000 crustaceans in the oceans worldwide, only a handful of species in the *Synalpheus gambarelloides* group are highly social. Yet, this group of approximately 45 species of obligate sponge-dwellers is extremely socially diverse, ranging from eusocial colonies, to cooperatively breeding groups, to communal associations, to simple pairs of males and females. Despite being poorly studied compared to most other vertebrate and invertebrate social lineages, we are beginning to learn a great deal about the biology, life history, and behavior of snapping shrimps. In particular, many eusocial species share both individual and group traits with other social vertebrates and insects, making them an ideal system to study social evolution in a comparative context.

References

Aviles, L. & Purcell, J. (2012) The evolution of inbred social systems in spiders and other organisms: From short-term gains to long-term evolutionary dead ends? *Advances in the Study of Behavior*, **44**, 99–133.

Banner, D. M. & Banner, A. H. (1975) The alpheid shrimp of Australia. Part 2: The genus *Synalpheus*. *Records of the Australian Museum*, **29**, 267–389.

(1981) Annotated checklist of the alpheid shrimp of the Red Sea and Gulf of Aden. *Zoologische Verhandelingen*, **190**, 1–99.

(1983) An annotated checklist of the alpheid shrimp from the Western Indian Ocean. *Travaux et Documents de l'ORSTOM*, **158**, 2–164.

Bauer, R. T. (2011) Chemical communication in decapod shrimps: The influence of mating and social systems on the relative importance of olfactory and contact pheromones. In: Breithaupt, T. & Thiel, M. (eds.) *Chemical Communication in Crustaceans*. New York: Springer, pp. 277–296.

Bertness, M., Garrit, S., & Levings, S. (1981) Predation pressure and gastropod foraging: A tropical-temperate comparison. *Evolution*, **35**, 995–1007.

Boomsma, J. J. (2007) Kin selection vs. sexual selection: Why the ends do not meet. *Current Biology*, **17**, R673-R683.

(2009) Lifetime monogamy and the evolution of eusociality. *Philosophical Transactions of the Royal Society of London B*, **364**, 3191–3207.

(2013) Beyond promiscuity: Mate-choice commitments in social breeding. *Philosophical Transactions of the Royal Society of London B*, **368**, 20120050.

Bourke, A. (1999) Colony size, social complexity and reproductive conflict in social insects. *Journal of Evolutionary Biology*, **12**, 245–257.

Bruce, A. (1988) *Synalpheus dorae*, a new commensal alpheid shrimp from the Australian Northwest shelf. *Proceedings of the Biological Society of Washington*, **101**, 843–852.

Chace, F. A. J. (1972) The shrimps of the Smithsonian-Bredin Caribbean Expeditions with a summary of the West Indian shallow-water species (Crustacea: Decapoda: Natantia). *Smithsonian Contributions to Zoology*, **98**, 1–179.

Chak, T. C. S, Duffy, J. E., & Rubenstein, D. R. (2015a) Reproductive skew drives patterns of sexual dimorphism in sponge-dwelling snapping shrimps. *Proceedings of the Royal Society of London B*, **282**, 20150342.

Chak, T. C. S, Rubenstein, D. R., & Duffy, J. E. (2015b) Social control of reproduction and breeding monopolization in the eusocial snapping shrimp *Synalpheus elizabethae*. *The American Naturalist*, **186**, 660–668.

Chak, S. C., Bauer, R., & Thiel, M. (2015c) Social behaviour and recognition in decapod shrimps, with emphasis on the Caridea. In: Aquiloni, L. & Tricarico, E. (eds.) *Social Recognition in Invertebrates*, Switzerland: Springer International Publishing, pp. 57–84.

Cockburn, A. (2003) Cooperative breeding in oscine passerines: Does sociality inhibit speciation? *Proceedings of the Royal Society of London B*, **270**, 2207–2214.

Coutière, H. (1909) The American species of snapping shrimps of the genus *Synalpheus*. *Proceedings of the United States National Museum*, **36**, 1–93.

Crespi, B. J. & Mound, L. A. (1997) Ecology and evolution of social behavior among Australian gall thrips and their allies. In: Choe, J. C. & Crespi, B. J. (eds.) *The Evolution of Social Behavior in Insects and Arachnids*. Cambridge: Cambridge University Press, pp. 166–180.

Dardeau, M. (1984) *Synalpheus* shrimps (Crustacea: Decapoda: Alpheidae). I. The Gambarelloides group, with a description of a new species. *Memoirs of the Hourglass Cruises*, **7** (part 2), 1–125.

Didderen, K., Fransen, C., & deVoogd, N. (2006) Observations on sponge-dwelling colonies of *Synalpheus* (Decapoda, Alpheidae) of Sulawesi, Indonesia. *Crustaceana*, **79**, 961–975.

Diesel, R. (1997) Maternal control of calcium concentration in the larval nursery of the bromeliad crab, *Metopaulias depressus* (Grapsidae). *Proceedings of the Royal Society of London B*, **264**, 1403–1406.

Dobkin, S. (1965) The first post-embryonic stage of *Synalpheus brooksi* Coutière. *Bulletin of Marine Science*, **15**, 450–462.

(1969) Abbreviated larval development in caridean shrimps and its significance in the artificial culture of these animals. *FAO Fisheries Reports*, **57**, 935–946.

Duffy, J. E. (1993) Genetic population-structure in two tropical sponge-dwelling shrimps that differ in dispersal potential. *Marine Biology*, **116**, 459–470.

(1996a) Eusociality in a coral-reef shrimp. *Nature*, **381**, 512–514.

(1996b) Species boundaries, specialization, and the radiation of sponge-dwelling Alpheid shrimp. *Biological Journal of the Linnean Society*, **58**, 307–324.

(1996c) Resource-associated population subdivision in a symbiotic coral-reef shrimp. *Evolution*, **50**, 360–373.

(2003) The ecology and evolution of eusociality in sponge-dwelling shrimp. In: T. Kikuchi, T., Higashi, S. and Azuma, N. (eds.) *Genes, Behaviour and Evolution in Social Insects*. Sapporo, Japan: University of Hokkaido Press, pp. 1–38.

(2007) Ecology and evolution of eusociality in sponge-dwelling shrimp. In: Duffy, J. E. & Thiel, M. (eds.) *Evolutionary Ecology of Social and Sexual Systems: Crustaceans as Model Organisms*. New York: Oxford University Press, pp. 387–409.

(2010) Social biology of crustacea. In: Breed, M. & Moore, J. (eds.) *Encyclopedia of Animal Behavior*. Oxford: Elsevier, pp. 421–429.

Duffy, J. E. & Thiel, M. (eds). (2007) *Evolutionary Ecology of Social and Sexual Systems: Crustaceans as Model Organisms*. New York: Oxford University Press.

Duffy, J. E. & Macdonald, K. (1999) Colony structure of the social snapping shrimp *Synalpheus filidigitus* in Belize. *Journal of Crustacean Biology*, **19**, 283–292.

Duffy, J. E. & Macdonald, K. S. (2010) Kin structure, ecology and the evolution of social organization in shrimp: A comparative analysis. *Proceedings of the Royal Society of London B*, **277**, 1–13.

Duffy, J. E., Morrison, C., & Rios, R. (2000) Multiple origins of eusociality among sponge-dwelling shrimps (*Synalpheus*). *Evolution*, **54**, 503–516.

Duffy, J. E., Morrison, C., & Macdonald, K. (2002) Colony defense and behavioral differentiation in the eusocial shrimp *Synalpheus regalis*. *Behavioral Ecology and Sociobiology*, **51**, 488–495.

Duffy, J. E., Macdonald, K. S., III, Hultgren, K. M., *et al.* (2013) Decline and local extinction of Caribbean eusocial shrimp. *PLOS ONE*, **8**, e54637.

Emlen, S. T. (1982) The evolution of helping. 1. An ecological constraints model. *The American Naturalist*, **119**, 29–39.

Freestone, A. L., Osman Richard, W., Ruiz, G. M., & Torchin, M. E. (2011) Stronger predation in the tropics shapes species richness patterns in marine communities. *Ecology*, **92**, 983–993.

Greenwood, P. J. (1980) Mating systems, philopatry and dispersal in birds and mammals. *Animal Behaviour*, **28**, 1140–1162.

Hamilton W. D. (1964) The genetical evolution of social behaviour. II. *Journal of Theoretical Biology*, **7**, 17–52.

Hartnoll, R. G. (2001) Growth in crustacea. *Hydrobiologia*, **449,** 111–122.

Hernáez, P., Martínez-Guerrero, B., Anker, A., & Wehrtmann, I. S. (2010) Fecundity and effects of bopyrid infestation on egg production in the Caribbean sponge-dwelling snapping shrimp *Synalpheus yano* (Decapoda: Alpheidae). *Journal of the Marine Biological Association of the United Kingdom*, **90**, 691–698.

Hughes, M. (1996a) Size assessment via a visual signal in snapping shrimp. *Behavioral Ecology and Sociobiology*, **38**, 51–57.

(1996b) The function of concurrent signals: Visual and chemical communication in snapping shrimp. *Animal Behaviour*, **52**, 247–257.

Hughes, M., Williamson, T., Hollowell, K., & Vickery, R. (2014) Sex and weapons: Contrasting sexual dimorphisms in weaponry and aggression in snapping shrimp. *Ethology*, **120**, 982–994.

Hultgren, K. M. (2014) Variable effects of symbiotic snapping shrimps on their sponge hosts. *Marine Biology* **161**, 1217–1227.

Hultgren, K. M. & Brandt, A. (2015) Taxonomy and phylogenetics of the *Synalpheus paraneptunus*-species-complex (Decapoda: Alpheidae), with a description of two new species. *Journal of Crustacean Biology*, **35**, 547–558.

Hultgren, K. M. & Duffy, J. E. (2011) Multi-locus phylogeny of sponge-dwelling snapping shrimp (Caridea: Alpheidae: *Synalpheus*). supports morphology-based species concepts. *Journal of Crustacean Biology*, **31**, 352–360.

(2010) Sponge host characteristics shape the community structure of their shrimp associates. *Marine Ecology Progress Series*, **407**, 1–12.

(2012) Phylogenetic community ecology and the role of social dominance in sponge-dwelling shrimp. *Ecology Letters*, **15**, 704–713.

Hultgren, K., Macdonald, K. S., & Duffy, J. E. (2010) Sponge-dwelling snapping shrimps of Curaçao, with descriptions of three new species, *Zootaxa*, **2372**, 221–262.

Hultgren, K. M., Macdonald, K. S., & Duffy, J. E. (2011) Sponge-dwelling snapping shrimps (Alpheidae: *Synalpheus*) of Barbados, West Indies, with a description of a new eusocial species. *Zootaxa*, **2834**, 1–16.

Hultgren, K. M., Hurt, C., & Anker, A. (2014) Phylogenetic relationships within the snapping shrimp genus *Synalpheus* (Decapoda: Alpheidae). *Molecular Phylogenetics and Evolution*, **77**, 116–125.

Jeffrey N. W., Hultgren, K. M., Chak, T. C. S, Gregory, T. R., & Rubenstein, D. R. (2016) Patterns of genome size variation in snapping shrimps. *Genome*, **59**, 393–402.

Kakui, K. & Hiruta, C. (2013) Selfing in a malacostracan crustacean: Why a tanaidacean but not decapods. *Naturwissenschaften*, **100**, 891–894.

Karplus, I. & Thompson, A. (2011) The partnership between gobiid fishes and burrowing alpheid shrimps. In: Patzner, R. A., Van Tassell, J. L., Kovacic, M., & Kapoor, B. G. (eds.) *The Biology of Gobies*. Boca Raton: Science Publishers, pp. 559–608.

Keller, L. & Genoud, M. (1997) Extraordinary lifespans in ants: A test of evolutionary theories of ageing. *Nature*, **389,** 958–960.

Keller, L. & Perrin, N. (1995) Quantifying the level of eusociality, *Proceedings of the Royal Society of London B*, **260**, 311–315.

Knowlton, N. (1980) Sexual selection and dimorphism in two demes of a symbiotic, pair-bonding snapping shrimp, *Evolution*, **34**, 161–173.

Koenig, W. D., Pitelka, F. A., Carmen, W. J., Mumme, R. L., & Stanback, M. T. (1992) The evolution of delayed dispersal in cooperative breeders. *Quarterly Review of Biology*, **67**, 111–150.

Korb, J. (2008) The ecology of social evolution in termites. In: Korb, J. & Heinze, J. (eds.) *Ecology of Social Evolution*. Berlin: Springer-Verlag, pp. 151–174.

Kough, A. S., Paris, C. B., & Butler M. J. IV. (2013) Larval connectivity and the international management of fisheries. *PLOS ONE*, **8**, e64970.

Linsenmair, K. E. (1987) Kin recognition in subsocial arthropods, in particular in the desert isopod *Hemilepistus reaumuri*. In: Fletcher, D. & Michener, C. (eds.) *Kin Recognition in Animals*. Chichester: John Wiley & Sons, Ltd., pp. 121–207.

Macdonald, K. S., Rios, R., & Duffy, J. E. (2006) Biodiversity, host specificity, and dominance by eusocial species among sponge-dwelling alpheid shrimp on the Belize Barrier Reef, *Diversity and Distributions*, **12**, 165–178.

Macdonald, K. S., Hultgren, K., & Duffy, J. E. (2009) The sponge-dwelling snapping shrimps (Crustacea, Decapoda, Alpheidae, *Synalpheus*) of Discovery Bay, Jamaica, with descriptions of four new species. *Zootaxa*, **2199**, 1–57.

Martin, J. & Davis, G. (2001) *An Updated Classification of the Recent Crustacea*. Los Angeles: Natural History Museum of Los Angeles County.

Mathews, L. (2002) Tests of the mate-guarding hypothesis for social monogamy: Does population density, sex ratio, or female synchrony affect behavior of male snapping shrimp (*Alpheus angulatus*)? *Behavioral Ecology and Sociobiology*, **51**, 426–432.

McMurry, S. E., Blum, J. E., & Pawlik, J. R. (2008) Redwood of the reef: Growth and age of the giant barrel sponge *Xestospongia muta* in the Florida Keys. *Marine Biology*, **155**, 159–171.

Morrison, C., Rios, R., & Duffy, J. E. (2004) Phylogenetic evidence for an ancient rapid radiation of Caribbean sponge-dwelling snapping shrimps (*Synalpheus*). *Molecular Phylogenetics and Evolution*, **30**, 563–581.

Nolan, B. & Salmon, N. (1970) The behavior and ecology of snapping shrimp (Crustacea: *Alpheus heterochaelis* and *Alpheus normanni*). *Forma et Functio*, **2**, 289–335.

Obermeier, M. & Schmitz, B. (2003) Recognition of dominance in the big-clawed snapping shrimp (*Alpheus heterochaelis* Say 1818), Part I: Individual or group recognition? *Marine and Freshwater Behavior and Physiology*, **36**, 1–16.

Ory, N. C., Dudgeon, D., Duprey, N., & Thiel, M. (2014) Effects of predation on diel activity and habitat use of the coral-reef shrimp *Cinetorhynchus hendersoni* (Rhynchocinetidae). *Coral Reefs*, **33**, 639–650.

Pawlik, J., Chanas, B., Toonen, R., & Fenical, W. (1995) Defenses of Caribbean sponges against predatory reef fish. 1. Chemical deterrency. *Marine Ecology Progress Series*, **127**, 183–194.

Queller, D. & Strassmann, J. (1998) Kin selection and social insects, *Bioscience*, **48**, 165–175.

Ríos, R. & Duffy, J. E. (2007) A review of the sponge-dwelling snapping shrimp from Carrie Bow Cay, Belize, with description of *Zuzalpheus*, new genus, and six new species (Crustacea: Decapoda: Alpheidae). *Zootaxa*, **1602**, 1–89

Roberts, C.M. (1997) Connectivity and management of Caribbean coral reefs. *Science*, **278**, 1454–1456.

Rubenstein, D. R. (2012) Sexual and social competition: Broadening perspectives by defining female roles. *Philosophical Transactions of the Royal Society B-Biological Sciences*, **367**, 2248–2252.

Rubenstein, D. R., McCleery, B., & Duffy, J. E. (2008) Microsatellite development suggests evidence of polyploidy in the social sponge-dwelling snapping shrimp *Zuzalpheus brooksi*. *Molecular Ecology Resources*, **8**, 890–894.

Shellman-Reeve, J. S. (1997) The spectrum of eusociality in termites. In: Choe, J. C. & Crespi, B. J. (eds.) *The Evolution of Social Behavior in Insects and Arachnids.* Cambridge: Cambridge University Press, pp. 52–93.

Sherman, P. W., Lacey, E. A., Reeve, H. K., & Keller, L. (1995) The eusociality continuum. *Behavioral Ecology*, **6**, 102–108.

Shuster, S. M. & Wade, M. J. (1991) Equal mating success among male reproductive strategies in a marine isopod. *Nature*, **350**, 608–610.

Spanier, E., Cobb, J. S., & James, M. J. (1993) Why are there no reports of eusocial marine crustaceans? *Oikos*, **67**, 573–576.

Stern, D. L. & Foster, W. A. (1997) The evolution of sociality in aphids: A clone's-eye view. In: Choe, J. C. & Crespi, B. J. (eds.) *The Evolution of Social Behavior in Insects and Arachnids.* Cambridge: Cambridge University Press, pp. 150–165.

Sun, S.-J., Rubenstein, D.R., Liu, J.-N., *et al.* (2014) Climate-mediated cooperation promotes niche expansion in burying beetles. *eLife*, **3**, e02440.

Tóth, E. & Bauer, R. T. (2007) Gonopore sexing technique allows determination of sex ratios and helper composition in eusocial shrimps. *Marine Biology*, **151**, 1875–1886.

(2008) *Synalpheus paraneptunus* (Crustacea: Decapoda: Caridea) populations with intersex gonopores: A sexual enigma among sponge-dwelling snapping shrimps. *Invertebrate Reproduction and Development*, **51**, 49–59.

(2005) Coordinated group response to nest intruders in social shrimp. *Biology Letters*, **1**, 49–52.

(2008) Influence of sociality on allometric growth and morphological differentiation in sponge-dwelling alpheid shrimp. *Biological Journal of the Linnean Society*, **94**, 527–540.

Vehrencamp, S. (1983) Optimal degree of skew in cooperative societies. *American Zoologist*, **23**, 327–335.

Versluis, M., Schmitz, B., Heydt, von der, A., & Lohse, D. (2000) How snapping shrimp snap: Through cavitating bubbles. *Science*, **289**, 2114–2117.

Vollmer, S. V. & Palumbi, S. R. (2007) Restricted gene flow in the Caribbean staghorn coral *Acropora cervicornis*: Implications for the recovery of endangered reefs. *Journal of Heredity*, **98**, 40–50.

Wilson, E. (1971) *The Insect Societies.* Cambridge, MA: Belknap Press of Harvard University.

Woodward, G., B. Ebenman, M. Emmerson, J., *et al.* (2005) Body size in ecological networks. *Trends in Ecology and Evolution*, **20**, 402–409.

Part II

Vertebrates

9 Sociality in Primates

Joan B. Silk and Peter M. Kappeler

Overview

Primates are a large and diverse order of mammals. They play an important role in studies of the evolution of sociality because there is considerable diversity in social organization and mating systems across the order, and because primates have evolved particularly complex forms of social behavior that rely on individual recognition, well-differentiated relationships, and well-developed social cognition. The first field studies of primates were conducted in the first half of the twentieth century by Clarence Ray Carpenter who observed mantled howler monkeys (*Alouatta palliata*, Carpenter, 1934), white-handed gibbons (*Hylobates lar*, Carpenter, 1940) and rhesus monkeys (*Macaca mulatta*, Carpenter, 1942) in their natural habitats. In the 1940s, Kinji Imanishi began studying free-ranging macaques (*Macaca fuscata*) in Japan, and developed a set of techniques that would become the foundation of future research on wild primates: habituation, individual recognition, and long-term studies (Matsuzawa & McGrew, 2008). In the past 60 years, researchers have amassed a large body of information about the social behavior and ecology of many of the more than 400 primate species, and as a result, today more is known about sociality in primates than in any other mammalian order.

I SOCIAL DIVERSITY

9.1 How Common is Sociality in Primates?

To answer this question in a precise way, we need to define a set of terms to describe the dimensions of sociality, and relate them to terms commonly used in the behavioral ecology literature. This is not a trivial exercise in semantics. Precise definitions facilitate cross-taxonomic comparisons and help to identify targets of selection and their interdependencies. Our task is complicated by the fact that biologists use numerous descriptive labels for different aspects of sociality

We thank Marina Cords, Charles Janson, and the editors for many helpful comments on earlier versions of this chapter, and the contributors to *The Evolution of Primate Societies*, whose work we relied on to write this chapter. Ulrike Walbaum provided valuable help in formatting the manuscript.

(Chapter 1). In particular *social system, social structure, social living* and *social organization* are often used interchangeably, and mating system terminology has frequently been used to describe how a population is sub-structured into social units, thereby hampering comparisons across primates and comparisons among primates and other taxa (Rowell, 1993; Müller & Thalmann 2000; Shultz, *et al.*, 2011; Swedell, 2012). However, a conceptual dissection of social systems and the associated terminology proposed by Kappeler & van Schaik (2002) has enjoyed growing acceptance among primatologists (Cords, 2007), and has also been applied to other mammals (Kappeler, *et al.*, 2013).

According to the framework developed by Kappeler & van Schaik (2002), primate social systems are defined by and encompass three interrelated, but heuristically distinct, components: (1) social organization; (2) mating system; and (3) social structure. *Social organization* refers to who lives with whom and is defined as the size, composition, cohesion, and genetic structure of a social unit. Three fundamental types of social organization can be distinguished: (1) solitary: adults do not coordinate their activities with conspecifics and spend most of their time alone (or with dependent offspring); (2) pair-living: adults associate and coordinate their activities with one member of the opposite sex for extended periods of time; (3) non-paired group living: adults associate and coordinate their activities with at least two other conspecifics, and form groups that can be further categorized in terms of their size and composition (e.g. single-male vs. multi-male groups). Although group living is often equated with sociality, it is important to emphasize that solitary species can display some of the same kinds of social patterns that we see in group living species, such as non-random patterns of contacts, individual recognition, cooperative predator mobbing, and communal breeding (Kappeler, 2012).

Mating system refers to who mates with whom and how often. Because there is not necessarily a strict correspondence between who lives with whom and who mates with whom, genetic analyses play an important role in the analysis of mating systems. Apart from providing information on the main categories of mating systems (e.g. monogamy, polygyny, polyandry, polygynandry), these data also allow for estimates of the degree of reproductive skew within each sex. Mating system terminology should not be used to characterize different types of social organization because they are not necessarily congruent. For example, individuals living in pairs (a type of social organization) can have a monogamous mating system or participate in extra-pair matings, as occurs among birds (Griffith, *et al.*, 2002).

Social structure is defined as the set of all dyadic social relationships (except mating) among group members, and constitutes an emergent property of the group. According to Hinde (1976), social relationships represent the outcome of a contingent series of social interactions between two individuals. The pattern and frequency of interactions within a dyad will be influenced by the traits of each individual, such as their age, sex, and physical condition, and by properties of the dyad, such as their degree of relatedness. The pattern of relationships among all of the individuals in the group shapes the social structure. Hinde's abstract construct has been difficult to operationalize, but social network analysis provides a valuable set of tools to visualize and quantify group-level properties of the social structure (Sueur, *et al.*, 2011). For example, it is possible to assess the density of direct and indirect connections among individuals within groups and the extent to which groups are subdivided into cliques. Primatologists are just beginning to explore how the topography of social networks varies across species, across groups within species, and across time within groups.

Having defined the elements of social systems, we can now return to the question of how social are primates. Only 5 percent of extant primates are classified as solitary (Table 9.1). However, the proportion of solitary species will probably rise as we learn more about the many strepsirrhine species and several tarsier species whose social organization has not yet been described. If all these unknown species were solitary, the proportion of solitary species would increase to 21 percent. Thus, 79 to 95 percent of all primate species form groups, and all but two of the 275 species of Old World monkeys, New World monkeys, and apes are group living.

It is important to point out that group living is not exclusively synonymous with being social. Some solitary primates form day-time sleeping groups, some tolerate same-sex conspecifics within their ranges, and some use (ultrasonic) vocal and olfactory signals along with direct interactions to manage their relationships with kin, mates, and neighbors (Schülke & Ostner, 2005; Nekaris & Bearder, 2011; Kappeler, 2012). Furthermore, the spatial distribution of some solitary primates is structured by kinship (Kappeler, *et al.*, 2002), and this may have favored the evolution of forms of sociality that are not usually associated with a solitary lifestyle, such as temporary sleeping associations (Dammhahn & Kappeler, 2009; van Noordwijk, *et al.*, 2012) and communal nursing (Eberle & Kappeler, 2006).

Table 9.1 Sociality across major groups of primates. Number of species exhibiting one of the three major types of social organization. Each species listed in Mitani, *et al.* (2012) was classified according to its modal type of social organization. Many species of lemurs, lorises, and tarsiers (sometimes collectively also referred to as prosimians, even though this is not a monophyletic lineage) have not been studied yet, and were therefore classified as "unknown". Even though not all species of platyrrhine (New World) and catarrhine (Old World) monkeys have been studied yet, we assumed they would exhibit the same type of social organization as other members of their genus.

Infraorder	Common name	Solitary	Pair-living	Group-living	Unknown	Total
Lemuriformes	Lemurs	9	14	26	44	93
Lorisiformes	Lorises	8	3	0	16	27
Tarsiformes	Tarsiers	2	2	1	6	11
Platyrrhini	New World monkeys	0	45	94	-	139
Catarrhini	Old World monkeys and apes	2	16	118	-	136
Total		**21**	**80**	**239**	**66***	**406**

* The majority of these species is expected to be solitary or pair-living.

9.2 Forms of Sociality in Primates

There is considerable diversity in the social organization of primates. Some primate species are solitary, others live in pairs, some live in groups with multiple adult females and one or many adult males, and some form large, multi-level societies (Table 9.1). All social primate species form stable bisexual social units that include at least one adult male and one adult female (van Schaik & Kappeler, 1997), although in a handful of Old World primate species – such as grey langurs, *Semnopithecus entellus* (Rajpurohit, *et al.*, 1995) and blue monkeys, *Cercopithecus mitis* (Cords, 2000) – all-male groups exist along with bisexual groups. Thus sexual segregation is much less common primates than it is in other mammals (Ruckstuhl & Neuhaus, 2002).

9.2.1 Solitary Species

Apart from the two species of orangutan, *Pongo* spp., solitary primates are restricted to members of the lemurs, lorises, and tarsiers (Table 9.1). Most solitary primates are nocturnal, and the difficulties of group coordination, enhanced crypsis, or a lack of anti-predator benefits derived from grouping, together with a heavy reliance on unpredictably distributed, non-sharable resources (e.g. insects) are thought to explain the absence of group formation in these species (Kappeler, 1997).

9.2.2 Pair-Living Species

Pair-living has evolved in all the major groups of primates. At least 20 percent of the 406 species of primates live in pairs, and this proportion may rise as we learn more about the social organization of a number of strepsirrhine and tarsier species. Among mammals, only the small order of elephant shrews (Macroscelidae) has a larger proportion of pair-living species than primates. Pair-living species typically live in groups of less than five individuals (Patterson, *et al.*, 2014), and generally include one adult male, one adult female, and several immatures.

9.2.3 Group Living Species

Group living characterizes most of the New World monkeys, Old World monkeys, and apes. Only cetaceans and bats have larger proportions of group living species (Lukas & Clutton-Brock, 2013). Group living primates can live in groups that contain one male and multiple females (one-male groups), multiple males and multiple females (multi-male groups), or more rarely, a single female and multiple males. Some Old World primates – such as snub-nosed monkeys, *Rhinopithecus* spp., hamadryas baboons, *Papio hamadryas*, and geladas, *Theropithecus gelada* – live in multi-level societies in which the primary unit is the one-male group. Multiple one-male groups (and all male groups) regularly associate together and form clans or bands, and these units may be further combined into even larger aggregations (Grueter, *et al.*, 2012). Group size data are notoriously inaccurate, but the most careful estimates indicate that for species that live in multi-male, multi-female groups, group size typically ranges from 50 to 70 (Patterson, *et al.*, 2014).

9.2.4 Flexibility in Forms of Sociality

Although primatologists typically define social organization in the way that we do, the boundaries between categories sometimes blur. For example, as we noted earlier, many solitary primates forage alone, but form daytime sleeping groups. Gibbons, *Hylobates* spp., typically form groups that include only one adult male and one adult female, but in some populations of white-handed gibbons nearly one quarter of the groups include two adult males and one adult female (Reichard, *et al.*, 2012). Many pair-living primates are characterized by close association, coordination, and bonding of pair-partners (Fernandez-Duque, *et al.*, 2012), but some lemur species are best characterized as "dispersed pairs" (van Schaik & Kappeler, 2003). In these taxa, male and female territories overlap extensively, but pair-partners rarely interact or actively avoid each other (Schülke & Kappeler, 2003; Fichtel, *et al.*, 2011). Taken to extremes, males and females of one species (white-footed sportive lemur, *Lepilemur leucopus*) were never seen grooming in year-round observations, and they had their only physical contact during the one night of the year when they mated (Dröscher & Kappeler, 2013). In some species that typically form one-male, multi-female groups, there are influxes of nonresident males during the mating season (e.g. blue monkeys, Cords, 2000). In other species that form one male, multi-female groups, adult males sometimes tolerate subordinate "followers" who may be former residents in the group (e.g. geladas, Snyder-Mackler, *et al.*, 2012) or younger natal males (e.g. mountain gorillas, *Gorilla gorilla berengei*, Yamagiwa, *et al.*, 2003).

The degree of spatial cohesion among group members also varies considerably, complicating simple descriptions of group composition. Chimpanzees, bonobos, and some New World monkeys (spider monkeys, *Ateles* spp. and muriquis, *Brachyteles* spp.) are characterized by a fission-fusion dynamic. In these species, as in some bats, dolphins and hyenas, a set of individuals forms a community and shares a collective range, but all members of the group rarely travel together as a cohesive unit. Instead, they form temporary subgroups (or parties) that change in size and composition over the course of a day (Aureli, *et al.*, 2008).

9.3 Why Primates Form Social Groups

Of the numerous benefits of living in groups (Alexander, 1974; Krause & Ruxton, 2002), two advantages seem to play the most important roles for primates: (1) communal resource defense; and (2) anti-predator benefits. In evaluating the relative importance of these and other factors, it is often difficult to distinguish between the selective factors that promoted evolutionary transitions to group living and the advantages that accrue once groups exist. The benefits contribute to the maintenance of groups despite the centrifugal forces generated by the costs of group living. We summarize these benefits and costs in this section.

9.3.1 Resource Acquisition and Use

Wrangham (1980) first proposed that sociality in primates evolved as a means to enhance females' access to food resources. He argued that food is the key resource limiting female reproductive success, and females benefit from collectively defending food patches against neighboring groups. These benefits will accrue only if food occurs in discrete, defensible patches of high quality, where patch size determines the number of individuals that can feed together. Wrangham (1980) also emphasized the importance of fallback foods (i.e. abundant low quality items that are eaten when more preferred foods are scarce), which would enable females to maintain bonds in the face of seasonal changes in food availability. He further reasoned that groups would be composed of related females because long-term relationships and kinship would facilitate cooperation in intergroup conflicts. These "female-bonded" species were expected to rely upon clumped resources – like fruit, which could be defended efficiently – while the non-female-bonded species without pronounced between-group competition would feed primarily upon more dispersed resources, such as leaves. Wrangham's hypothesis is hard to reconcile with the fact that group living primates have extremely variable diets, and not all female-bonded species collectively defend access to food resources. Moreover, van Schaik (1983, 1989) argued that the benefits that females might gain from between-group competition would be offset by the costs incurred from increased within-group competition, and he emphasized the impact of female–female competition on the evolution of primate social organization.

9.3.2 Predator Avoidance

As small to medium-sized mammals inhabiting tropical and subtropical habitats, primates face a variety of predators, including snakes, raptors, carnivores, and other primates (Fichtel, 2012). Predation rates are difficult to quantify because predator attacks are not often observed directly. Predation is more often inferred when healthy animals disappear abruptly, and alternate explanations (such as dispersal) can be confidently eliminated. Studies of habituated chimpanzees at Gombe National Park in Tanzania, which hunt red colobus monkeys (*Procolobus badius*) provide an exception. Chimpanzees killed 15–53 percent of the red colobus monkey within the park population each year (Teelen, 2008). Vigilance among diurnal species is ubiquitous and increases with perceived predation risk (Teichroeb & Sicotte, 2012). Predation risk decreases with increased body size and varies between habitats (Janson & Goldsmith, 1995; Hill & Lee, 1998). Primates exhibit numerous behavioral adaptations to reduce predation risk, including alarm calling, mobbing, and formation of mixed-species associations (Isbell, 1994; Stojan-Dolar & Heymann, 2010; Fichtel, 2012). Predation has undoubtedly played a prominent role in the course of primate evolution, and predation avoidance has been widely accepted as the ultimate reason why virtually all diurnal primates live in groups (van Schaik, 1983).

Predation risk has two fundamental impacts on primate social organization. First, group size and cohesion vary as a function of predation risk. In the absence of predators, group size decreases and the distance among individuals within groups increases (van Schaik & van Noordwijk, 1985; Kappeler & Fichtel, 2012; Bettridge & Dunbar, 2012). Second, increased predation risk favors an increase in the number of adult males per group. This effect is evident in both inter-specific (van Schaik & Hörstermann, 1994; Hill & Lee, 1998) and intra-specific comparisons (Stanford, 1998; Bettridge & Dunbar, 2013). Additional males contribute to group augmentation (Kokko, *et al.*, 2001) and adult males may be best suited to confront predators because of their superior size and weaponry. Predation risk could also affect primate mating systems by favoring multi-male societies and providing opportunities for more than one male to mate with females (van Schaik & Hörstermann, 1994; Port & Kappeler, 2010). Thus, predation avoidance has been an important force in primate evolution with far-ranging consequences for all aspects of their social behavior.

9.3.3 Homeostasis

Although group-level benefits may promote the evolution of certain aspects of sociality in many social insects (Oldroyd & Fewell, 2007), such benefits of grouping have not been discussed in the primate literature. There are numerous links between aspects of the social system of a species and the homeostasis of individuals, but these processes do not seem to have acted as selective advantages promoting sociality in primates. Life in groups facilitates thermoregulatory behaviors like huddling in primates inhabiting cool or seasonally cold habitats (Ostner, 2002), but it is more likely that social tolerance in group living species allowed some primates to successfully occupy temperate habitats than *vice versa*. Permanent association and interaction with others may also create stress for some individuals and promote the spread of pathogens (Freeland, 1976; Sapolsky, 2004), thus actually leading to deviations from homeostasis.

9.3.4 Mating

Mating opportunities may influence primate sociality in some ways. Lukas & Clutton-Brock's (2013) comparative study suggested that the formation of pairs among primates is primarily due to mate defense, although other explanations focusing on infanticide avoidance have been proposed as well (Opie, *et al.*, 2013). Although males often guard access to groups of females in species that live in larger social groups, male mate guarding is unlikely to be the explanation for why females live together. It seems more likely that females benefit from living in groups because they are safer from predators or have better access to food. Males join groups to gain access to females, and their presence in groups may help to protect females and their offspring from predators and infanticidal threats.

Most pair-living primates are socially, but not genetically, monogamous (e.g. western fat-tailed dwarf lemur, *Cheirogaleus medius*, Fietz, *et al.*, 2000; fork-marked lemur, *Phaner furcifer*, Schülke, *et al.*, 2004; white-handed gibbons, Barelli, *et al.*, 2013) as they are among birds (Griffith, *et al.*, 2002). However, genetic analyses confirm that nocturnal owl monkeys, *Aotus azarae*, have a monogamous mating system (Huck, *et al.*, 2014).

Species that form multi-male and multi-female groups show considerable diversity in mating systems. Most members of the subfamily Callitrichinae live in groups with multiple adult males and females, but breed cooperatively. Current evidence indicates that the dominant male and female monopolize reproduction in most cases (Anzenberger & Falk, 2012). In other species that live in multi-male, multi-female groups, all females are reproductively active, although some females may reproduce more successfully than others (Pusey, 2012). Among males, there is considerable variation in the extent of reproductive skew across species (Port & Kappeler, 2010). In some species, like Northern muriquis, *Brachyteles hypoxanthus*, paternity is fairly evenly distributed among male group members and the most successful male sires only 18 percent of the infants (Strier, *et al.*, 2011). In contrast, in white-fronted capuchins, *Cebus capucinus*, and yellow baboons, *Papio cynocephalus*, the most successful males sire about 80 percent of all infants (Altmann, *et al.*, 1996; Alberts, *et al.*, 2006; Muniz, *et al.*, 2010), and in Verreaux's sifakas, *Propithecus verreauxi*, even more than 90 percent (Kappeler & Schäffler, 2008).

9.3.5 Offspring Care

In many social insects and cooperatively breeding birds, groups form as a result of extended care for offspring and their subsequent retention in natal groups with overlapping generations (Alexander, 1974; Emlen, 1994). Cooperative breeding among primates is rare, however, and has evolved only once within the primate order in the subfamily Callitrichinae, which includes the tamarins and marmosets, presumably from pair-living ancestors (Dunbar, 1995; Lukas & Clutton-Brock, 2013). Associations among otherwise solitary females in mouse lemurs, *Microcebus murinus*, provide mothers with benefits from communal care of offspring (Eberle & Kappeler, 2006), providing a potential mechanism for the promotion of female gregariousness. This pattern is not widespread among primates and is unlikely to be a primary factor favoring sociality across the primate order (van Noordwijk, *et al.*, 2012). However, a growing body of evidence suggest that benefits derived from male paternal care may favor the unusual prevalence of the enduring bisexual associations that distinguish primates from most other mammalian orders. Although direct male care is uncommon in primates, particularly those that do not live in pairs, males may play an important role in reducing infanticide risk in primates (van Schaik & Kappeler, 1997). In addition, several studies of yellow baboons and chacma baboons, *Papio ursinus*, which live in multi-male, multi-female groups have shown that males provide benefits to their offspring even when infanticide risk is low (Buchan, *et al.*, 2003; Charpentier, *et al.*, 2008; Nguyen, *et al.*, 2009; Huchard, *et al.*, 2010).

9.4 The Role of Ecology in Shaping Sociality in Primates

As in most social vertebrates, ecology plays an important role in shaping primate social organization. Primates occur in a range of habitats, they spend their lives on the ground or up in the canopy, and they occur across the tropics on a number of continents.

9.4.1 Habitat and Environment

Primates occupy a diverse set of habitats, ranging from tropical rainforests to woodland savannahs to arid deserts. Habitat may influence sociality through its influence on predation risk and food distribution. Terrestrial animals living in more open habitats are vulnerable to a wider range of predators than arboreal animals living in more wooded areas are, and as a consequence terrestrial species typically live in larger groups than arboreal species of the same body size (Janson & Goldsmith, 1995). As we noted earlier, predation risk also influences the number of males per group, and this influences both social organization and mating system.

Primate social organization is thought to be primarily affected by the number of co-occurring predators and their hunting styles (see Fichtel, 2012), as well as the distribution and abundance of key resources, such as food and safe shelters (Schülke & Ostner, 2012). Resource distribution played a key role in verbal models of primate sociality, which had an influential impact on primate behavioral ecology in the 1990's (van Schaik, 1989; Isbell, 1991; Sterck, *et al.*, 1997). These models predicted that the extent of between-group versus within-group competition for resources would influence the value of alliances among females, and this in turn would influence dispersal patterns and a number of social traits. Van Schaik (1989) suggested that within-group contest competition (WGC), within-group scramble competition (WGS), and between-group contest competition (BGC) would have different effects on social structure. When females compete over access to resources with group members, the outcome will be a function of individual dominance rank. Thus, he predicted that WGC would be associated with the formation of strict linear dominance hierarchies. Because dominance rank is likely to influence females' access to resources and ultimately to affects their fitness, he also predicted that females would form alliances to support close kin and related females would come to occupy adjacent positions in the dominance hierarchy. This, in turn, would create strong selective pressures favoring female philopatry. In contrast, when WGS is the primary form of resource competition, a female's reproductive success will depend upon the size of her social group, not her own dominance rank. This will weaken selective pressures favoring the formation of dominance hierarchies, nepotistic alliances, and female philopatry. A combination of WGS and weak BGC will have similar effects.

This basic scheme, which has been elaborated upon by several other researchers (Isbell, 1991; Sterck, *et al.*, 1997), seemed to fit with much of what was known about primate feeding ecology and social organization in the 1990's, and it became possible to assign a number of species to particular quadrants in the socioecological scheme.

Primate behavioral ecologists devoted considerable effort to investigating the socio-ecological correlates of social organization and behavior (Janson, 2000). Because it is difficult to experimentally manipulate key ecological variables and measure social responses of primates (but see Corbin & Schmid, 1995; Janson, 1998, 2012; Janson & Byrne, 2007), natural experiments and comparative studies provided the primary methods for evaluating predictions derived from these models. For example, species with large geographical ranges are likely to be exposed to different types of habitats or climatic variability in different parts of their ranges. Such ecological variation seems to influence group size, group composition, and competitive regimes in a number of species, including gorillas, baboons, and grey langurs (Newton, 1988; Barton, *et al.*, 1996; Yamagiwa, *et al.*, 2003), as the models predict. Mitchell, *et al.* (1991) compared two closely related species of squirrel monkeys, *Saimiri* spp., in Costa Rica and Peru that had similar diets, lived in groups of similar sizes, experienced similar levels of predation, but differed in the extent of WGC. As predicted by socioecological models, the species that experienced high WGC formed stable, linear dominance hierarchies, kin-based coalitions, and exhibited strict female philopatry. In the species that experienced low levels of WGC, a dominance hierarchy was not detected among females, female coalitions were not observed, and females sometimes left their natal groups.

However, as more analyses were conducted, researchers found that standard dietary classifications (e.g. frugivore, folivore) did map on to competitive regimes in a simple way (Janson, 2000; Snaith & Chapman, 2007), complicating efforts to apply the model. It also became clear that there were examples of groups of closely related species that have similar forms of social organization across quite different kinds of habitats. Perhaps the most striking examples of this are the Cercopithecinae, a subfamily of Old World monkeys that includes many species of macaques, baboons, vervets, *Chlorocebus* spp., and mangabeys, *Lophocebus* spp. These species exhibit marked uniformity in patterns of their social organization despite their distribution across Africa and Asia (Di Fiore & Randall, 1994). Similarly, different species of lemurs, *Eulemur* spp., inhabiting very different habitats throughout Madagascar exhibit no relationship between social organization and environment (Ossi & Kamilar, 2006). Thus, there is a growing consensus that socioecological models do not fully account for observed variation in social organization and behavior in extant primate species (Janson, 2000; Clutton-Brock & Janson, 2012) and that other factors, such as the relationship between group size and infanticide risk or phylogenetic history, may also play important roles.

9.4.2 Biogeography

Primates are mainly found in tropical regions of the Americas, Africa, and Asia, but (non-human) primates are absent from Australia and New Zealand. The monkeys of the New World belong to the infraorder Platyrrhini, and range from northern Mexico to northern Argentina. While biogeographic factors seem to have an important impact on species richness and diversity in primates as they do in other taxa (Lehman & Fleagle, 2005), there is no clear link between primate biogeography and sociality.

9.4.3 Niches

As a taxonomic group, primates rely upon a wide range of food resources, including invertebrates and vertebrates, fruit, leaves, seeds, gum and saps, as well as various other minor items (e.g. mushrooms, bark, soil). These food items vary greatly in energy content, digestibility, processing requirements, and toxicity (Chapman, *et al.*, 2012). Furthermore, food items and food patches vary in size, density, and seasonal availability (Koenig, 2002). Most primates eat a variety of kinds of foods, although the proportion of fruit, leaves, and animal prey in the diet varies considerably across taxa (Chapman, *et al.*, 2012). Animals like the bamboo lemurs, *Hapalemur* spp., which feed almost exclusively on bamboo (Tan, 1999) and spectral tarsiers, *Tarsier spectrum*, whose diet is composed entirely of animals (Gursky, 2002), are exceptions to the rule. As noted earlier, there is not a clear link between what primates eat and the kinds of groups they live in, and the niche concept does not seem to play an important role in understanding primate sociality.

Circadian activity acts as a strong constraint on primate social evolution (Terborgh & Janson, 1986; Shultz, *et al.*, 2011). Group living, in particular, is strongly associated with diurnal activity, and only a few species of owl monkeys, lemurs, and bamboo lemurs, *Hapalemur* spp., that can be active both day and night (Donati, *et al.*, 2009; Parga, 2011), exhibit some nocturnal activity in groups. The factors that favor cathemeral activity remain in the dark, as do the advantages of diurnal activity, which has evolved at least three times in the course of primate evolution (Hill, 2006; Griffin, *et al.*, 2012).

9.5 The Role of Evolutionary History in Shaping Sociality in Primates

There is a growing appreciation of the role that phylogenetic history plays in the evolution of primate social organization (Di Fiore & Rendall, 1994; Shultz, *et al.*, 2011; Opie, et al., 2012; Kappeler, *et al.*, 2013; Thierry, 2013). It has long been recognized that closely related primate species exhibit very similar types of social organization. For example, all species of the Hylobatidae, which includes gibbons and siamangs, are pairliving. Comparative behavioral studies of macaques revealed differences in social structure among ecologically very similar species that can best be explained by their phylogenetic affiliation (Balasubramaniam, *et al.*, 2012).

Formal phylogenetic analyses also suggest that evolutionary history has an important impact on primate social organization. Shultz, *et al.* (2011) analyzed the phylogeny and social organization of more than 200 primate species. Their analyses suggest that there were a series of unilateral transitions in primate social organization over the last 65 million years. About 52 million years ago, there were the first transitions from solitary life to group living. These analyses also indicate that the first groups were loose aggregations of multiple males and females. From there, more stable multi-male, multi-female groupings arose. And from the stable groups, both pair-bonded and one-male harems emerged roughly 16 million years ago. In contrast, Lukas & Clutton-Brock

(2013) found a direct transition from solitary to pair-living in primates and other mammalian taxa. The discrepancy in these findings may be related to the way social organization, particularly pair-living, was categorized (Lukas & Clutton-Brock, 2013), and underscores the importance of clear definitions and consistent terminology.

II SOCIAL TRAITS

Primates lead rich social lives. They recognize other group members and some neighbors as individuals, participate in a diverse range of social interactions, and develop well-differentiated social relationships. The patterning of their social interactions is often structured by kinship, age, sex, and dominance rank.

9.6 Traits of Social Species

9.6.1 Cognition and Communication

Patterns of variation in cognition, brain size, and brain organization are uniquely important in understanding primate sociality. According to the social brain hypothesis, selection favored enhanced cognitive abilities in primates to help individuals cope with the challenges of life in large and complex social groups (Dunbar, 2003). For example, to compete effectively with other group members, primates benefit from sophisticated social cognition. In contrast, others have argued that non-social challenges, such as processing foods that are inaccessible to most potential consumers, have favored enhanced cognitive abilities in primates. More generally, cognitive complexity appears to facilitate innovation and social learning (Reader & Laland, 2001, 2002).

Primates, as an order, have larger brains than expected for their body sizes (Jerrison, 1973; Passingham, 1981) or basal metabolic rates (Isler & van Schaik, 2006), with humans representing a particularly pronounced outlier. Within the primates, there have also been several graded shifts, with most of the haplorhine primates having relatively larger brains than strepsirrhines, and the great apes having relatively larger brains than most of the monkeys (Rilling, 2006). Brain organization exhibits similar trends (Radinsky, 1975; Finlay & Darlington, 1995). For example, various measures of relative brain size (e.g. the size of the neocortex in relation to the rest of the brain) are correlated with several proxies of social complexity, including social group size (Sawaguchi, 1990; Sawaguchi & Kudo, 1990; Dunbar 1992), size of grooming cliques (Kudo & Dunbar, 2001), frequency of tactical deception (Byrne & Corp, 2004), and fission-fusion social organization (Barrett, *et al.*, 2003; Aureli, *et al.*, 2008). Moreover, measures of relative brain size are also associated with several measures linked to foraging behavior, such as frugivory, extractive foraging, tool use, social learning, and the length of the juvenile period (Joffe & Dunbar, 1997; Reader & Laland, 2001, 2002; Walker, *et al.*, 2006). For example, tool use has been reported for a variety of monkey and ape species (reviewed in Perry, 2006; Whiten, 2012). In nearly all cases, tools are used for food processing, although chimpanzees also use tools for other purposes, such as wiping debris from their bodies. Primatologists have compiled detailed lists of innovations (Reader & Laland, 2001, 2002). Most innovations are linked to foraging, but there are some exceptions. For example, Perry (2011) has documented a number of novel social behaviors in white-faced capuchins, such as tail-sucking, or inserting fingers in others' eyes, mouth, or nostrils.

As the social brain hypothesis implies, monkeys have well developed knowledge about the other members of their group (reviewed in Seyfarth & Cheney, 2012). They recognize other group members as individuals, and know something about their own kinship and dominance relationships to other group members. Observational studies and experimental playback studies conducted on several haplorhine primates indicate that they also know something about the kinds of relationships that exist among others, which is referred to as third-party knowledge. Third-party knowledge extends to kinship relationships (Cheney & Seyfarth 1980, 1986, 1989; Kitchen, *et al.*, 2005), dominance relationships (Silk, 1999; Perry, *et al.*, 2004; Slocombe & Zuberbühler, 2007), and relationship quality (Bachmann & Kummer, 1980; Perry, *et al.*, 2004, 2008; Wittig & Boesch, 2010).

Communication and individual recognition in primate groups involve a combination of olfactory, visual, and auditory modalities. Olfactory signals play a role in sexual behavior, mate choice, territorial defense, individual and species recognition, mother-infant bonding, and foraging (Zuberbühler, 2012). Scent marking behavior occurs in strepsirrhine primates and some New World monkeys, and some of them anoint themselves with urine and other scented matter. In some strepsirrhine primates, mates with heterozygous or dissimilar MHC alleles are preferred (Western fat-tailed dwarf lemur and grey mouse lemur, *M. murinus*, Schwensow, *et al.*, 2008a,b) and this information may also affect mate choice decisions in haplorhines (Setchell & Huchard, 2010; but see Huchard, *et al.*, 2010 for chacma baboons). Visual signals, including facial expressions and gestures, also play important roles in primates, particularly the monkeys and apes. Facial expressions are highly uniform across individuals, develop early in life, and appear in individuals reared in isolation, suggesting that they are under strong genetic control (Arbib, *et al.*, 2008). The diversity of visual signals in primates may be related to the extent of social complexity. A phylogenetic analysis of 12 haplorhine primate species showed that the number of facial expressions are positively associated with group size (Dobson, 2009), a crude measure of social complexity. Vocal communication has been studied more intensively than visual or olfactory signals, in part because vocalizations lend themselves to acoustic analyses of structure and experimental analyses of meaning. Some primates produce a range of acoustically distinct call types that are used in specific contexts, such as predator detection, foraging, and socio-sexual interactions. The size of call repertoires is correlated with two measures of social complexity: group size and the percentage of time spent grooming (McComb & Semple, 2005). Anecdotal evidence indicates that primates may use vocal signals deceptively to gain strategic advantages (Byrne & Whiten, 1990; Gouzoules, *et al.*, 1996), much as some avian species do (Munn, 1986; Møller, 1990).

9.6.2 Lifespan and Longevity

Longevity is relevant for comparative studies of primate social behavior because it correlates positively with opportunities for reciprocity and the development of social bonds. As expected for large-brained mammals of their body size, primates have relatively long lifespans (Lindstedt & Calder, 1981; van Schaik & Isler, 2012; but see Austad & Fischer, 1992). For example, the maximum lifespan for strepsirrhine primates ranges from 12 to 37 years, while the maximum life spans for great apes range up to 60 years. Within the primate order, larger-bodied and larger-brained primates generally live longer than other primates do (Austad & Fischer, 1992; Judge & Carey, 2000; van Schaik & Isler, 2012).

Increases in longevity are generally associated with decreases in extrinsic sources of mortality, particularly predation. Terrestrial animals are vulnerable to a wider range of predators than arboreal animals, and have more limited means to escape from predatory attacks. This helps to explain why flying animals, such as birds and bats, live longer than terrestrial animals of similar body size, and arboreal mammals live longer than terrestrial mammals of similar body size (Prothero & Jürgens, 1987; Austad & Fischer, 1992; Shattuck & Williams, 2010; van Schaik & Isler, 2012). The same relationship may hold for primates (van Schaik & Isler, 2012; but see Shattuck & Williams, 2010).

9.6.3 Fecundity

Some strepsirrhine primates give birth to twins and occasionally triplets (van Schaik & Isler, 2012). Most of the haplorhine primates typically produce singletons, but cooperatively breeding callitrichids consistently produce litters of two or three infants. In general, larger-bodied species have longer gestation periods and wean their infants later than smaller-bodied species, and fertility rates are thus inversely related to body size (van Schaik & Isler, 2012).

9.6.4 Age at First Reproduction

Age at first reproduction ranges from less than 1 year in some of the strepsirrhines to nearly 61 years in orangutans. Across primates, there is a close relationship between age at first reproduction and body size, with large bodied species maturing later than small bodied species (van Schaik & Isler, 2012).

9.6.5 Dispersal

In all group living primate species, members of one or both sexes disperse from their natal groups as they reach the age of sexual maturity (Pusey, 1987). As in a number of other mammalian taxa, the primary function of dispersal in primates seems to be inbreeding avoidance (Pusey, 1987; Pusey & Wolf, 1996). Dispersal by members of both sexes is characteristic of pair-bonded primates, and is the ancestral pattern for primates living in larger social groups (Schülke & Ostner, 2012).

In species that are characterized by dispersal by only one sex, male dispersal and female philopatry are considerably more common than female dispersal and male philopatry. The male bias in dispersal is generally attributed to the benefits that philopatric females gain from forming alliances with kin (Wrangham, 1980).

9.7 Traits of Social Groups

9.7.1 Genetic Structure

Although the development of noninvasive methods for collecting genetic information makes it possible to describe the genetic structure of free-ranging groups, such analyses remain relatively limited for primates, thereby precluding comprehensive analyses of how social organization, male reproductive skew, group fissioning, or sex-biased dispersal patterns influence genetic structure (reviewed by Di Fiore, 2012).

Sex-biased dispersal patterns are expected to influence population structuring in the portion of the genome inherited from only one parent, i.e. the Y-chromosome for males and mtDNA for females. While dispersal shuffles genetic material across groups, sex biases in philopatry have the opposite effect. When females are philopatric, there is expected to be more sub-structuring of variation in mtDNA than Y-chromosome or autosomal DNA; similarly, when males are philopatric, there is expected to be more sub-structuring in Y-chromosome variation than in mtDNA or autosomal DNA. These predictions are largely supported by studies of grey mouse lemurs (Wimmer, *et al.*, 2002), several species of cercopithecine monkeys (Melnick & Hoelzer 1992, 1996; Shimada, *et al.*, 2000; Hapke, *et al.*, 2001; Hammond, *et al.*, 2006), red woolly monkeys, *Lagothrix poeppigii* (Di Fiore, 2009), chimpanzees, *Pan troglodytes* (Morin, *et al.*, 1994), bonobos, *Pan paniscus* (Gerloff, *et al.*, 1999; Eriksson, *et al.*, 2006), orangutans, *Pongo borneo* (Nietlisbach, *et al.*, 2012), and western lowland gorillas, *Gorilla gorilla* (Inoue, *et al.*, 2013).

Sex-biased dispersal patterns are also expected to create higher degrees of relatedness among members of the philopatric sex than members of the dispersing sex. However, relatedness among philopatric females is expected to be somewhat higher than among philopatric males (Lukas, *et al.*, 2005). This is partly because in species with female philopatry the offspring of two females may be related through both maternal and paternal lines, while in species with male philopatry the offspring of two females will be related through the paternal line only. In line with this, substantially higher degrees of relatedness among females than males have been reported for some species in which females are philopatric, including in yellow baboons (Altmann, *et al.*, 1996), long-tailed macaques, *Macaca fasicularis* (de Ruiter & Geffen, 1998), Verreaux's sifaka (Lawler, *et al.*, 2003), and red-fronted lemurs, *Eulemur rufifrons* (Wimmer & Kappeler, 2002). In contrast, male philopatry is not associated with substantially higher levels of relatedness among males than females in chimpanzees (Vigilant, *et al.*, 2001; Lukas, *et al.*, 2005), bonobos (Gerloff, *et al.*, 1999), red woolly monkeys, or white-bellied spider monkeys,

Ateles belzebuth (Di Fiore, 2009; Di Fiore, *et al.*, 2009). Similar patterns characterize a range of mammal and bird species (Lukas, *et al.*, 2005).

Group fission may increase the degree of relatedness within groups if groups divide along genetic lines. Groups sometimes divide along matrilineal lines in macaques (Missakian, 1973; Chepko-Sade & Sade, 1979; Dittus, 1988; Ménard & Vallet, 1993; Kuester & Paul, 1997; Koyama, 2003; Widdig, *et al.*, 2006), olive baboons, *Papio anubis* (Nash, 1976), wedge-capped capuchins, *Cebus olivaceus* (Robinson, 1988), and ring-tailed lemurs, *Lemur catta* (Ichino, 2006; Ichino & Koyama, 2006), and genetic studies indicate that relatedness within groups is sometimes higher after fission (Cheverud, *et al.*, 1978; Olivier, *et al.*, 1981; Van Horn, *et al.*, 2007). However, female-bonded groups do not always fission cleanly along matrilineal lines (Melnick & Kidd, 1983; Ron, 1996; Widdig, *et al.*, 2006; Van Horn, *et al.*, 2007; Cords, 2012), and the genetic consequences of fission may be less pronounced in these cases.

High reproductive skew, which is characteristic of many primate groups (reviewed by Alberts, 2012) is also expected to influence the genetic structure of groups and may influence the patterning of social behavior. As Altmann (1979) originally pointed out, when there is high male skew in multi-male groups and individual males monopolize reproduction for relatively short periods of time, members of the same birth cohort are likely to be paternal half siblings and age proximity may be an accurate proxy for paternal relatedness. In some taxa, females selectively associate with paternal half-siblings (rhesus macaques, Schülke, *et al.*, 2013; Widdig, *et al.*, 2001, 2002, 2006; yellow baboons, Silk, *et al.*, 2006, Smith, *et al.*, 2003; mandrills, *Mandrillus sphinx*, Charpentier, *et al.*, 2007), while in others paternal siblings are not preferred partners (white-faced capuchins, Perry, *et al.*, 2008; sun-tailed monkeys, *Cercopithecus solatus*, Charpentier, *et al.*, 2008).

9.7.2 Group Structure, Breeding Structure and Sex Ratio

As noted earlier, there is considerable diversity in social organization across primates, and thus considerable variation in adult sex ratios. In species that form pair bonds, sex ratios are generally even. In species that form larger social groups, sex ratios can vary from 1:1 (Kappeler, 2000) to more than 10:1 (F:M; Nunn, 1999). There is a strong linear relationship between the number of females and number of males in social groups (Nunn, 1999; Lindenfors, *et al.*, 2004), and there does not seem to be a clear difference in the sex ratios of species that form one-male and multi-male groups

III SOCIAL SYNTHESIS

Primates are eminently social creatures. In the vast majority of species, individuals are permanently associated with at least one conspecific of the opposite sex. They exhibit virtually all mating systems known from other vertebrates and they have multiple differentiated social relationships that can extend over decades. Sociality is best explained as a response to predation risk, but variation in the size and composition of social groups is probably linked to both phylogenetic history and ecological conditions.

9.8 A Summary of Primate Sociality

Because virtually all haplorhine primates live in stable social groups, and the solitary strepsirrhines are harder to study and less well-known, it is difficult to draw strong conclusions about the impact of sociality within the primate order. However, according to proponents of the social brain hypothesis, primates living in larger groups have relatively larger brains than species living in smaller groups, and relative brain size is, in turn, linked to several proxies of social complexity. This suggests that selective pressures favoring increases in group size have also favored the development of more complex cognition, but the causal arrows could be reversed. Animals that evolved relatively large brains to solve ecological problems might be better able to cope with the challenges of living in larger groups.

9.9 Comparative Perspectives on Primate Sociality

With their large brains and close phylogenetic relationship to humans, there has been a strong tendency to think of primates as "exceptional" animals. This is at least partly because many primatologists are trained in anthropology or psychology programs, and are prone to think of primates in terms of their similarities to humans, not their similarities to other animals. Researchers from the biological sciences tend to avoid studying primates because they live too long to measure important variables such as lifetime fitness, are not readily amenable to most types of field experiments, and they are difficult (and expensive) to bring into the laboratory. As a consequence, the primate literature is not fully integrated with the larger literature in behavioral ecology even though we rely on a common body of theory. Primate researchers are often criticized (or gently mocked) by biologists for their insularity and their penchant for publishing in taxon-specific journals and attending taxon-specific conferences.

Claims of primate exceptionalism are progressively harder to defend in the face of evidence that many traits that were thought to be unique to primates, and closely tied to their large brains and sophisticated cognitive abilities, can be found in other taxa. Thus, a variety of organisms use tools or construct elaborate structures (Hansell, 2007). Corvids have well-developed social cognition (Clayton, *et al.*, 2007), fish display social learning (Brown & Laland, 2003), and meerkats, *Suricata suricatta*, teach immatures how to handle dangerous prey (Thornton & McAuliffe, 2006). All of these capacities are found within the primate order, but not all have yet been found within any single primate species.

Claims of primate exceptionalism would be justified if primates followed different "rules" than other organisms, or constituted exceptions to well-established patterns of adaptation. There is not much evidence that this is the case. For example, the sexually selected infanticide hypothesis, which was originally developed to explain male behavior in primate groups, seems to apply to a wide range of mammalian taxa (Hrdy & Hausfater, 1984). Similarly, the link between sexual dimorphism and intrasexual selection seems to follow the same patterns in primates as it does in pinnipeds, ungulates, carnivores, and birds (Alexander, *et al.*, 1979; Rubenstein & Lovette, 2009).

More recently, researchers have begun to examine the generality of the social brain hypothesis, which was originally developed to explain the evolution of large brains and expansion of the neocortex in primates. As in primates, brain size is positively associated with group size in dolphins (Marino, 1996), carnivores (Peréz-Barbería, *et al.*, 2007; Dunbar & Bever, 2010), ungulates (Peréz-Barbería, *et al.*, 2007), and some groups of insectivores (Dunbar & Bever, 2010). However, there are hints that the selective forces influencing the relationship between brain and group size may operate differently in primates than in other orders. Evolutionary transitions from solitary to pair-living are associated with an increase in brain size in bats, artiodactyl ungulates, and carnivores, as well as in birds (Emery, 2004; Emery, *et al.*, 2007; Shultz & Dunbar, 2007). However, in these taxa, pair-living species have larger brains than species living in larger groups. Primates represent a conspicuous exception to this pattern: pair-living primates have smaller brains in relation to body size than those that live in larger groups.

To explain the variation in the relationship between group size, pair-living, and relative brain size across taxa, it may be important to consider the characteristics of relationships that animals form, not just the size of the groups in which they live (Holekamp, 2007). Evolution may have favored increases in brain size in pair-living mammals and birds in response to the cognitive demands of closely coordinating daily activities and movements and resolving conflicts of interest that arise among pair-bonded partners (Emery, *et al.*, 2007; Shultz & Dunbar, 2007). In most mammals and birds, larger social units are less stable and less tightly coordinated, and may impose weaker cognitive demands on group members. Shultz & Dunbar (2007) suggest that "...individuals in transitory groups may not need to invest in cognitive resources for cataloguing previous experiences (or identifying cheats), judging relative resource holding potential, manipulating the behavior of other individuals or recognizing relatedness." Using crude proxies for social bonding among females (i.e. time devoted to social activity, female philopatry, or formation of stable kin groups), Shultz & Dunbar (2007) found no differences between pair-living primates and pair-living ungulates. However, monkeys and apes that live in larger social units are much more likely to form "bonded" groups than ungulates or carnivores, and also devote more time to social activity than ungulates do.

9.10 Concluding Remarks

In summary, it is possible that (some) primates differ from (most) other mammals and birds in their capacity to form stable bisexual social units, establish social bonds that extend beyond mates and immature offspring, and maintain extended social networks. Primate cognition seems well designed to cope with the opportunities and challenges of social life: identifying kin, choosing allies, remembering past interactions, monitoring rank and reproductive status, tracking third party relationships, perceiving others' intentions, and so on. Moreover, in some primate species, the quality of social bonds is associated with fitness outcomes (reviewed by Silk, 2012). It is not yet clear whether primates are unique in the capacity to form these kinds of extended connections. Spotted hyenas, *Crocuta crocuta*, seem uncannily like baboons in their social lives (Holekamp, 2007), and associations between social relationships and female fitness have been found in several other mammalian taxa, including horses, *Equus equus* (Cameron, *et al.*, 2009), bottlenose dolphins, *Tursiops truncates* (Frère, *et al.*, 2010), and house mice, *Mus musculus* (Weidt, *et al.*, 2008). It may turn out to be the case that the patterns that we see in primates may also characterize a wide range of large-brained, long-living animals that live in stable groups. Thus, primates may be among a select group of exceptional animals, in which natural selection has favored complex sociality.

References

Alberts, S. C. (2012) Magnitude and sources of variation in male reproductive performance. *In*: Mitani, J. C., Call, J. Kappeler, P. M., Palombit, R. A., & Silk, J. B. (eds.) *The Evolution of Primate Societies*. Chicago: University of Chicago Press, pp. 412–431.

Alberts, S. C., Buchan, J. C., & Altmann, J. (2006) Sexual selection in wild baboons: From mating opportunities to paternity success. *Animal Behaviour*, **72**, 1177–1196.

Alexander, R. D. (1974) The evolution of social behavior. *Annual Review of Ecology and Systematics*, **5**, 325–383.

Alexander, R. D., Hoogland, J. L., Howard, R. D., Noonan K. M., & Sherman, P. W. (1979) Sexual dimorphisms and breeding systems in pinnipeds, ungulates, primates, and humans. *In*: Chagnon, N. A. & Irons, W. (eds.) *Evolutionary Biology and Human Social Behavior: An Anthropological Perspective*. North Scituate/Ma: Duxbury, pp. 402–435.

Altmann, J. (1979) Age cohorts as paternal sibships. *Behavioral Ecology and Sociobiology*, **6**, 161–4.

Altmann, J., Alberts, S. C., Haines, S. A., et al. (1996) Behavior predicts genetic structure in a wild primate group. *Proceeding of the National Academy of Science of the United States of America*, **93**, 5797–5801.

Anzenberger, G. & Falk, B. (2012) Monogamy and family life in callitrichid monkeys: Deviations, social dynamics and captive management. *International Zoo Yearbook*, **46**, 109–122.

Arbib, M. A., Liebal, K., & Pika, S. (2008) Primate vocalization, gesture, and the evolution of human language. *Current Anthropology*, **49**, 1053–1076.

Aureli, F., Schaffner, C. M., Boesch, C., et al. (2008) Fission–fusion dynamics: New research frameworks. *Current Anthropology*, **49**, 627–654.

Austad, S. N. & Fischer, K. E. (1992) Primate longevity: Its place in the mammalian scheme. *American Journal of Primatology*, **28**, 251–261.

Bachmann, C. & Kummer, H. (1980) Male assessment of female choice in hamadryas baboons. *Behavioral Ecology and Sociobiology*, **6**, 315–321.

Balasubramaniam, K. N., Dittmar, K., Berman, C. M., et al. (2012) Hierarchical steepness and phylogenetic models: Phylogenetic signals in *Macaca. Animal Behaviour*, **83**, 1207–1218.

Barelli, C., Matsudaira, K., Wolf, T., et al. (2013) Extra-pair paternity confirmed in wild white-handed gibbons. *American Journal of Primatology*, **75**, 1185–1195.

Barrett, L., Henzi, S. P., & Dunbar, R. I. M. (2003) Primate cognition: From 'what now?' to 'what if?'. *Trends in Cognitive Sciences*, **7**, 494–497.

Barton, R. A., Byrne, R. W., & Whiten, A. (1996) Ecology, feeding competition and social structure in baboons. *Behavioral Ecology and Sociobiology*, **38**, 321–329.

Bettridge, C. M. & Dunbar, R. I. M. (2013) Predation as a determinant of minimum group size in baboons. *Folia Primatologica*, **83**, 332–352.

Bradley, B. J., Robbins, M. M., Williamson, E. A., et al. (2005) Mountain gorilla tug-of-war: silverbacks have limited control over reproduction in multimale groups. *Proceedings of the National Academy of Sciences of the United States of America*, **102**, 9418–9423.

Brown, C. & Laland, K. N. (2003) Social learning in fishes: A review. *Fish and Fisheries*, **4**, 280–288.

Buchan, J. C., Alberts, S. C., Silk, J. B., & Altmann, J. (2003) True paternal care in a multi-male primate society. *Nature*, **425**, 179–181

Byrne, R. W. & Corp, N. (2004) Neocortex size predicts deception rate in primates. *Proceedings of the Royal Society of London B*, **271**, 1693–1699.

Byrne, R. W. & Whiten, A. (1990) Tactical deception in primates: The 1990 database. *Primate Report*, **27**, 1–101.

Cameron, E. Z., Setsaas, T. H., & Linklater, W. L. (2009) Social bonds between unrelated females increase reproductive success in feral horses. *Proceedings of the National Academy of Sciences of the United States of America*, **106**, 13850–138503.

Carpenter, C. R. (1934) A field study of the behavior and social relations of howling monkeys (*Alouatta palliata*). *Comparative Psychology Monographs*, **10**, 1–168.

(1940) A field study in Siam of the behavior and social relations of the gibbon (*Hylobates lar*). *Comparative Psychology Monographs*, **16**, 1–212.

(1942) Sexual behavior of free-ranging rhesus monkeys, *Macaca mulatta*. *Journal of Comparative Psychology*, **33**, 113–142.

Chapman, C. A., Rothman, J. M., & Lambert, J. E. (2012) Food as a selective force in primates. *In*: Mitani, J. C., Call, J., Kappeler, P. M., Palombit, R. A., & Silk, J. B. (eds.) *The Evolution of Primate Societies*. Chicago: University of Chicago Press, pp. 149–168.

Charpentier, M. J., Peignot, P., Hossaert–McKey, M., & Wickings, E. J. (2007) Kin discrimination in juvenile mandrills, *Mandrillus sphinx*. *Animal Behaviour*, **73**, 37–45.

Charpentier, M. J., Deubel, D., & Peignot, P. (2008) Relatedness and social behaviors in *Cercopithecus solatus*. *International Journal of Primatology*, **29**, 487–495.

Cheney, D. L. & Seyfarth, R. M. (1980) Vocal recognition in free-ranging vervet monkeys. *Animal Behaviour*, **28**, 362–367.

Cheney, D. L. & Seyfarth, R. M. (1986) The recognition of social alliances among vervet monkeys. *Animal Behaviour*, **34**, 1722–1731.

(1989) Redirected aggression and reconciliation among vervet monkeys, *Cercopithecus aethiops*. *Behaviour*, **110**, 258–275.

Chepko-Sade, B. D. & Sade, D. S. (1979) Patterns of group splitting within matrilineal kinship groups: A study of social group structure in *Macaca mulatta* (Cercopithecidae: Primates). *Behavioral Ecology and Sociobiology*, **5**, 67–86.

Cheverud, J. M., Buettner-Janusch, J., & Sade, D. (1978) Social group fission and the origin of intergroup genetic differentiation among the rhesus monkeys of Cayo Santiago. *American Journal of Physical Anthropology*, **49**, 449–456.

Clayton, N. S., Dally, J. M., & Emery, N. J., (2007) Social cognition by food-caching corvids. *The western scrub-jay as a natural psychologist. Philosophical Transactions of the Royal Society of London B*, **362**, 507–522.

Clutton-Brock, T. H. & Harvey, P. H. (1977) Primate ecology and social organization. *Journal of Zoology*, **183**, 1–39.

Clutton-Brock, T. H. & Janson, C. H. (2012) Primate socioecology at the crossroads: Past, present, and future. *Evolutionary Anthropology*, **21**, 136–150.

Corbin, G. D. & Schmid, J. (1995) Insect secretions determine habitat use patterns by a female lesser mouse lemur (*Microcebus murinus*). *American Journal of Primatology*, **37**, 317–324.

Cords, M. (2000) The number of males in guenon groups. *In*: P. M. Kappeler (ed.) *Primate Males: Causes and Consequences of Variation in Group Composition*. Cambridge: Cambridge University Press, pp. 84–96.

(2007) Primates in perspective. *International Journal of Primatology*, **28**, 497–500.

(2012) The Thirty-year blues: What we know and don't know about life history, group size, and group fission of blue monkeys in the Kakamega Forest, Kenya. *In*: Kappeler, P. M. & Watts, D. P. (eds.) *Long-term field studies of primates*. Springer: Berlin, pp. 289–311.

Dammhahn, M. & Kappeler, P. M. (2009) Females go where the food is: Does the socio–ecological model explain variation in social organisation of solitary foragers? *Behavioral Ecology and Sociobiology*, **63**, 939–952.

de Ruiter, J. D. & Geffen, E. (1998) Relatedness of matrilines, dispersing males and social groups in long-tailed macaques (*Macaca fascicularis*). *Proceedings of the Royal Society of London B*, **265**, 79–87.

Di Fiore, A. (2009) Genetic approaches to the study of dispersal and kinship in New World primates. *In*: *Garber, P. A., Estrada, A., Bicca-Marques, J. C.*, Heymann, E. W. & Strier, K. B. (eds.) *South American Primates*. New York: Springer, pp. 211–50.

(2012) Genetic consequences of primate social organization. Mitani, J. C., Call, J., Kappeler, P. M., Palombit, R. A., & Silk, J. B. (eds.) *The Evolution of Primate Societies*. Chicago: University of Chicago Press, pp. 269–92.

Di Fiore, A. & Rendall, D. (1994) Evolution of social organization: A reappraisal for primates by using phylogenetic methods. *Proceedings of the National Academy of Sciences of the United States of America*, **91**, 9941–9945.

Di Fiore, A., Link, A., Schmitt, C. A., & Spehar, S. N. (2009) Dispersal patterns in sympatric woolly and spider monkeys: Integrating molecular and observational data. *Behaviour*, **146**, 437–470.

Dittus, W. P. (1988) Group fission among wild toque macaques as a consequence of female resource competition and environmental stress. *Animal Behaviour*, **36**, 1626–1645.

Dobson, S. D. (2009) Socioecological correlates of facial mobility in nonhuman anthropoids. *American Journal of Physical Anthropology*, **139**, 413–420.

Donati, G., Baldi, N., Morelli, V., Ganzhorn, J. U. and Borgognini-Tarli, S. M. (2009) Proximate & ultimate determinants of cathemeral activity in brown lemurs. *Animal Behaviour*, **77**, 317–325.

Dröscher, I. & Kappeler, P. M. (2013) Defining the low end of primate social complexity: The social organization of the nocturnal white-footed sportive lemur (*Lepilemur leucopus*). *International Journal of Primatology*, **34**, 1225–1243.

Dunbar, R. I. M. (1992) Neocortex size as a constraint on group size in primates. *Journal of Human Evolution*, **22**, 469–493.

(1995) The mating system of callitrichid primates. I. Conditions for the coevolution of pair bonding and twinning. *Animal Behaviour*, **50**, 1057–1070.

(2003) The social brain: Mind, language, and society in evolutionary perspective. *Annual Review of Anthropology*, **32**, 163–181.

Dunbar, R. I. M. and Bever J. (2010) Neocortex size predicts group size in carnivores and some insectivores. *Ethology*, **104**, 695–708.

Eberle, M. & Kappeler, P. M. (2006) Family insurance: Kin selection and cooperative breeding in a solitary primate (*Microcebus murinus*). *Behavioral Ecology and Sociobiology*, **60**, 582–588.

Emery, N. J. (2004) Are corvids 'feathered apes'? Cognitive evolution in crows, jays, rooks, and jackdaws. *In*: Watanabe, S. (ed.) *Comparative Analysis of Minds*. Tokyo: Keio University Press, pp. 181–213.

Emery, N. J., Seed, A. M., von Bayern, A. M. P., & Clayton, N. S. (2007) Cognitive adaptations of social bonding in birds. *Philosophical Transactions of the Royal Society of London B*, **362**, 489–505.

Emlen, S. T. (1994) Benefits, constraints and the evolution of the family. *Trends in Ecology and Evolution*, **9**, 282–284.

Eriksson, J., Siedel, H., Lukas, D., et al. (2006) Y–chromosome analysis confirms highly sex–biased dispersal and suggests a low male effective population size in bonobos (*Pan paniscus*) *Molecular Ecology*, **15**, 939–949.

Fernandez-Duque, E., Di Fiore, A., & Huck, M. (2012) The behavior, ecology, and social evolution of New World monkeys. *In*: Mitani, J. C., Call, J., Kappeler, P. M., Palombit, R. A., & Silk, J. B. (eds.) *Evolution of Primate Societies*. Chicago: University of Chicago Press, pp. 43–64.

Fichtel, C. (2012) Predation. *In*: Mitani, J. C., Call, J., Kappeler, P. M., Palombit, R. A. & Silk, J. B. (eds.) *Evolution of Primate Societies*. Chicago: University of Chicago Press, pp. 169–94.

Fichtel, C., Zucchini, W., & Hilgartner, R. (2011) Out of sight but not out of mind? Behavioral coordination in red-tailed sportive lemurs (*Lepilemur ruficaudatus*). *International Journal of Primatology*, **32**, 1383–1396.

Fietz, J., Zischler, H., Schwiegk, C., Tomiuk, J., Dausmann, K. H., & Ganzhorn, J. U. (2000) High rates of extra-pair young in the pair–living fat-tailed dwarf lemur, *Cheirogaleus medius*. *Behavioral Ecology and Sociobiology*, **49**, 8–17.

Finlay, B. L. & Darlington, R. B. (1995) Linked regularities in the development and evolution of mammalian brains. *Science*, **268**, 1578–1584.

Freeland, W. J. (1976) Pathogens and the evolution of primate sociality. *Biotropica*, **8**, 12–24.

Frère, C. H., Krützen, M., Mann, J., Connor, R. C., Bejder, L., & Sherwin, W. B. (2010) Social and genetic interactions drive fitness variation in a free–living dolphin population. *Proceedings of the National Academy of Sciences of the United States of America*, **107**, 19949–19954.

Gerloff, U., Hartung, B., Fruth, B., Hohmann, G., & Tautz, D. (1999) Intracommunity relationships, dispersal pattern and paternity success in a wild living community of bonobos (*Pan paniscus*) determined from DNA analysis of faecal samples. *Proceedings of the Royal Society of London B*, **266**, 1189–1195.

Gouzoules, H., Gouzoules, S., & Miller, K. (1996) Skeptical responding in rhesus monkeys (*Macaca mulatta*). *International Journal of Primatology*, **17**, 549–68.

Griffith, S. C., Owens, I. P. F., & Thuman, K. A. (2002) Extra pair paternity in birds: A review of interspecific variation and adaptive function. *Molecular Ecology*, **11**, 2195–2212.

Griffin, R. H., Matthews, L. J., & Nunn, C. L. (2012) Evolutionary disequilibrium and activity period in primates: A bayesian phylogenetic approach. *American Journal of Physical Anthropology*, **147**, 409–416.

Grueter, C. C., Chapais, B., & Zinner, D. (2012) Evolution of multilevel social systems in nonhuman primates and humans. *International Journal of Primatology*, **33**, 1002–1037.

Gursky, S. (2002) Determinants of gregariousness in the spectral tarsier (Prosimian: *Tarsius spectrum*). *Journal of Zoology*, **256**, 401–410.

Hammond, R. L., Handley, L. J. L., Winney, B. J., Bruford, M. W., & Perrin, N. (2006) Genetic evidence for female-biased dispersal and gene flow in a polygynous primate. *Proceedings of the Royal Society of London B*, **273**, 479–484.

Hansell, M. (2007) *Built by Animals: The Natural History of Animal Architecture*. Oxford: Oxford University Press.

Hapke, A., Zinner, D., & Zischler, H. (2001) Mitochondrial DNA variation in Eritrean hamadryas baboons (*Papio hamadryas hamadryas*): Life history influences population genetic structure. *Behavioral Ecology and Sociobiology*, **50**, 483–492.

Hill, R. A. (2006) Why be diurnal? Or, why not be cathemeral? *Folia Primatologica*, **77**, 72–86.

Hill, R. A. & Lee, P. C. (1998) Predation risk as an influence on group size in cercopithecoid primates: Implications for social structure. *Journal of Zoology*, **245**, 447–456.

Hinde, R. A. (1976) Interactions, relationships and social structure. *Man*, **11**, 1–17.

Holekamp, K. E. (2007) Questioning the social intelligence hypothesis. *Trends in Cognitive Sciences*, **11**, 65–69.

Hrdy, S. B. & Hausfater, G. (eds.) (1984) *Infanticide: Comparative and Evolutionary Perspectives*. New York: Aldine.

Huchard, E., Knapp, L.A., Wang, J., Raymond, M., & Cowlishaw, G. (2010) MHC, mate choice and heterozygote advantage in a wild social primate. *Molecular Ecology* **19**, 2545–2561

Huck, M., Fernandez-Duque, E., Babb, P., & Schurr, T. (2014) Correlates of genetic monogamy in socially monogamous mammals: Insights from Azara's owl monkeys. *Proceedings of the Royal Society of London B*, **281**, 20140195.

Ichino, S. (2006) Troop fission in wild ring-tailed lemurs (*Lemur catta*) at Berenty, Madagascar. *American Journal of Primatology*, **68**, 97–102.

Ichino, S. & Koyama, N. (2006) Social changes in a wild population of ringtailed lemurs (*Lemur catta*) at Berenty, Madagascar. *In*: Jolly, A., Sussman, R. W., Koyama, N. & Rasamimanana, H. R. (eds.) *Ringtailed Lemur Biology: Lemur catta in Madagascar*. New York: Springer, pp. 233–44.

Inoue, E., Akomo-Okoue, E. F., Ando, C., et al. (2013) Male genetic structure and paternity in western lowland gorillas (*Gorilla gorilla gorilla*). *American Journal of Physical Anthropology*, **151**, 583–588.

Isbell, L. A. (1991) Contest and scramble competition: Patterns of female aggression and ranging behavior among primates. *Behavioral Ecology*, **2**, 143–155.

(1994) Predation on primates: Ecological patterns and evolutionary consequences. *Evolutionary Anthropology*, **3**, 61–71.

Isler, K. & van Schaik, C. P. (2006) Metabolic costs of brain size evolution. *Biology Letters*, **2**, 557–560.

Janson, C. H. (1998) Testing the predation hypothesis for vertebrate sociality: Prospects and pitfalls. *Behaviour*, **135**, 389–410.

(2000). Primate socioecology: The end of a golden Age. *Evolutionary Anthropology*. **9**, 73–86.

Janson, C. H. (2012) Reconciling rigor and range: Observations, experiments, and quasi-experiments in field primatology. *International Journal of Primatology*, **33**, 520–541.

Janson, C. H. & Byrne, R. (2007) What wild primates know about resources: Opening up the black box. *Animal Cognition*, **10**, 357–367.

Janson, C. H. & Goldsmith, M. L. (1995) Predicting group size in primates: Foraging costs and predation risks. *Behavioral Ecology*, **6**, 326–336.

Jerison, H. J. (1973) *Evolution of the Brain and Intelligence*. New York: Academic Press.

Joffe, T. H. & Dunbar, R. I. M. (1997) Visual and socio-cognitive information processing in primate brain evolution. *Proceedings of the Royal Society of London B*, **264**, 1303–1307

Judge, D. A. & Carey, J. R. (2000) Postreproductive life predicted by primate patterns. *Journal of Gerontology A*, **55**, B201–B2019.

Kappeler, P. M. (1997) Determinants of primate social organization: Comparative evidence and new insights from Malagasy lemurs. *Biological Reviews of the Cambridge Philosophical Society*, **72**, 111–151.

(2000) Causes and consequences of unusual sex ratios among lemurs. *In*: Kappeler, P. (ed.) *Primate Males: Causes and Consequences of Variation in Group Composition*. Cambridge: Cambridge University Press, pp. 55–63.

(2012) The behavioral ecology of strepsirrhines and tarsiers. *In*: Mitani, J. C., Call, J., Kappeler, P. M., Palombit, R. A., & Silk, J. B. (eds.) *Evolution of Primate Societies*. Chicago: University of Chicago Press, pp. 17–42.

Kappeler, P. M. & Fichtel, C. (2012) A fifteen-year perspective on the social organization and life history of sifaka in Kirindy Forest. *In*: Kappeler, P. M. & Watts, D. P. (eds.) *Long-Term Field Studies of Primates*. Heidelberg: Springer, pp. 101–121.

Kappeler, P. M. & Schäffler, L. (2008) The lemur syndrome unresolved: Extreme male reproductive skew in sifakas (*Propithecus verreauxi*), a sexually monomorphic primate with female dominance. *Behavioral Ecology and Sociobiology*, **62**, 1007–1015.

Kappeler, P. M. & van Schaik, C. P. (2002) Evolution of primate social systems. *International Journal of Primatology*, **23**, 707–740.

Kappeler, P. M., Wimmer, B., Zinner, D. P., & Tautz, D. (2002) The hidden matrilineal structure of a solitary lemur: Implications for primate social evolution. *Proceedings of the Royal Society of London B*, **269**, 1755–1763.

Kappeler, P. M., Barrett, L., Blumstein, D. T., & Clutton-Brock, T. H. (2013) Constraints and flexibility in mammalian social behaviour: Introduction and synthesis. *Philosophical Transactions of the Royal Society B*, **368**, 20120337.

Kenyon, M., Roos, C., Binh, V. T., & Chivers, D. (2011) Extrapair paternity in golden-cheeked gibbons (*Nomascus gabriellae*) in the secondary lowland forest of Cat Tien National Park, Vietnam. *Folia Primatologica*, **82**, 154–164.

Kitchen, D. M., Cheney, D. L., & Seyfarth, R. M. (2005) Male chacma baboons (*Papio hamadryas ursinus*) discriminate loud call contests between rivals of different relative ranks. *Animal Cognition*, **8**, 1–6.

Koenig, A. (2002) Competition for resources and its behavioral consequences among female primates. *International Journal of Primatology*, **23**, 759–783.

Kokko, H., Johnstone, R. A., & Clutton-Brock, T. H. (2001) The evolution of cooperative breeding during group augmentation. *Proceedings of the Royal Society of London B*, **268**, 187–196.

Koyama, N. F. (2003) Matrilineal cohesion and social networks in *Macaca fuscata*. *International Journal of Primatology*, **24**, 797–811.

Krause, J. & Ruxton, G. D. (2002) *Living in Groups*. Oxford: Oxford University Press.

Kudo, H. & Dunbar, R. I. M. (2001) Neocortex size and social network size in primates. *Animal Behaviour*, **62**, 711–722.

Kuester, J. & Paul, A. (1997) Group fission in Barbary macaques (*Macaca sylvanus*) at Affenberg Salem. *International Journal of Primatology*, **18**, 941–966.

Launhardt, K., Borries, C., Hardt, C., Epplen, J. T., & Winkler, P. (2001) Paternity analysis of alternative male reproductive routes among the langurs (*Semnopithecus entellus*) of Ramnagar. *Animal Behaviour*, **61**, 53–64.

Lawler, R. R., Richard, A. F., & Riley, M. A. (2003) Genetic population structure of the white sifaka (*Propithecus verreauxi verreauxi*) at Beza Mahafaly Special Reserve, southwest Madagascar (1992–2001). *Molecular Ecology*, **12**, 2307–2317.

Lehman, S. M. & Fleagle, J. G. (2005) Biogeography and primates: A review. *In*: Lehman, S. M., and Fleagle, J. G (eds.) *Primate Biogeography*. Springer, US. pp. 1–58.

Lindenfors, P., Fröberg, L., & Nunn, C. L. (2004) Females drive primate social evolution. *Proceedings of the Royal Society of London B*, **271**(Suppl 3), S101–S103.

Lindstedt, S. L. & Calder III, W. A. (1981) Body size, physiological time, and longevity of homeothermic animals. *Quarterly Review of Biology*, **56**, 1–16.

Lukas, D. & Clutton-Brock, T. H. (2013) The evolution of social monogamy in mammals. *Science*, **341**, 526–530.

Lukas, D., Reynolds, V., Boesch, C., & Vigilant, L. (2005) To what extent does living in a group mean living with kin? *Molecular Ecology*, **14**, 2181–2196.

Marino, L. (1996) What can dolphins tell us about primate evolution? *Evolutionary Anthropology*, **5**, 81–85.

Matsuzawa, T. & McGrew, W. C. (2008) Kinji Imanishi and Sixty years of Japanese primatology. *Current Biology*, **18**, R587–R591.

McComb, K. & Semple, S. (2005) Coevolution of vocal communication and sociality in primates. *Biology Letters*, **1**, 381–385.

Melnick, D. J. & Hoelzer, G. A. (1992) Differences in male and female macaque dispersal lead to contrasting distributions of nuclear and mitochondrial DNA variation. *International Journal of Primatology*, **13**, 379–393.

(1996) The population genetic consequences of macaque social organization and behaviour. *In*: Fa, J. E. & Lindburg, D. G. (eds.) *Evolution and Ecology of Macaque Societies*. Cambridge: Cambridge University Press, pp. 413–443.

Melnick, D. J. & Kidd, K. K. (1983) The genetic consequences of social group fission in a wild population of rhesus monkeys (*Macaca mulatta*). *Behavioral Ecology and Sociobiology*, **12**, 229–236.

Ménard, N. & Vallet, D. (1993) Dynamics of fission in a wild Barbary macaque group (*Macaca sylvanus*). *International Journal of Primatology*, **14**, 479–500.

Missakian, E. A. (1973) The timing of fission among free-ranging rhesus monkeys. *American Journal of Physical Anthropology*, **38**, 621–624.

Mitchell, C.L., Boinski, S., & van Schaik, C.P. (1991) Competitive regimes and female bonding in two species of squirrel monkeys (*Saimiri oerstedi* and *S. sciureus*). *Behavioral Ecology and Sociobiology*, **28**, 55–60

Møller, A. P. (1990) Deceptive use of alarm calls by male swallows, *Hirundo rustica*: A new paternity guard. *Behavioral Ecology*, **1**, 1–6.

Morin, P. A., Moore, J. J., Chakraborty, R., Jin, L., Goodall, J., & Woodruff, D. S. (1994) Kin selection, social structure, gene flow, and the evolution of chimpanzees. *Science*, **265**, 1193–1201.

Müller, A. E. & Thalmann, U. (2000) Origin and evolution of primate social organisation: A reconstruction. *Biological Reviews*, **75**, 405–435.

Muniz, L., Perry, S., Manson, J. H., Gilkenson, H., Gros–Louis, J., & Vigilant, L. (2010) Male dominance and reproductive success in wild white-faced capuchins (*Cebus capucinus*) at Lomas Barbudal, Costa Rica. *American Journal of Primatology*, **72**, 1118–1130.

Munn, C. A. (1986) Birds that 'cry wolf'. *Nature*, **319**, 143–145.

Nash, L. T. (1976) Troop fission in free-ranging baboons in the Gombe Stream National Park, Tanzania. *American Journal of Physical Anthropology*, **44**, 63–77.

Nekaris, K. A. I. & Bearder, S. K. (2011) The Lorisiform primates of Asia and mainland Africa: Diversity shrouded in darkness. *In*: Campbell, C. J., Fuentes, A., MacKinnon, K. C., Bearder, S. K., & Stumpf, R. M. (eds.) *Primates in Perspective*. Oxford: Oxford University Press, pp. 34–54.

Newton, P. N. (1988) The variable social organization of hanuman langurs (*Presbytis entellus*), infanticide, and the monopolization of females. *International Journal of Primatology*, **9**, 59–77.

Nguyen, N., Van Horn, R.C., Alberts, S.C., & Altmann, J. (2009) "Friendships" between new mothers and adult males: Adaptive benefits and determinants in wild baboons (*Papio cynocephalus*). *Behavioral Ecology and Sociobiology*, **63**, 1331–1344

Nietlisbach, P., Arora, N., Nater, A., Goossens, B., van Schaik, C. P., & Krützen, M. (2012) Heavily male-biased long-distance dispersal of orang-utans (genus: *Pongo*), as revealed by Y-chromosomal and mitochondrial genetic markers. *Molecular Ecology*, **21**, 3173–3186.

Nsubuga, A. M., Robbins, M. M., Boesch, C., & Vigilant, L. (2008) Patterns of paternity and group fission in wild multimale mountain gorilla groups. *American Journal of Physical Anthropology*, **135**, 263–274.

Nunn, C. L. (1999) The number of males in primate social groups: A comparative test of the socioecological model. *Behavioral Ecology and Sociobiology*, **46**, 1–13.

Oldroyd, B. P. & Fewell, J. H. (2007) Genetic diversity promotes homeostasis in insect colonies. *Trends in Ecology and Evolution*, **22**(8), 408–413.

Olivier, T. J., Ober, C., Buettner–Janusch, J., & Sade, D. S. (1981) Genetic differentiation among matrilines in social groups of rhesus monkeys. *Behavioral Ecology and Sociobiology*, **8**, 279–285.

Opie, C., Atkinson, Q. D., & Shultz, S. (2012) The evolutionary history of primate mating systems. *Communicative and Integrative Biology*, **5**, 458–461.

Opie, C., Atkinson, Q. D., Dunbar, R. I. M., & Shultz, S. (2013) Male infanticide leads to social monogamy in primates. *Proceedings of the National Academy of Sciences of the United States of America*, **110**, 13328–13332.

Ossi, K. & Kamilar, J. M. (2006) Environmental and phylogenetic correlates of *Eulemur* behavior and ecology (Primates: Lemuridae). *Behavioral Ecology and Sociobiology*, **61**, 53–64.

Ostner, J. (2002) Social thermoregulation in redfronted lemurs (*Eulemur fulvus rufus*). *Folia Primatologica*, **73**, 175–180.

Parga, J. A. (2011) Nocturnal ranging by a diurnal primate: Are ring–tailed lemurs (*Lemur catta*) cathemeral? *Primates*, **52**, 201–205.

Passingham, R. E. (1981) Primate specializations in brain and intelligence. *Symposia of the Zoological Society of London*, **46**, 361–388.

Patterson, S. K., Sandel, A. A., Miller, J. A., & Mitani, J. C. (2014) Data quality and the comparative method: The case of primate group size. *International Journal of Primatology*, **25**, 990–1003.

Pérez-Barbería, F. J., Shultz, S., & Dunbar, R. I. M. (2007) Evidence for coevolution of sociality and relative brain size in three orders of mammals. *Evolution*, **61**, 2811–2821.

Perry, S. (2006) What cultural primatology can tell anthropologists about the evolution of culture. *Annual Review of Anthropology*, **35**, 171–190.

(2011) Social traditions and social learning in capuchin monkeys (*Cebus*) *Philosophical Transactions of the Royal Society of London B*, **366**, 988–996.

Perry, S., Barrett, H. C., & Manson, J. H. (2004) White-faced capuchin monkeys show triadic awareness in their choice of allies. *Animal Behaviour*, **67**, 165–170.

Perry, S., Manson, J. H., Muniz, L., Gros-Louis, J., & Vigilant, L. (2008) Kin–biased social behaviour in wild adult female white–faced capuchins, *Cebus capucinus*. *Animal Behaviour*, **76**, 187–199.

Port, M. & Kappeler, P. M. (2010) The utility of reproductive skew models in the study of male primates, a critical evaluation. *Evolutionary Anthropology*, **19**, 46–56.

Prothero, J. & Jürgens, K. D. (1987) Scaling of maximal lifespan in mammals: A review. *In*: Woodhead, A. D. & Thompson, K. H. (eds.) *Evolution of Longevity in Animals*. New York: Springer, pp. 49–74.

Pusey, A. (2012) Magnitude and sources of variation in female reproductive performance. *In*: Mitani, J. C., Call, J., Kappeler, P. M., Palombit, R. A., & Silk, J. B. (eds.) *The Evolution of Primate Societies*. Chicago: University of Chicago Press, pp. 343–366.

Pusey, A. E. (1987) Sex-biased dispersal and inbreeding avoidance in birds and mammals. *Trends in Ecology and Evolution*, **2**, 295–299.

Pusey, A., and Wolf, M. (1996) Inbreeding avoidance in animals. *Trends in Ecology and Evolution*, **11**, 201–206.

Radinsky, L. (1975) Primate Brain Evolution: Comparative studies of brains of living mammal species reveal major trends in the evolutionary development of primate brains, and analysis of endocasts from fossil primate braincases suggests when these specializations occurred. *American Scientist*, 63, 656–663.

Rajpurohit, L. S., Sommer, V., & Mohnot, S. M. (1995) Wanderers between harems and bachelor bands: Male hanuman langurs (*Presbytis entellus*) at Jodhpur in Rajasthan. *Behaviour*, **132**, 255–299.

Reader, S. M. & Laland, K. N. (2001) Primate innovation: Sex, age and social rank differences. *International Journal of Primatology*, **22**, 787–805.

(2002) Social intelligence, innovation and enhanced brain size in primates. *Proceedings of the National Academy of Sciences of the United States of America*, **99**, 4436–4441.

Reichard, U. H., Ganpanakngan, M., & Barelli, C. (2012) White–handed gibbons of Khao Yai: Social flexibility, complex reproductive strategies, and a slow life history. *In*: P. M. Kappeler and D. P. Watts (eds.) *Long-term Field Studies of Primates*. Heidelberg: Springer, pp. 237–258.

Rilling, J. K. (2006) Human and nonhuman primate brains: Are they allometrically scaled versions of the same design? *Evolutionary Anthropology*, **15**, 65–77.

Roberts, S. J., Nikitopoulos, E., & Cords, M. (2014) Factors affecting low resident male siring success in one-male groups of blue monkeys. *Behavioral Ecology*, **25**, 852–861

Robinson, J. G. (1988) Demography and group structure in wedgecapped capuchin monkeys, *Cebus olivaceus*. *Behaviour*, **104**, 202–232.

Ron, T. (1996) Who is responsible for fission in a free-ranging troop of baboons? *Ethology*, **102**, 128–133.

Rowell, T. E. (1993) Reification of social systems. *Evolutionary Anthropology*, **2**(4), 135–137.

Rubenstein, D. R. & Lovette, I. J. (2009) Reproductive skew and selection on female ornamentation in social species. *Nature*, **462**, 786–789.

Ruckstuhl, K. E. & Neuhaus, P. (2002) Sexual segregation in ungulates: A comparative test of three hypotheses. *Biological Reviews*, **77**, 77–96.

Sapolsky, R. M. (2004) Social status and health in humans and other animals. *Annual Review of Anthropology*, **33**, 393–418.

Sawaguchi, T. (1990) Relative brain size, stratification, and social structure in anthropoids. *Primates*, **31**, 257–272.

Sawaguchi, T. and Kudo, H. (1990) Neocortical development and social structure in primates. *Primates*, **31**(2), 283–289.

Schülke, O. & Kappeler, P. M. (2003) So near and yet so far: Territorial pairs but low cohesion between pair-partners in a nocturnal lemur, *Phaner furcifer*. *Animal Behaviour*, **65**, 331–343.

Schülke, O., & Ostner, J. (2005) Big times for dwarfs: Social organization, sexual selection, and cooperation in the Cheirogaleidae. *Evolutionary Anthropology: Issues, News, and Reviews*, 14, 170–185.

(2012) Ecological and social influences on sociality. *In*: J. C. Mitani, J. Call, P. M. Kappeler, R. A. Palombit and J. B. Silk (eds.) *The Evolution of Primate Societies*. Chicago: University of Chicago Press, pp. 195–219.

Schülke, O., Kappeler, P. M., & Zischler, H. (2004) Small testes size despite high extra-pair paternity in the pair-living nocturnal primate *Phaner furcifer*. *Behavioral Ecology and Sociobiology*, **55**, 293–301

Schülke, O., Wenzel, S., & Ostner, J. (2013) Paternal relatedness predicts the strength of social bonds among female rhesus macaques. *PLoS One*, **8**, e59789.

Schwensow, N., Fietz, J., Dausmann, K. H., & Sommer, S. (2008a) MHC-associated mating strategies and the importance of overall genetic diversity in an obligate pair-living primate. *Evolutionary Ecology*, **22**, 617–636.

Schwensow, N., Eberle, M., & Sommer, S. (2008b) Compatibility counts: MHC-associated mate choice in a wild promiscuous primate. *Proceedings of the Royal Society of London B*, **275**, 555–564.

Setchell, J. M. & Huchard, E. (2010) The hidden benefits of sex: Evidence for MHC-associated mate choice in primate societies. *BioEssays*, **32**, 940–948.

Snaith, T. V. & Chapman, C. A. (2007) Primate group size and interpreting socioecological models: Do folivores really play by different rules? *Evolutionary Anthropology: Issues, News, and Reviews*, **16**, 94–106.

Seyfarth, R. M. & Cheney, D. L. (2012) Knowledge of social relations. *In*: Mitani, J. C., Call, J., Kappeler, P. M., Palombit, R. A., & Silk, J. B. (eds.) *The Evolution of Primate Societies*. Chicago: University of Chicago Press, pp. 628–642.

Shattuck, M. R. & Williams, S. A. (2010) Arboreality has allowed for the evolution of increased longevity in mammals. *Proceedings of the National Academy of Sciences of the United States of America*, **107**, 4635–4639.

Shimada, M. K. (2000) Geographic distribution of mitochondrial DNA variations among grivet (*Cercopithecus aethiops aethiops*) populations in central Ethiopia. *International Journal of Primatology*, **21**, 113–129.

Shultz, S. & Dunbar, R. I. M. (2007) The evolution of the social brain: Anthropoid primates contrast with other vertebrates. *Proceedings of the Royal Society of London B*, **274**, 2429–2436.

Shultz, S., Opie, C., & Atkinson, Q. D. (2011) Stepwise evolution of stable sociality in primates. *Nature*, **479**, 219–222.

Silk, J. B. (1999) Male bonnet macaques use information about third-party rank relationships to recruit allies. *Animal Behaviour*, 58, 5–51.

(2012) The adaptive value of sociality. *In*: Mitani, J. C., Call, J., Kappeler, P. M., Palombit, R. A., & Silk, J. B. (eds.) *The Evolution of Primate Societies*. Chicago: University of Chicago Press, pp. 552–564.

Silk, J. B., Altmann, J., & Alberts, S. C. (2006) Social relationships among adult female baboons (*Papio cynocephalus*) I. Variation in the strength of social bonds. *Behavioral Ecology and Sociobiology*, **61**, 183–195.

Slocombe, K. E. & Zuberbühler, K. (2007) Chimpanzees modify recruitment screams as a function of audience composition. *Proceedings of the National Academy of Sciences of the United States of America*, **104**, 17228–17233.

Smith, K. L., Alberts, S. C., & Altmann, J. (2003) Wild female baboons bias their social behaviour towards paternal half-sisters. *Proceedings of the Royal Society of London B*, **270**, 503–510.

Snyder-Mackler, N., Alberts, S. C., & Bergman, T. J. (2012) Concessions of an alpha male? Cooperative defence and shared reproduction in multi–male primate groups. *Proceedings of the Royal Society of London B*, **279**, 3788–3795.

Stanford, C. B. (1998) Predation and male bonds in primate societies. *Behaviour*, **135**, 513–533.

Sterck, E. H. M., Watts, D. P., & van Schaik, C. P. (1997) The evolution of female social relationships in nonhuman primates. *Behavioral Ecology and Sociobiology*, **41**, 291–309.

Stojan-Dolar, M. & Heymann, E. W. (2010) Vigilance of mustached tamarins in single-species and mixed–species groups: The influence of group composition. *Behavioral Ecology and Sociobiology*, **64**, 325–335.

Strier, K. B., Chaves, P. B., Mendes, S. L., Fagundes, V., & Di Fiore, A. (2011) Low paternity skew and the influence of maternal kin in an egalitarian, patrilocal primate. *Proceedings of the National Academy of Sciences of the United States of America*, **108**, 18915–18919.

Sueur, C., Jacobs, A., Amblard, F., Petit, O., & King, A. J. (2011) How can social network analysis improve the study of primate behavior? *American Journal of Primatology*, **73**, 703–719.

Swedell, L. (2012) Primate sociality and social systems. *Nature Education Knowledge*, **3**, e84.

Tan, C. L. (1999) Group composition, home range size, and diet of three sympatric bamboo lemur species (genus *Hapalemur*) in Ranomafana National Park, Madagascar. *International Journal of Primatology*, **20**, 547–566.

Teelen, S. (2008) Influence of chimpanzee predation on the red colobus population at Ngogo, Kibale National Park, Uganda. *Primates*, **49**, 41–49.

Teichroeb, J. A. & Sicotte, P. (2012) Cost-free vigilance during feeding in folivorous primates? Examining the effect of predation risk, scramble competition, and infanticide threat on vigilance in ursine colobus monkeys (*Colobus vellerosus*). *Behavioral Ecology and Sociobiology*, **66**, 453–466.

Terborgh, J. & Janson, C. H. (1986) The socioecology of primate groups. *Annual Review of Ecology and Systematics*, **17**, 111–135.

Thierry, B. (2013) Identifying constraints in the evolution of primate societies. *Philosophical Transactions of the Royal Society B*, **368**, 20120342.

Thornton, A. & McAuliffe, K. (2006) Teaching in wild meerkats. *Science*, **313**, 227–229.

Van Horn, R. C., Buchan, J. C., Altmann, J., & Alberts, S. C. (2007) Divided destinies: Group choice by female savannah baboons during social group fission. *Behavioral Ecology and Sociobiology*, **61**, 1823–1837.

van Noordwijk, M. A., Arora, N., Willems, E. P., et al. (2012) Female philopatry and its social benefits among Bornean orangutans. *Behavioral Ecology and Sociobiology*, **66**, 823–834.

van Schaik, C. P. (1983) Why are diurnal primates living in groups? *Behaviour*, **87**, 120–144.

(1989) The ecology of social relationships amongst female primates. *In*: V. Standen and R. A. Foley (eds.) *Comparative Socioecology*. Oxford: Blackwell, pp. 195–218.

van Schaik, C. P. & Hörstermann, M. (1994) Predation risk and the number of adult males in a primate group: A comparative test. *Behavioral Ecology and Sociobiology*, **35**, 261–272.

van Schaik, C. P. & Isler, K. (2012) Life–history evolution in primates. *In*: Mitani, J. C., Call, J., Kappeler, P. M., Palombit, R. A., & Silk, J. B. (eds.) *The Evolution of Primate Societies*. Chicago: University of Chicago Press, pp. 220–244.

van Schaik, C. P. & Kappeler, P. M. (1997) Infanticide risk and the evolution of male–female association in primates. *Proceedings of the Royal Society of London B*, **264**, 1687–1694.

(2003) The evolution of social monogamy in primates. *In*: U. H. Reichard & Boesch, C. (eds.) *Monogamy: Mating Strategies and Partnerships in Birds, Humans and Other Mammals*. Cambridge: Cambridge University Press, pp. 59–80.

van Schaik, C. P. & van Noordwijk, M. A. (1985) Evolutionary effect of the absence of felids on the social organization of the macaques on the island of Simeulue (*Macaca fascicularis*, Miller 1903). *Folia Primatologica*, **44**, 138–147.

Vigilant, L., Hofreiter, M., Siedel, H., & Boesch, C. (2001) Paternity and relatedness in wild chimpanzee communities. *Proceedings of the National Academy of Sciences of the United States of America*, **98**, 12890–12895.

Walker, R., Burger, O., Wagner, J., & von Rueden, C.R. (2006) Evolution of brain size and juvenile periods in primates. *Journal of Human Evolution*, **51**, 480–489.

Weidt, A., Hofmann, S. E., & König, B. (2008) Not only mate choice matters: Fitness consequences of social partner choice in female house mice. *Animal Behaviour*, **75**, 801–808.

Whiten, A. (2012) Social learning, traditions, and culture. *In*: Mitani, J. C., Call, J., Kappeler, P. M., Palombit, R. A., & Silk, J. B. (eds.) *The Evolution of Primate Societies*. Chicago: University of Chicago Press, pp. 681–699.

Widdig, A., Nürnberg, P., Krawczak, M., Streich, W. J., & Bercovitch, F. B. (2001) Paternal relatedness and age proximity regulate social relationships among adult female rhesus macaques. *Proceedings of the National Academy of Sciences of the United States of America*, **98**, 13769–13773.

Widdig, A., Nürnberg, P., Krawczak, M., Streich, W. J., & Bercovitch, F. B. (2002) Affiliation and aggression among adult female rhesus macaques: A genetic analysis of paternal cohorts. *Behaviour*, **139**, 371–391.

Widdig, A., Streich, W. J., Nürnberg, P., Croucher, P. J. P., Bercovitch, F. B., & Krawczak, M. (2006) Paternal kin bias in the agonistic interventions of adult female rhesus macaques (*Macaca mulatta*). *Behavioral Ecology and Sociobiology*, **61**, 205–214.

Wimmer, B. & Kappeler, P. M. (2002) The effects of sexual selection and life history on the genetic structure of redfronted lemur, *Eulemur fulvus rufus*, groups. *Animal Behaviour*, **63**, 557–568.

Wimmer, B., Tautz, D., & Kappeler, P. M. (2002) The genetic population structure of the gray mouse lemur (*Microcebus murinus*), a basal primate from Madagascar. *Behavioral Ecology and Sociobiology*, **52**, 166–175.

Wittig, R. M. & Boesch, C. (2010) Receiving post-conflict affiliation from the enemy's friend reconciles former opponents. *PLoS One*, **5**, e13995.

Wrangham, R. W. (1980) An ecological model of female-bonded primate groups. *Behaviour*, **75**, 262–300.

Yamagiwa, J., Kahekwa, J., & Basabose, A. (2003) Intra-specific variation in social organization of gorillas: Implications for their social evolution. *Primates*, **44**, 359–369.

Zuberbühler, K. (2012) Communication strategies. *In*: Mitani, J. C., Call, J., Kappeler, P. M., Palombit, R. A., & Silk, J. B. (eds.) *The Evolution of Primate Societies*. Chicago: University of Chicago Press, pp. 643–663.

10 Sociality in Non-Primate Mammals

Jennifer E. Smith, Eileen A. Lacey, and Loren D. Hayes

Overview

Broadly defined, mammalian social behavior includes all activities that individuals engage in when interacting with conspecifics (Eisenberg, 1966). Such interactions may be agonistic or affiliative and reflect a wide array of functional contexts, including – but not limited to – foraging, predator defense, mate choice, reproductive competition, and parental care. Indeed, these social interactions are a central feature of the biology of all mammals (Lukas & Clutton-Brock, 2013). Sociality – defined as the degree to which individuals live with conspecifics in groups or societies (Eisenberg, 1966) – varies markedly across mammalian species. Sociality shapes multiple aspects of the mammalian phenotype, including mating and breeding success (Silk, 2007), physiology (Creel, 2001), and neurobiology (Fleming, *et al.*, 1999; Carter, 2003; Hofmann, *et al.*, 2014; Young & Wang 2004). These effects are evident over ecological and evolutionary time scales, indicating that social behavior is a powerful selective pressure.

Sociality is widespread, occurring in all major lineages of non-primate mammals. Among metatherians (marsupials), sociality is found primarily in the Diprotodontia (kangaroos, wallabies, and wombats) (Jarman, 1991). Among eutherians (placentals), sociality is well-documented in the Cetacea (whales and dolphins, Connor, 2000), Hyracoidea (hyraxes, Hoeck, *et al.*, 1982), Proboscidea (elephants, Moss, *et al.*, 2011), Perissodactyla (odd-toed ungulates, Cameron, *et al.*, 2009; Sundaresan, *et al.*, 2007), Cetartiodactyla (even-toed ungulates, Clutton-Brock & Guinness, 1982; Coté & Festa-Bianchet, 2001; Jarman, 1974), and Lagomorpha (rabbits, hares, and pikas; Chapman & Flux, 1990). Although members of these taxa have greatly contributed to our understanding of mammalian cooperation, communication, grouping and dominance, here we focus primarily upon the social members of three prominent orders of non-primate mammals: the Carnivora, the Rodentia, and the Chiroptera. First, collectively these three orders represent more than 70 percent of extant mammal species (Wilson & Reeder, 2005). Second, because these lineages include many of the best-studied species of social

We thank the editors, Dustin Rubenstein and Patrick Abbot, for inviting us to contribute to this book and for their careful guidance and patience at all stages of this project. Jennifer E. Smith was supported by Faculty Development Funds from the Provost's Office at Mills College and by the Barrett Foundation. Eileen A. Lacey was supported by the Gary and Donna Freedman Chair in Undergraduate Education and the Museum of Vertebrate Zoology at UC Berkeley. Loren D. Hayes was supported by US NSF grants 0553910 and 0853719 and the University of Tennessee at Chattanooga.

mammals, they offer rich opportunities to explore variation in social structure. Third, because each of these orders contains multiple social species, they allow for informative comparisons of behavior across multiple taxonomic scales.

To characterize sociality in these lineages, we begin by reviewing the leading hypotheses developed to explain why groups form. We then explore variation in the life history traits characterizing social non-primate mammals to explore whether group living is associated with particular suites of such variables. Finally, we outline the major empirical and conceptual advances emerging from studies of these social systems. For each topic, we summarize existing data and identify areas in which our understanding of mammalian social structure is lacking to stimulate further study of the causes and consequences of variation in mammalian sociality.

I SOCIAL DIVERSITY

Non-primate mammals vary enormously with respect to the nature and frequency of interactions with conspecifics, with social species as well as those that form only temporary aggregations or that lack structured social relationships (Eisenberg, 1966). We focus upon species in which individuals belong to groups and interact with group members in ways that are distinct from interactions with non-group mates. Within groups, individuals tend to communicate extensively with one another and to engage in specialized forms of cooperation and conflict such as alloparental care and reproductive competition, respectively. Living with conspecifics is expected to generate a distinct suite of environmental conditions that may impact multiple aspects of individual phenotypes. Accordingly, we consider not only the nature of social structure but also the selective forces giving rise to and resulting from the tendency for members of some species of carnivorans, rodents, and chiropterans to live in groups.

10.1 How Common is Sociality in Non-Primate Mammals?

Social structure has been characterized for roughly 2,500 (46 percent) of the approximately 5,400 species of extant mammals, including primates (Wilson & Reeder, 2005). The majority (approximately 70 percent) of mammalian species are solitary, meaning that adults tend to live alone and that interactions with conspecifics are limited primarily to relatively brief encounters associated with reproduction. The remaining approximately 30 percent of extant mammals are social, meaning that adults engage in some form of group living. Social species are distributed throughout the mammalian radiation and include some of the most extreme forms of social structure known among vertebrates (Sherman, *et al.*, 1995). Given its prevalence, phylogenetic distribution, and evolutionary significance, it is clear that sociality is a fundamental component of the biology of many non-primate mammals.

The Carnivora consists of approximately 300 extant species that, collectively, are global in distribution and occupy a vast range of terrestrial habitats and ecological niches (Wilson & Reeder, 2005). Group living evolved multiple times from a solitary ancestral state, with recent estimates of up to 25 percent of extant carnivoran species recognized as social (Dalerum, 2007). Because social systems are complex, Smith, *et al.* (2012a) developed composite "cooperation scores" to provide a species-level index of the occurrence of none (score = 0) to all (score = 5) of the following forms of cooperation: (1) alloparental care; (2) group hunting; (3) intra-group coalitions; (4) inter-group coalitions; and (5) predator protection. Based upon this analysis, carnivorans engage in 0 to 5 forms of cooperation (mean score = 0.8 ± 0.1; N = 87 species), with alloparental care being most common (Figure 10.1). The occurrence of cooperation varies among families. For example, cooperation is rare in ursids (bears), mustelids (weasels), and most felids (cats), with a few notable exceptions such as the African lion (*Panthera leo*) (Figure 10.2). Cooperative behavior also occurs in hyenids (hyenas) and is particularly pronounced among canids (dogs) and herpestids (mongooses), with the African wild dog (*Lycaon pictus*) engaging in all five forms of cooperation.

The Rodentia includes approximately 2,000 extant species, making it the largest order of mammals (Wilson & Reeder, 2005). Accordingly, the Rodentia encompasses a vast array of social systems. Examples range from strictly solitary pocket gophers (Geomyidae) to eusocial naked mole-rats (Bathyergidae) (Lacey & Sherman, 2007). More generally, sociality among rodents ranges from short-term seasonal aggregations to life-long social groups. Social structure has been studied for at least a few species in each major sub-lineage of rodents, with the best-studied taxa being the ground-dwelling sciurids (ground squirrels) and bathyergids (African mole-rats). Although an index comparable to that of Smith, *et al.* (2012a) has not been applied to rodents, Blumstein & Armitage (1998) established a metric of social complexity for sciurids based on the number of age-sex roles within social groups such that species with the greatest diversity of roles are most complex. Values for this metric varied considerably within and among the genera *Cynomys* (prairie dogs: 1.05 ± 0.06, from 0.84 to 1.23) and *Marmota* (marmots: 1.12 ± 0.15, from 0.27 to 1.46) as well as the genera *Spermophilus* and *Otospermophilus* (ground squirrels, formally known simply as *Spermophilus*: 0.42 ± 0.03, from 0.26 to 0.65). Ongoing research is revealing remarkable diversity in the social structures of the Caviomorpha, a large, highly diverse sub-order of rodents that occurs in South America and parts of Africa; recent estimates indicate that more than 50 percent of the approximately 240 recognized species of caviomorphs are social.

The Chiroptera is second only to the Rodentia in taxonomic breadth, with more than 1,200 species currently recognized (Wilson & Reeder, 2005). Although poorly studied, bats are thought to be among the most socially complex lineages of non-primate mammals (Kerth, 2008). While no quantitative overview of the nature or frequency of sociality is available for this order, group living appears to be widespread. Group sizes range from dozens of individuals to several million bats per roost or colony, although the latter arguably represent aggregations rather than behaviorally cohesive social groups (Kunz, 1982). There is a clear need for additional study of the social structures of roosting colonies, particularly given that one of the most striking aspects of chiropteran social structure is the tendency for individuals to engage in complex forms of cooperation despite spending substantial time away from group mates (Kerth, 2008).

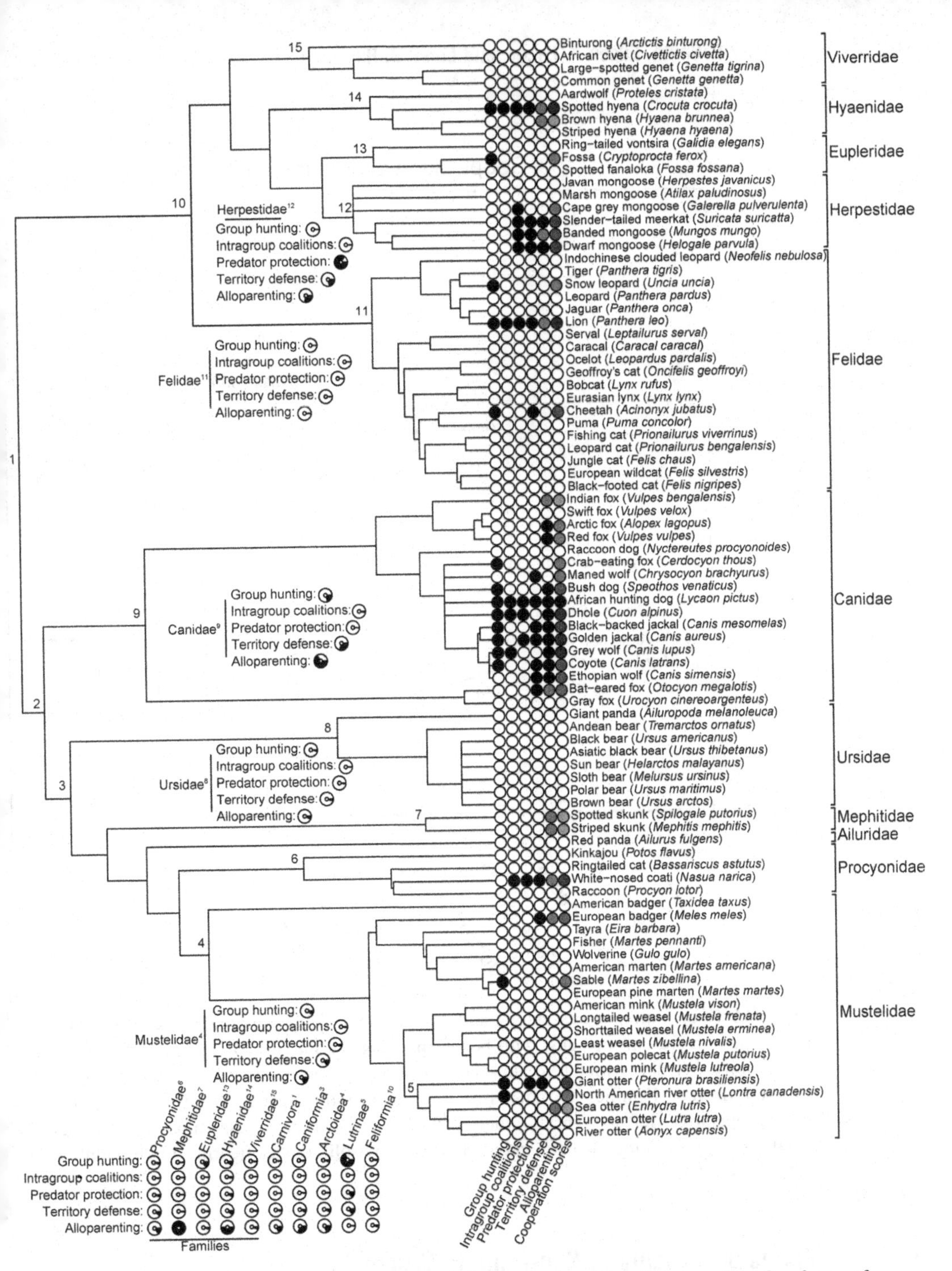

Figure 10.1 Phylogeny of extant carnivoran species for which data regarding five forms of cooperation were available. Tip labels display species names. At tips of the phylogeny are circles shaded to varying degrees indicating for each cooperation variable whether or not the species exhibits the trait. Black indicates trait presence, whereas white indicates that the species does not exhibit that trait. For alloparenting, grey indicates communal denning. Circles for the continuous variable of overall cooperation scores are shaded, with darker shading indicating higher cooperation scores for the species. Pie charts are given for each binary variable for select taxonomic groups giving the probabilities that a common ancestor was uncooperative (white) or cooperative (black). Superscripts for each taxonomic grouping refer to labeled nodes on phylogeny (reprinted from Smith, *et al.*, 2012).

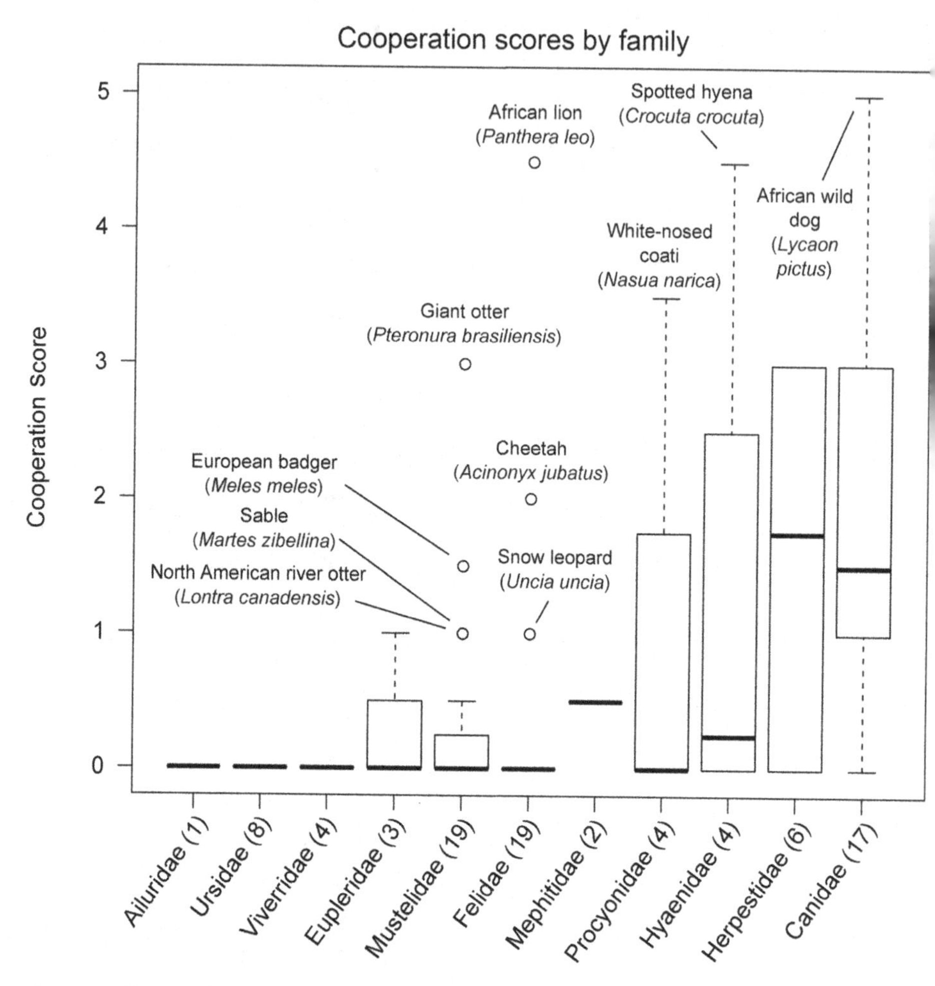

Figure 10.2 Cooperation scores among and within families of Carnivora. Dark horizontal lines in box plots represent medians, with boxes spanning the middle 50 percent of the data for each family. Whiskers stretch to any values that are outside boxes but within 2.5 quartiles from the median. Families are ranked by mean cooperation scores from least cooperative to most cooperative (left to right). Outliers within each family are those values greater than 2.5 quartiles away from the median, and are indicated by solid circles (reprinted from Smith, *et al.*, 2012).

10.2 Forms of Sociality in Non-Primate Mammals

Mammalian social groups vary considerably with respect to multiple parameters, including size, longevity, sex and age composition, kin structure, breeding structure, and degree of cooperation among group members (Krause & Ruxton, 2002; Ebensperger & Hayes, 2008). With respect to breeding structure, non-primate mammals exhibit singular breeding (i.e. one female monopolizes direct reproduction and non-breeders provide alloparental care) and plural breeding (i.e. multiple females in a group breed), each occurring with and without communal care of offspring (Silk, 2007; Chapter 1). These elements of behavior tend to be correlated, resulting in evolutionarily important emergent trends in social structure (Lacey & Sherman, 2007). For example, large groups that form when individuals follow temporally variable food resources; tend to be short-lived and to be composed of aggregations of unrelated individuals. Within these aggregations, all adults are reproductively competent (i.e. no reproductive skew *sensu* Vehrencamp, 1983) and individuals rarely cooperate with one another (e.g. blue wildebeest, *Connochaetes taurinus*, Thirgood, *et al.*, 2004; otherwise solitary species of carnivorans, MacDonald, 1983). In contrast, groups that form when ecological conditions preclude natal dispersal tend to be considerably smaller, to persist longer, and to be composed of close kin; within these groups, individuals often cooperate extensively and reproduction may be restricted to only a subset of adults of each sex (e.g. meerkats, *Suricata suricatta*, Clutton-Brock, *et al.*, 2001). Societies with fission-fusion dynamics are characterized by individuals or subsets of group-mates that regularly merge and subdivide (Smith, *et al.*, 2008); social relationships endure without constant close proximity among group members. In general, the breeding structures of social non-primate mammals are assigned to one of four categories based on the combination of degree of reproductive skew within groups and the presence of alloparental care (Hayes, 2000; Silk, 2007; Ebensperger, *et al.*, 2012; Lukas & Clutton-Brock, 2012; Chapter 1). The resulting categories are as follows:

10.2.1 Plural Breeders without Communal Care

In plural breeding societies without communal care of offspring, all group members reproduce but care of offspring is not shared among group mates (i.e. no alloparental care occurs). This breeding structure appears to be relatively rare among social mammals, having been reported for fewer than a dozen non-primate species, including yellow-bellied marmots (*Marmota flaviventris*, Blumstein & Armitage, 1999) and spotted hyenas (*Crocuta crocuta*, Holekamp, *et al.*, 1996). For example, yellow-bellied marmot groups consist of several related breeding females and their offspring (Blumstein & Armitage, 1999). Although communal nursing occurs in some groups of marmots, communal care is not a consistent feature of the breeding structure of these animals.

10.2.2 Plural Breeders with Communal Care

In plural breeding societies with communal care, all group members reproduce and care of offspring is routinely shared among group mates. This is relatively common, occurring in more than 30 species. Among carnivorans, this list includes Artic foxes (*Alopex lagopus semenovi*), European badgers (*Meles meles*), banded mongooses (*Mungos mungo*), lions, and white-nosed coati (*Nasua narica*) (Silk, 2007; Ebensperger, *et al.*, 2012; Lukas & Clutton-Brock, 2012). Among rodents, plural breeding often involves communal nursing, in which mothers share milk with non-descendent offspring (Hayes, 2000); this type of social structure occurs in phylogenetically diverse species found in Africa (common gundi, *Ctenodactylus gundi*), Asia (great gerbil, *Rhombomys opimus*), Europe (dormouse, *Glis glis*), North America (black-tailed prairie dog, *Cynomys ludovicianus*; Gunnison's prairie dog, *C. gunnisoni*; arctic ground squirrel, *S. parryii*) and South America (cavies, *C. porcellus, Microcavia australis;* colonial tuco-tucos, degus). Because groupmates in many of these societies share a communal nest or den site, other potential forms of alloparental care include thermoregulation and defense of young from predators; among lions, alloparenting includes protection against infanticide by conspecific males (Packer, *et al.*, 1990, 2001). Although parentage assignment studies for bats lags behind those for rodents and carnivorians, data from several well-studied species such as Mexican free-tailed bats, *Tadarida brasiliensis*, and evening bats, *Nycticeius humeralis*, suggest plural breeding and communal nursing in these animals (Burland & Worthington, 2001).

10.2.3 Singular Breeders

Reproduction within groups of singular breeders is restricted to a subset of individuals of each sex – typically one male and one female – with alloparental care provided by non-breeding group members. Among non-primate mammals, this type of breeding structure is observed primarily in carnivorans and rodents (Lukas & Clutton-Brock, 2012). Among carnivorans, singular breeding occurs in canids and herpestids; in these species, non-breeding helpers of both sexes attend the den and provision the offspring of dominant breeders (Moehlman, 1986). Arguably, the best-studied singular breeder is the meerkat, in which the pronounced reproductive skew between breeding and non-breeding group members is maintained by evicting non-breeders that attempt to reproduce directly (Clutton-Brock, *et al.*, 2001, 2008; Griffin, *et al.*, 2003). In rodents, singular breeding occurs in voles (Arvicolinae), beavers (Castoridae), African mole rats (Bathyergidae), porcupines (Hystricidae), marmots (Sciuridae), Mongolian gerbils (*Meriones unguiculatus*), and several species of New World mice and rats (Sigmodontinae and Murinae) (Hayes, 2000; Ebensperger, *et al.*, 2012; Lukas & Clutton-Brock, 2012). In most singular breeding rodents, both sexes are philopatric (McGuire, *et al.*, 1993; Lacey & Sherman, 2007), with the result that females and males may act as non-breeding alloparents.

10.2.4 Eusocial Breeders

In eusocial species, reproduction within groups is restricted to a subset of individuals of each sex, typically one male and one female, with alloparental care provided by non-breeding group members. The distinction between eusociality and singular breeding is debated (e.g. Sherman, *et al.*, 1995 versus Crespi & Yanega, 1995), with potential differences between these breeding structures tied to the degree to which non-breeding alloparents retain the physiological capacity to reproduce. Only two mammal species are consistently recognized as eusocial: the naked mole-rat (*Heterocephalus glaber*) and the Damaraland mole-rat (*Fukomys damarensis*) (Jarvis, *et al.*, 1994; Bennett, *et al.*, 2000). Although both of these species are members of the rodent family Bathyergidae, phylogenetic analyses indicate that eusociality has arisen independently in each species (Rowe & Honeycutt, 2002). Non-breeding mole-rats appear to retain the ability to reproduce throughout their lifetimes (Lacey & Sherman, 2007), although ecological data suggest that most individuals will never produce offspring (Lacey & Sherman, 2007).

10.2.5 Intraspecific Variation in Social Organization

Some social species of non-primate mammals do not fit readily into any single breeding structure, instead displaying intraspecific variation in social group structure. Such variation has been documented for African striped mice, Rhabdomys pumilio, Gunnison's prairie dogs, European badgers, and some species of *Peromyscus* mice (Schradin, 2013). The prairie vole (*Microtus ochrogaster*) is a well-documented example of a species with intraspecific variation in its social and breeding structure; this variation is evident within (McGuire, *et al.*, 2002; Solomon & Crist, 2008) and among populations (Streatfeild, *et al.*, 2011). This species is particularly notable because it displays plural breeding with communal care under some conditions, but singular breeding with alloparental care under others (Hayes & Solomon, 2004; 2006). Such species – those that engage in both breeding systems – represent particularly important targets for future study.

10.3 Why Non-Primate Mammals Form Social Groups

Social behavior has significant fitness consequences for individuals. Sociality is expected to occur when the benefits of living with conspecifics exceed the associated costs (Alexander, 1974; Krause & Ruxton, 2002). Benefits of group formation are numerous but generally encompass elements of resource acquisition, predator avoidance, offspring care, or physiological homeostasis. Typical costs include resource competition and susceptibility to pathogen transmission. The specific nature and exact magnitude of each clearly varies among taxa. Given the extended period of maternal care that characterizes many mammals, milk "theft" and misdirected care distinguish this clade from other animals (Hayes, 2000; Bordes, *et al.*, 2007). Such costs are critical in determining the consequences of group living. Because they are expected to vary with behavior, demography, and ecology, their impacts on social structure warrant careful study. Below, we consider the proposed benefits of sociality to mammals in more detail.

10.3.1 Resource Acquisition and Use

As originally described by Alexander (1974), group living is predicted to increase an individual's access to critical resources when the spatial distribution of those items requires that conspecifics aggregate in the environment. More generally, access to spatially or temporally variable resources plays an important role in the formation of social groups in numerous mammal species; this idea has been formalized as the "resource dispersion hypothesis" (MacDonald, 1983). Specifically, when key trophic resources are patchily distributed, mammals such as European badgers, might share portions of the habitat with little or no net cost due to competition for that resource. Similarly, clumped resources promote group living at roost sites in some species of bats (Kerth, 2008), select pupping beaches in some marine carnivorans (e.g. Galápagos sea lions, *Zalophus wollebaeki*, Wolf, *et al.*, 2007) food resources in white-nosed coatis (Gompper, 1996), watering holes in capybara, Hydrochoerus hydrochaeris, (Herrera, *et al.*, 2011), and African elephants, *Loxodonta africana* (Grainger, *et al.*, 2005). The specific resources in question vary but members of these species share the tendency to benefit from group formation at key resources.

Resource acquisition, rather than resource dispersion *per se*, may also favor sociality in non-primate mammals. For example, in taxa that rely on food resources that are particularly challenging to secure, cooperative hunting or foraging are often important benefits to group living. Cooperative foraging is distinct from the earlier examples of resource acquisition in that the benefits to individuals require the presence and actions of conspecifics. Forms of cooperative foraging such as group hunting have been well-studied in species of carnivorans such as lions (Packer, *et al.*, 1990), African wild dogs (Creel & Creel, 1995), spotted hyenas (Holekamp, *et al.*, 1996), and wolves, *Canis lupus* (Mech, *et al.*, 2015). In each of these taxa, cooperative hunting allows group members to capture larger prey items than would be possible for lone individuals (Smith, *et al.*, 2012). The role of cooperative foraging in promoting sociality among rodents has received less attention, although this lineage includes at least one notable example, the naked mole-rat, in which members of a social group cooperate to dig the subterranean tunnels required to reach food resources (Brett, 1991). In an apparently unique twist, vampire bats engage in reciprocal food sharing of blood meals (Wilkinson, 1984; Carter & Wilkinson, 2013).

10.3.2 Predator Avoidance

Living with conspecifics often decreases the risk of predation by increasing the likelihood that threats are detected and creating opportunities for cooperative defense against predators (Ebensperger, 2001). As with access to spatially or temporally patchy resources, the reduced risk that any given individual will be depredated when in a large group (i.e. the "dilution effect," Foster & Treherne, 1981) requires only the presence of conspecifics. In contrast, benefits resulting from cooperative defense and, in some cases, increased predator detection, require the active participation of group mates. For example, cooperative defense by small carnivores involves coordinated deterrence of predators (Smith, *et al.*, 2012). In multiple species of rodents, improved predator detection involves vigilance and alarm calling (e.g. Blumstein, 2007). In some taxa, this enhanced vigilance has been shown not only to increase predator detection but also to allow group members to allocate more time to other direct fitness enhancing activities such as foraging (FitzGibbon, 1989). Demonstrating that vigilance is cooperative can be challenging. Participation in predator detection is clearly shared among members of some species such as meerkats, in which group members take turns acting as sentinels (Clutton-Brock, *et al.*, 1999). In Belding's' ground squirrels, *Urocitellus beldingi* (Sherman, 1977), however, alarm calling may simply be a function of which individual first detects a threat (Blumstein, 2007). Because improved predator detection often impacts both direct and indirect fitness – including simultaneously enhancing both (e.g. Sherman, 1977) – determining the exact inclusive fitness benefits of this behavior is challenging. Clearly, however, predator detection and anti-predator defense of conspecifics represent potentially important benefits to sociality in multiple non-primate mammals.

10.3.3 Homeostasis

Group living may promote physiological homeostasis, which refers to an organism's ability to maintain bodily functions within a species-typical range of values. Although all mammals generate internal heat to thermoregulate, the traits they use to conserve metabolically generated heat vary. In some species, huddling by members of social groups enhances thermoregulation; this is a benefit of group formation for multiple species of rodents (McShea & Madison, 1984; Blumstein, *et al.*, 2004) and some species of temperate-zone bats (Kerth, 2008; Ebensperger, 2001). This benefit may be particularly important to the survival of offspring in taxa with altricial young that are born helpless and requiring extensive care. Communal rearing typically decreases the amount of time that offspring are left unattended. For example, communal care allows some adult prairie voles to forage while others remain with young and help them to maintain a stable thermal environment (Hayes & Solomon, 2006), which contributes to the more rapid development of communally-reared offspring (Hayes & Solomon, 2004). Similarly, a primary factor favoring sociality in bats is their extremely demanding thermoregulatory physiology, particularly at night and during hibernation (Kerth, 2008). Many species of bats are thought to benefit from roosting socially to maintain body heat (Eisenberg, 1966). Although thermal benefits have received less attention in studies of carnivoran societies, these animals often benefit from group living in terms of increased homeostasis in communal dens (Creel, 2001).

10.3.4 Mating

Individuals that live in groups may benefit via improved access to mates; a shortage of potential reproductive partners may favor natal philopatry and thus the formation of social groups (Emlen & Oring, 1977). Mammalian mating systems are typically characterized based on the number of reproductive partners per individual and include monogamy (i.e. one male mates with one female), polygyny (i.e. one male mates with multiple females), polyandry (i.e. one female mates with multiple males), and polygynandry (i.e. multiple partners for both sexes) (Clutton-Brock, 1988). Which of these systems a species exhibits reflects a combination of ecological, evolutionary, and life history parameters that, collectively, determine the number of mates that an individual can gain access to (Emlen & Oring, 1977). For species in which all group members reproduce (i.e. plural breeders, low reproductive skew), membership in the social group may ensure access to mates (e.g. shortage of partners elsewhere in the habitat), to a larger number of mates (e.g. male defense of multiple females), or to better quality mates (e.g. males that win competitive encounters to secure access to a group of females). For species in which breeding is limited to a subset of adults per group (i.e. singular breeding, high reproductive skew), membership in the group and the associated long-term social bonds that develop among group mates may increase the chances that an individual will eventually become a breeder within the group. Reproductive "queuing" has been suggested for dwarf mongooses, *Helogale parvula* (Creel & Waser, 1994), meerkats (Clutton-Brock, *et al.*, 2001), and naked mole-rats (Lacey & Sherman, 2007).

10.3.5 Offspring Care

In most mammals, care of offspring is performed primarily – if not exclusively – by the female parent. In some social species, however, other adults contribute to care of young; helpers may include other reproductive females or non-breeding "helpers" of one or both sexes (Hayes, 2000; Silk, 2007). Such alloparental care occurs in multiple forms. These behaviors may be divided into two types: direct care (i.e. activities that are clearly targeted toward young) and indirect care (i.e. activities that benefit all group members, including young). Examples of the former include guarding, grooming, and feeding young; examples of the latter include predator detection, cooperative foraging, and thermoregulatory benefits associated with use of a communal nest or den site (Creel & Creel, 1991; Solomon & Hayes, 2009; Lukas & Clutton-Brock, 2012). In groups containing multiple breeding females, care of non-offspring may also include allonursing of young, which has been reported for multiple species of carnivorans, chiropterans, and rodents (e.g. Packer, *et al.*, 1992; Hayes, 2000; Kerth, 2008). In many cases, however, it is unclear whether this behavior occurs because females allow non-offspring to nurse or because females lack the ability to discriminate between their own young from the offspring of others (König, 1994; Hayes & Solomon, 2004, 2006).

More generally, assessing the benefits of alloparental care requires consideration of fitness consequences for both parents and non-parents. Alloparental care is expected to enhance the direct fitness of parents and, depending upon patterns of kinship, the indirect fitness of other group members. These predicted relationships are most clearly upheld in singular breeding groups containing non-breeding alloparents; in these species, the direct fitness of breeding animals tends to increase with the number of non-breeders providing care (Silk, 2007) and alloparents tend to be closely related to the young that they assist, indicating that indirect fitness benefits are possible (Clutton-Brock, 2009). In some social species, however, these predicted fitness outcomes may be complicated by other aspects of group living, such that per capita direct fitness decreases with group size (e.g. colonial tuco-tucos and black tailed prairie dogs, Hoogland, 1995; Lacey, 2004). Moreover, not all carnivorans that engage in alloparental behavior are related to the offspring receiving care (e.g. dwarf mongooses, Creel & Waser, 1994). Thus, while care of offspring is an important potential benefit of group living, the specific nature and magnitude of the fitness consequences of this behavior vary substantially among non-primate mammals.

10.4 The Role of Ecology in Shaping Sociality in Non-Primate Mammals

Ecology is a critical determinant of multiple aspects of social structure. Perhaps most conspicuous is the fundamental role that ecology plays in establishing the costs and benefits of group living. This includes not only competition for critical resources, but also aspects of predator defense and pathogen exposure (Alexander, 1974).

10.4.1 Habitat and Environment

Empirical studies that examine interactions between ecology and social behavior have been slow to emerge. Groups structured by fission-fusion dynamics represent one form of mammalian sociality that may prove particularly informative regarding interactions between ecology and behavior because group members regularly separate and reunite (Smith, *et al.*, 2008). Local ecological conditions both promote and constrain grouping patterns, particularly within the Carnivora; most social species of carnivorans forage in small parties to reduce feeding competition (Smith, *et al.*, 2008). Among the Rodentia, evidence that ecological factors contribute to individual dispersal decisions is limited to largely qualitative comparisons of habitat conditions in social versus solitary taxa (e.g. colonial tuco-tucos, Lacey & Wieczorek, 2004). For others, intraspecific variation in ecology and social structure reveals key ecological variables associated with natal philopatry (e.g. great gerbils, Randall, *et al.*, 2005; degus, Ebensperger, *et al.*, 2012; African striped mice, Schradin, *et al.*, 2012). The most compelling studies are those that experimentally manipulate resource availability to characterize interactions between ecology and natal philopatry (e.g. prairie voles, Smith & Batzli, 2006; Lucia, *et al.*, 2008), but such experimental studies are particularly rare. Because of the challenges associated with field manipulation of habitat parameters, carefully designed intra- and inter-specific comparisons are likely the most tractable way forward for understanding the complex relationships among ecology, dispersal patterns, and social structure.

10.4.2 Biogeography

Carnivorans, rodents, and bats are speciose groups that occur worldwide, with social taxa found throughout the geographic distribution for each lineage. Carnivorans are nearly cosmopolitan; this includes several specific families of Carnivora (e.g. Felidae, Canidae, Phocidae) that are nearly global in distribution. Geographic "hot spots" for endemic species of carnivorans include Borneo, Java and Sumatra, Madagascar, Mesoamerica, Western Ghats, and Sri Lanka, and the Guinean Forests of West Africa (Sechrest, *et al.*, 2002). Rodents are also geographically widespread, occurring on all continents except Antarctica (Honeycutt, *et al.*, 2007). Although a few families (e.g. Sciuridae) are effectively cosmopolitan, in general individual rodent clades tend to be associated with particular geographic regions. Rodent families endemic to sub-Saharan Africa are among the oldest members of this order (Honeycutt, *et al.*, 2007), with the extensive caviomorph clade representing the oldest radiation in South America. Interestingly, 57 percent of extant rodent species are endemic to Australia, despite the general association between this continent and marsupial diversity (Honeycutt, *et al.*, 2007). The biogeography of chiropterans is similar in that, collectively, members of this order are nearly cosmopolitan. Although one family, the Vespertilionidae, spans much of the geographic distribution of the order, other families tend to be more restricted to either specific landmasses or ecogeographic regions (e.g. tropical habitats) of the world. In comparison to carnivorans and rodents, the fossil record for bats is relatively sparse, but a well-resolved molecular phylogeny for all extant families suggests that megabats are nested among four major lineages of microbats (Teeling, *et al.*, 2005). Sociality – specifically cooperative breeding – among birds is particularly prevalent in northern South America, sub-Saharan Africa, and Australia (Jetz & Rubenstein, 2011); in contrast, although no formal quantitative analyses have been conducted, sociality among mammals does not seem to be similarly associated with a particular subset of geographic regions.

10.4.3 Niches

Given their taxonomic and ecological diversity, carnivorans, rodents and bats occupy a wide range of ecological niches. Because carnivoran diets are typically comprised primarily of meat, carnivorans are important in many ecosystems as top predators that regulate the abundance of other animals (e.g. Mech, *et al.*, 2015). The highly social mongooses are a notable exception because members of this clade feed primarily on invertebrates and thus, ecologically, can be characterized as insectivorous. Social rodents are ecologically much more diverse, occurring in all terrestrial biomes and occupying subterranean, aquatic, and arboreal habitats. Rodents are primarily herbivorous and include grazers, browsers, frugivores, and root or bulb specialists. In addition to serving as important prey, some social rodents serve as ecosystem engineers by substantially altering the structure of the habitats in which they occur; social species that fill this role include beavers (*Castor canadensis*) and multiple species of burrowing squirrels. Bats occupy numerous trophic roles, including those of insectivores, carnivores, nectarivores, frugivores, and sanguinivores (Neuweiler, 1989). Accordingly, bats fulfill multiple critical ecological functions as pollinators, seed dispersers, and predators.

10.5 The Role of Evolutionary History in Shaping Sociality in Non-Primate Mammals

In addition to ecology, evolutionary history is expected to be a significant determinant of social structure in non-primate mammals. Carnivorans, rodents, and bats represent markedly divergent lineages of mammals, each of which has followed a distinct evolutionary trajectory within the larger framework of their shared membership in the mammalian clade. Consideration of phylogenetic history has revealed that patterns of social behavior vary within each of these lineages. In the carnivorans sociality generally appears to be rare among ursids, mustelids, and felids, although several notable exceptions occur in the latter two families (Smith, *et al.*, 2012). In contrast, most canids (dogs), herpestids (mongooses), and hyenids (hyenas) are social and tend to engage in cooperative behavior (Smith, *et al.*, 2012). Among rodents, sociality is particularly prevalent within the Caviomorpha (Hayes & Ebensperger, 2011) and some families of the Sciuromorpha (e.g. sciurids, Murie & Michener, 1984; Blumstein & Armitage, 1999) but conspicuously absent in others (e.g. geomyid pocket gophers, Lacey, 2000). Within the bats, phylogenetic history fails to predict call characters used for sophisticated laryngeal echolocation; these traits are apparently highly convergent across the chiropterans (Eick, *et al.*, 2005). In contrast, the negative evolutionary relationship between male testes size and brain size apparently drives mating systems within chiropterans; female promiscuity is most common in species with the smallest brains (Pitnick, *et al.*, 2016). These strong phylogenetic patterns suggest that evolutionary history shapes social structure in concert with current environmental conditions and interspecific differences in ecology.

Evolutionary history also shapes mating structure within groups. For example, singular breeding groups evolved from monogamous ancestors (Lukas & Clutton-Brock, 2012; 2013), with the resulting close kinship among group mates enhancing the potential for indirect fitness benefits. Although it has been argued that singular and plural breeding systems represent endpoints along a continuum of reproductive structures (Sherman, *et al.*, 1995), a quantitative review of relationships between group size and breeding structure suggests that singular and plural breeding instead represent distinct forms of sociality (Rubenstein, *et al.*, 2016). Taken together, these findings imply that evolutionary history has contributed significantly to the extent to which individuals forego direct production of offspring in species of non-primate mammals.

To better understand how evolutionary history and current ecology interact to shape sociality, a growing number of studies use comparative, phylogenetically informed analyses to examine correlates of variation in social behavior. For instance, Lukas & Clutton-Brock (2012) argue that cooperative (singular) breeding is evolutionarily derived from social monogamy. This form of sociality has evolved at least fourteen times in mammals, with thirteen of the evolutionary transitions from monogamy to cooperative breeding occurring in rodents and carnivorans. Whereas sociality likely evolved within African mole-rats (Bathyergidae) due to ecological constraints on dispersal imposed by arid habitats and patchily distributed food resources (Rowe & Honeycutt, 2002), sociality in caviomorph rodents coevolved with increased body sizes and use of subterranean burrows (Ebensperger & Cofre, 2001). Unfortunately, because these studies focus on different rodent taxa, it is not possible to determine the extent to which the ecological factors identified as important to each clade (e.g. aridity, distribution of food, burrow use) may contribute to apparent differences in general patterns of social structure between bathyergid and caviomorph rodents.

II SOCIAL TRAITS

Natural selection has created correlations between multiple life history traits and elements of social structure in non-primate mammals. Here, we identify a suite of traits (e.g. forms of communication, reproductive attributes, dispersal patterns) that appear to differ systematically between social and non-social species of carnivorans, rodents, and bats.

10.6 Traits of Social Species

10.6.1 Cognition and Communication

Communication contributes to numerous elements of mammalian social structure, including social bonding, predator defense, cooperative foraging, dominance hierarchies, and reproduction. Although all mammals engage in multiple modes of communication, chemical and vocal signals are particularly important in most non-primates (Bradbury & Vehrencamp, 2011). While many solitary species have complex forms of communication (e.g. Peters & Wozencraft, 1989), social species are distinguished by the tendency to have a suite of signals that are used only during interactions with group mates (Bradbury & Vehrencamp, 2011). Given the complexity and short-term unpredictability of social relationships, communication is generally expected to be more nuanced and thus potentially more cognitively demanding in social compared to non-social species (Adolphs, 2001). Accordingly, reviewing patterns of communication in group living carnivorans, rodents, and bats should yield important insights into not only the social structures of these animals but also the selective pressures acting on their neurobiology.

Social carnivorans routinely use vocal and olfactory signals to communicate with group members (Peters & Wozencraft, 1989; Archie & Theis, 2011); contexts for these signals include recognizing individuals, assessing kinship, and fostering group cohesion, all of which are critical to the maintenance of within-group social relationships. For example, spotted hyenas distinguish between the calls produced by one versus multiple intruding conspecifics as well as to distinguish among potential partners in cooperative interactions (Smith, *et al.*, 2007, 2010, 2011; Benson-Amram, *et al.*, 2011). Members of this species also use olfactory cues present in secretions from their anal glands to convey information about sex, reproductive status, and individual identity (Drea, *et al.*, 2002). Although particularly well studied in spotted hyenas, both modes of communication are expected to be common among social carnivores (e.g. Gorman & Trowbridge, 1989).

Among social rodents, chemical signals, particularly those in urine, play important roles in territory defense, regulation of reproductive activity, and decisions regarding social and reproductive partners (Wolff & Sherman, 2007). Olfactory information may be used to recognize individuals and to discriminate kin from non-kin (Smith, 2014).

Vocal communication is also widespread among rodents and occurs in multiple contexts. For example, many species emit alarm calls in response to predators; such calls may vary depending on the level of risk and type of predator (e.g. aerial versus terrestrial predators), although it is often unclear whether calls function primarily to signal predators (e.g. pursuit deterrent signals that warn predators of detection) or to warn conspecifics (Blumstein, 2007). Vocalizations can also be crucial to mother-offspring interactions and may influence the neurobehavioral development of young (Branchi, *et al.*, 2001). For both olfactory and vocal communication, it seems likely that many examples of relevant signals have yet to be described. For example, a growing body of evidence indicates that the use of ultrasonic calls may be important during courtship and aggressive encounters (Portfors, 2007). Thus, olfactory and vocal communication play critical and apparently underappreciated roles in multiple aspects of rodent social behavior.

The cognitive demands – the underlying neurobiological complexity and processing capabilities – of communication in social carnivorans and rodents are not well understood. With regard to olfactory signals, chemical cues thought to be involved in kin recognition appear to be processed only by the olfactory bulb and thus may be cognitively less demanding than other forms of communication that require the integration of multiple brain regions (e.g. Haxby, *et al.*, 1996). To date, however, the relative cognitive requirements of processing olfactory versus auditory or other types of information remain unclear.

Among social Chiroptera, information transfer via vocal communication is particularly important and used widely for mother-offspring recognition (Safi & Kerth, 2007). In common vampire bats, *Desmodus rotundus*, vocal communication is also used among non-kin to recognize and locate the most generous food-sharing partners (Carter & Wilkinson 2016). Olfactory information may also be used to attract mates (McCracken & Wilkinson, 2000). Among species of microbats, there is a positive relationship between social complexity, defined simply as group size, and relative brain size (Dunbar & Shultz, 2007). Although many bats live and forage in challenging physical conditions, it is the complexity of the social environment that appears to be most strongly associated with potential cognitive abilities in these animals (Kerth, 2008). Indeed, the relative difference in residual brain volume (a measure that controls for differences in body size) between bats that form pair bonds and those with other mating systems is the most striking known among vertebrates (Pitnick, *et al.*, 2006), suggesting that sociality in the form of reproductive pair bonds is cognitively demanding.

10.6.2 Lifespan and Longevity

Phylogenetically informed comparisons of carnivorans suggest that multiple life history traits are correlated with sociality and cooperation in these animals (Beckhoff, *et al.*, 1984; Gittleman, 1986; Smith, *et al.*, 2012); key variables include reduced sexual dimorphism, increased age at weaning, and increased age of first reproduction. Demographic and ecological factors that likely interact with these life history parameters include population density, degree of social group cohesion, and prolonged hunting efforts in open habitats. Interestingly, however, neither longevity nor mass-corrected basal metabolic rate is correlated with the degree of cooperation in these taxa (Smith, *et al.*, 2012). These patterns are intriguing and efforts to determine the direction of causality of these relationships should generate exciting insights into the role of lifespan in the evolution of social structure in carnivorans.

Among rodents, sociality and communal breeding may be favored in multiple species due to high rates of mortality coupled with the often pronounced need for parental care. In particular, taxa in which young individuals face harsh seasonal conditions may benefit from the extended care made possible by group living (Arnold, 1990). With regard to reproductive parameters, Blumstein & Armitage (1998) found that social complexity (based on number of age-sex categories in a group) in ground-dwelling sciurids is linked to increased age at first reproduction, reduced litter size and increased offspring survival. With regard to body size, although Ebensperger (2001) predicted that sociality should be more pronounced among small-bodied species, an empirical review (Ebensperger & Blumstein, 2006) revealed that large-bodied species of hystricognath rodents tend to be more social. As evident from this brief synopsis, such analyses have typically not included all rodent taxa and thus it remains unclear how the general differences in life history strategies characterizing sciuromorph versus hystricorph rodents, the two most speciose suborders, may contribute to patterns of sociality across this lineage.

A primary factor thought to favor sociality among chiropterans is their remarkable longevity (Kerth, 2008). Bats are extraordinarily long-lived compared to other mammals of similar body sizes (Figure 10.3), with average life spans of 5 to 10 years and maximum longevities of up to 30 or more years (Barclay & Harder, 2003; Kerth, 2008). Maternal investment is also often prolonged, with females nursing young until the latter are roughly the size of adults. This combination of factors is expected to lead to extended social relationships between adults and young, which may favor natal philopatry and thus facilitate the formation of social groups comprised of multiple generations of related females (Eisenberg, 1966).

Clearly, life history attributes play important roles in the evolution of sociality among non-primate mammals. To date, however, efforts to quantify these relationships have tended to focus on traits that are distinctive to specific sub-clades of mammals, such as the striking longevity of chiropterans. Moving forward, studies that examine the same suites of parameters across diverse mammalian taxa should prove useful in elucidating the impacts of larger patterns of life history variation in mammals. For example, recent comparative analyses of cooperatively breeding rodents, carnivorans, and primates (Lukas & Clutton-Brock, 2012) indicate that this form of sociality is associated with polytocy (i.e. production of litters of multiple young) but is not consistently linked with longevity or other forms of reproductive investment in these lineages. The apparent contradiction between this result and the findings for bats described above illustrates the potential complexity of interactions between life history traits and behavior. Together, these results underscore the need for comparative studies conducted across multiple taxonomic scales.

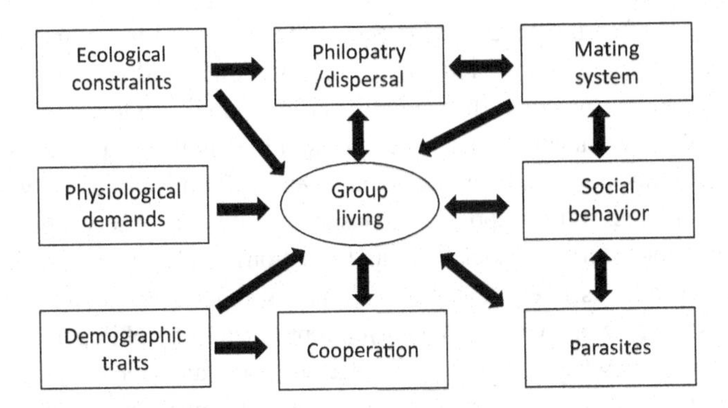

Figure 10.3 Relationships among the causes and consequences of group living in bats (redrawn from Kerth, 2008).

10.6.3 Fecundity

Fecundity, defined as the number of offspring produced by females, is an important component of fitness in all mammals. A growing body of literature suggests that individual females living in large groups and those that are more socially connected (i.e. have relationships with a greater number of conspecifics) within their social group tend to be more fecund than females that live in smaller groups or are less connected with group mates, although exceptions to these patterns do occur (e.g. badgers, Cresswell, 1992). The effects of these social parameters are layered on top of life history traits that may be strongly influenced by phylogenetic history. For example, in multiple plural (communal) breeding species of carnivorans (e.g. banded mongooses, Gilchrist, *et al.*, 2004; lions, Packer, *et al.*, 2001), age and body size are more important predictors of female fecundity than social environment. Demography may also be important, with high population density increasing within-group differences in female fecundity in some species (e.g. red deer, Clutton-Brock, *et al.*, 1987). In sum, female fecundity is clearly shaped by multiple factors, including elements of social structure, life history, and demography. Assessing the relative contributions of each of these parameters to patterns of fecundity and fitness in individual species represents an important line of research for future studies of non-primate mammals.

10.6.4 Age at First Reproduction

The age at which females first reproduce is expected to significantly impact lifetime reproductive success, since a longer reproductive lifespan should increase the number of offspring produced. Harvey & Zammuto (1985) demonstrated that the age of first reproduction is strongly influenced by life expectancy at birth. As a result, mortality schedules for juveniles are a critical determinant of several important components of female fitness, including age-specific costs and benefits of producing young and investing in parental care (Promislow & Harvey, 1990). For example, litter size and neonatal survival of offspring generally increases with maternal age in carnivorans (e.g. common raccoons, *Procyon lotor*, red foxes, *Vulpes vulpes*) and rodents (e.g. white-footed mice, *Peromyscus leucopus*, beavers). Because group living tends to reduce rates of juvenile mortality, members of social species often initiate their reproductive careers at earlier ages than do individuals in non-social species (Promislow & Harvey, 1990). Although this outcome at first appears inconsistent with the observation that natal philopatry (a common form of social group formation) tends to be associated with delayed reproduction, it is likely that this discrepancy reflects the inclusion of plural breeding species in some analyses since delayed reproduction by philopatric animals is most typical of the subset of social species that are singular breeders (Promislow & Harvey, 1990). Thus, social structure appears to play an important role in the reproductive schedules of group living species, suggesting future studies should examine the impacts of sociality on this aspect of offspring production.

10.6.5 Dispersal

Patterns of dispersal, or conversely patterns of natal philopatry, are central to social structure in all of the mammalian taxa considered here. In most mammals, males disperse more often and tend to move greater distances than do females (Greenwood, 1980). Ecology shapes social structure through its effects on natal dispersal. Specifically, when ecological conditions render dispersal prohibitively costly, individuals may remain in their natal area, leading to the formation of social groups (Koenig, *et al.*, 1992; Hayes, 2000; Lucia, *et al.*, 2008). Ecological parameters that may contribute to high costs of dispersal include risk of predation, habitat availability, limited access to mates, and the probability of successfully rearing young in a new (i.e. non-natal) portion of the habitat (Emlen, 1994). While groups that form when conspecifics cluster around critical resources may or may not contain related individuals, groups that arise due to natal philopatry are almost necessarily composed of kin (Eisenberg, 1966; Emlen, 1994; Smith, 2014). Ecological factors are clearly central in both of these cases, but the specific mechanisms by which ecology causes conspecifics to congregate are thought to be fundamental to determining the kin structure of a group and thus the potential for direct versus indirect fitness benefits to group members (Emlen, 1994; Clutton-Brock, 2002).

Most carnivorans follow the typical mammalian pattern of male-biased dispersal. The exact conditions favoring natal philopatry and, more generally, interspecific variation in patterns and rates of movement are often poorly understood. Potential contributing factors include mating system (Greenwood, 1980; Dobson, 1982), the difficulty of hunting alone, and the extended parental care required by generally altricial carnivoran young (e.g. African wild dogs, Creel & Creel, 1995). Given the tendency for dispersal to be male-biased, social groups of carnivorans are most often comprised of closely related adult females, their offspring, and unrelated, immigrant adult males. Even in species in which some males remain in their natal group (e.g. dwarf mongooses), the extent of philopatry and thus the degree of kinship within groups tend to be greater among females, although immigrants of both sexes may occur (Lucas, *et al.*, 1994).

Dispersal among rodents is also typically male-biased (Greenwood, 1980; Dobson, 1982). Natal dispersal may be adaptive in reducing inbreeding depression, which can have negative fitness consequences, as has been reported for Belding's ground squirrels (Holekamp & Sherman, 1989) and yellow-bellied marmots (Olson, *et al.*, 2012). In the former species, male dispersal is triggered by the combination of a minimum level of body fat and circulating levels of testosterone (Nunes, *et al.*, 1998; 1999). Male-biased dispersal also occurs in most other species of ground squirrels, although it is less likely when a close female relative (e.g. mother or sibling) is absent from the natal area (Armitage, *et al.*, 2011; Hoogland, 2013), providing further suggestion that inbreeding avoidance may be important. In those species characterized by natal philopatry, ecological constraints on dispersal reflect access to critical resources, predation pressure, and the potential for cooperation to resolve ecological challenges (Lacey & Sherman, 2007, Figure 10.4). Although ecological drivers of sociality have been studied extensively for subterranean species, including African mole-rats (Jarvis, *et al.*, 1994) and tuco-tucos (Lacey, 2000), surprisingly few studies provide empirical data regarding relationships between ecology and dispersal (Lacey, 2016). Thus, future studies that quantify interactions between environmental conditions and patterns of movement to elucidate the causes of dispersal should prove informative.

As in carnivorans and rodents, dispersal in bats tends to be male-biased (Burland & Worthington, 2001). This pattern appears to be particularly characteristic of European and North American species, with more variable patterns of sex-specific dispersal occurring among tropical species (Kerth, 2008). The ecological bases for these patterns appear to vary. For example, female philopatry in Bechstein's bats (*Myotis bechsteinii*) has been linked to fitness benefits resulting from communal breeding and reduced parasite transmission between neighboring colonies (Kerth, *et al.*, 2002). In contrast, Kerth (2008) posited that male dispersal in tropical species is best explained by the benefits of avoiding resource competition among kin. These explanations are not mutually exclusive and the relative importance of factors such as parasite load and competition likely vary among taxa and geographic regions. Clearly, considerable additional research is needed to evaluate the nature, frequency, and adaptive bases for natal dispersal and natal philopatry in chiropterans.

10.7 Traits of Social Groups

10.7.1 Genetic Structure

Inclusive fitness theory predicts that animals gain important fitness benefits from living and cooperating with close kin (Hamilton, 1964). Not surprisingly, a major focus of research on sociality has been to document the kin structure of groups across diverse taxa. In general, kin structure in mammalian groups reflects the prevalence and duration of natal philopatry (Lukas & Clutton-Brock, 2012; Smith, 2014). At one extreme, groups that form when individuals cluster around critical resources may contain few related individuals, resulting in little genetic kinship within these aggregations (Ebensperger & Hayes,

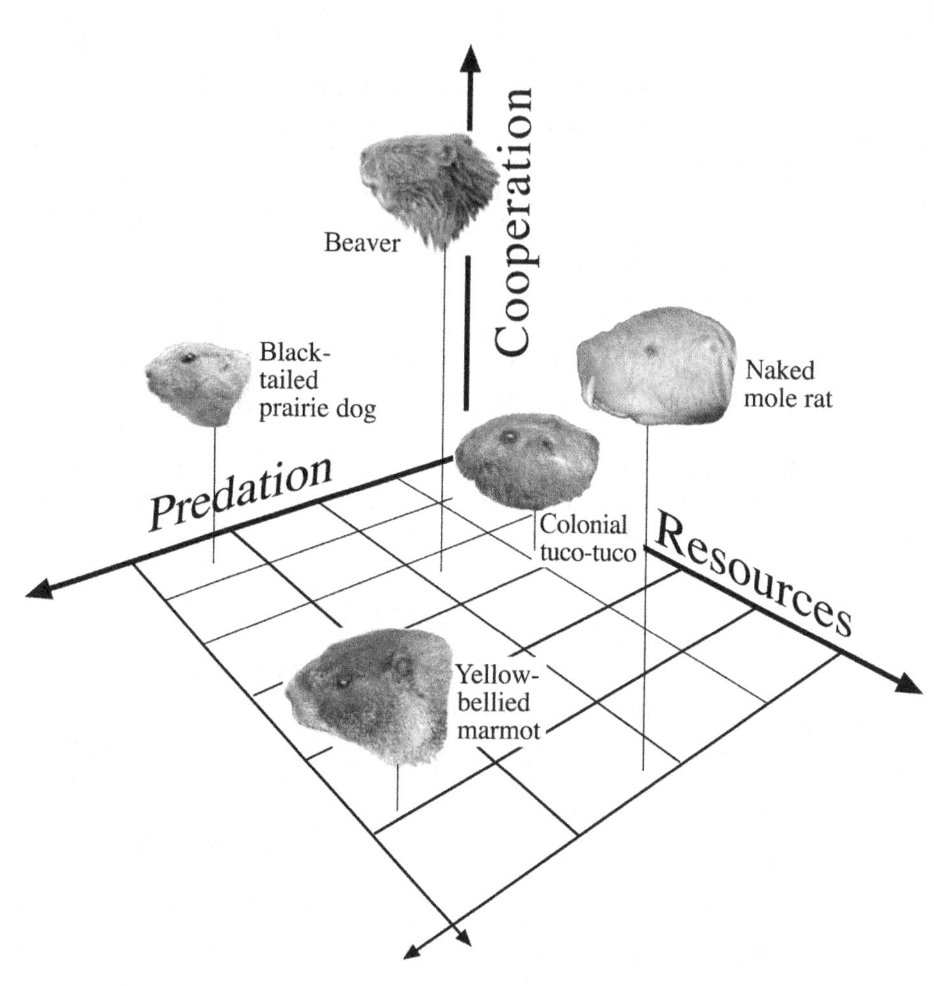

Figure 10.4 Three-dimensional conceptual model of the ecological bases for sociality. Postulated locations for several well-studied species are shown to indicate how data for these species would be represented in the model (adapted from Lacey & Sherman, 2007).

2008; Clutton-Brock & Lukas, 2012). At the other extreme, individuals may spend their entire lives in their natal group, resulting in highly kin-structured social systems, as has been suggested for naked mole-rats (Lacey & Sherman, 1997). Most species of non-primate mammals fall between these endpoints, resulting in patterns and levels of kinship vary (Smith, 2014). Although rare, the absence of kin structure within social groups has been reported for some social non-primate mammals including degus, raccoons, sperm whales, *Physeter microcephalus*, and river otters, *Lontra canadensis* (Smith, 2014). More generally, despite female philopatry, social species of bats tend to have relatively low levels of within-colony relatedness; this pattern likely reflects the presence of multiple breeding males within social groups (Burland & Worthington, 2001).

Because the majority of social non-primate mammals occur in groups containing closely related individuals of one or both sexes (e.g. Creel & Creel, 1991; Clutton-Brock, 2002;

Wilkinson & Baker, 1988; Faulkes, *et al.*, 1997), many individuals in these species interact frequently, if not primarily, with kin. This tendency is expected to significantly shape the nature of social relationships among group mates (Smith, 2014). Members of these species generally prefer to socialize with kin versus non-kin, exhibiting social preferences that parallel those in many species of primates (Chapter 9). For example, in social groups comprised of adult females that vary with regard to relatedness, females often direct more cooperative and affiliative behaviors toward their closest kin (e.g. African elephants, Archie, *et al.*, 2006; spotted hyenas, Smith, *et al.*, 2010; Indo-Pacific bottlenose dolphins, Moeller, 2012; yellow-bellied marmots, Smith, *et al.*, 2013). Interestingly, in some species, a small number of immigrant (i.e. unrelated) individuals also participate in cooperative interactions (Clutton-Brock, 2002), leading to intriguing questions regarding the interplay between direct and indirect fitness benefits in these societies. Although kin selection likely contributes substantially to the behavior of these species, future studies will benefit from expanded consideration of the range of potential adaptive benefits accruing to members of social groups characterized by variable levels of kinship.

10.7.2 Group Structure, Breeding Structure and Sex Ratio

The demographic composition, dominance structure, and social cohesion of groups have significant implications for social structure. Demographically, groups may be composed of varying ratios of males to females and may include one or multiple generations of individuals. Clearly, these elements of group structure are closely tied to the mechanisms by which groups form, notably the extent to which males and females engage in natal philopatry. Demographic structure, in turn, is critical in shaping multiple aspects of social relationships among group members, including breeding structure, dominance interactions, and the potential for cooperative care of young. Among social non-primate mammals, multi-generational groups are common, with such groups most often consisting primarily of female kin groups.

Within groups, one often conspicuous element of social structure is the presence of a dominance hierarchy. The degree to which groupmates may be ranked in a linear dominance hierarchy based on the outcomes of agonistic interactions is highly variable across non-primate mammals. For those species with pronounced dominance hierarchies, high social status often has substantial fitness benefits for carnivorans (e.g. Holekamp, *et al.*, 1996) and rodents (e.g. Huang, *et al.*, 2011), but these effects in bats, if they exist, are not well understood. Because dominance rank is often predicted by the relative age and size of individuals, the demographic structure of a group may be an important determinant of which individuals accrue such benefits. The tendency for males to have larger body sizes than females leads to the expectation that, between the sexes, males should be socially dominant over females. One striking exception occurs in spotted hyenas, in which adult females clearly dominate immigrant males (Kruuk, 1972; Frank, 1986). Female hyenas acquire social status via associative learning, also referred to as "maternal rank inheritance"; this process closely resembles rank acquisition in some Old World monkeys but appears to be unique among non-primate mammals (Engh, *et al.*, 2000; Holekamp, *et al.*, 2007).

Social cohesion among group mates, defined as the degree to which individuals within a group associate spatially and behaviorally, varies enormously among the species considered here. Group structure is spatially and temporally stable for species in which group mates are almost always together (e.g. naked molerats, Lacey & Sherman, 2007; African wild dogs, Creel & Creel, 1995) but is considerably more

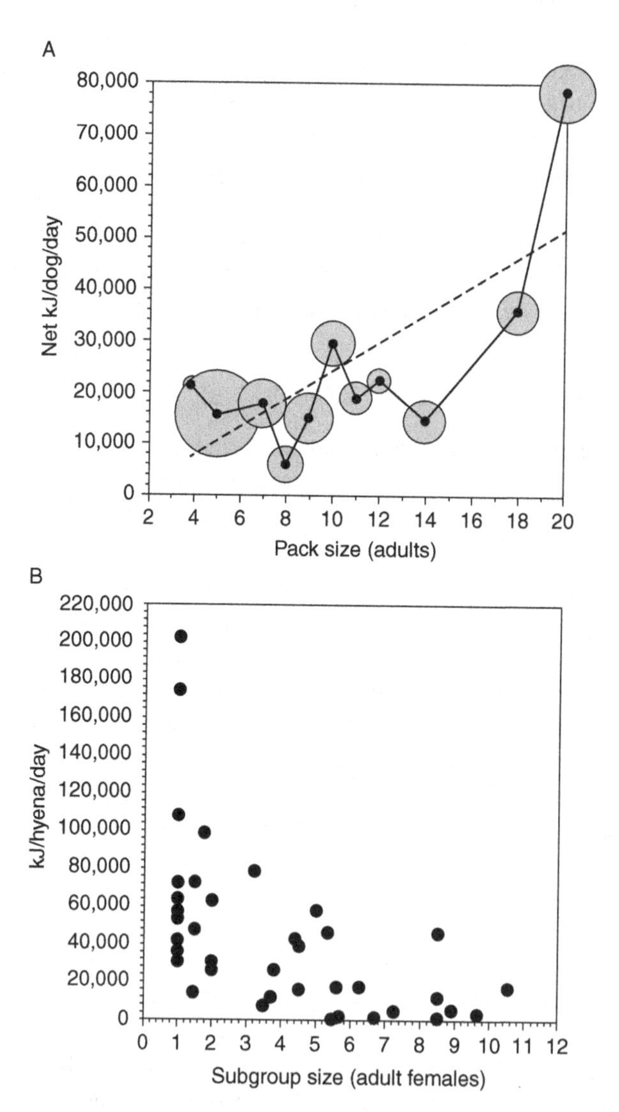

Figure 10.5 Per capita energy intake as a function of foraging group size among adult: (A) African wild dogs (reprinted from Creel, 1997) and (B) female spotted hyenas (reprinted from Smith, *et al.*, 2008). Individual points in (A) represent mean pack sizes, with sizes of surrounding gray circles proportional to the number of observations; the dashed line represents the linear regression between group size and energy intake. Points in (B) represent individual adult female hyenas observed foraging together in a subgroup after a kill has been acquired during hunting or scavenging.

fluid in species characterized by fission-fusion dynamics (e.g. spotted hyenas, Smith, *et al.*, 2008). Whereas wild dogs increase per capita energy gain from hunting in large packs, individuals within most species of social carnivorans instead often hunt alone or separate from group-mates into smaller foraging parties to reduce feeding competition at kills (Figure 10.5). Because some taxa displaying fission-fusion dynamics are often characterized by predictable, stable subgroups of individuals (e.g. elephants, Wittemyer, *et al.*, 2005), the difference in social cohesion between these and other social non-primate mammals may be more a question of scale than actual absence of cohesion. This is in apparent contrast to short-term aggregations that form around specific resources, in which no persistent relationships among group members are expected. To our knowledge, efforts to compare degrees of social cohesion beyond carnivorans (Smith, *et al.*, 2012) using a common set of metrics have not been undertaken and thus it is often unclear at what spatial and temporal levels cohesion among group mates occurs.

The sex ratio of adults in a social group tends to be closely linked to the mating system, with strongly female-biased groups often representing harem units while groups with more equitable sex ratios tend to be polygynandrous (plural breeding) or monogamous (singular breeding) (Clutton-Brock, 1988). As a result, information regarding sex ratio may generate important insights into patterns of reproductive behavior among group members. Clearly, the sex ratio of a social group is influenced by patterns of natal philopatry and group formation, with greater bias toward female philopatry expected to increase the probability of the formation of harem groups Clutton-Brock & Iason, 1986). As a result, adult sex ratio provides an important tie between the reasons for group living and the fitness consequences to group members of living with conspecifics. The "Trivers-Willard hypothesis" (Trivers & Willard, 1973) posits that females may alter the sex ratio of young in response to differences in the costs of producing sons versus daughters and the asymmetry in breeding success of low versus high quality offspring of each sex. Although evidence for this hypothesis varies among mammals, a comprehensive meta-analysis revealed that studies focusing on female condition near the time of conception provide strong support for the prediction that mothers in good condition should bias their litters towards sons (Cameron, 2004). Given that group living may enhance the direct fitness of at least some group members, it is possible that patterns of sex ratio allocation differ between solitary and social species with otherwise similar breeding structures. Future studies that explore this theme in greater detail should generate important insights into this aspect of mammalian sociality.

Mammals exhibit a wide range of mating systems including polygyny, polyandry, polygynandry, and both social and genetic monogamy. In striking contrast to birds, less than 10 percent of mammalian species are monogamous (Lukas & Clutton-Brock, 2013), indicating that in the vast majority of the animals that are the focus of this chapter, adults of one or both sexes mate with multiple partners. Across all non-primate mammals, monogamy is particularly prevalent among elephant shrews (Macroscelidea, 14/15 or 93 percent of species), carnivorans (32/202 or 16 percent of species), and diprotodont marsupials (Diprotodontia, 2/15 or 13 percent of species). In contrast, monogamy occurs in less than 10 percent of extant species of chiropterans, rodents, artiodactyls (even-toed ungulates), lagomorphs (rabbits, hares, pikas), and soricomorphs (shrews, Lukas & Clutton-Brock, 2013). Mating systems are known for at least 963 species of chiropterans; the vast majority live in polygynous groups with only 17 (1.8 percent) species described as monogamous (McCracken & Wilkinson, 2000). Among social mammals more generally, singular breeding is believed to have evolved from monogamy (Lukas & Clutton-Brock, 2012) and, not surprisingly, singular breeding species tend to be monogamous. Less obvious is why monogamy is generally rare among other mammals; considerable research has focused upon at identifying the ultimate and proximate bases for monogamy in the few species in which it occurs (Clutton-Brock, 1988). Given the apparent link between monogamy and singular breeding, understanding the adaptive bases for monogamy would likely generate important insights into the reasons for variation in breeding structure and correlated aspects of social structure in non-primate mammals.

III SOCIAL SYNTHESIS

Despite extensive research on a phylogenetically diverse array of species, there is still much that remains to be learned regarding the social lives of non-primate mammals. Here, we summarize what is known and outline important gaps in our understanding of the social structures of these animals, including taxa for which relevant data are particularly lacking. To conclude, we outline several ways to address these shortcomings and emphasize the need for long-term field studies (Armitage, 1991; Clutton-Brock & Sheldon, 2010), integration of proximate and ultimate information (Hofmann, *et al.*, 2014), and inclusion of new model organisms (Taborsky, *et al.*, 2015).

10.8 A Summary of Non-Primate Mammalian Sociality

Because individuals must interact to reproduce and offspring must nurse from their mothers, all species of mammals are to some degree social. However, the form and degree of sociality vary widely both within and among species. Current ecology plays a central role in this variation, influencing the tendency for groups to form, remain cohesive or to fragment. Because evolutionary history influences demographic and life history traits, phylogenetically-based analyses reveal general patterns. Given the wealth of demographic, life history, ecological, and behavioral information now available for mammals and the increasingly powerful phylogenetic tools that can be used to explore these data, the range of possible comparative studies of the evolution of social behavior in non-primate mammals is tremendous.

One critical direction for future studies of mammalian social structure is to determine the nature and magnitude of the fitness consequences associated with group living. Such analyses can yield important insights into the selective factors favoring sociality (Silk, 2007) and thus can guide efforts to link patterns of social structure to specific environmental factors, life history traits, and evolutionary parameters. For example, in species in which group living is associated with increased per capita direct fitness (e.g. yellow-bellied marmots, Armitage & Schwartz, 2000; African wild dogs, Creel & Creel, 1995), the benefits of sociality are assumed to outweigh those associated solitary life; in species in which per capita direct fitness decreases within groups (e.g. colonial tuco-tucos, Lacey, 2004), ecological or other costs to dispersal are thought to predominate. At the same time, analyses of the fitness consequences of group living can facilitate understanding of variation in fundamental within-group social traits such as breeding structure. For example, while the relationship between per capita direct fitness and group size tends to be neutral or negative in plural breeding species, this relationship tends to be positive – at least for reproductive individuals – in singular breeding species (Ebensperger, *et al.*, 2012). Collectively, these contrasts suggest that further comparative analyses of patterns of direct fitness will yield critical insights into the causes and consequences of sociality in non-primate mammals.

The majority of information summarized here comes from studies of a relatively limited number of carnivorans and rodents, with information from social chiropterans being particularly underrepresented. Long-term field studies conducted in natural settings are essential for linking patterns of behavior to the selective environments in which they occur, as has been attempted for carnivorans (Smith, *et al.*, in press), rodents (Hayes, *et al.*, in press), and bats (Kerth & Van Schaik, 2012). Equally importantly, long-term studies can reveal significant patterns of variation that are not evident over short time scales (Clutton-Brock & Sheldon, 2010), such as the lifetime fitness trajectories of individual dwarf mongooses (Creel & Waser, 1994), the impact of inter-annual environmental variation on reproductive success among female degus (Ebensperger, *et al.*, 2014), the social and fitness consequences of subordinate breeding in meerkats (Young, *et al.*, 2006), the lifetime impacts of inherited social rank in spotted hyenas (Holekamp, *et al.*, 2012), and the implications of decadal patterns of environmental change on the demography and behavior of yellow-bellied marmots (Blumstein, 2013). In addition to increasing the number and diversity of mammalian taxa for which information is available, we advocate the collection of behavioral, demographic, and ecological data across multiple generations of free-living individuals to allow for greater understanding of relationships among environmental conditions, adaptive consequences, and patterns of social structure.

10.9 Comparative Perspectives on Non-Primate Mammalian Sociality

This review identifies emergent patterns of mammalian social structure. Expanding the scope of these comparisons to include primates as well as members of other animal lineages reveals intriguing similarities and differences in social behavior. Species considered here are similar to primates in that cognitive abilities appear to play an important role in shaping patterns of nepotism and communication. Social dominance hierarchies and fission-fusion dynamics within many of the social non-primate mammals resemble those of primates. With regard to other vertebrates, group defense and predator detection reviewed here mirror those of some birds and fish. Patterns of dispersal and parental care are in direct contrast to predominant patterns for birds (Chapter 11). More generally, some mammals cooperate to acquire resources and to rear young. As in most other vertebrates, but in striking contrast to the caste structure found in many insect societies, non-primate mammals apparently lack morphological and physiological specializations for reproductive division of labor. This raises intriguing questions about the ecological and evolutionary bases for this apparent difference between social vertebrates and invertebrates. It also raises important questions regarding the differences within vertebrates that have led to remarkable parallels between social insects and a very limited number of (eu)social rodents. In sum, we expect that future broad-scale comparative studies that explore these types of similarities and differences in greater detail will prove highly informative regarding the ecological and evolutionary bases for variation in animal social structure.

10.10 Concluding Remarks

The study of mammalian social behavior has reached an exciting juncture due to emerging opportunities to integrate extensive data sets from long-term field studies with rapidly expanding molecular and neurobiological technologies capable of revealing the proximate substrates for social interactions (Blumstein, *et al.*, 2010; Hofmann, *et al.*, 2014). In the decades since the publication of Eisenberg's (1966) seminal review of the social behavior of mammals, multiple long-term (and often ongoing) studies of social structure have been initiated for free-living populations of carnivorans, rodents, and bats (e.g., Kerth & Van Schaik, 2012; Hayes, *et al.*, in press; Smith, *et al.*, in press). These efforts continue to reveal new and intriguing relationships among social behavior, ecology, demography, and life history traits. To capitalize upon these efforts and to better delineate the role of evolutionary history in shaping such relationships, the taxonomic coverage of field research programs needs to be expanded, with particular attention to studies of social chiropterans. At the same time, the ever-changing technological landscape is improving our ability to examine physiological and genetic mechanisms of social behavior in free-living animals and these emerging opportunities promise to yield important new insights into variation in social structure.

We hope that the overview of non-primate mammalian social structure provided here – including the gaps in understanding and taxonomic coverage that we have highlighted – will serve to stimulate considerable additional research on all aspects of the social behavior of these fascinating animals. As field biologists, we strongly encourage efforts to characterize patterns of behavior as they occur in natural selective environments. As active participants in collaborative studies of neurobiological, endocrine and genetic underpinnings for social behavior, we advocate integrative efforts to understand variation in social structure. Ideally, readers of this chapter – including both established scientists and future experts in behavior – will be intrigued by the emergent trends in mammalian sociality that we have described and will build upon this foundation to undertake new and exciting studies of social structure in this important clade.

References

Adolphs, R. (2001) The neurobiology of social cognition. *Current Opinion in Neurobiology*, **11**, 231–239.

Alexander, R. D. (1974) The evolution of social behavior. *Annual Review of Ecology and Systematics*, **5**, 325–383.

Archie, E. A. & Theis, K. R. (2011) Animal behaviour meets microbial ecology. *Animal Behaviour*, **82**, 425–436.

Archie, E. A., Moss, C. J., & Alberts, S. C. (2006) The ties that bind: Genetic relatedness predicts the fission and fusion of social groups in wild African elephants. *Proceedings of the Royal Society of London B*, **273**, 513–522.

Armitage, K. B. (1991) Social and population dynamics of yellow-bellied marmots: Results from long-term research. *Annual Review of Ecology and Systematics*, **22**, 379–407.

Armitage, K. B. & Schwartz, O. A. (2000) Social enhancement of fitness in yellow-bellied marmots. *Proceedings of the National Academy of Sciences*, **97**, 12149–12152.

Armitage, K. B., Vuren, D. H. V., Ozgul, A., & Oli, M. K. (2011) Proximate causes of natal dispersal in female yellow-bellied marmots, *Marmota flaviventris*. *Ecology*, **92**, 218–227.

Arnold, W. (1990) The evolution of marmot sociality: II. Costs and benefits of joint hibernation. *Behavioral Ecology and Sociobiology*, **27**, 239–246.

Barclay, R. M. R. & Harder, L. D. (2003) *Bat Ecology*. Chicago: University of Chicago Press.

Beckhoff, M., Daniels, T. J., & Gittleman, J. L. (1984) Life history patterns and the comparative social ecology of carnivores. *Annual Review of Ecology and Systematics*, **15**, 191–232.

Bennett, N. C., Faulkes, C. G., & Molteno, A. J. (2000) Reproduction in subterranean rodents. In: Lacey, E. A. Patton, J. L. Cameron, G. N. (eds.) *Life Underground: The Biology Of Subterranean Rodents*. Chicago, IL: University of Chicago Press, pp. 145–178.

Benson-Amram, S., Heinen, V. K., Dryer, S. L., & Holekamp, K. E. (2011) Numerical assessment and individual call discrimination by wild spotted hyaenas, *Crocuta crocuta*. *Animal Behaviour*, **82**, 743–752.

Blumstein, D. T. (2007) The evolution of alarm communication in rodents: Structure, function, and the puzzle of apparently altruistic calling. In: Wolff, J. O. Sherman, & P. W. (eds.) *Rodent Societies: An Ecological and Evolutionary Perspective*. Chicago: University of Chicago Press, pp. 317–327.

Blumstein, D. T. (2013). Yellow-bellied marmots: Insights from an emergent view of sociality. *Philosophical Transactions of the Royal Society B*, **368**, 20120349.

Blumstein, D. T. & Armitage, K. B. (1998) Life history consequences of social complexity: A comparative study of ground-dwelling sciurids. *Behavioral Ecology*, **9**, 8–19.

(1999) Cooperative breeding in marmots. *Oikos*, **84**, 369–382.

Blumstein, D. T., Im, S., Nicodemus, A., & Zugmeyer, C. (2004) Yellow-bellied marmots (*Marmota flaviventris*) hibernate socially. *Journal of Mammalogy*, **85**, 25–29.

Blumstein, D. T., Ebensperger, L. A., Hayes, L. D., *et al.* (2010) Toward an integrative understanding of social behavior: New models and opportunities. *Frontiers in Behavioral Neuroscience*, **4,** 1–9.

Bordes, F., Blumstein, D. T., & Morand, S. (2007) Rodent sociality and parasite diversity. *Biology Letters*, **3**, 692–694.

Bradbury, J. W. & Vehrencamp, S. L. (2011) *Principles of Animal Communication*. Sunderland, MA: Sinauer Associates, Inc.

Branchi, I., Santucci, D., & Alleva, E. (2001) Ultrasonic vocalisation emitted by infant rodents: A tool for assessment of neurobehavioural development. *Behavioural Brain Research*, **125**, 49–56.

Brett, R. A. (1991) The ecology of naked mole-rat colonies: Burrowing, food, and limiting factors. In: Sherman, P. W., Jarvis, J. U. M., & Alexander, R. D. (eds.) *The Biology of the Naked Mole-rat*, Princeton: Princeton University Press, pp. 137–184.

Burland, T. M. & Worthington, W., J. (2001) Seeing in the dark: Molecular approaches to the study of bat populations. *Biological Reviews of the Cambridge Philosophical Society*, **76**, 389–409.

Cameron, E. Z. (2004) Facultative adjustment of mammalian sex ratios in support of the Trivers-Willard hypothesis: Evidence for a mechanism. *Proceedings of the Royal Society of London, Series B: Biological Sciences*, **271**, 1723–1728.

Cameron, E. Z., Setsaas, T. H., & Linklater, W. L. (2009) Social bonds between unrelated females increase reproductive success in feral horses. *Proceedings of the National Academy of Sciences*, **106**, 13850–13853.

Carter, C. S. (2003) Developmental consequences of oxytocin. *Physiology and Behavior*, **79**, 383–397.

Carter, G. G. & Wilkinson, G. S. (2013) Food sharing in vampire bats: Reciprocal help predicts donations more than relatedness or harassment. *Proceedings of the Royal Society of London B*, **280**, 20122573.

Carter, G. G. & Wilkinson, G. S. (2016) Common vampire bat contact calls attract past food-sharing partners. *Animal Behaviour*, **116**, 45–51.

Chapman, J. A. & Flux, J. E. (1990) Introduction and overview of the lagomorphs. In: Chapman, J. A. & Flux, J. E. C. (eds.) *Rabbits, Hares and Pikas, Status Survey and Conservation Action Plan*. Oxford: The World Conservation Union, pp. 1–6.

Clutton-Brock, T. H. (2009). Cooperation between non-kin in animal societies. *Nature*, **462**, 51–57.

Clutton-Brock, T. H. (ed.) (1988) *Reproductive Success: Studies of Individual Variation In Contrasting Breeding Systems*. Chicago: University of Chicago Press.

Clutton-Brock, T. H. (2002) Breeding together: Kin selection and mutualism in cooperative vertebrates. *Science*, **296**, 69–72.

Clutton-Brock, T. H. & Guinness, F. E. (1982) *Red Deer: Behavior and Ecology of Two Sexes*. Chicago: University of Chicago Press.

Clutton-Brock, T. H. & Iason, G. R. (1986) Sex ratio variation in mammals. *Quarterly Review of Biology*, **61**, 339–374.

Clutton-Brock, T. H. & Lukas, D. (2012) The evolution of social philopatry and dispersal in female mammals. *Molecular Ecology*, **21**, 472–492.

Clutton-Brock, T. H. & Sheldon, B. C. (2010). Individuals and populations: The role of long-term, individual-based studies of animals in ecology and evolutionary biology. *Trends in Ecology & Evolution*, **25**, 562–573.

Clutton-Brock, T. H., Albon, S. D., & Guinness, F. E. (1987) Interactions between population density and maternal characteristics affecting fecundity and juvenile survival in red deer. *The Journal of Animal Ecology*, **56**, 857–871.

Clutton-Brock, T. H., O'Riain, M. J., Brotherton, P. N. M., *et al.* (1999) Selfish sentinels in cooperative mammals. *Science*, **284**, 1640–1644.

Clutton-Brock, T. H., Russell, A. F., Sharpe, L. L., *et al.* (2001) Effects of helpers on juvenile development and survival in meerkats. *Science*, **293**, 2446–2449.

Clutton-Brock, T. H., Hodge, S. J., & Flower, T. P. (2008) Group size and the suppression of subordinate reproduction in Kalahari meerkats. *Animal Behaviour*, **76**, 689–700.

Connor, R. C. (2000) Group living in whales and dolphins. In: J. Mann, R. C., Connor, P. L. Tyack, & Whitehead, H. (eds.) *Cetacean Societies: Field Studies of Dolphins and Whales.* Chicago: University of Chicago Press, pp. 199–218.

Côté, S. D. & Festa-Bianchet, M. (2001) Reproductive success in female mountain goats: The influence of age and social rank. *Animal Behaviour*, **62**, 173–181.

Creel, S. (1997) Cooperative hunting and group size: Assumptions and currencies. *Animal Behaviour*, **54**, 1319–1324.

(2001) Social dominance and stress hormones. *Trends in Ecology and Evolution*, **16**, 491–497.

Creel, S. & Creel, N. M. (1991) Energetics, reproductive suppression and obligate communal breeding in carnivores. *Behavioral Ecology and Sociobiology*, **28**, 263–270.

(1995) Communal hunting and pack size in African wild dogs, *Lycaon pictus*. *Animal Behaviour*, **50**, 1325–1339.

Creel, S. R. & Waser, P. M. (1994) Inclusive fitness and reproductive strategies in dwarf mongooses. *Behavioral Ecology*, **5**, 339–348.

Crespi, B. J. & Yanega, D. (1995) The definition of eusociality. *Behavioral Ecology*, **6**, 109–115.

Cresswell, W. J., Harris, S., Cheeseman, C. L., & Mallinson, P. J. (1992) To breed or not to breed: An analysis of the social and density-dependent constraints on the fecundity of female badgers (*Meles meles*) *Philosophical Transactions of the Royal Society B*, **338**, 393–407.

Dalerum, F. (2007) Phylogenetic reconstruction of carnivore social organizations. *Journal of Zoology*, **273**, 90–97.

Dobson, F. S. (1982) Competition for mates and predominant juvenile male dispersal in mammals. *Animal Behaviour*, **30**, 1183–1192.

Drea, C. M., Vigniere, S. N., Kim, H. S., Weldele, M. L., & Glickman, S. E. (2002) Responses to olfactory stimuli in spotted hyenas (*Crocuta crocuta*): II. *Discrimination of conspecific scent. Journal of Comparative Psychology*, **116**, 342–349.

Dunbar, R. I. & Shultz, S. (2007) Evolution in the social brain. *Science*, **317**, 1344–1347.

Ebensperger, L. A. (2001) A review of the evolutionary causes of rodent group living. *Acta Theriologica*, **46**, 115–144.

Ebensperger, L. A. & Blumstein, D. T. (2006) Sociality in New World hystricognath rodents is linked to predators and burrow digging. *Behavioral Ecology*, **17**, 410–418.

Ebensperger, L. A. & Cofré, H. (2001) On the evolution of group living in the New World cursorial hystricognath rodents. *Behavioral Ecology*, **12**, 227–236.]

Ebensperger, L. A. & Hayes, L. D. (2008) On the dynamics of rodent social groups. *Behavioural Processes*, **79**, 85–92.

Ebensperger, L. A., Rivera, D. S., & Hayes, L. D. (2012) Direct fitness of group living mammals varies with breeding strategy, climate and fitness estimates. *Journal of Animal Ecology*, **81**, 1013–1023.

Ebensperger, L. A., Villegas, Á., Abades, S., & Hayes, L. D. (2014) Mean ecological conditions modulate the effects of group living and communal rearing on offspring production and survival. *Behavioral Ecology*, **25**, 862–870.

Eick, G. N., Jacobs, D. S., & Matthee, C. A. (2005). A nuclear DNA phylogenetic perspective on the evolution of echolocation and historical biogeography of extant bats (Chiroptera). *Molecular Biology and Evolution*, **22**, 1869–1886..

Eisenberg, J. F. (1966) The social organization of mammals. *Handbuch der Zoologie*, **10**, 1-92.

Emlen, S. T. (1994) Benefits, constraints and the evolution of the family. *Trends in Ecology and Evolution*, **9**, 282–285.

Emlen, S. T. & Oring, L. W. (1977) Ecology, sexual selection, and the evolution of mating systems. *Science*, **197**, 215–223.

Engh, A. L., Esch, K., Smale, L., & Holekamp, K. E. (2000) Mechanisms of maternal rank "inheritance" in the spotted hyaena, *Crocuta crocuta*. *Animal Behaviour*, **60**, 323–332.

Faulkes, C. G., Abbott, D. H., O'Brien, H. P., *et al.* (1997) Micro- and macrogeographical genetic structure of colonies of naked mole-rats *Heterocephalus glaber*. *Molecular Ecology*, **6**, 615–628.

FitzGibbon, C. D. (1989) A cost to individuals with reduced vigilance in groups of Thomson's gazelles hunted by cheetahs. *Animal Behaviour*, **37**, 508–510.

Fleming, A. S., O'Day, D. H., & Kraemer, G. W. (1999) Neurobiology of mother–infant interactions: Experience and central nervous system plasticity across development and generations. *Neuroscience and Biobehavioral Reviews*, **23**, 673–685.

Foster, W. A. & Treherne, J. E. (1981) Evidence for the dilution effect in the selfish herd from fish predation on a marine insect. *Nature*, **293**, 466–467.

Frank, L. G. (1986) Social organization of the spotted hyaena (*Crocuta crocuta*). II. Dominance and reproduction. *Animal Behaviour*, **35**, 1510–1527.

Gittleman, J. L. (1986) Carnivore brain size, behavioral ecology, and phylogeny. *Journal of Mammalogy*, **67**, 23–36.

Gilchrist, J. S., Otali, E., & Mwanguhya, F. (2004) Why breed communally? Factors affecting fecundity in a communal breeding mammal: The banded mongoose (*Mungos mungo*). *Behavioral Ecology and Sociobiology*, **57**, 119–131.

Gompper, M. E. (1996) Sociality and asociality in white-nosed coatis (*Nasua narica*): Foraging costs and benefits. *Behavioral Ecology*, **7**, 254–263.

Gorman, M. L. & Trowbridge, B. J. (1989) The role of odor in the social lives of carnivores. In: Gittleman, J. L. *Carnivore Behavior, Ecology, and Evolution*. New York: Springer, pp. 57–88

Grainger, M., Aarde, R., & Whyte, I. (2005) Landscape heterogeneity and the use of space by elephants in the Kruger National Park, South Africa. *African Journal of Ecology*, **43**, 369–375.

Greenwood, P. J. (1980) Mating systems, philopatry and dispersal in birds and mammals. *Animal Behaviour*, **28**, 1140–1162.

Griffin, A. S., Pemberton, J. M., Brotherton, P. N., *et al.* (2003) A genetic analysis of breeding success in the cooperative meerkat (*Suricata suricatta*). *Behavioral Ecology*, **14**, 472–480.

Hamilton, W. D. (1964) The genetical evolution of social behavior, I and II. *Journal of Theoretical Biology*, **7**, 1–52.

Harvey, P. H. & Zammuto, R. M. (1985) Patterns of mortality and age at first reproduction in natural populations of mammals. *Nature*, **315**, 319–320.

Haxby, J. V., Ungerleider, L. G., Horwitz, B., *et al.* (1996) Face encoding and recognition in the human brain. *Proceedings of the National Academy of Sciences*, **93**, 922–927.

Hayes, L. D. (2000) To nest communally or not to nest communally: A review of rodent communal nesting and nursing. *Animal Behaviour*, **59**, 677–688.

Hayes, L. D. & Ebensperger, L. A. (2011) Caviomorph rodent social systems: An introduction. *Journal of Mammalogy*, **92**, 1–2.

Hayes, L. D. & Solomon, N. G. (2004) Costs and benefits of communal rearing to female prairie voles (*Microtus ochrogaster*). *Behavioral Ecology and Sociobiology*, **56**, 585–593.

(2006) Mechanisms of maternal investment by communal prairie voles, *Microtus ochrogaster*. *Animal Behaviour*, **72**, 1069–1080.

Hayes, L. D., Ebensperger, L. A., Kelt, D. A., Meserve, P. L., Pillay, N., Viblanc, V. A., & Schradin, C. (in press). Long-term field studies in rodents. *Journal of Mammalogy*.

Herrera, E. A., Salas, V., Congdon, E. R., Corriale, M. J., & Tang-Martínez, Z. (2011) Capybara social structure and dispersal patterns: Variations on a theme. *Journal of Mammalogy*, **92**, 12–20.

Hoeck, H. N., Klein, H., & Hoeck, P. (1982) Flexible social organization in Hyrax. *Zeitschrift für Tierpsychologie*, **59**, 265–298.

Hofmann, H. A., Beery, A. K., Blumstein, D. T., *et al.* (2014) An evolutionary framework for studying mechanisms of social behavior. *Trends in Ecology and Evolution*, **29**, 581–589.

Holekamp, K. E. & Sherman, P. W. (1989) Why male ground squirrels disperse: A multilevel analysis explains why only males leave home. *American Scientist*, **77**, 232–239.

Holekamp, K. E., Smale, L., & Szykman, N. (1996) Rank and reproduction in the female spotted hyaena. *Journal of Reproduction and Fertility*, **108**, 229–237.

Holekamp, K. E., Sakai, S. T., & Lundrigan, B. L. (2007) Social intelligence in the spotted hyena (*Crocuta crocuta*). *Philosophical Transactions of the Royal Society of London B: Biological Sciences*, **362**, 523–538.

Holekamp, K. E., Smith, J. E., Strelioff, C. C., Van Horn, R. C., & Watts, H. E. (2012) Society, demography and genetic structure in the spotted hyena. *Molecular Ecology*, **21**, 613–632.

Honeycutt, R. L., Frabotta, L. J., & Rowe, D. L. (2007) Rodent evolution, phylogenetics, and biogeography. In: Wolff, J. O. & Sherman, P. W. (eds.) *Rodent Societies: An Ecological and Evolutionary Perspective*. Chicago: University of Chicago Press, pp. 8–23.

Hoogland, J. L. (1995) *The Black-Tailed Prairie Dog: Social Life of a Burrowing Mammal*. University of Chicago Press.

(2013) Prairie dogs disperse when all close kin have disappeared. *Science*, **339**, 1205–1207.

Huang, B., Wey, T.W., & Blumstein, D. T. (2011) Correlates and consequences of dominance in a social rodent. *Ethology*, **117**, 573–585.

Jarman, P. J. (1974) The social organisation of antelope in relation to their ecology. *Behaviour*, **48**, 215–267.

(1991) Social behavior and organization in the Macropodoidea. *Advances in the Study of Behavior*, **20**, 1–50.

Jarvis, J. U. M., O'Riain, M. J., Bennett, N. C., & Sherman, P. W. (1994) Mammalian eusociality: A family affair. *Trends in Ecology and Evolution*, **9**, 47–51.

Jetz, W. & Rubenstein, D. R. (2011) Environmental uncertainty and the global biogeography of cooperative breeding in birds. *Current Biology*, **21**, 72–78.

Kerth, G. (2008) Causes and consequences of sociality in bats. *Bioscience*, **58**, 737–746.

Kerth, G., Safi, K., & König, B. (2002) Mean colony relatedness is a poor predictor of colony structure and female philopatry in the communally breeding Bechstein's bat (*Myotis bechsteinii*). *Behavioral Ecology and Sociobiology*, **52**, 203–210.

Kerth, G. & Van Schaik, J. (2012) Causes and consequences of living in closed societies: Lessons from a long-term socio-genetic study on Bechstein's bats. *Molecular Ecology*, **21**, 633–646.

Koenig, W. D., Pitelka, F. A., Carmen, W. J., Mumme, R. L., & Stanback, M. T. (1992) The evolution of delayed dispersal in cooperative breeders. *Quarterly Review of Biology*, **67**, 111–150.

König, B. (1994) Fitness effects of communal rearing in house mice: The role of relatedness versus familiarity. *Animal Behaviour*, **48**, 1449–1457.

Krause, J. & Ruxton, G. D. (2002) *Living in Groups*. Oxford: Oxford University Press.

Kruuk, H. (1972) *The Spotted Hyena: A Study Of Predation And Social Behavior.* Chicago: University of Chicago Press.

Kunz, T. H. (1982) Roosting ecology of bats. In: T. H. Kunz (ed.) *Ecology of Bats*. United States: Springer, pp. 1–55.

Lacey, E. A. (2016) Dispersal in caviomorph rodents. In: Hayes, L. D. (ed.) *Sociobiology of Caviomorph Rodents: An Integrative View.* New Jersey: Wiley & Associates, pp. 119–146.

(2000) Spatial and social systems of subterranean rodents. . In: Lacey, E.A. Patton, J. L. Cameron, G.N. (eds.) *Life Underground: The Biology Of Subterranean Rodents.* Chicago, IL: University of Chicago Press, pp. 257–296.

(2004) Sociality reduces individual direct fitness in a communally breeding rodent, the colonial tuco-tuco (*Ctenomys sociabilis*). *Behavioral Ecology and Sociobiology*, **56**, 449–457.

Lacey, E. A. & Sherman, P. W. (1997) Cooperative breeding in naked mole-rats: Implications for vertebrate and invertebrate sociality. In: Solomon, N. G. & French, J. A. (eds.) *Cooperative Breeding in Mammals.* New York: Cambridge University Press, pp. 267–301.

(2007) The ecology of sociality in rodents. In: Wolff, J. O. & Sherman, P. W. (eds.) *Rodent Societies: An Ecological and Evolutionary Perspective*, Chicago: University of Chicago Press, pp. 243–254.

Lacey, E. A. & Wieczorek, J. R. (2004) Kinship in colonial tuco-tucos: Evidence from group composition and population structure. *Behavioral Ecology*, **15**, 988–996.

Lucas, J. R., Waser, P. M., & Creel, S. R. (1994) Death and disappearance: Estimating mortality risks associated with philopatry and dispersal. *Behavioral Ecology*, **5**, 135–141.

Lucia, K. E., Keane, B., Hayes, L. D., *et al.* (2008) Philopatry in prairie voles: An evaluation of the habitat saturation hypothesis. *Behavioral Ecology*, **19**, 774–783.

Lukas, D. & Clutton-Brock, T. H. (2012) Cooperative breeding and monogamy in mammalian societies. *Proceedings of the Royal Society of London B*, **279**, 2151–2156.

(2013) The evolution of social monogamy in mammals. *Science*, **341**, 526–530.

Macdonald, D. W. (1983) The ecology of carnivore social behaviour. *Nature*, **301**, 379–384.

McCraken, G. F. & Wilkinson, G. S. (2000) Bat mating systems. *Reproductive Biology of Bats*, **12**, 3–5.

McGuire, B., Getz, L. L., Hofmann, J. E., Pizzuto, T., & Frase, B. (1993) Natal dispersal and philopatry in prairie voles (*Microtus ochrogaster*) in relation to population density, season, and natal social environment. *Behavioral Ecology and Sociobiology*, **32**, 293–302.

McGuire, B., Getz, L. L., & Oli, M. K. (2002) Fitness consequences of sociality in prairie voles, *Microtus ochrogaster*: Influence of group size and composition. *Animal Behaviour*, **64**, 645–654.

McShea, W. J. & Madison, D. M. (1984) Communal nesting between reproductively active females in a spring population of *Microtus pennsylvanicus*. *Canadian Journal of Zoology*, **62**, 344–346.

Mech, L. D., Smith, D. W., & MacNulty, D. (2015) *Wolves on the Hunt: The Behavior of Wolves Hunting Wild Prey*. Chicago: University of Chicago Press.

Moehlman, P. D. (1986) Ecology and cooperation in canids. In: Rubenstein, D. I. & Wrangham, R. W. (eds.) *Ecological Aspects of Social Evolution: Birds and Mammals*. Princeton: Princeton University Press, pp. 64–86.

Moeller, L. M. (2012) Sociogenetic structure, kin associations and bonding in delphinids. *Molecular Ecology*, **21**, 745–764.

Moss, C. J., Croze, H., & Lee, P. C. (2011) *The Amboseli elephants: A Long-term perspective on a long-lived mammal*. Chicago: University of Chicago Press.

Murie, J.O. & Michener, G.R. (eds.). (1984) *Biology of Ground-dwelling Squirrels: Annual Cycles, Behavioral Ecology and Sociality*. Lincoln: University of Nebraska Press.

Neuweiler, G. (1989) Foraging ecology and audition in echolocating bats. *Trends in Ecology and Evolution*, **4**, 160–166.

Nunes, S., Ha, C. T., & Garrett, P. J., *et al.* (1998) Body fat and time of year interact to mediate dispersal behaviour in ground squirrels. *Animal Behaviour*, **55**, 605-614.

Nunes, S., Duniec, T. R., Schweppe, S. A., & Holekamp, K. E. (1999) Energetic and endocrine mediation of natal dispersal behavior in Belding's ground squirrels. *Hormones and Behavior*, **35**, 113–124.

Olson, L. E., Blumstein, D. T., Pollinger, J. R., & Wayne, R. K. (2012) No evidence of inbreeding avoidance despite demonstrated survival costs in a polygynous rodent. *Molecular Ecology*, **21**, 562–571.

Packer, C., Scheele, D., & Pusey, A. E. (1990) Why lions form groups: Food is not enough. *The American Naturalist*, **136**, 1–19.

Packer, C., Lewis, S., & Pusey, A. (1992) A comparative analysis of non-offspring nursing. *Animal Behaviour*, **43**, 265–281.

Packer, C., Pusey, A. E. & Eberly, L. E. (2001) Egalitarianism in female African lions. *Science*, **293**, 690–693.

Peters, G. & Wozencraft, W. C. (1989) Acoustic communication by fissiped carnivores. In: Gittleman, J. L. (ed.) *Carnivore Behavior, Ecology, and Evolution*. New York: Springer, pp. 14–56.

Pitnick, S., Jones, K. E., & Wilkinson, G. S. (2006) Mating system and brain size in bats. *Proceedings of the Royal Society of London B: Biological Sciences*, **273**, 719–724.

Portfors, C. V. (2007) Types and functions of ultrasonic vocalizations in laboratory rats and mice. *Journal of the American Association for Laboratory Animal Science*, **46**, 28–34.

Promislow, D. E. & Harvey, P. H. (1990) Living fast and dying young: A comparative analysis of life-history variation among mammals. *Journal of Zoology*, **220**, 417–437.

Randall, J. A., Rogovin, K., Parker, P. G., & Eimes, J. A. (2005) Flexible social structure of a desert rodent, *Rhombomys opimus*: Philopatry, kinship, and ecological constraints. *Behavioral Ecology*, **16**, 961–973.

Rowe, D. L. & Honeycutt, R. L. (2002) Phylogenetic relationships, ecological correlates, and molecular evolution within the Cavioidea (Mammalia, Rodentia). *Molecular Biology and Evolution*, **19**, 263–277.

Rubenstein, D. R., Botera, C. A., & Lacey E. A.. (2016) Discrete but variable structure of animal societies leads to the false perception of a social continuum. *Royal Society Open Science*, **3**, 160147.

Safi, K. & Kerth, G. (2007) Comparative analyses suggest that information transfer promoted sociality in male bats in the temperate zone. *The American Naturalist*, **170**, 465–472.

Schradin, C. (2013) Intraspecific variation in social organization by genetic variation, developmental plasticity, social flexibility or entirely extrinsic factors. *Philosophical Transactions of the Royal Society B*, **368**, 20120346.

Schradin, C., Lindholm, A. K., Johannesen, J. E. S., *et al.* (2012) Social flexibility and social evolution in mammals: A case study of the African striped mouse (*Rhabdomys pumilio*). *Molecular Ecology*, **21**, 541–553.

Sechrest, W., Brooks, T. M., da Fonseca, G. A., *et al.* (2002) Hotspots and the conservation of evolutionary history. *Proceedings of the National Academy of Sciences USA*, **99**, 2067–2071.

Sherman, P. W. (1977) Nepotism and the evolution of alarm calls. *Science*, **197**, 1246–1253.

Sherman, P. W., Lacey, E. A., Reeve, H. K., & Keller L. (1995) Forum: The eusociality continuum. *Behavioral Ecology*, **6**, 102–108.

Silk, J. B. (2007) The adaptive value of sociality in mammalian groups. *Philosophical Transactions of the Royal Society B*, **362**, 539–559.

Smith, J. E. (2014) Hamilton's legacy: Kinship, cooperation and social tolerance in mammalian groups. *Animal Behaviour*, **92**, 291–304.

Smith, J. E. & Batzli, G. O. (2006) Dispersal and mortality of prairie voles (*Microtus ochrogaster*) in fragmented landscapes: A field experiment. *Oikos*, **112**, 209–217.

Smith, J. E., Lehmann, K. D. S., Montgomery, T. M., Strauss, E. D. and K. E. Holekamp. (in press). Insights from long-term field studies of mammalian carnivores. *Journal of Mammalogy*.

Smith, J. E., Memenis, S. K., & Holekamp, K. E. (2007) Rank-related partner choice in the fission-fusion society of spotted hyenas (*Crocuta crocuta*). *Behavioral Ecology and Sociobiology*, **61**, 753–765.

Smith, J. E., Kolowski, J. M., Graham, K. E., Dawes, S. E., & Holekamp, K. E. (2008) Social and ecological determinants of fission-fusion dynamics in the spotted hyaena. *Animal Behaviour*, **76**, 619–636.

Smith, J. E., Van Horn, R. C., Powning, K. S., *et al.* (2010) Evolutionary forces favoring intragroup coalitions among spotted hyenas and other animals. *Behavioral Ecology*, **21**, 284–303.

Smith, J. E., Powning, K., Dawes, S., *et al.* (2011) Greetings promote cooperation and reinforce social bonds among spotted hyaenas. *Animal Behaviour*, **81**, 401–415.

Smith, J. E., Swanson, E. M., Reed, D., & Holekamp, K. E. (2012a) Evolution of cooperation among mammalian carnivores and its relevance to hominin evolution. *Current Anthropology*, **53**, S436-S452.

Smith, J. E., Chung, L. K., & Blumstein, D. T. (2013) Ontogeny and symmetry of social partner choice among free-living yellow-bellied marmots. *Animal Behaviour*, **85**, 715–725.

Solomon, N. G. & Crist, T. O. (2008) Estimates of reproductive success for group living prairie voles, *Microtus ochrogaster*, in high-density populations. *Animal Behaviour*, **76**, 881–892.

Solomon, N. G. & Hayes, L. D. (2009) The biological basis of alloparental behaviour in mammals. In: Bentley, G., & Mace, R. (eds.) *Substitute Parents: Biological and Social Perspectives on Alloparenting in Human Societies*. New York: Berghahn Books, pp. 13–49.

Streatfeild, C. A., Mabry, K. E., Keane, B., Crist, T. O., & Solomon, N. G. (2011) Intraspecific variability in the social and genetic mating systems of prairie voles, *Microtus ochrogaster*. *Animal Behaviour*, **82**, 1387–1398.

Sundaresan, S. R., Fischhoff, I. R., Dushoff, J., & Rubenstein, D. I. (2007) Network metrics reveal differences in social organization between two fission–fusion species, Grevy's zebra and onager. *Oecologia*, **151**, 140–149.

Taborsky, M., Hofmann, H. A., Beery, A. K., *et al.* (2015) Taxon matters: Promoting integrative studies of social behavior. *Trends in Neurosciences*, **38**, 189–191.

Teeling, E. C., Springer, M. S., Madsen, O., Bates, P., O'Brien, S. J., & Murphy, W. J. (2005) A molecular phylogeny for bats illuminates biogeography and the fossil record. *Science*, **307**, 580–584.

Thirgood, S., Mosser, A., Tham, S., *et al.* (2004) Can parks protect migratory ungulates? The case of the Serengeti wildebeest. *Animal Conservation*, **7**, 113–120.

Trivers, R. L. & Willard, D. E. (1973) Natural selection of parental ability to vary the sex ratio of offspring. *Science*, **179**, 90–92.

Vehrencamp, S. L. (1983) Optimal degree of skew in cooperative societies. *American Zoologist*, **23**, 327–335.

Wilkinson, G. S. (1984) Reciprocal food sharing in the vampire bat. *Nature*, **308**(5955), 181–184.

Wilkinson, G. S. & Baker, A. E. M. (1988) Communal nesting among genetically similar house mice. *Ethology*, **77**, 103–114

Wilson, D. E. & Reeder, D. M. (eds.) (2005) *Mammal Species of the World: A Taxonomic and Geographic Reference,* John Hopkins University Press, Baltimore, MA.

Wittemyer, G., Douglas-Hamilton, I., & Getz, W. M. (2005) The socioecology of elephants: Analysis of the processes creating multitiered social structures. *Animal Behaviour*, **69**, 1357–1371.

Wolf, J. B. W., Mawdsley, D., Trillmich, F., & James, R. (2007) Social structure in a colonial mammal: Unravelling hidden structural layers and their foundations by network analysis. *Animal Behaviour*, **74**, 1293–1302.

Wolff, J. O. & Sherman, P. W. (2007) Rodents societies as model systems. *In*: Wolff, J. O. & Sherman, P. W. (eds.) *Rodent Societies: An Ecological and Evolutionary Perspective*, Chicago: University of Chicago Press, pp. 3–7.

Young, A. J., Carlson, A. A., Monfort, S. L., Russell, A. F., Bennett, N. C., & Clutton-Brock, T. (2006). Stress and the suppression of subordinate reproduction in cooperatively breeding meerkats. *Proceedings of the National Academy of Sciences*, **103**(32), 12005–12010.

Young, L. J. & Wang, Z. (2004) The neurobiology of pair bonding. *Nature Neuroscience*, **7**, 1048–1054.

11 Sociality in Birds

Andrew Cockburn, Ben J. Hatchwell, and Walter D. Koenig

Overview

Birds differ from the other groups discussed in this book not via any single feature, but by often combining four unusual traits. First, most birds fly, which enhances their ability to choose where they breed and forage, activities that they therefore can, at least potentially, do in very different areas. For example, some albatross travel more than 10,000 kilometers to gain food to provision their young. Second, birds are homeotherms and maintain a very high body temperature, so the metabolic investment that is required to rear offspring to independence is extremely high. Third, unlike most of their homeothermic mammalian counterparts, males and females can share much of this expensive parental care, as both sexes can incubate and provision offspring. Finally, again unlike most mammals, in birds male offspring are more likely to be philopatric than females (Greenwood, 1980; Mabry, *et al.*, 2013), and hence more likely to spend much of their lives in close proximity to their relatives. At least in part because of this combination of features, birds live social lives to an extent that is rare among animals. Birds cooperate to obtain mates, to obtain food, to fly more efficiently, and to evade predators. In addition, most species form pair-bonds, and males and females rear young collaboratively, which itself is a form of relatively advanced social organization (Lack, 1968). Because a review of all aspects of sociality, including the evolution of pair-breeding, is beyond the space allocated to this chapter, here we concentrate primarily on an advanced form of sociality: cooperative breeding, in which more than a pair of individuals cooperate in some way to raise young.

There is some disagreement in the literature over how avian cooperative breeding should be defined. One definition restricts the term to the case where some group members contribute parental care but do not themselves gain parentage

Collectively the authors of this chapter have studied cooperative breeding for almost 90 years. It is fair to say that we do not agree on many issues. Hence while the primary aim here was to describe cooperative breeding in birds in a way that facilitates comparison with the other taxa in the book, it is equally the case that our disagreements may prove to be profitable avenues for future research, and we have highlighted rather than obscured differing opinions for that reason. Our research on cooperative breeding has been funded by grants from the Australian Research Council to AC, the National Environment Research Council to BJH, and the US National Science Foundation to WDK. We thank Andy Russell for getting us to write this chapter and the editors for their persistent encouragement.

(Ligon & Burt, 2004). This definition is problematic. First, it is generally not possible to judge from simple behavioral observations whether helpers are attempting to reproduce or not, nor whether they actually achieve parentage of any offspring (Cockburn, 1998; Cockburn, 2004). This is particularly true in long-lived species with low fecundity, where it may take many years before the true degree of reproductive sharing can be discerned (Faaborg, *et al.*, 1995; Haydock & Koenig, 2002). Second, the definition may also restrict consideration of the factors that could potentially lead to the formation of breeding groups (Cockburn, 2010). Here we define any social unit where three or more birds provision at a single nest as cooperative breeding (Ekman, *et al.*, 2004; Cockburn, 2006; Riehl, 2013). Our definition therefore excludes cases of polygyny or polyandry where the young are reared in different nests and only one or two individuals contribute to care at any one of the nests.

I SOCIAL DIVERSITY

11.1 How Common is Sociality in Birds?

Recent estimates of avian relationships suggest that there are more than 10,500 extant species of birds, belonging to about 40 orders and 240 families, following Gill & Donsker's (2014) categorization of families and species, in combination with the recent higher taxon arrangements proposed by Jarvis, *et al.* (2014). Cooperative breeding was initially estimated to be a relatively rare behavior among birds (Brown, 1987), and hence amenable to analysis as a trait that always occurred as a response to local ecological conditions. However, it is now thought to be much more common, occurring in at least 9 percent of all species, and in more than 10 percent of the birds where both males and females provision young (Cockburn, 2006). Cooperative breeding has been repeatedly gained and lost through the evolutionary history of birds, and can be ubiquitous in some families of birds with wide geographic and ecological distributions (Cockburn, *et al.*, 2013).

The phylogenetic distribution of cooperative breeding is puzzling, but of considerable interest. On one hand cooperative breeding is broadly scattered across the phylogenetic tree of birds, having been recorded in more than 100 families and in more than half (63/98, 64 percent) of species-rich families containing more than 20 species. However, 85 percent (n = 815) of known or suspected cooperative species occur in just two clades with broad geographic distributions: (1) the Coraciimorphae, which includes kingfishers, rollers, bee-eaters, and woodpeckers and their allies, where almost 25 percent of species breed cooperatively; and (2) the oscine passerines, the largest clade of birds (comprising more than 5,000 species), where the incidence is about 13 percent (Table 11.1). There are 34 species-rich families where cooperative breeding occurs in at least 10 percent of the species with bi-parental care. Twenty-four of these are oscines (of 49 oscine families with more than 20 species; 49 percent), six are coraciimorph (of 8; 75 percent), and just four occur in the rest of the birds (of 42; 9 percent). The exceptions are the Falconidae

Table 11.1 The distribution of cooperative breeding in the two clades of birds containing most cooperative species, compared with other birds[a,b,c].

	Coraciimorphae[b]	Oscine passerines	All other birds
Number of species with biparental care[d]	581	4452	3755
Number of species that breed cooperatively	126	570	119
Proportion of cooperative breeders	0.22	0.13	0.03
Number of species-rich families (>20 spp)	8	49	42
With at least 10% of spp. cooperative	6	24	4
Proportion of families cooperative	0.75	0.49	0.09
Proportion of cooperative breeders in Africa[e]	0.39	0.18	0.04
Proportion of cooperative breeders in AustraloPapua[e]	0.24	0.18	0.04
Proportion of cooperative breeders in Indomalaya[e]	0.16	0.08	0.03
Proportion of cooperative breeders in Neotropics[e]	0.18	0.15	0.03

[a] All proportions are expressed as the proportion of species with biparental care that breed cooperatively.
[b] The classification of higher taxa, and the identification of sister taxa, follows Jarvis, *et al.* (2014).
[c] Number of species and assignment to families follows Gill & Donsker (2014).
[d] Parental care assigned to species follows Cockburn (2006), amended and corrected where necessary, and realigned to the new species delimitation.
[e] Species that span more than one zoogeographic region are excluded.

(falcons), Musophagidae (turacos), Rallidae (rails and gallinules) and Apodidae (swifts), which share very few features of ecology or breeding biology.

One approach to understanding the bias towards cooperative breeding in some clades of birds is to compare them with their sister group. The sister groups are not cooperative for both the Coraciimorphae (Strigiformes; 0/235 species, 0 percent) and the oscines (suboscines; 21/1167, 2 percent). The same is true for the swifts, rallids, and turacos. Falcons have oscines as part of their sister clade. The difference between the oscine passerines and their suboscine sister group is most dramatic, as both of these passerine groups have undergone a vast radiation in the Neotropics, where they overlap more or less completely in distribution, diet, life history, and indeed in all other features of their biology except for the incidence of cooperative breeding. Given that the evolutionary transition to cooperative breeding among birds has been the focus of so much research effort, we agree on the significance of these stark differences, and that the resolution of why clades with very similar ecology do or do not develop cooperative breeding is the most important problem in fully understanding cooperative breeding in birds (Cockburn, 2006). Possible explanations require exploration of the biogeography, ecology, and life history traits among cooperative and non-cooperative breeders, which we discuss in greater detail below.

11.2 Forms of Sociality in Birds

The variety of forms of cooperative breeding among birds is bewildering. Most variation is expressed along five axes. First, there is variation in the number of nests attended by the breeding group at any time. Second, there is variation in whether supernumerary (or helper) males, females, or both contribute to care. Third, there is variation in the degree to which reproduction is shared or dominated by a dominant male and female (reproductive skew). Fourth, there is variation in the extent to which extra-group mating occurs. Finally, group members may or may not be related to each other. Some of this variation sets preconditions on the expression of variation along other axes. For example, there can be no reproductive sharing among females if there is only one female attending the nest. However, there is currently a frustrating lack of evidence for consistent correlations between patterns of variation along the five axes (Cockburn, 2004, 2013), and growing evidence that variation along each axis represents a continuum rather than a series of discrete states. These factors preclude a simple classification of types of social organization in cooperatively breeding birds.

In the majority of cooperatively breeding species, the group attends a single nest, and usually all group members contribute to feeding the young and predator defense (singular breeding). However, in some cases multiple nests can be active at the same time within a territory (plural breeding). In the simplest cases of plural breeding, such as in the splendid fairy-wren, *Malurus splendens*, secondary females initiating additional nests are not assured of care by group members (Rowley, *et al.*, 1989). However in other cases such as the Mexican jay, *Aphelocoma wollweberi*, the support of helpers is virtually guaranteed (Brown & Brown, 1990). In extreme cases, such as in the *Manorina* miners, each nest can be attended by 20 or more birds, and most birds, including dominants, feed at multiple nests (Wright, *et al.*, 2010). Complex arrangements can be facilitated by breeding in colonies, as occurs in white-fronted bee-eaters, *Merops bullockoides* (Emlen & Wrege, 1992), or where communal nest structures are constructed allowing many groups to breed in close proximity, as in sociable weavers, *Philetairus socius* (Covas, *et al.*, 2004; van Dijk, *et al.*, 2014). Although rarer than cases of singular breeding, the patterns of care in some of these plural breeding societies can be very complex. In colonies of white-fronted bee-eaters, for example, parents can recruit helpers by interfering with attempts by their offspring to establish independent nests and thereby preventing them from breeding independently (Emlen & Wrege, 1992).

The second primary axis of variation is whether both sexes contribute help, or whether just male or female non-breeders contribute care. We lack a full understanding of this variation because cooperative breeders are often sexually monomorphic (Rubenstein & Lovette, 2009), and data from many species have been collected without any capacity to test which sexes are involved. Nonetheless, it is clear that female-only care is less common than care by males or both sexes.

A critical question is whether reproduction is shared by group members, or whether it is dominated by just one individual, an axis of variation called reproductive skew.

In addition to plural breeding, species in which two or more females contribute care include cases where more than one female lays in the nest. Such "joint-nesting" is rare among cooperative breeders and involves two or more females from the same group contributing to the same clutch and rearing offspring cooperatively. It is patchily distributed across the avian phylogeny, occurring among both non-passerines (including ratites, some rails, the crotophagid cuckoos, and acorn woodpeckers, *Melanerpes formicivorus*), but also among passerines (Seychelles warbler, *Acrocephalus sechellensis*, yellow-billed shrike *Corvinella corvina*, and Taiwan yuhina, *Yuhina brunneiceps*). A major disincentive to joint-nesting is perturbation to the optimum clutch size, which may diminish the efficiency of incubation (Koenig, *et al.*, 1992; Chao 1997). Competition among females is widely reported in such systems, often in the form of egg destruction, which can entail significant costs in terms of both time and energy (Koenig, *et al.*, 1995; Vehrencamp & Quinn, 2004). Importantly, the limitations to nest-sharing posed by perturbation of the optimum clutch size do not apply to males. In addition, an extra egg in the nest is likely to be obvious in many cases, whereas extra-pair mating can be acquired furtively.

In some cooperatively breeding species unassisted pairs are incapable of rearing young (obligate cooperative breeding) because the adults cannot supply the young with sufficient food (Grimes, 1980; Heinsohn, 1992; Boland, *et al.*, 1997a). In some cooperative breeders, it is difficult to test for a critical group size because the birds always breed in groups, though further evidence for obligate cooperative breeding comes from brown jays, *Psilorhinus morio*, where the population has expanded rapidly, but new groups always form through fission from large groups rather than by pairs budding off to attempt independent reproduction (Williams, *et al.*, 1994).

Most cooperatively breeding social groups involving nonbreeding helpers consist of related individuals, a point we will discuss more extensively later. Many cooperatively breeding groups involving cooperative polyandry or polygyny, however, consist largely of unrelated individuals (Riehl, 2013). This enormous variation in social organization among cooperative breeders remains one of the primary impediments to understanding its phylogenetic and biogeographic distribution.

11.3 Why Birds Form Social Groups

Cooperative breeding in birds arises most commonly when offspring delay dispersal and remain associated with, and assist reproduction of, the adults that reared them. Dispersal can be delayed long after the young have achieved reproductive maturity and could potentially breed on their own. For example, in the Florida scrub-jay, *Aphelocoma coerulescens*, young of both sexes typically remain with their parents on their natal territory for years, during which time they are "helpers-at-the-nest," provisioning younger siblings at subsequent nests of their parents and helping to defend the territory and scan for predators (Woolfenden & Fitzpatrick, 1984).

11.3.1 Resource Acquisition and Use

Living in groups is a double-edged sword. It can improve the efficiency of resource acquisition in a number of ways, but it also leads to competition between group members for resources (Alexander, 1974). There is no clear evidence that the procurement of food resources is the primary cause of cooperative breeding in birds. However, as is true for some mammals such as lions, coalition formation can be necessary in order to procure a breeding vacancy. For example, Galapagos hawks, *Buteo galapagoensis*, live on oceanic islands where the resources necessary to sustain reproduction are confined to territories in narrow coastal strips. When young birds become independent they leave their natal territory and forage in groups in marginal habitat in the center of the island. During this period, males – who are often unrelated – form coalitions that are large enough in size to drive out owners of existing territories (Faaborg & Bednarz, 1990).

One interesting case of resource manipulation that appears to facilitate cooperative breeding in some species is the construction of a compound nest, where many nesting cavities are contained within a large superstructure that is built by several breeding pairs. Brown (1987) predicted that cooperative breeding would be common in these cases, and it increasingly appears that cooperative breeding does occur in each of the species that construct these superstructures (Cockburn, 1996). A parallel situation involving food, rather than nesting, occurs in acorn woodpeckers, where family groups defend "granaries" potentially consisting of thousands of small holes in dead limbs and thick bark of trees in which the birds communally store acorns harvested each autumn from nearby oaks. Granaries are a limiting resource that, along with suitable nest cavities, restrict the ability of birds to breed independently, thereby facilitating delayed dispersal and potentially cooperative polygamy (Koenig, *et al.*, 1992).

11.3.2 Predator (and Brood Parasite) Avoidance

Predation risk promotes group formation in many species of birds. Group membership decreases the individual probability of predation, enables mobbing of ground predators, confuses aerial predators, improves predator detection through greater corporate vigilance, and consequently allows individuals to reduce their vigilance without increasing their likelihood of predation (Bertram, 1978; Pulliam & Caraco, 1984). Such advantages have the potential to be refined in stable social groups, and are commonly gained in cooperative breeders by means of sentinel behavior. Sentinels can greatly improve foraging efficiency of the group (Bell, *et al.*, 2009), summon group members to mob predators (Fraga, 1991), and facilitate monitoring of the nest, thereby preventing provisioners from alerting predators to its presence when they feed young (Yasukawa & Cockburn, 2009). Sentinel behavior may be risky (Ridley, *et al.*, 2013), and is most likely to be undertaken by individuals that have already satiated themselves (Bednekoff & Woolfenden, 2003). Cooperative defense against brood parasitism provides similar advantages to defense against predators (Poiani & Elgar, 1994; Feeney, *et al.*, 2013; Feeney & Langmore, 2015).

While the benefits of nest defense and reduced predation and brood parasitism are very strong, there are problems in trying to separate cause and effect. It would be of great interest if reduced predation and brood parasitism promote the evolution of cooperative breeding, but it is also possible that cooperative breeding promotes the evolution of defensive behavior. For example, it is difficult to see how costly sentinel behavior could evolve except in the context of cooperation among kin and prolonged care of offspring. Such problems of the directionality of causation recur throughout the study of cooperative breeding, where it is difficult to tease apart the benefits of living in groups from those caused by helping behavior because it is difficult to manipulate one without changing the other (Cockburn, 1998).

Indeed, it has also been suggested that the risk of predation can select against cooperative breeding. Some birds live in stable groups throughout the breeding season, yet unlike many of the birds discussed in this chapter, only the dominant pair feed the young. In the best-studied case, the *Perisoreus* jays, young can remain in their natal group for substantial periods. Philopatry confers considerable benefits to the young during the harsh winter, as their parents ensure that their young receive privileged access to scarce and scattered food resources (Ekman, *et al.*, 2000). However, in the Siberian jay, *P. infaustus*, young never help young raised by their parents and in the gray jay, *P. canadensis*, young are not allowed to approach the nest, but they do help feed fledglings. By contrast, young birds do provide help at the nest in the Sichuan jay *P. internigrans*. This variation has been plausibly related to differences in the type of predators in each habitat and the possibility that inexperienced feeders would expose nestlings and fledglings to greater risk of predation (Strickland & Waite, 2001; Jing, *et al.*, 2009).

11.3.3 Homeostasis

Thermoregulation has been studied in few cooperative breeders. The principal context for investigation has been communal roosting, either in nest cavities or in huddles. Ligon, *et al.* (1988) argued that green woodhoopoes, *Phoeniculus purpureus*, are forced to roost in tree cavities because of their inability to thermoregulate at low ambient temperatures, and thus that the shortage of suitable cavities selects for philopatry and cooperative breeding. Woodhoopoes gain substantial energetic benefits from roosting communally (du Plessis & Williams, 1994), but single birds can easily maintain their body temperature at very low ambient temperatures (Williams, *et al.*, 1991), countering the hypothesis that thermoregulatory needs drive cooperative breeding. Acorn woodpeckers also gain thermoregulatory benefits from roosting communally in cavities (du Plessis, *et al.*, 1994), and in sociable weavers, the massive communal nests buffers extreme ambient temperatures, an effect that is enhanced by the presence of conspecifics (White, *et al.*, 1975) and in more central nest chambers (van Dijk, *et al.*, 2013).

Roosting huddles in long-tailed tits, *Aegithalos caudatus*, and bushtits, *Psaltriparus minimus*, have a similar function of reducing energetic costs; in the former species, access to preferred positions is determined by competitive interactions among flock members and there is no link between helping and access to roosts (Chaplin, 1982;

McGowan, *et al.*, 2006; Hatchwell, *et al.*, 2009; Napper, *et al.*, 2013). Indeed, communal roosting is widespread among birds (Beauchamp 1999) and evidence for a direct link between thermoregulatory needs and cooperative breeding is equivocal (du Plessis, 2004), especially given the paucity of cooperative breeding in northern temperate regions. A likely exception to this general conclusion is in the Coliiformes (mousebirds), an order comprising six species, all of which are cooperative. Mousebirds have an extraordinary thermoregulatory system that does not appear to be closely regulated about a set point, instead being subject to ambient temperature, food intake, metabolic suppression (torpor), and huddling (McKechnie & Lovegrove 2001a,b). Mousebirds frequently bask, but predictable access to huddles during the day and particularly at night is also likely to be critical and may have selected for sociality.

11.3.4 Mating

Mating relationships among cooperative breeders are often very difficult to observe. For example, copulations are rarely seen in either Florida scrub-jays or in acorn woodpeckers, two cooperative breeders that have been the subject of long-term studies. As a consequence, there have in some cases – although not in these two species – been stark disparities between the conclusions inferred from field-based behavioral studies and the actual patterns revealed by molecular determination of parentage. It is now clear that the true mating system among cooperative breeders is enormously variable (Cockburn, 2004).

Cooperative breeders range from the least faithful to the most faithful of all birds. The range of habits is well illustrated by three species of plural breeders. In the *Manorina* miners the dominant male at each nest gains 100 percent of parentage (Põldmaa, *et al.*, 1995; Conrad, *et al.*, 1998). By contrast, in the Mexican jay, subordinates sire 40 percent of young (Li & Brown, 2000). Finally, in plural breeding populations of the Australian magpie there are also very high levels of infidelity, but it is expressed by mating outside the group (82 percent of young, Hughes, *et al.*, 2003). None of this variation was anticipated by knowledge of ecology or observations of behavior. Similar variation occurs among singular breeders, where species with only male helpers vary from virtual complete fidelity (e.g. Florida scrub-jay Quinn, *et al.*, 1999) to extreme infidelity (e.g. superb fairy-wrens, Mulder, *et al.*, 1994).

In addition to this variation among socially monogamous species, polyandrous, polygynous, and polygynandrous social arrangements can occur, and in some species these can occur within a single population. Examples include acorn woodpeckers, where coalitions of related males compete for matings within a group for one to three related joint-nesting females, and accentors (Prunellidae) such as the dunnock, *Prunella modularis*, in which males and female defend space independently from each other (Davies, 1992; Davies, *et al.*, 1995). Dunnock females can defend large territories, resulting in both polyandrous and polygynandrous associations. Females benefit from living with several males from which they can extract care, thus increasing their reproductive output. Males, on the other hand, enhance their fitness through copulatory access to multiple females. This leads to complicated behavioral counterploys between males and females, with females soliciting frequent copulations from multiple males and cloaca-pecking by males to encourage the female to eject the sperm of the male with which she has mated most recently. An equally complex example occurs in the eclectus parrot, *Eclectus roratus*, where extreme competition for high quality nests forces females to stay within nest cavities for most of the year, during which time they are completely dependent on males for food. Females that display at high quality nests attract more male feeders and copulate with all of them (Heinsohn & Legge, 2003; Heinsohn, *et al.*, 2007; Heinsohn, 2008)

11.3.5 Offspring Care

We have defined cooperative breeding as the case where multiple birds care for a single brood. Some early discussion of cooperative breeding argued that feeding by supernumerary individuals was an inevitable response to proximity to begging offspring, and hence did not need an adaptive explanation (Jamieson & Craig 1987; Jamieson 1991). This view has been largely discredited through repeated demonstration that the amount of alloparental care varies greatly between individuals in ways that reflect the costs and benefits of care (Dickinson & Hatchwell, 2004). On the other hand, there is great variation across species in the amount of care that helpers invest in broods that remains largely unexplained (Hatchwell, 1999; Heinsohn, 2004).

The failure of some supernumeraries to provide care has attracted some attention, but is in general poorly understood. We have already discussed the possibility that inexperienced feeders may increase the risk of predation of offspring, and hence any contribution should be selected against. Attention has also been focused on the surprising phenomenon of "false feeding", where supernumerary birds carry food to the nest but then consume it themselves. This could be a response to nestlings being satiated, or a response to uncertainty over whether it is safe to feed the nestlings (McDonald, *et al.*, 2007). For example, in bell miners false feeds decline when the demands of the brood are low, but increase when observers are too close. However, there is evidence from at least one species that feeding has a signaling function (Boland, *et al.*, 1997b), as originally argued by Zahavi (1977). First, false feeding declines to very low levels when food is supplemented, though supplementation should reduce nestling demand. Second, helpers do not swallow the food until the previous sentinel at the nest has departed, which greatly prolongs begging. Demonstrating reliability may be unusually important in this obligate cooperative breeding species, since at least four birds are needed for reproduction and thus birds rely on coalition partners in order to form new breeding groups (Rowley, 1978; Heinsohn, 1992).

11.4 The Role of Ecology in Shaping Sociality in Birds

The original interest in cooperative breeding in birds was focused on population ecology, reflecting the importance attributed to constraints on dispersal. This has fostered one of the notable research approaches that has dominated the study of cooperative breeding: long-term studies of individual species (Stacey & Koenig, 1990; Koenig & Dickinson, 2016). Some of these studies have run for decades, and include some of the most thorough investigations of the ecology of single species ever undertaken. However, this empirical approach has been less successful at the comparative level, and has not led to effective prediction of how other species should behave (Cockburn, 2013).

11.4.1 Habitat and Environment

Some decades ago, Emlen (1982) suggested that cooperative breeding was favored in two contrasting ecological contexts. In the first, cooperative breeding was hypothesized to be found among species living in stable environments with specialized ecological requirements. Such "ecological constraints" have been experimentally demonstrated to be important drivers of sociality in several species. For example, in red-cockaded woodpeckers, *Picoides borealis*, a monogamous species with helpers-at-the-nest in the southeastern United States experiments have shown that roosting cavities limit the ability of birds to disperse from their natal territory (Walters, *et al.*, 1992). Even more dramatic are observations and experiments on the Seychelles warbler, where habitat destruction at one point confined the species to very low numbers on a single small island. After careful management led to recovery, population growth initially occurred through the formation of new breeding territories. However, once a certain of number of territories was reached, subsequent population growth occurred only through recruitment of supernumeraries (Komdeur, 1992). The same result was obtained when the birds were introduced to a second island (Komdeur, *et al.*, 1995), clearly implicating resource availability, specifically resources defining high-quality habitat (Komdeur, 1992). These results thus support Emlen's idea that when all such habitat is occupied, young individuals may be better off delaying dispersal and remaining in their natal territory, an idea frequently referred to as the "habitat saturation" hypothesis.

Emlen (1982) also suggested, however, that additional assistance with the cost of reproduction might be most helpful (and independent reproduction most fraught with risk) in unpredictable and fluctuating environments – essentially the opposite conditions as those defining habitat saturation. Comparative tests with appropriate phylogenetic controls that attempt to distinguish these alternative possibilities have produced equivocal results. Cooperative breeding is correlated with environmental variability in African starlings (Lovette & Rubenstein, 2007), but with environmental stability in the hornbills (Gonzalez, *et al.*, 2013), with no obvious reason for this difference. In a global analysis, Jetz & Rubenstein (2011) also found an association with variability, but the effects were marginal relative to the phylogenetic effects, and in a problem that is likely to recur with global analyses, it is not possible to dissociate whether the effect of climatic variability is driven by the loss of cooperative breeding among Holarctic families or helps explain variation among tropical and southern taxa (Cockburn & Russell, 2011; Cockburn, 2013).

Early discussions of natal philopatry emphasized remaining on the natal territory because of ecological constraints in the environment that limited dispersal (Selander,

1964; Emlen, 1982; Koenig, *et al.*, 1992). However, it is increasingly clear than many cooperative breeders do not maintain year-round territories like the Florida scrub-jay, yet limited dispersal is still crucial. Such cases include species where individuals initially attempt independent reproduction but assist at another nest if their own nest fails, as in long-tailed tits (Hatchwell & Sharp, 2006); those that spend most of their lives in mixed-species feeding flocks but break away to breed, as in the Gola malimbe, *Malimbus ballmanni* (Gatter & Gardner, 1993); and some species where belonging to a group is the primary criterion of success and territory defense does not occur, as in white-winged choughs, *Corcorax melanorhamphos* (Heinsohn, 1992). Despite compelling empirical evidence from several species, an operational and predictive definition of habitat saturation has proved troublesome (Pen & Weissing, 2000; Kokko & Lundberg, 2001; Kokko & Ekman, 2002).

11.4.2 Biogeography

Four patterns emerge from recent global analyses of the geographic distribution of cooperative breeding. First, cooperative breeding occurs primarily in diurnal landbirds, being rare in the clades of birds that exploit aquatic habitats (Pelecanimorphae, Procellarimorphae, Gaviiformes, Phoenicopterimorphae, Anseriformes, 7/953 species, less than 1 percent) (Cockburn, 2006), and the two unrelated clades of nightbirds (Strigiformes, Caprimulgiformes; 0/366 species, 0 percent) (Cockburn, 2006). Second, cooperative breeding among diurnal landbirds declines in prevalence with latitude to the north but not to the south of the equator (Cockburn, 2003; Jetz & Rubenstein, 2011). The decline in sociality at the northerly latitudes of the Holarctic has recurred repeatedly in some families of birds (Ekman & Ericson, 2006). Although ecological causes are therefore strongly implicated, most aspects of the environment change with latitude, and the plethora of correlated explanatory factors have made cause and effect of latitudinal gradients notoriously difficult to dissect in all branches of ecology. Indeed, understanding the stark difference in latitudinal changes in behavior and life history in birds living north and south of the equator is likely to offer great insights, but has proved difficult to disentangle, although considerable progress has been made by studies that compare phylogenetically-matched suites of species at a number of sites in different continents (Martin, 2002; Martin, *et al.*, 2007; Chalfoun & Martin, 2007; Lloyd, *et al.*, 2014). Third, cooperation often evolves on islands in taxa that are not otherwise cooperative (Covas, 2012). Island faunas are often species-poor compared to their mainland counterparts, and their constituent species may experience ecological release and achieve high densities. This result is therefore compatible with the notion of habitat saturation. Finally, among tropical and southern regions, cooperative breeding is more prevalent in sub-Saharan Africa and Australia than in Indomalaya or the Neotropics (Jetz & Rubenstein, 2011). The causes here are also elusive, and some care is required in assessing the source of these differences. For example, the Neotropics are rich in species with female-only care, particularly among frugivores and the speciose nectarivorous hummingbirds. In order to understand the evolution of cooperative breeding, it is therefore better to focus upon species where both males and females provide parental care that breed cooperatively. Among the taxa with a high incidence of cooperative breeding, cooperation is most prevalent among Coraciiformes in Africa, while it is least prevalent among oscines in the Indomalayan region (Table 11.1). However, it is equally clear that these species-rich taxa have very high rates of cooperative breeding across their geographic range. The difference among these biogeographic regions is therefore likely driven by the mixture of cooperative and non-cooperative clades rather than variation within the clades themselves. In particular, the incidence of cooperative breeding in the Neotropics is driven down by the large radiation of suboscine birds where cooperation is very rare.

11.4.3 Niches

It has proven difficult to identify a particular foraging niche associated with cooperative breeding. A potential factor identified in several studies is higher prevalence in open habitats such as savannas compared to that found in forests and other dense habitats. This has been hypothesized to result from both higher (Gaston, 1978; du Plessis, *et al.*, 1995; Rubenstein & Lovette, 2007) and lower (Ford, *et al.*, 1988) inter- and intra-season variation of savanna habitats relative to forest habitats. By contrast, Bell (1985) argued that group cohesion might be more difficult to maintain in visually occluded habitats, and Cockburn (1996) extended that idea to suggest that social conflict over reproduction might prevent group living where the behavior of reproductive competitors could not be monitored.

An alternative hypothesis is that cooperative breeding may have been reported less frequently in rainforest because of the difficulty in observing birds (Brown, 1987; Ford, *et al.*, 1988). Although one of us originally argued against that view (Cockburn, 2003), the improved data reported by Cockburn (2006) and subsequent reports suggest that cooperation in rainforest habitats may have been greatly underestimated, particularly among Indomalayan babblers and Neotropical tanagers. These groups are primarily denizens of rainforest, and may be the two clades with the greatest number of cooperative species.

11.5 The Role of Evolutionary History in Shaping Sociality in Birds

We have seen earlier that cooperative breeding is particularly common in some clades of birds. There is just one species-rich family of birds that is exclusively cooperative (Maluridae), though some small families also fall into this category (e.g. Opisthocomidae, in which the only species is highly cooperative). Such social variability, as found in many groups of birds, allows examination of the evolutionary transitions within families. Mapping of trait distributions on to well-defined phylogenies remains in its infancy, but it is clear that, just as cooperative breeding frequently evolves independently in different groups, in some families reversion to pair breeding also occurs (Berg, *et al.*, 2012; Ekman & Ericson, 2006; Nicholls, *et al.*, 2000;

Rubenstein & Lovette, 2007). Both evolutionary gains and losses are critical to understanding the distribution of sociality in birds.

Cockburn (2003) proposed a macroevolutionary hypothesis for the global distribution of cooperative breeding among the oscine passerines, pointing out that clades of cooperative breeders were often highly successful within their continent of origin, but species-poor relative to closely related clades that were predominantly pair-breeding. The difference could be accounted for by the absence in cooperative clades of species that have crossed major biogeographic barriers such as Wallace's line, crossed major barriers to migration such as the Himalayan plateau, or colonized remote oceanic islands. Cockburn (2003) further proposed that this failure stems from high rates of natal philopatry by at least one sex that is common among passerine cooperative breeders. This argument has been supported using more fully resolved phylogenies that allow the rate of speciation to be properly estimated (Marki, *et al.*, 2015), but not by a comparative analysis of vagrancy among North American species that failed to support the hypothesis of reduced long-distance dispersal among cooperative breeders (Rusk, *et al.*, 2013).

II SOCIAL TRAITS

Thanks to careful observations by natural historians like Alexander Skutch (1935, 1961), we know a great deal about the life histories of cooperatively breeding birds. Below we discuss the traits of social species, and then the traits of social groups of birds.

11.6 Traits of Social Species

11.6.1 Cognition and Communication

Some primatologists have argued that the challenges of living in social groups drive the evolution of bigger brains and more sophisticated cognitive abilities (social intelligence hypothesis, Humphrey, 1976; Dunbar & Shultz, 2007; Chapter 9), and this idea has been elaborated to suggest that cooperative breeding in particular demands complex cognitive abilities and larger brains (Burkart, *et al.*, 2009). Whether this idea can be applied to birds has recently been reviewed by Thornton & McAuliffe (2015), who found little convincing evidence for this path of causation.

Auditory communication has been studied extensively in avian cooperative breeders, especially in the context of recognition, although the strength of selection on the communication of social identity within and between groups is likely to vary greatly among species in relation to their social organization (Cornwallis, *et al.*, 2009). There is good experimental evidence for discrimination of group from non-group members using vocalizations in several species, including rufous-crowned babblers, *Pomatostomus ruficeps* (Crane, *et al.*, 2015), green woodhoopoes (Radford, 2005), and superb starlings, *Lamprotornis superbus* (Keen, *et al.*, 2013). Such group-specific calls may also be used as coordinated displays in territorial contests among neighboring groups (Radford, 2003; Seddon & Tobias, 2003) in much the same way that lions and various primate species signal to neighbors vocally. Playback experiments show that communication of social identity also occurs at a finer resolution, with discrimination among individuals within groups in apostlebirds, *Struthidea cinerea* (Warrington, *et al.*, 2015), and kin discrimination in long-tailed tits (Sharp, *et al.*, 2005). In the latter case, vocal cues are learned during early development and so can be characterized as environmental rather than genetic cues to kinship that are learned through association.

It is also notable that some of the best examples in birds of complex female song are found among cooperative breeders. In social groups with multiple female reproductives, females may compete for breeding opportunities or for valuable male care. Thus, in alpine accentors, *Prunella collaris*, and dunnocks, females sing to attract males to copulate, thereby securing their care (Langmore, *et al.*, 1996; Langmore & Davies, 1997), while in African starlings (Sturnidae) female-female competition has apparently selected for ornamentation and complex song in females, resulting in little sexual dimorphism in either form of signal (Rubenstein & Lovette, 2009; Pilowsky & Rubenstein, 2013). Finally, it is intriguing that female song is also associated with duetting (i.e. behavior associated with territory and mate defense) in the Icteridae (Odom, *et al.*, 2015), and that duetting and cooperative breeding appear to be correlated with several common biogeographic and life-history traits (Hall, 2009; Logue & Hall, 2014). However, the possibility that cooperative species are more likely to exhibit vocal duetting remains to be tested.

11.6.2 Lifespan and Longevity

It has become axiomatic to assume that cooperative breeders are long-lived and have low fecundity relative to their pair-breeding counterparts, and a number of theoretical models have been developed to demonstrate how these associations could come about (Hardling & Kokko, 2003; Ridley, *et al.*, 2005). A causative effect of life history is attractive because it draws together many of the known correlates of cooperative breeding, such as the decline in incidence in the northern but not the southern hemisphere as life histories speed up; the frequent evolution of cooperative breeding on islands where life histories are slow; and an explicit theoretical link between life history characteristics and the concept of habitat saturation, as breeding vacancies will arise less frequently as lifespan increases (Kokko, *et al.*, 2001, 2002).

A widely cited series of comparative analyses proposed that life history variation had greater power in explaining the incidence of cooperative breeding than any "ecological" variable (Arnold & Owens, 1998, 1999). These studies relied on earlier results that suggested that life history traits are highly conserved in birds, so there is little intra-familial life history variation (Owens & Bennett, 1995). Such conservatism might help explain the strong phylogenetic bias in cooperative breeding. Indeed, the analysis treated taxonomic "family" as the appropriate unit for comparative analysis, and showed that incidence of cooperative breeding is highest in families with slow life histories, long lifespans, and small clutches. Unfortunately, these conclusions are based on dubious assumptions. Most important, there were great biases in the species selected to characterize families (Cockburn, 2003). By and large, families of passerines that occur in the Holarctic were characterized only by species living at high latitudes. For example, the parid tits were typified by Palearctic species that are short-lived, have very large clutch sizes, and are pair-breeders, while their southern counterparts have small clutch sizes and are often cooperative breeders (Shaw, *et al.*, 2015). By contrast, data from cooperative breeders were over-represented among families confined to tropical and southern latitudes because of the intense research effort prompted by cooperation.

A more recent analysis using phylogenetically-paired comparisons to test for an association between life history traits and cooperative breeding in which the pairs selected for analysis were taken from the same geographic region and body size was controlled found no difference between cooperative and pair breeders for maximum longevity over a large number of paired comparisons (Beauchamp, 2014). Similar results were obtained in a geographically restricted analysis of North American birds (Blumstein & Møller, 2008). Beauchamp (2014) did, however, find that cooperative species enjoyed a 10 percent increase in annual survival of adults. Whether these results provide support for the life history hypothesis or not is therefore questionable.

11.6.3 Fecundity

Beauchamp (2014) also showed that cooperative species do not differ in fecundity from pair-breeding species (see also a geographically restricted analysis by Poiani & Jermiin, 1994). Indeed, while cooperatively breeding birds can respond to the presence of helpers by using the additional care to enhance their annual fecundity, many species use the alternative strategy of reducing their own contribution to the brood (load-lightening) in order to enhance their own survival (Hatchwell, 1999; Legge, 2000). For these species, increased adult survival is predicted to be a consequence rather a cause of cooperative breeding, and the increased adult survival reported by Beauchamp (2014) could result from this effect. Thus, while a link between life history and the decline of cooperative breeding at northern latitudes is of great interest, there is currently no convincing evidence that life history differences among taxa explain the considerable variation in the incidence of cooperative breeding between species living in tropical/south temperate habitats.

11.6.4 Age at First Reproduction

All avian cooperative breeders are totipotent, which is to say that they are physiologically capable of reaching reproductive maturity. Unlike some other social taxa, including mammals, there is no evidence of any species having made the transition to true eusociality, in which some individuals are permanently sterile. There is also no convincing evidence of specialization into different castes within birds, despite some remarkable observations of "dwarf" helpers in New Caledonian parakeets, *Cyanoramphus saisseti* (Theuerkauf, *et al.*, 2009). While these latter observations are provocative and demand further study, helpers in this case tried to mate with the breeding female, and are thus unlikely to indicate the existence of a sterile caste.

While they are all totipotent, cooperative breeders differ considerably in the age at which they first reproduce. In some cases there are long periods – up to 5 years – during which younger group members are physiologically incapable of reproduction (Kemp, 1995). In other species reproductive totipotency is already present by the first breeding season (Dickinson, *et al.*, 1996; Hatchwell, *et al.*, 2004), although one-year-old birds may fail to reproduce because of lack of opportunity or competition with older birds.

Possible physiological bases for delayed reproduction have been carefully discussed by Schoech, *et al.* (2004). Failure to attain reproductive maturity could reflect slow development for reasons unrelated to breeding competition. The southern ground hornbill, *Bucorvus leadbeateri*, is probably the extreme case, as young can still be receiving food from the group at 2 years of age (Kemp, 1995). In white-winged choughs, another bird with slow development, young continue to develop foraging skills over the first four years of life, and it is unlikely that they would have sufficient skill to rear a brood before that time (Heinsohn, *et al.*, 1988; Heinsohn, 1991). Again, cause and effect is problematical. Cockburn (2013), for example, has suggested a model where the near inevitable presence of helpers in such species has allowed young to spend more time acquiring complex foraging skills, which in turn has forced a greater and ultimately irreversible dependence upon cooperation.

Of particular interest in understanding cooperative breeding in birds is reproductive suppression by dominants that renders subordinates temporarily or facultatively sterile. Physiological suppression could be caused by either same-sex or opposite-sex dominants (Schoech, *et al.*, 2004), the latter most likely to arise because of incest avoidance (Koenig & Haydock, 2004). In many cooperatively breeding societies, the only mates available within the group may be immediate relatives, which may lower reproductive interest and affect physiological maturation. We know that these effects are potent because they can also limit reproduction of the dominants in the group (Koenig, *et al.*, 2004). For example, when related acorn woodpeckers assume the senior positions in the group, they can nonetheless exhibit reproductive restraint despite severe demographic penalties (Koenig, *et al.*, 1999). At a proximate level, male helpers in red-cockaded woodpeckers have much lower testosterone levels when the breeding female is a relative (Khan, *et al.*, 2001), confirming that suppression is driven by intrasexual interactions and incest avoidance.

Demonstrating an unequivocal role for intrasexual suppression outside of incest avoidance is more difficult. It can be excluded in many cooperative breeders, where reproductive sharing within groups is common provided that there is no incest constraint (Cockburn, 2004). In the Florida scrub-jay, for example, young do not contribute to reproduction while they are in the helper role, but can attain reproductive maturity rapidly if they are promoted to the dominant position. Dissecting the physiological mechanism of suppression and distinguishing cause from effect, however, has turned out to be more complicated than in cooperatively breeding mammals (Schoech, *et al.*, 2004).

A possible source of variability in susceptibility to reproductive suppression arises because in some species subordinates can gain reproductive success outside the social group, and hence should become sexually mature even if they are related to all the opposite-sex adults in their group. In superb fairy-wrens, where parentage is dominated by extra-group matings (Mulder, *et al.*, 1994), subordinates living with attractive dominants gain more extra-group success than unattractive dominants because they are able to achieve extra-group matings parasitically (Double & Cockburn, 2003; Cockburn, *et al.*, 2009; Cockburn, 2013; Cockburn, *et al.*, 2016). Hence it is likely that, at least in this species, the opportunity for success outside the group exerts strong selective pressure for reproductive maturity regardless of the relatedness of subordinates to other members of their social group.

11.6.5 Dispersal

Dispersal in general, and sex-biased dispersal in particular, plays a fundamental role in understanding cooperative breeding in birds. Philopatry promotes interactions between close relatives and allows kin selection to occur. However, limited dispersal can also have costs, such as local resource and mate competition between kin, and inbreeding depression (West, *et al.*, 2002). Like most birds, avian cooperative breeders often show female-biased dispersal, and this sex bias remains the most plausible general explanation for the bias towards male help among avian cooperative breeders (Cockburn, 1998).

Despite significant investment using methods such as radiotelemetry in some species (Koenig, *et al.*, 2000), sample sizes for dispersal studies in most avian cooperative species remain limited and biased to observations of short-distance movements. Outstanding sample sizes have been obtained in red-cockaded woodpeckers, but in this species there is no difference in dispersal distance between males and females (Kesler, *et al.*, 2010) and, perhaps as a consequence, no evidence that dispersal is effective in reducing inbreeding (Schiegg, *et al.*, 2006) despite strong inbreeding depression (Daniels & Walters, 2000). It should be noted that contemporary dispersal patterns are difficult to interpret in this species because the preferred habitat is highly fragmented (Schiegg, *et al.*, 2002), and the large sample size reflects conservation concerns for its fate. Similarly, dispersal distances have also been well characterized for Seychelles warblers, because originally the entire world population was confined to a single, small island (Eikenaar, *et al.*, 2008a). In this species, early results suggested that females were most likely to delay dispersal, but this pattern disappeared as the population grew as a result of conservation efforts (Eikenaar, *et al.*, 2010) and, as for red-cockaded woodpeckers, there is no strong evidence that dispersal facilitates inbreeding avoidance (Eikenaar, *et al.*, 2008b). Unfortunately, neither of these species shows sex-biased dispersal, and a research priority must be to achieve data of comparable quality on species that exhibit the more common pattern of sex-biased dispersal.

11.7 Traits of Social Groups

11.7.1 Genetic Structure

The prevalence of family-based social organization in avian cooperative breeding systems has been recognized since the early studies of Skutch (1961) and Selander (1964), a characterization supported by subsequent reviews (Brown, 1987) and the fact that delayed natal dispersal plays a key role in the formation of social groups in many cooperative species (Emlen, 1982; Koenig, *et al.*, 1992). The theoretical framework provided by inclusive fitness theory (Hamilton, 1964) also prompted a widespread acceptance of kin selection as a critical process in the evolution of helping behavior (Emlen, 1994), reinforcing the assumption that relatedness within cooperative groups is generally high. The advent of novel techniques to determine parentage and other genetic relationships in natural populations in the past 25 years has revolutionized the capacity of researchers to determine patterns of kinship, but, despite some remarkable revelations about paternity (Mulder, *et al.*, 1994), maternity (Berg, *et al.*, 2009) or both (Richardson, *et al.*, 2001), the general conclusion that cooperative breeding typically occurs among members of kin-groups has remained largely unchanged, though such groups can often include unrelated members (Cockburn, 1998).

Although family-living among cooperatively breeding birds is the typical pattern, there are also many intriguing exceptions in which cooperation occurs principally among non-kin. Thus, there is a continuum of social organization between conventional "helper-at-the-nest" systems and communal or joint-nesting species and cooperative polygamous species (Cockburn, 1998) across which kinship among group members varies considerably. Most challenging to social evolution theory is cooperative behavior that occurs entirely within non-kin groups. Unrelated helpers occur in social groups otherwise composed principally of kin (Riehl, 2013), and such individual can be expected to derive some direct fitness benefit from their behavior (Cockburn, 1998). True communal nesting systems are relatively rare and highly variable in terms of the inter- and intra-sexual strategies that are expressed. In several such species, however, conspecific brood parasitism appears to be one of the main ways through which joint nesting is initiated, which can then become stable if females benefit from joining another female, male parental care is present, and other factors increase the cost of leaving or the benefit of staying (Vehrencamp & Quinn, 2004).

Two recent studies have attempted to quantify the proportion of avian cooperative breeders whose social groups include close kin. Hatchwell (2009) characterized avian families containing species with cooperative brood care as having social groups composed of kin or non-kin, finding that 80 percent of the 55 families exhibiting cooperative breeding that could be classified typically consisted of kin groups. Extending these assignments by phylogenetic inference to all the species contained within those families, 91 percent ($n = 612$) of species could be provisionally described as cooperating predominantly within kin groups. This analysis must be qualified with various caveats, however, as there are few families where the prevalence of cooperative breeding can be assessed completely. Riehl (2013) used an alternative approach to address the same question, classifying the pattern of relatedness for cooperative species, and reaching broadly similar conclusions. A large majority of species (85 percent; $n = 213$) bred either in nuclear family groups (55 percent) or in mixed groups of kin and non-kin (30 percent), with just 15 percent of species living primarily with non-kin.

Genealogical relationships among kin may change through time in avian social groups as breeder turnover occurs, resulting in extended families of variable composition. Retained kin helpers in such systems may be augmented by unrelated helpers that have either dispersed into a social group or whose relatives have been replaced as breeders by non-kin on their natal territory (Reyer, 1990; Piper, *et al.*, 1995). Related helpers may obviously gain indirect and/or direct fitness benefits, while unrelated immigrant helpers may be selected to help non-kin if they gain direct fitness benefits that outweigh the costs of helping (Cockburn, 1998; Dickinson & Hatchwell, 2004).

Dispersal does not necessarily preclude subsequent kin-directed helping behavior if relatives disperse together in coalitions, or if dispersal is spatially limited so that kinship ties are retained. In acorn woodpeckers and long-tailed tits, for example, relatives may disperse together and subsequently cooperate either as helpers or co-breeders when they settle together in a new territory or location (Koenig, *et al.*, 2000; Sharp, *et al.*, 2008). More commonly, if dispersal distances are limited, then kin neighborhoods may result in which close kin may continue to interact cooperatively even though they no longer live in discrete family groups (Dickinson, *et al.*, 2004). Such kin-structured populations are often associated with redirected helping, where most individuals may attempt to breed, but failed breeders may become helpers at the nests of nearby relatives (Emlen, *et al.*, 1992; Dickinson, *et al.*, 1996; MacColl & Hatchwell, 2002). In these social networks of kin, the permissive conditions for the expression of cooperative behavior are often more diverse than in the stable nuclear families described above in which offspring typically assist their parents. Instead, helpers may be offspring, siblings, or parents of the breeders whose brood they care for, or they may be failed breeders, non-breeders, or even "helper-breeders" that breed at their own nest while simultaneously helping elsewhere (Dickinson & Akre, 1998; Nam, *et al.*, 2010; Preston, *et al.*, 2013).

In summary, the distinction between cooperative and non-cooperative species, in terms of their social and population genetic structure, is not clear-cut. Family-living is more widespread than cooperative breeding across the avian phylogeny (Ekman, *et al.*, 2004; Covas & Griesser, 2007), in some cases with retained offspring living on their natal territory but not acting as helpers when the opportunity arises, as in Siberian jays (Ekman, *et al.*, 2001). Similarly, there is substantial overlap in dispersal strategies, with fine-scale kin structure (i.e. kin neighborhoods) occurring in both cooperative and non-cooperative species (Hatchwell, 2010). Understanding the selection pressures that result in cooperation in some species and not others therefore remains a significant challenge.

11.7.2 Group Structure, Breeding Structure and Sex Ratio

As we have already seen, the composition of social groups and the issue of which birds within them can breed vary widely both within and among species. In the great majority of cases, cooperation is facultative, although in a few species it is obligate. In facultative cooperative breeders, groups generally occupy stable territories that persist throughout the year; such year-round residency is consistent with the fact that very few migratory species exhibit cooperative breeding (Arnold & Owens, 1998), although again there are exceptions, including European bee-eaters, *Merops apiaster* (Lessells, 1990) and rainbow bee-eaters, *M. ornatus* (Boland, 2004).

Social group structure is closely linked to reproductive activity, but the degree of cooperation and competition over reproduction is variable to an extent that almost defies generalization, as we discussed earlier. One of the most exciting pieces of work recently emerging from the study of social insects has been the demonstration that the evolution of eusociality originates in groups that express genetic monogamy, which is consistent with the idea that increased relatedness that occurs among offspring where the female does not mate multiply is an important precondition for advanced sociality (Boomsma, 2007; Hughes, *et al.*, 2008; Boomsma, *et al.*, 2011). Comparative analyses have been used to extend the generality of this argument to cooperative breeding in mammals (Lukas & Clutton-Brock, 2013), and birds, where it was argued that low rates of extra-pair paternity increase relatedness among group members and facilitate the evolution of cooperative breeding (Cornwallis, *et al.*, 2010).

While these results promise a common mechanism for the evolutionary origins of sociality, there are several caveats to the application of these ideas to birds, where as we have seen, eusociality has yet to arise. First, in order to calculate an index of infidelity in birds (Cornwallis, *et al.*, 2010), it was necessary to confine the analysis to family groups, which excludes cases where some or all of the group members are unrelated, and hence analyses have an intrinsic bias toward finding family associations to be a precursor of sociality. In those species where family groups are augmented by unrelated immigrants, it is common for parentage to be shared among group members; such species include the carrion crow, *Corvus corone* (Baglione, *et al.*, 2002) and white-browed scrubwren, *Sericornis frontalis* (Whittingham, *et al.*, 1997). Second, there are some spectacular examples of promiscuous mating behavior having profound implications for reproductive success and kinship within avian cooperative breeders. The only species-rich family in which every species is a cooperative breeder (the Maluridae) also includes some of the species which exhibit the highest rates of infidelity, though this is a derived trait within a cooperative clade (Cockburn, *et al.*, 2013). Third, the monogamy hypothesis works well in the social insects, where queen longevity greatly exceeds that of workers, so that workers raise their brothers and sisters. This situation contrasts with the social birds, where helpers typically raise later broods, and so it is their relatedness to those offspring rather than to their brood-mates that matters, and this will be influenced by many factors besides parental infidelity (Kramer & Russell, 2014). For example, because young birds are more likely to be cuckolded than older individuals, high rates of extra-group paternity can actually result in sons being *more* closely related to young in their parent's nest than young in their own nest (Dickinson, *et al.*, 1996; Kramer, *et al.*, 2014).

Living in close proximity to kin of both sexes, as is typical of avian cooperative breeders, also incurs a risk of inbreeding. However, incest appears to be rare in social birds, indicating the existence of effective mechanisms for inbreeding avoidance

(Koenig, *et al.*, 2004). Sex-biased dispersal before the onset of reproduction is one such mechanism common to both cooperative and non-cooperative species (Greenwood, 1980), and the efficacy of this strategy is reinforced by reproductive skew within groups, with reproduction typically monopolized by older individuals within age-related dominance hierarchies (Magrath, *et al.*, 2004). This could result from helpers simply refraining from reproduction, but it can take the form of active eviction of potentially reproductive kin, as in acorn woodpeckers (Koenig, *et al.*, 1998). Thus, a combination of delayed reproduction by subordinates, reproductive suppression by dominants, and promiscuous mating beyond the social group appears to result in avoidance of inbreeding despite the substantial risk in highly kin-structured populations.

The sex ratio of helpers in avian cooperative breeding systems generally reflects the adult sex ratio through its effect on reproductive opportunities and sex-biased dispersal, especially when helpers accrue indirect fitness benefits. First, adult sex ratios in birds tends to be male-biased, a pattern unsurprisingly associated with a tendency for females to experience higher mortality rates than males (Szekely, *et al.*, 2014). Among cooperative breeders, a male-bias among helpers is associated with a strongly male-biased adult sex ratio and hence limited male reproductive opportunities in species including the superb fairy-wren (100 percent male helpers, 1.8:1 M:F adult sex ratio, Cockburn & Double, 2008) and pied kingfisher, *Ceryle rudis* (100 percent male helpers, 1.6 to 1 sex ratio, Reyer, 1990).

In general, however, it appears that it is the identity of the dispersing sex and the timing of dispersal events that determine the sex of helpers. Across a sample of 34 species that breed cooperatively with extensive observations of provisioning data and the sex of the provisioners, the mean sex ratio of helpers was 76 percent male (range: 12 to 100 percent). Just two species in this sample exhibited unequivocal female-biased helping: white-throated magpie jays, *Calocitta formosa* (12 percent male, Langen, 1996; Berg, *et al.*, 2009) and American crows, *Corvus brachyrhynchos* (28 percent male, Caffrey, 1992). In both species, dispersal is male-biased, an unusual pattern among birds (Greenwood, 1980). In the remaining species, the sex ratio of helpers was either more or less even (43 to 67 percent, n = 12 species), a pattern typically associated with helping by juveniles and/or delayed dispersal by both sexes, or it was strongly male-biased (72 to 100 percent, n = 20 species; Green, et al. 2016). In the latter category, males typically delay dispersal (and independent breeding) for longer than females and so have greater opportunity to help. Kin neighborhood cooperators, such as western bluebirds, *Sialia mexicana* (100 percent male helpers, Dickinson, *et al.*, 1996), long-tailed tits (86 percent, Hatchwell, *et al.*, 2004) and *Climacteris* treecreepers (77 to 97 percent, Luck, 2001; Doerr & Doerr, 2006), also fall within this strongly male-biased helper category, a finding that is consistent with the fact that in these species kin neighborhoods arise via limited dispersal of males rather than females.

Cooperatively breeding birds have also proved a popular testing ground for hypotheses regarding the production of biased offspring sex ratios. Specifically, the local resource enhancement hypothesis predicts that females should bias sex ratio towards the helping (philopatric) sex, while the local resource competition hypothesis predicts that females should bias production towards the dispersing sex. The evidence in support of these alternative hypotheses is equivocal (Koenig & Walters, 1999; Komdeur, 2004; West, *et al.*, 2005). There is compelling evidence from several studies for facultative sex ratio adjustment (Bednarz & Hayden, 1991; Komdeur, *et al.*, 1997; Dickinson, 2004); however, population sex-ratio patterns predicted by either the local resource enhancement or the local resource competition hypotheses have not been consistently observed (Cockburn, *et al.*, 2008; Koenig & Dickinson, 1996), often despite large sample sizes derived from long-term studies.

III SOCIAL SYNTHESIS

11.8 A Summary of Avian Sociality

As we have seen in the review, there are a variety of hypotheses that attempt to deal with the repeated transitions from pair to cooperative breeding and vice versa, and growing interest in how the diversity of avian cooperative breeding systems emerge. The numerous evolutionary transitions to and from sociality within birds stand in stark contrast to other vertebrate taxa, making comparative analyses and attempts to develop predictive hypotheses attractive. This opportunity is enriched because the natural history on birds, despite numerous deficiencies, dwarf what is available for most other taxa. But, has all of our effort delivered the promised rewards? "Yes" in the sense that many authors have identified factors promoting the transition to cooperation within their species, but an emphatic "no" when it comes to the question of why some avian lineages exhibit sociality and cooperative breeding while others do not.

There are several reasons for this lack of success. Cooperative breeding is often treated as a unitary phenomenon, whereas, as we have seen, there are several very different phenomena to be explained. While we have an outline of an explanation for the decline of cooperative breeding in the Holarctic, there are no predictive models that explain its distribution among families and biogeographic areas in the tropics and southern temperate regions beyond the observed phylogenetic bias, which has descriptive value but is uninformative unless we can predict why cooperative breeding proliferates in some clades but not in others with similar ecology.

11.9 Comparative Perspectives on Avian Sociality

Much of the conceptual/theoretical framework for thinking about the evolution of cooperative breeding in vertebrates has resulted from studies of avian systems. However, ornithologists have been receptive to ideas developed through studies of social insects, particularly in the areas of sex allocation theory, mating patterns, and reproductive skew, although the advances made in this regard are limited because of the fundamental differences in control over sex allocation. Moreover, ornithologists have been less effective in studying the proximate mechanisms underlying cooperation and conflict than students of social insects, mammals, and fishes. This is in part because it is harder to conduct controlled experiments on key parameters in birds. For example, the key role of philopatry is impossible to study in captive populations, and it is difficult to have aviary configurations that allow more than one group to interact. In the wild, it is often difficult to conduct experiments that do not simultaneously manipulate multiple parameters. For example, experimental removal of helpers simultaneously manipulates the benefits of group living and the ability of the helper to rear young, thereby triggering cascading effects that are almost impossible to interpret (Dunn & Cockburn, 1996; Jamieson & Quinn, 1997). A further complication is the desirability of not experimentally manipulating populations that have been followed for many years, since such work can interfere with long-term studies.

One of the field tools that has been put to good use by mammalogists has been habituation of study groups to the presence of the investigator, which has allowed highly informative short-term measures of fitness such as condition and foraging success. With the (very successful) exception of workers studying *Turdoides* babblers (Ridley & Raihani, 2007a, 2007b; Bell, *et al.*, 2009), this technique has not been applied widely, and as a result ornithologists lag behind in understanding individual variation in state and the consequences for decision-making. Ultimately, marrying short-term and longer-term perspectives of fitness will be needed for progress in these areas.

Finally, in some respects, primatologists have adopted a research approach that is most similar to that of ornithologists, often conducting field studies that last many decades. However, there has been a tendency for researchers in the two disciplines to develop non-overlapping lexicons. Furthermore, while primatologists have often had a strong focus on comparative analysis, ornithologists have made more concerted efforts to engage with measures of the fitness consequences of cooperation, thereby testing social evolution theory. More recently, studies of other social mammals, particularly mongooses, have followed a research agenda closer to that of ornithologists. Such studies facilitate comparisons across taxa and promise a greater understanding of the similarities, as well as the distinctions, between them.

11.10 Concluding Remarks

The high incidence of cooperative breeding in birds, relative to other vertebrate taxa, is probably attributable to the prevalence of pair-bonding – itself a form of sociality – and biparental care in the taxon. As a consequence, cooperative breeding has evolved numerous times in multiple lineages, facilitating analysis of its evolutionary origins. The dominant research paradigm in the study of avian cooperative breeding systems is to use long-term studies of marked individuals, an approach that has resulted in many meticulously detailed studies of the direct and indirect fitness consequences of helping behavior and of the conflict inherent within such systems. Such work provides some of the strongest evidence supporting a key role for kin selection in the evolution of social behavior and has done much to further our understanding of the complexities inherent in social evolution. Nevertheless, despite detailed understanding of the causes and consequences of cooperation within species, many questions remain to be answered concerning the evolution of behavior in these intensely interesting and complex species.

References

Alexander, R. D. (1974) The evolution of social behavior. *Annual Review of Ecology and Systematics*, **5**, 325–383.

Arnold, K. E. & Owens, I. P. F. (1998) Cooperative breeding in birds: A comparative test of the life history hypothesis. *Proceedings of the Royal Society B*, **265**, 739–745.

(1999) Cooperative breeding in birds: The role of ecology. *Behavioral Ecology*, **10**, 465–471.

Baglione, V., Marcos, J. M., Canestrari, D., & Ekman, J. (2002) Direct fitness benefits of group living in a complex cooperative society of carrion crows, *Corvus corone corone*. *Animal Behaviour*, **64**, 887–893.

Beauchamp, G. (1999) The evolution of communal roosting in birds: Origin and secondary losses. *Behavioral Ecology*, **10**, 675–687.

(2014) Do avian cooperative breeders live longer? *Proceedings of the Royal Society B*, **281**: 20140844.

Bednarz, J. C. & Hayden, T. J. (1991) Skewed brood sex-ratio and sex-biased hatching sequence in Harris's hawks. *The American Naturalist*, **137**, 116–132.

Bednekoff, P. A. & Woolfenden, G. E. (2003) Florida scrub-jays (*Aphelocoma coerulescens*) are sentinels more when well-fed (even with no kin nearby). *Ethology*, **109**, 895–903.

Bell, H. L. (1985) The social organization and foraging behaviour of three syntopic thornbills *Acanthiza* spp. *In:* Keast, A., Recher, H. F., Ford, H., & Saunders, D. (eds.) *Birds of Eucalypt Forests and Woodlands: Ecology, Conservation, Management.* Sydney: RAOU and Surrey Beatty & Sons, pp. 151–163.

Bell, M. B. V., Radford, A. N., Rose, R., Wade, H. M., & Ridley, A. R. (2009) The value of constant surveillance in a risky environment. *Proceedings of the Royal Society B*, **276**, 2997–3005.

Berg, E. C., Eadie, J. M., Langen, T. A., & Russell; A. F. (2009) Reverse sex-biased philopatry in a cooperative bird: Genetic consequences and a social cause. *Molecular Ecology*, **18**, 3486–3499.

Berg, E. C., Aldredge, R. A., Peterson, A. T., & McCormack, J. E. (2012) New phylogenetic information suggests both an increase and at least one loss of cooperative breeding during the evolutionary history of *Aphelocoma* jays. *Evolutionary Ecology*, **26**, 43–54.

Bertram, B. C. R. (1978) Living in groups: Predators and prey. *In:* Krebs, J. R. & Davies, N. B. (eds.) *Behavioral Ecology: An Evolutionary Approach.* Oxford: Blackwell Scientific Publications, pp. 23–63.

Blumstein, D. T. & Møller, A. P. (2008) Is sociality associated with high longevity in North American birds? *Biology Letters*, **4**, 146–148.

Boland, C. R. J. (2004) Breeding biology of rainbow bee-eaters (*Merops ornatus*): A migratory, colonial, cooperative bird. *Auk*, **121**, 811–823.

Boland, C. R. J., Heinsohn, R., & Cockburn, A. (1997a) Experimental manipulation of brood reduction and parental care in cooperatively breeding white-winged choughs. *Journal of Animal Ecology*, **66**, 683–691.

(1997b) Deception by helpers in cooperatively breeding white-winged choughs and its experimental manipulation. *Behavioral Ecology and Sociobiology*, **41**, 251–256.

Boomsma, J. J. (2007) Kin selection versus sexual selection: Why the ends do not meet. *Current Biology*, **17**, R673-R683.

Boomsma, J. J., Beekman, M., Cornwallis, C. K., *et al.* (2011) Only full-sibling families evolved eusociality. *Nature*, **471**, E4-E5.

Brown, J. L. (1987) *Helping and Communal Breeding in Birds: Ecology and Evolution*. Princeton: Princeton University Press.

Brown, J. L. & Brown, E. R. (1990) Mexican jays: Uncooperative breeding. *In:* Stacey, P. B. & Koenig, W. D. (eds.) *Cooperative Breeding in Birds: Long-Term Studies of Ecology and Behavior*. Cambridge: Cambridge University Press, pp. 267–288.

Burkart, J. M., Hrdy, S. B., & Van Schaik, C. P. (2009) Cooperative breeding and human cognitive evolution. *Evolutionary Anthropology*, **18**, 175–186.

Caffrey, C. (1992) Female-biased delayed dispersal and helping in American crows. *Auk*, **109**, 609–619.

Chalfoun, A. D. & Martin, T. E. (2007) Latitudinal variation in avian incubation attentiveness and a test of the food limitation hypothesis. *Animal Behaviour*, **73**, 579–585.

Chao, L. (1997) Evolution of polyandry in a communal breeding system. *Behavioral Ecology*, **8**, 668–674.

Chaplin, S. B. (1982) The energetic significance of huddling behavior in common bushtits (*Psaltriparus minimus*). *Auk*, **99**, 424–430.

Cockburn, A. (1996) Why do so many Australian birds cooperate? Social evolution in the Corvida *In:* Floyd, R. B., Sheppard, A. W., & De Barro, P. J. (eds.) *Frontiers of Population Ecology*. East Melbourne: CSIRO, pp. 451–472.

(1998) Evolution of helping behaviour in cooperatively breeding birds. *Annual Review of Ecology and Systematics*, **29**, 141–177.

(2003) Cooperative breeding in oscine passerines: Does sociality inhibit speciation? *Proceedings of the Royal Society of London Series B*, **270**, 2207–2214.

(2004) Mating systems and sexual conflict. *In:* Koenig, W. D. & Dickinson, J. L. (eds.) *Ecology and Evolution of Cooperative Breeding in Birds*. Cambridge: Cambridge University Press, pp. 81–101.

(2006) Prevalence of different modes of parental care in birds. *Proceedings of the Royal Society B*, **273**, 1375–1383.

(2010) Oh sibling, who art thou? *Nature*, **466**, 930–931.

(2013) Cooperative breeding in birds: Towards a richer conceptual framework. *In:* Sterelny, K., Joyce, R., Calcott, B., & Fraser, B. (eds.) *Cooperation and Its Evolution*. Cambridge, MA: MIT Press, pp. 223–245.

Cockburn, A. & Double, M. C. (2008) Cooperatively breeding superb fairy-wrens show no facultative manipulation of offspring sex ratio despite plausible benefits. *Behavioral Ecology and Sociobiology*, **62**, 681–688.

Cockburn, A. & Russell, A. F. (2011) Cooperative breeding: A question of climate? *Current Biology*, **21**, R195–R197.

Cockburn, A., Sims, R. A., Osmond, H. L., Green, D. J., Double, M. C., & Mulder, R. A. (2008) Can we measure the benefits of help in cooperatively breeding birds? The case of superb fairy-wrens *Malurus cyaneus*, *Journal of Animal Ecology*, **77**, 430–438.

Cockburn, A., Dalziell, A. H., Blackmore, C. J., *et al.* (2009) Superb fairy-wren males aggregate into hidden leks to solicit extragroup fertilizations before dawn. *Behavioral Ecology*, **20**, 501–510.

Cockburn, A., Brouwer, L., Double, M., Margraf, N., & van de Pol, M. (2013) Evolutionary origins and persistence of infidelity in *Malurus*: The least faithful birds. *Emu*, **113**, 208–217.

Cockburn, A., Brouwer, L., Margraf, N., Osmond, H. L., & van de Pol, M. (2016) Superb fairy-wrens: making the worst of a good job. *In:* Koenig, W. D. & Dickinson, J. L. (eds.) *Cooperative Breeding in Vertebrates: Studies of Ecology, Evolution, and Behavior.* Cambridge: Cambridge University Press, pp. 133–149.

Conrad, K. F., Clarke, M. F., Robertson, R. J., & Boag, P. T. (1998) Paternity and the relatedness of helpers in the cooperatively breeding bell miner. *Condor*, **100**, 343–349.

Cornwallis, C. K., West, S. A., & Griffin, A. S. (2009) Routes to indirect fitness in cooperatively breeding vertebrates: Kin discrimination and limited dispersal. *Journal of Evolutionary Biology*, **22**, 2445–2457.

Cornwallis, C. K., West, S. A., Davis, K. E., & Griffin, A. S. (2010) Promiscuity and the evolutionary transition to complex societies. *Nature*, **466**, 969–972.

Covas, R. (2012) Evolution of reproductive life histories in island birds worldwide. *Proceedings of the Royal Society B*, **279**, 1531–1537.

Covas, R. & Griesser, M. (2007) Life history and the evolution of family living in birds. *Proceedings of the Royal Society B*, **274**, 1349–1357.

Covas, R., Doutrelant, C., & du Plessis, M. A. (2004) Experimental evidence of a link between breeding conditions and the decision to breed or to help in a colonial cooperative bird. *Proceedings of the Royal Society B*, **271**, 827–832.

Crane, J. M. S., Pick, J. L., Tribe, A. J., Vincze, E., Hatchwell, B. J., & Russell, A. F. (2015) Chestnut-crowned babblers show affinity for calls of removed group members: A dual playback without expectancy violation. *Animal Behaviour*, **104**, 51–57.

Daniels, S. J. & Walters, J. R. (2000) Inbreeding depression and its effects on natal dispersal in Red-cockaded Woodpeckers. *Condor*, **102**, 482–491.

Davies, N. B. (1992) *Dunnock Behaviour and Social Evolution*, Oxford: Oxford University Press.

Davies, N. B., Hartley, I. R., Hatchwell, B. J., Desrochers, A., Skeer, J., & Nebel, D. (1995) The polygynandrous mating system of the alpine accentor, *Prunella collaris*. 1. Ecological causes and reproductive conflicts. *Animal Behaviour*, **49**, 769–788.

Dickinson, J. L. (2004) Facultative sex ratio adjustment by western bluebird mothers with stay-at-home helpers-at-the-nest. *Animal Behaviour*, **68**, 373–380.

Dickinson, J. L. & Akre, J. J. (1998) Extrapair paternity, inclusive fitness, and within-group benefits of helping in western bluebirds. *Molecular Ecology*, **7**, 95–105.

Dickinson, J. L. & Hatchwell, B. J. (2004) Fitness consequences of helping. *In:* Koenig, W. D., & Dickinson, J. L. (eds.) *Ecology and Evolution of Cooperative Breeding in Birds*. Cambridge: Cambridge University Press, pp. 48–66.

Dickinson, J. L., Koenig, W. D., & Pitelka, F. A. (1996) Fitness consequences of helping behavior in the western bluebird. *Behavioral Ecology*, **7**, 168–177.

Doerr, E. D. & Doerr, V. A. J. (2006) Comparative demography of treecreepers: Evaluating hypotheses for the evolution and maintenance of cooperative breeding. *Animal Behaviour*, **72**, 147–159.

Double, M. C. & Cockburn, A. (2003) Subordinate superb fairy-wrens (*Malurus cyaneus*) parasitize the reproductive success of attractive dominant males. *Proceedings of the Royal Society of London Series B*, **270**, 379–384.

Dunbar, R. I. M. & Shultz, S. (2007) Evolution in the social brain. *Science*, **317**, 1344–1347.

Dunn, P. O. & Cockburn, A. (1996) Evolution of male parental care in a bird with almost complete cuckoldry. *Evolution*, **50**, 2542–2548.

du Plessis, M. A. (2004) Physiological ecology. *In:* Koenig, W. D., & Dickinson, J. L. (eds.) *Ecology and Evolution of Cooperative Breeding in Birds*. Cambridge: Cambridge University Press, pp. 117–127.

du Plessis, M. A. & Williams, J. B. (1994) Communal cavity roosting in green woodhoopoes: Consequences for energy expenditure and the seasonal pattern of mortality. *Auk*, **111**, 292–299.

du Plessis, M. A., Weathers, W. W., & Koenig, W. D. (1994) Energetic benefits of communal roosting by acorn woodpeckers during the nonbreeding season. *Condor*, **96**, 631–637.

du Plessis, M. A., Siegfried, W. R., & Armstrong, A. J. (1995) Ecological and life-history correlates of cooperative breeding in South African birds. *Oecologia*, **102**, 180–188.

Eikenaar, C., Richardson, D. S., Brouwer, L., & Komdeur, J. (2008a) Sex biased natal dispersal in a closed, saturated population of Seychelles warblers *Acrocephalus sechellensis*. *Journal of Avian Biology*, **39**, 73–80.

Eikenaar, C., Komdeur, J., & Richardson, D. S. (2008b) Natal dispersal patterns are not associated with inbreeding avoidance in the Seychelles warbler. *Journal of Evolutionary Biology*, **21**, 1106–1116.

Eikenaar, C., Brouwer, L., Komdeur, J., & Richardson, D. S. (2010) Sex biased natal dispersal is not a fixed trait in a stable population of Seychelles warblers. *Behaviour*, **147**, 1577–1590.

Ekman, J. & Ericson, P. G. P. (2006) Out of Gondwanaland; The evolutionary history of cooperative breeding and social behaviour among crows, magpies, jays, and allies. *Proceedings of the Royal Society B*, **273**, 1117–1125.

Ekman, J., Bylin, A., & Tegelström, H. (2000) Parental nepotism enhances survival of retained offspring in the Siberian jay. *Behavioral Ecology*, **11**, 416–420.

Ekman, J., Eggers, S., Griesser, M., & Tegelström, H. (2001) Queuing for preferred territories: Delayed dispersal of Siberian jays. *Journal of Animal Ecology*, **70**, 317–324.

Ekman, J., Dickinson, J. L., Hatchwell, B. J., & Griesser, M. (2004) Delayed dispersal. *In:* Koenig, W. D. & Dickinson, J. L. (eds.) *Ecology and Evolution of Cooperative Breeding in Birds*. Cambridge: Cambridge University Press, pp. 35–47.

Emlen, S. T. (1982) The evolution of helping. I. An ecological constraints model. *The American Naturalist*, **119**, 29–39.

(1994) Benefits, constraints and the evolution of the family. *Trends in Ecology and Evolution*, **9**, 282–285.

Emlen, S. T. & Wrege, P. H. (1992) Parent–offspring conflict and the recruitment of helpers among bee-eaters. *Nature*, **356**, 331–333.

Faaborg, J. & Bednarz, J. C. (1990) Galápagos and Harris' hawks: Divergent causes of sociality in two raptors. *In:* Stacey, P. B. & Koenig, W. D. (eds.) *Cooperative Breeding in Birds: Long-Term Studies of Ecology and Behavior*. Cambridge: Cambridge University Press, pp. 359–383.

Faaborg, J., Parker, P. G., DeLay, L., *et al.* (1995) Confirmation of cooperative polyandry in the Galápagos hawk (*Buteo galapagoensis*). *Behavioral Ecology and Sociobiology*, **36**, 83–90.

Feeney, W. E. & Langmore, N. E. (2015) Superb Fairy-wrens (*Malurus cyaneus*) increase vigilance near their nest with the perceived risk of brood parasitism. *Auk*, **132**, 359–364.

Feeney, W. E., Medina, I., Somveille, M., *et al.* (2013) Brood parasitism and the evolution of cooperative breeding in birds. *Science*, **342**, 1506–1508.

Ford, H. A., Bell, H., Nias, R., & Noske, R. (1988) The relationship between ecology and the incidence of cooperative breeding in Australian birds. *Behavioral Ecology and Sociobiology*, **22**, 239–249.

Fraga, R. M. (1991) The social system of a communal breeder, the bay-winged cowbird *Molothrus badius*. *Ethology*, **89**, 195–210.

Gaston, A. J. (1978) The evolution of group territorial behavior and cooperative breeding. *The American Naturalist*, **112**, 1091–1100.

Gatter, W. & Gardner, R. (1993) The biology of the Gola malimbe *Malimbus ballmanni* Wolters 1974. *Bird Conservation International*, **3**, 87–103.

Gill, F. & Donsker, D. (eds.) (2014) *IOC World Bird List (v.4.3)*.

Gonzalez, J.-C. T., Sheldon, B. C., & Tobias, J. A. (2013) Environmental stability and the evolution of cooperative breeding in hornbills. *Proceedings of the Royal Society B*, **280**, 20131297.

Green, J. P., Freckleton, R. P., & Hatchwell, B. J. (2016). Variation in helper investment among cooperatively breeding bird species is consistent with Hamilton's rule. *Nature Communications*, 10.1038/ncomms12663.

Greenwood, P. J. (1980) Mating systems, philopatry and dispersal in birds and mammals. *Animal Behavior*, **28**, 1140–1162.

Grimes, L. G. (1980) Observations of group behaviour and breeding biology of the yellow-billed shrike *Corvinella corvina*. *Ibis*, **122**, 166–192.

Hall, M. L. (2009) A review of vocal duetting in birds. *Advances in the Study of Behavior*, Vol. 40. pp. 67–121.

Hamilton, W. D. (1964) The genetical evolution of social behaviour. *I. Journal of Theoretical Biology*, **7**, 1–16.

Härdling, R. & Kokko, H. (2003) Life-history traits as causes or consequences of social behaviour: Why do cooperative breeders lay small clutches? *Evolutionary Ecology Research*, **5**, 691–700.

Hatchwell, B. J. (1999) Investment strategies of breeders in avian cooperative breeding systems. *The American Naturalist*, **154**, 205–219.

(2009) The evolution of cooperative breeding in birds: Kinship, dispersal and life history. *Philosophical Transactions of the Royal Society B*, **364**, 3217–3227.

(2010) Cryptic kin selection: Kin structure in vertebrate populations and opportunities for kin-directed cooperation. *Ethology*, **116**, 203–216.

Hatchwell, B. J. & Sharp, S. P. (2006) Kin selection, constraints, and the evolution of cooperative breeding in long-tailed tits. *Advances in the Study of Behavior*, **36**, 355–395.

Hatchwell, B. J., Russell, A. F., MacColl, A. D. C., Ross, D. J., Fowlie, M. K., & McGowan, A. (2004) Helpers increase long-term but not short-term productivity in cooperatively breeding long-tailed tits. *Behavioral Ecology*, **15**, 1–10.

Hatchwell, B. J., Sharp, S. P., Simeoni, M., & McGowan, A. (2009) Factors influencing overnight loss of body mass in the communal roosts of a social bird. *Functional Ecology*, **23**, 367–372.

Haydock, J. & Koenig, W. D. (2002) Reproductive skew in the polygynandrous acorn woodpecker. *Proceedings of the National Academy of Sciences USA*, **99**, 7178–7183.

Heinsohn, R. (2008) The ecological basis of unusual sex roles in reverse-dichromatic eclectus parrots. *Animal Behaviour*, **76**, 97–103.

Heinsohn, R. & Legge, S. (2003) Breeding biology of the reverse-dichromatic, co-operative parrot *Eclectus roratus*. *Journal of Zoology*, **259**, 197–208.

Heinsohn, R., Ebert, D., Legge, S., & Peakall, R. (2007) Genetic evidence for cooperative polyandry in reverse dichromatic *Eclectus* parrots. *Animal Behaviour*, **74**, 1047–1054.

Heinsohn, R. G. (1991) Slow learning of foraging skills and extended parental care in cooperatively breeding white-winged choughs. *The American Naturalist*, **137**, 864–881.

(1992) Cooperative enhancement of reproductive success in white-winged choughs. *Evolutionary Ecology*, **6**, 97–114.

(2004) Parental care, load-lightening, and costs. *In:* Koenig, W. D. & Dickinson, J. L. (eds.) *Ecology and Evolution of Cooperative Breeding in Birds*. Cambridge: Cambridge University Press, pp. 67–80.

Heinsohn, R. G., Cockburn, A., & Cunningham, R. B. (1988) Foraging, delayed maturation, and advantages of cooperative breeding in white-winged choughs, *Corcorax melanorhamphos*. *Ethology*, **77**, 177–186.

Hughes, J. M., Mather, P. B., Toon, A., Ma, J., Rowley, I., & Russell, E. (2003) High levels of extra-group paternity in a population of Australian magpies *Gymnorhina tibicen*: Evidence from microsatellite analysis. *Molecular Ecology*, **12**, 3441–3450.

Hughes, W. O. H., Oldroyd, B. P., Beekman, M., & Ratnieks, F. L. W. (2008) Ancestral monogamy shows kin selection is key to the evolution of eusociality. *Science*, **320**, 1213–1216.

Humphrey, N. K. (1976) The social function of intellect. *In:* Bateson, P. P. G. & Hinde, R. A. (eds.) *Growing Points in Ethology*. Cambridge: Cambridge University Press, pp. 303–317.

Jamieson, I. G. (1991) The unselected hypothesis for the evolution of helping behavior: Too much or too little emphasis on natural selection? *The American Naturalist*, **138**, 271–282.

Jamieson, I. G. & Craig, J. L. (1987) Critique of helping behavior in birds: A departure from functional explanations. *In:* Bateson, P. P. G. & Klopfer, P. (eds.) *Perspectives in Ethology*. Vol. 7. New York: Plenum, pp. 79–98.

Jamieson, I. G. & Quinn, J. S. (1997) Problems with removal experiments designed to test the relationship between paternity and parental effort in a socially polyandrous bird. *Auk*, **114**, 291–295.

Jarvis, E. D., Mirarab, S., Aberer, A. J., *et al.* (2014) Whole-genome analyses resolve early branches in the tree of life of modern birds. *Science*, **346**, 1320–1331.

Jetz, W. & Rubenstein, D. R. (2011) Environmental uncertainty and the global biogeography of cooperative breeding in birds. *Current Biology*, **21**, 72–78.

Jing, Y., Fang, Y., Strickland, D., Lu, N., & Sun, Y.-H. (2009) Alloparenting in the rare Sichuan jay (*Perisoreus internigrans*). *Condor*, **111**, 662–667.

Keen, S. C., Meliza, C. D., & Rubenstein, D. R. (2013) Flight calls signal group and individual identity but not kinship in a cooperatively breeding bird. *Behavioral Ecology*, **24**, 1279–1285.

Kemp, A. C. (1995) *The Hornbills*, Oxford: Oxford University Press.

Kesler, D. C., Walters, J. R., & Kappes, J. J., Jr. (2010) Social influences on dispersal and the fat-tailed dispersal distribution in red-cockaded woodpeckers. *Behavioral Ecology*, **21**, 1337–1343.

Khan, M. Z., McNabb, F. M. A., Walters, J. R., & Sharp, P. J. (2001) Patterns of testosterone and prolactin concentrations and reproductive behavior of helpers and breeders in the cooperatively breeding red-cockaded woodpecker (*Picoides borealis*). *Hormones and Behavior*, **40**, 1–13.

Koenig, W., D. & Dickinson, J. L. (eds.) (2004) *Ecology and Evolution of Cooperative Breeding in Birds*, Cambridge: Cambridge University Press.

(2016) *Cooperative Breeding in Vertebrates: Studies of Ecology, Evolution and Behavior*. Cambridge: Cambridge University Press.

Koenig, W. D. & Dickinson, J. L. (1996) Nestling sex-ratio variation in western bluebirds. *Auk*, **113**, 902–910.

Koenig, W. D. & Haydock, J. (2004) Incest and incest avoidance. *In:* Koenig, W. D. & Dickinson, J. L. (eds.) *Ecology and Evolution of Cooperative Breeding in Birds*. Cambridge: Cambridge University Press, pp. 142–156.

Koenig, W. D. & Walters, J. R. (1999) Sex-ratio selection in species with helpers at the nest: The repayment model revisited. *The American Naturalist*, **153**, 124–130.

Koenig, W. D., Pitelka, F. A., Carmen, W. J., Mumme, R. L., & Stanback, M. T. (1992) The evolution of delayed dispersal in cooperative breeders. *Quarterly Review of Biology*, **67**, 111–150.

Koenig, W. D., Mumme, R. L., Stanback, M. T. & Pitelka, F. A. (1995) Patterns and consequences of egg destruction among joint-nesting acorn woodpeckers. *Animal Behaviour*, **50**, 607–621.

Koenig, W. D., Haydock, J., & Stanback, M. T. (1998) Reproductive roles in the cooperatively breeding acorn woodpecker: Incest avoidance versus reproductive competition. *The American Naturalist*, **151**, 243–255.

Koenig, W. D., Stanback, M. T., & Haydock, J. (1999) Demographic consequences of incest avoidance in the cooperatively breeding acorn woodpecker. *Animal Behaviour*, **57**, 1287–1293.

Koenig, W. D., Hooge, P. N., Stanback, M. T., & Haydock, J. (2000) Natal dispersal in the cooperatively breeding acorn woodpecker. *Condor*, **102**, 492–502.

Kokko, H. & Ekman, J. (2002) Delayed dispersal as a route to breeding: Territorial inheritance, safe havens, and ecological constraints. *The American Naturalist*, **160**, 468–484.

Kokko, H. & Lundberg, P. (2001) Dispersal, migration, and offspring retention in saturated habitats. *The American Naturalist*, **157**, 188–202.

Komdeur, J. (1992) Importance of habitat saturation and territory quality for evolution of cooperative breeding in the Seychelles warbler. *Nature*, **358**, 493–495.

(2004) Sex-ratio manipulation. *In:* Koenig, W. D. & Dickinson, J. L. (eds.) *Ecology and Evolution of Cooperative Breeding in Birds*. Cambridge: Cambridge University Press, pp. 102–116.

Komdeur, J., Huffstadt, A., Prast, W., Castle, G., Mileto, R. & Wattel, J. (1995) Transfer experiments of Seychelles warblers to new islands: Changes in dispersal and helping behaviour. *Animal Behaviour*, **49**, 695–708.

Komdeur, J., Daan, S., Tinbergen, J., & Mateman, C. (1997) Extreme adaptive modification in sex ratio of the Seychelles warbler's eggs. *Nature*, **385**, 522–525.

Kramer, K. L. & Russell, A. F. (2014) Kin-selected cooperation without lifetime monogamy: Human insights and animal implications. *Trends in Ecology and Evolution*, **29**, 600–606.

Lack, D. (1968) *Ecological Adaptations for Breeding in Birds*, London: Chapman & Hall.

Langen, T. A. (1996) The mating system of the white-throated magpie-jay *Calocitta formosa* and Greenwood's hypothesis for sex-biased dispersal. *Ibis*, **138**, 506–513.

Langmore, N. E. & Davies, N. B. (1997) Female dunnocks use vocalizations to compete for males. *Animal Behaviour*, **53**, 881–890.

Langmore, N. E., Davies, N. B., Hatchwell, B. J., & Hartley, I. R. (1996) Female song attracts males in the alpine accentor *Prunella collaris*. *Proceedings of the Royal Society B*, **263**, 141–146.

Legge, S. (2000) Helper contributions in the cooperatively breeding laughing kookaburra: Feeding young is no laughing matter. *Animal Behaviour*, **59**, 1009–1018.

Lessells, C. M. (1990) Helping at the nest in European bee-eaters: Who helps and why? *In:* Blondel, J., Gosler, A., Lebreton, J.-D., & McLeery, R. (eds.) *Population Biology of Passereine Birds: An Integrated Approach*. Berlin: Springer-Verlag, pp. 357–368.

Li, S.-H. & Brown, J. L. (2000) High frequency of extrapair fertilization in a plural breeding bird, the Mexican jay, revealed by DNA microsatellites. *Animal Behaviour*, **60**, 867–877.

Ligon, J. D. & Burt, D. B. (2004) Evolutionary origins. *In:* Koenig, W. D. & Dickinson, J. L. (eds.) *Ecology and Evolution of Cooperative Breeding in Birds*. Cambridge: Cambridge University Press, pp. 5–34.

Ligon, J. D., Carey, C., & Ligon, S. H. (1988) Cavity roosting, philopatry, and cooperative breeding in the green woodhoopoe may reflect a physiological trait. *Auk*, **105**, 123–127.

Lloyd, P., Abadi, F., Altwegg, R., & Martin, T. E. (2014) South temperate birds have higher apparent adult survival than tropical birds in Africa. *Journal of Avian Biology*, **45**, 493–500.

Logue, D. M. & Hall, M. L. (2014) Migration and the evolution of duetting in songbirds. *Proceedings of the Royal Society B*, **281**, 20140103.

Lovette, I. J. & Rubenstein, D. R. (2007) A comprehensive molecular phylogeny of the starlings (Aves: Sturnidae) and mockingbirds (Aves: Mimidae): Congruent mtDNA and nuclear trees for a cosmopolitan avian radiation. *Molecular Phylogenetics and Evolution*, **44**, 1031–1056.

Luck, G. W. (2001) The demography and cooperative breeding behaviour of the rufous treecreeper, *Climacteris rufa*. *Australian Journal of Zoology*, **49**, 515–537.

Lukas, D. & Clutton-Brock, T. H. (2013) The evolution of social monogamy in mammals. *Science*, **341**, 526–530.

Mabry, K. E., Shelley, E. L., Davis, K. E., Blumstein, D. T., & Van Vuren, D. H. (2013) Social mating system and sex-biased dispersal in mammals and birds: A phylogenetic analysis. *PLoS ONE*, **8**, e57980.

MacColl, A. D. C. & Hatchwell, B. J. (2002) Temporal variation in fitness payoffs promotes cooperative breeding in long-tailed tits *Aegithalos caudatus*. *The American Naturalist*, **160**, 186–194.

Magrath, R. D., Johnstone, R. A., & Heinsohn, R. G. (2004) Reproductive skew. *In:* Koenig, W. D. & Dickinson, J. L. (eds.) *Ecology and Evolution of Cooperative Breeding in Birds*. Cambridge: Cambridge University Press, pp. 157–176.

Marki, P. Z., Fabre, P.-H., Jønsson, K. A., Rahbek, C., Fjeldså, J., & Kennedy, J. D. (2015) Breeding system evolution influenced the geographic expansion and diversification of the core Corvoidea (Aves: Passeriformes). *Evolution*, **69**, 1874–1924.

Martin, T. E. (2002) A new view of avian life-history evolution tested on an incubation paradox. *Proceedings of the Royal Society B*, **269**, 309–316.

Martin, T. E., Auer, S. K., Bassar, R. D., Niklison, A. M., & Lloyd, P. (2007) Geographic variation in avian incubation periods and parental influences on embryonic temperature. *Evolution*, **61**, 2558–2569.

McDonald, P. G., Kazem, A. J. N., & Wright, J. (2007) A critical analysis of 'false-feeding' behavior in a cooperatively breeding bird: Disturbance effects, satiated nestlings or deception? *Behavioral Ecology and Sociobiology*, **61**, 1623–1635.

McGowan, A., Sharp, S. P., Simeoni, M., & Hatchwell, B. J. (2006) Competing for position in the communal roosts of long-tailed tits. *Animal Behaviour*, **72**, 1035–1043.

McKechnie, A. E. & Lovegrove, B. G. (2001a) Heterothermic responses in the speckled mousebird (*Colius striatus*). *Journal of Comparative Physiology B*, **171**, 507–518.

(2001b) Thermoregulation and the energetic significance of clustering behavior in the white-backed mousebird (*Colius colius*). *Physiological and Biochemical Zoology*, **74**, 238–249.

Mulder, R. A., Dunn, P. O., Cockburn, A., Lazenby-Cohen, K. A., & Howell, M. J. (1994) Helpers liberate female fairy-wrens from constraints on extra-pair mate choice. *Proceedings of the Royal Society of London B*, **255**, 223–229.

Mumme, R. L. (1992) Do helpers increase reproductive success? An experimental analysis in the Florida scrub jay. *Behavioral Ecology and Sociobiology*, **31**, 319–328.

Nam, K.-B., Simeoni, M., Sharp, S. P., & Hatchwell, B. J. (2010) Kinship affects investment by helpers in a cooperatively breeding bird. *Proceedings of the Royal Society B*, **277**, 3299–3306.

Napper, C. J., Sharp, S. P., McGowan, A., Simeoni, M., & Hatchwell, B. J. (2013) Dominance, not kinship, determines individual position within the communal roosts of a cooperatively breeding bird. *Behavioral Ecology and Sociobiology*, **67**, 2029–2039.

Nicholls, J. A., Double, M. C., Rowell, D. M., & Magrath, R. D. (2000) The evolution of cooperative and pair breeding in thornbills *Acanthiza* (Pardalotidae). *Journal of Avian Biology*, **31**, 165–176.

Odom, K. J., Omland, K. E., & Price, J. J. (2015) Differentiating the evolution of female song and male-female duets in the New World blackbirds: Can tropical natural history traits explain duet evolution? *Evolution*, **69**, 839–847.

Owens, I. P. F. & Bennett, P. M. (1995) Ancient ecological diversification explains life-history variation among living birds. *Proceedings of the Royal Society B*, **261**, 227–232.

Pen, I. & Weissing, F. J. (2000) Towards a unified theory of cooperative breeding: The role of ecology and life history re-examined. *Proceedings of the Royal Society B*, **267**, 2411–2418.

Pilowsky, J. A. & Rubenstein, D. R. (2013) Social context and the lack of sexual dimorphism in song in an avian cooperative breeder. *Animal Behaviour*, **85**, 709–714.

Piper, W. H., Parker, P. G., & Rabenold, K. N. (1995) Facultative dispersal by juvenile males in the cooperative stripe-backed wren. *Behavioral Ecology*, **6**, 337–342.

Poiani, A. & Elgar, M. A. (1994) Cooperative breeding in the Australian avifauna and brood parasitism by cuckoos (Cuculidae). *Animal Behaviour*, **47**, 697–706.

Poiani, A. & Jermiin, L. S. (1994) A comparative analysis of some life-history traits between cooperatively and non-cooperatively breeding Australian passerines. *Evolutionary Ecology*, **8**, 471–488.

Põldmaa, T., Montgomerie, R., & Boag, P. (1995) Mating system of the cooperatively breeding noisy miner *Manorina melanocephala*, as revealed by DNA profiling. *Behavioral Ecology and Sociobiology*, **37**, 137–143.

Preston, S. A. J., Briskie, J. V., Burke, T., & Hatchwell, B. J. (2013) Genetic analysis reveals diverse kin-directed routes to helping in the rifleman *Acanthisitta chloris*. *Molecular Ecology*, **22**, 5027–5039.

Pulliam, H. R. & Caraco, T. (1984) Living in groups: Is there an optimal group size? *In:* Krebs, J. R. & Davies, N. B. (eds.) *Behavioural Ecology: An Evolutionary Approach. 2nd ed.* Oxford: Blackwell Scientific Publications, pp. 122–147.

Quinn, J. S., Woolfenden, G. E., Fitzpatrick, J. W., & White, B. N. (1999) Multi-locus DNA fingerprinting supports genetic monogamy in Florida scrub-jays. *Behavioral Ecology and Sociobiology*, **45**, 1–10.

Radford, A. N. (2003) Territorial vocal rallying in the green woodhoopoe: Influence of rival group size and composition. *Animal Behaviour*, **66**, 1035–1044.

(2005) Group-specific vocal signatures and neighbour–stranger discrimination in the cooperatively breeding green woodhoopoe. *Animal Behaviour*, **70**, 1227–1234.

Reyer, H.-U. (1990) Pied kingfishers: Ecological causes and reproductive consequences of cooperative breeding. *In:* Stacey, P. B., & Koenig, W. D. (eds.) *Cooperative Breeding in Birds: Long-Term Studies of Ecology and Behavior*. Cambridge: Cambridge University Press, pp. 529–557.

Richardson, D. S., Jury, F. L., Blaakmeer, K., Komdeur, J., & Burke, T. (2001) Parentage assignment and extra-group paternity in a cooperative breeder: The Seychelles warbler (*Acrocephalus sechellensis*) *Molecular Ecology*, **10**, 2263–2273.

Ridley, A. R. & Raihani, N. J. (2007a) Facultative response to a kleptoparasite by the cooperatively breeding pied babbler. *Behavioral Ecology*, **18**, 324–330.

(2007b) Variable postfledging care in a cooperative bird: Causes and consequences. *Behavioral Ecology*, **18**, 994–1000.

Ridley, A. R., Nelson-Flower, M. J., & Thompson, A. M. (2013) Is sentinel behaviour safe? An experimental investigation. *Animal Behaviour*, **85**, 137–142.

Ridley, J., Yu, D. W. & Sutherland, W. J. (2005) Why long-lived species are more likely to be social: The role of local dominance. *Behavioral Ecology*, **16**, 358–363.

Riehl, C. (2013) Evolutionary routes to non-kin cooperative breeding in birds. *Proceedings of the Royal Society B*, **280**, 20132245.

Rowley, I. (1978) Communal activities among white-winged choughs *Corcorax melanorhamphos*. *Ibis*, **120**, 178–197.

Rowley, I., Russell, E., Payne, R. B., & Payne, L. L. (1989) Plural breeding in the splendid fairy-wren, *Malurus splendens* (Aves: Maluridae), a cooperative breeder. *Ethology*, **83**, 229–247.

Rubenstein, D. R. & Lovette, I. J. (2007) Temporal environmental variability drives the evolution of cooperative breeding in birds. *Current Biology*, **17**, 1414–1419.

(2009) Reproductive skew and selection on female ornamentation in social species. *Nature*, **462**, 786–789.

Rusk, C. L., Walters, E. L., & Koenig, W. D. (2013) Cooperative breeding and long-distance dispersal: A test using vagrant records. *PLoS ONE*, **8**, e58624.

Seddon, N. & Tobias, J. A. (2003) Communal singing in the cooperatively breeding subdesert mesite *Monias benschi*: Evidence of numerical assessment? *Journal of Avian Biology*, **34**, 72–80.

Schiegg, K., Walters, J. R., & Priddy, J. A. (2002) The consequences of disrupted dispersal in fragmented red-cockaded woodpecker *Picoides borealis* populations. *Journal of Animal Ecology*, **71**, 710–721.

Schiegg, K., Daniels, S. J., Walters, J. R., Priddy, J. A., & Pasinelli, G. (2006) Inbreeding in red-cockaded woodpeckers: Effects of natal dispersal distance and territory location. *Biological Conservation*, **131**, 544–552.

Schoech, S. J., Reynolds, S. J., & Boughton, R. K. (2004) Endocrinology. *In:* Koenig, W. D. & Dickinson, J. L. (eds.) *Ecology and Evolution of Cooperative Breeding in Birds*. Cambridge: Cambridge University Press, pp. 128–141.

Selander, R. K. (1964) Speciation in wrens of the genus *Campylorhychus*. *Univiversity of California Publication in Zoology*, **74**, 1–224.

Sharp, S. P., McGowan, A., Wood, M. J., & Hatchwell, B. J. (2005) Learned kin recognition cues in a social bird. *Nature*, **434**, 1127–1130.

Sharp, S. P., Simeoni, M., & Hatchwell, B. J. (2008) Dispersal of sibling coalitions promotes helping among immigrants in a cooperatively breeding bird. *Proceedings of the Royal Society B*, **275**, 2125–2130.

Shaw, P., Owoyesigire, N., Ngabirano, S., & Ebbutt, D. (2015) Life history traits associated with low annual fecundity in a central African parid: The stripe-breasted tit *Parus fasciiventer*. *Journal of Ornithology*, **156**, 209–221.

Strickland, D. & Waite, T. A. (2001) Does initial suppression of allofeeding in small jays help to conceal their nests? *Canadian Journal of Zoology*, **79**, 2128–2146.

Skutch, A. F. (1935) Helpers at the nest. *Auk*, **52**, 257–273.

(1961) Helpers among birds. *Condor*, **63**, 198–226.

Stacey, P. B. & Koenig, W. D. (eds.) (1990) *Cooperative Breeding in Birds: Long-Term Studies of Ecology and Behavior*, Cambridge: Cambridge University Press.

Stacey, P. B. & Ligon, J. D. (1991) The benefits-of-philopatry hypothesis for the evolution of cooperative breeding: Variation in territory quality and group size effects. *The American Naturalist*, **137**, 831–846.

Székely, T., Liker, A., Freckleton, R. P., Fichtel, C., & Kappeler, P. M. (2014) Sex-biased survival predicts adult sex ratio variation in wild birds. *Proceedings of the Royal Society B*, 281, 20140342.

Theuerkauf, J., Rouys, S., Mériot, J. M., Gula, R., & Kuehn, R. (2009) Cooperative breeding, mate guarding, and nest sharing in two parrot species of New Caledonia. *Journal of Ornithology*, **150**, 791–797.

Thornton, A. & McAuliffe, K. (2015) Cognitive consequences of cooperative breeding? A critical appraisal. *Journal of Zoology*, **295**, 12–22.

van Dijk, R. E., Kaden, J. C., Argüelles-Ticó, A., *et al.* (2013) The thermoregulatory benefits of the communal nest of sociable weavers *Philetairus socius* are spatially structured within nests. *Journal of Avian Biology*, **44**, 102–110.

(2014) Cooperative investment in public goods is kin directed in communal nests of social birds. *Ecology Letters*, **17**, 1141–1148.

Vehrencamp, S. L. & Quinn, J. S. (2004) Joint laying systems. *In:* Koenig, W. D. & Dickinson, J. L. (eds.) *Ecology and Evolution of Cooperative Breeding in Birds*. Cambridge: Cambridge University Press, pp. 177–196.

Walters, J. R., Copeyon, C. K., & Carter, J. H. III. (1992) Test of the ecological basis of cooperative breeding in red-cockaded woodpeckers. *Auk*, **109**, 90–97.

Warrington, M. H., McDonald, P. G., & Griffith, S. C. (2015) Within-group vocal differentiation of individuals in the cooperatively breeding apostlebird. *Behavioral Ecology*, **26**, 493–501.

West, S. A., Pen, I., & Griffin, A. S. (2002) Cooperation and competition between relatives. *Science*, **296**, 72–75.

West, S. A., Shuker, D. M., & Sheldon, B. C. (2005) Sex-ratio adjustment when relatives interact: A test of constraints on adaptation. *Evolution*, **59**, 1211–1228.

White, F. N., Bartholomew, G. A., & Howell, T. R. (1975) The thermal significance of the nest of the sociable weaver *Philetairus socius*: Winter observations. *Ibis*, **117**, 171–179.

Whittingham, L. A., Dunn, P. O., & Magrath, R. D. (1997) Relatedness, polyandry and extra-group paternity in the cooperatively-breeding white-browed scrubwren (*Sericornis frontalis*) *Behavioral Ecology and Sociobiology*, **40**, 261–270.

Williams, J. B., du Plessis, M. A., & Siegfried, W. R. (1991) Green woodhoopoes (*Phoeniculus purpureus*) and obligate cavity roosting provide a test of the thermoregulatory insufficiency hypothesis. *Auk*, **108**, 285–293.

Williams, D. A., Lawton, M. F., & Lawton, R. O. (1994) Population growth, range expansion, and competition in the cooperatively breeding brown jay, *Cyanocorax morio*. *Animal Behavior*, **48**, 309–322.

Woolfenden, G. E. & Fitzpatrick, J. W. (1984) *The Florida Scrub Jay: Demography of a Cooperative-Breeding Bird*, Princeton: Princeton University Press.

Wright, J., McDonald, P. G., te Marvelde, L., Kazem, A. J. N., & Bishop, C. M. (2010) Helping effort increases with relatedness in bell miners, but "unrelated" helpers of both sexes still provide substantial care. *Proceedings of the Royal Society B*, **277**, 437–445.

Yasukawa, K. & Cockburn, A. (2009) Antipredator vigilance in cooperatively breeding superb fairy-wrens (*Malurus cyaneus*). *Auk*, **126**, 147–154.

Zahavi, A. (1977) Reliability in communication systems and evolution of altruism. *In:* Stonehouse, B. & Perrins, C. M. (eds.) *Evolutionary Ecology*. London: Macmillan, pp. 253–259.

12 Sociality in Fishes

Michael Taborsky and Marian Wong

Overview

The reproductive biology of fish exemplifies the competition between conspecifics for the production of offspring (Taborsky 1994, 1997, 2008). This concerns intrasexual conflict, particularly in the male sex, but also the effects of the conflict of interest between the sexes (Henson & Warner, 1997; Maan & Taborsky, 2008; Taborsky & Neat, 2010). Nevertheless, there are remarkable examples of cooperation in fish reproduction, including some of the most highly advanced social systems and levels of cooperation known among vertebrates (Taborsky, 1987, 1994, 2016; Balshine & Buston, 2008). What at first glance may seem contradictory actually reflects alternative solutions to the same problem: individuals competing for resources such as food, shelter, or mates may succeed either through greater resource holding potential or by collaborating with (some of) their competitors. Resource competition may select either for high competence in conflict strategies (e.g. involving aggression or surreptitious exploitation) or for advanced social competence, group formation, and cooperation. Here we discuss which intrinsic characteristics of a taxon and which ecological conditions generate cooperation rather than conflict. Our emphasis will be on reproductive systems because social complexity is typically much greater in the context of reproduction than in other functional circumstances such as food acquisition or predator avoidance (Wilson, 1975; Trivers, 1985). In aquatic animals, the preconditions for advanced levels of sociality differ systematically between freshwater and marine environments, primarily due to the different mechanisms of dispersal. This has strong effects on the evolution of brood care and on the relatedness among conspecifics. Both factors are primary drivers of sociality; hence, the types of social systems that we find in freshwater and marine environments diverge substantially. Therefore, we shall deal with freshwater and marine systems separately where this seems appropriate, but we shall point out commonalities between freshwater and marine fishes wherever possible.

We are grateful to Dustin Rubenstein and Patrick Abbot for the invitation to contribute to this book, and for their and two referees' valuable comments on earlier drafts of this article. We thank Barbara Taborsky for help with the figures. M.T. was supported by SNSF grant 31003A_156152.

I SOCIAL DIVERSITY

12.1 How Common is Sociality in Fishes?

Even if social complexity is generally highest in the reproductive context, collective and cooperative behavior can also be observed in fish schools and shoals (i.e. a social group staying connected for some time; the term "school" refers to a shoal that shows coordinated movement regarding direction, swimming speed, and distance between its members). These more or less organized conspecific aggregations (Pitcher & Parrish, 1993; Delcourt & Poncin, 2012) are widely viewed as the most characteristic social organization of fish. They mainly serve an anti-predator function, but benefits may accrue also from collective food search and hunting (Partridge, *et al.*, 1983; Pitcher & Parrish, 1993), and from reduced energy consumption by joint locomotion (Videler, 1993; Johansen, *et al.*, 2010; Burgerhout, *et al.*, 2013). Such temporary aggregations are typically non-exclusive and "open" concerning membership, unstable with regard to size, and often exhibit fission-fusion dynamics. Complexity in these groups mainly relates to the behavioral coordination when moving, feeding and avoiding predators (Handegard, *et al.*, 2012; Tunstrom, *et al.*, 2013), the latter apparently being a direct evolutionary response to predation risk (Ioannou, *et al.*, 2012).

Collective behavior in fish assemblies may also include joint exploration of potential danger, where group members somewhat coordinate their acquisition of information (Milinski, *et al.* 1990; Dugatkin & Godin, 1992; Croft, *et al.* 2006). Grouping allows fish to obtain information from other group members ("public information", Danchin, *et al.*, 2004), which can be important not only to avoid risk, but also when searching for food (Coolen, *et al.*, 2003). Information exchange and behavioral coordination can occur also between fish species, as exemplified by the cooperative hunting of groupers and moray eels (e.g. Bshary, *et al.*, 2006) and the cleaning behavior exhibited mainly by marine wrasses (Labridae) and gobies (Gobiidae) (Cote, 2000; Bshary & Grutter, 2006).

When considering advanced levels of sociality, cooperative breeding arguably exemplifies the most complex social organization found in nature apart from eusocial insects. Cooperative breeding, where individuals help to care for offspring they have not produced themselves, is widespread in insects, mammals, and birds, and occurs also in some spiders and fishes (Taborsky & Limberger, 1981; Emlen, 1984; Solomon & French, 1997; Bourke, 2011; Biedermann, 2014). In fish, cooperative breeding has been observed in roughly 25 species of cichlids and a few other freshwater species (Taborsky, 1994; Heg & Bachar, 2006). It seems puzzling why cooperative breeding is so rare in fish (less than 0.1 percent of 32,700 known fish species). However, the majority of fish species show little or no brood care. Their specialty is to produce large offspring numbers rather than investing a lot of energy into few. In addition, most fish taxa with post-fertilization parental investment show uniparental care, often tending several clutches at a time (Breder & Rosen, 1966). As biparental care seems to be a precondition for the evolution of cooperative breeding (Komdeur & Ekman, 2010), the rather low number of cooperatively breeding fishes is less surprising. In addition, as doing research underwater is more difficult than on land, relatively few social and breeding systems of fish have yet been studied, thus there might be many cooperatively breeding fishes that have not yet been detected.

Remarkably, the majority of cooperatively breeding fishes occur in Lake Tanganyika, where roughly 70 species of cichlids are "substrate breeders" that attach eggs to the substrate and care for the brood by biparental guarding (Kuwamura, 1997). Brood care in these fish often involves keeping predators at bay by aggressive behavior, providing shelters for instance by digging out holes under stones or collecting empty gastropod shells, cleaning eggs and larvae, and supplying eggs with oxygen by fanning. About one-third of these biparental "substrate brooders" of Lake Tanganyika breed cooperatively (Taborsky, 1994; Heg & Bachar, 2006).

In marine fishes, social species forming stable, restricted-entry groups, in which group membership is regulated and individuals develop dominance hierarchies, have so far been identified in seven families of coral-reef fishes (Fricke, 1980; Clifton, 1990; Wong, *et al.*, 2007; Ang & Manica, 2010a). Unlike in freshwater fishes, there are no reported incidences of alloparental care in the sense that non-breeding helpers assist in the reproduction of breeders, although there have been reported cases of egg adoption (Thresher, 1985; Wisenden, 1999) and nest takeovers (e.g. Yanagisawa & Ochi, 1986; Taborsky, *et al.*, 1987; Bisazza, *et al.*, 1989) leading to alloparental care-like behavior in some marine fishes. There is, however, sometimes joint territory defense (Clifton, 1990; van Rooij, *et al.*, 1996). In addition to synergistic effects of joint defense, dominants could benefit from the presence of subordinates also from the reported positive effects on the growth rates of living coral and anemone hosts (Liberman, *et al.*, 1995; Roopin, *et al.*, 2011; Dixson & Hay, 2012). Even so, removal experiments have yet to reveal a significant benefit for dominants of tolerating subordinates (Ang & Manica, 2011). Therefore, existing evidence implies that in anemonefish, dominants suffer little cost (relative to evicting subordinates), and receive little benefit from the presence of subordinates.

12.2 Forms of Sociality in Fishes

In freshwater fishes, the lack of planktonic stages often results in relatively high relatedness levels locally (Sefc, *et al.*, 2012). In principle, this should favor the evolution of advanced levels of sociality and cooperation when compared to marine fish species. In addition, cooperative behavior in freshwater fishes often occurs between unrelated individuals, primarily in males (Taborsky, 1994). Male competitors may join forces for collaborative activities such as territory defense, nest building, courtship and even spawning (reviewed in Taborsky, 1994; Diaz-Munoz, *et al.*, 2014). Joint or alloparental brood care is of particular interest, as it results in the highest degrees of cooperation and social complexity known among fishes (Taborsky, 2016).

Alloparental care can result from several proximate mechanisms, including the stealing of eggs, adoption of foreign young, nest take-overs, and communal or cooperative breeding (Taborsky, 1987, 1994; Wisenden, 1999). The mixing of broods from different parents occurs frequently in brood-caring species of fish (e.g. Kellogg, *et al.*, 1998; Schaedelin, *et al.*, 2013). A common cause of brood mixing is the kidnapping of young by brood caring parents (Burchard, 1965; McKaye & McKaye, 1977), which may occur also at the egg stage (Jones, *et al.*, 1998). However, young may also join foreign parents on their own accord (Thresher, 1985; Wisenden, *et al.*, 2014). Alternatively, young may be consigned by their parent(s) to other, brood-caring parents' custody ("brood farming out", Yanagisawa, 1985; Wisenden & Keenleyside, 1992). Still other possibilities include "family conflux" (Taborsky, 1994) when brood-caring parents with their offspring join, and consequently the young no longer separate into distinct groups (Baylis, 1974; Wisenden & Keenleyside, 1992). Finally, alloparental care may result also from a nest take-over (Constantz, 1979; Daniels, 1979; Taborsky, *et al.*, 1987). Benefits to alloparents may result from dilution or confusion effects to reduce predation on own young (McKaye & McKaye, 1977; Wisenden & Keenleyside, 1992), increased attractiveness to potential mates (Kraak & Groothuis, 1994), and synergistic effects from joint brood care (Ward & Wyman, 1977; Ribbink, *et al.*, 1981; Wisenden & Keenleyside, 1992).

A particular condition leading to alloparental care is the dispersal delay of young, which typically causes overlapping generations of offspring sharing a territory, and some sort of cooperative brood care (Koenig, *et al.*, 1992). In freshwater fish, cooperatively breeding groups may form in two ways. First, offspring produced in territories may delay dispersal and help raise the offspring of related dominants, which corresponds to the most widespread pattern found in birds and mammals (Koenig & Dickinson, 2016). Alternatively, groups may consist of unrelated adults assembling for joint reproduction (Martin & Taborsky, 1997; Awata, *et al.*, 2005; Kohda, *et al.*, 2009). Combinations of both forms of group formation occur as well (Taborsky, 2009, 2016). For example, in the Lake Tanganyika cichlid, *Neolamprologus pulcher*, groups typically consist of immature and mature brood care helpers of both sexes that were born in the respective territory and have subsequently delayed their dispersal, plus a dominant male and female breeder that may be unrelated to the helpers due to the frequent replacement of breeders (Figure 12.1A). In addition, there are sometimes subordinate immigrants that enter the group and share in cooperative territory defense and maintenance, and in alloparental care (Dierkes, *et al.*, 2005; Zöttl, *et al.*, 2013a). A similar combination of related and unrelated cooperating group members partly participating in reproduction exists also in *Julidochromis ornatus* (Awata, *et al.*, 2005; Heg & Bachar, 2006). In *Julidochromis marlieri*, female breeders exceed male breeders in size and can be polyandrous, and these groups also contain brood care helpers (Figure 12.1B) (Taborsky, 1994; Yamagishi & Kohda, 1996; Sunobe, 2000). Still another variation is shown in *Neolamprologus obscurus*, where small, usually related helpers of both sexes and large, less related males form groups together with dominant breeders (Figure 12.1C) (Tanaka, *et al.*, 2015). In West African riverine cichlids *Pelvicachromis pulcher*, unrelated satellite males may join a pair and share in reproduction and brood care (Figure 12.1D) (Martin & Taborsky, 1997).

Like cooperatively breeding freshwater fishes, the basic form of some social groups in tropical marine fish is a dominant breeding pair that monopolizes reproduction at the

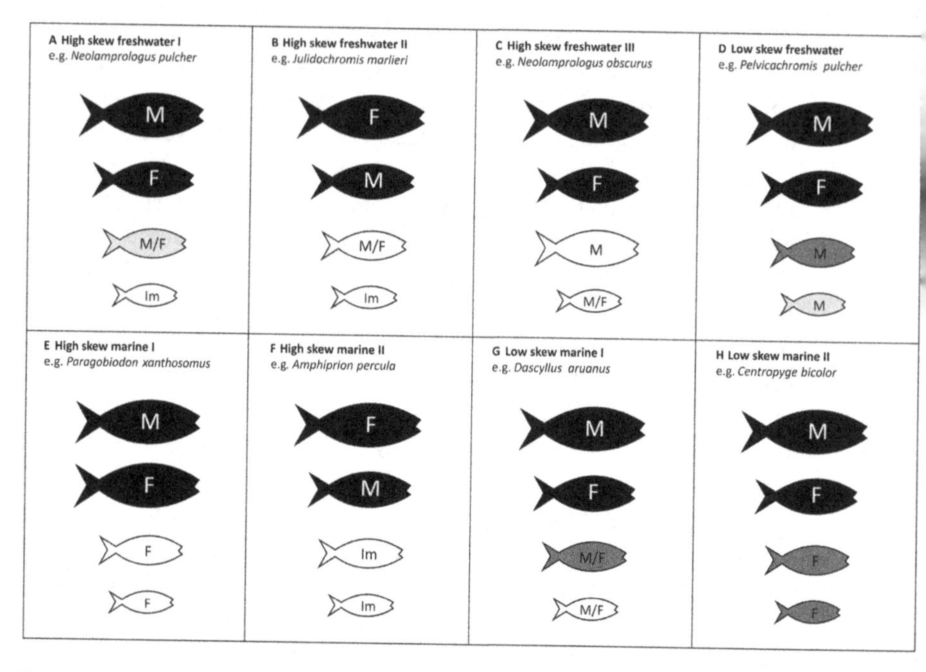

Figure 12.1 Typical group structures of social freshwater (top; A–D) and marine (bottom; I–H) fish species showing reproductive patterns (i.e. skew) and relative within-group body size distributions. Due to indeterminate growth in most fish species, body size is a prime determinant of social structure. M = Male, F = Female, Im = Immature. Black fish = dominant breeders, grey fish = subordinates highly (dark grey) or slightly (pale grey) participating in reproduction, white fish = non-reproductive subordinates. In freshwater species, subordinate group members typically show alloparental care; in marine species they do not.

expense of other more subordinate group members, commonly termed "despotic societies" (e.g. coral-gobies, Gobiidae, Wong, *et al.*, 2007) and anemone fishes (Pomacentridae, Allen, 1972), thereby exhibiting monogamy within the context of group living (Figure 12.1E & F). These subordinates are socially suppressed from reproduction (e.g. Wong, *et al.*, 2007), and in contrast to the mentioned cooperatively breeding freshwater fishes, they do not engage in alloparental care. In addition, socially more "egalitarian" marine fishes form groups in which reproduction is somewhat shared amongst dominants and subordinates, typically via a polygynous or polygynandrous mating system (e.g. Ang & Manica, 2010a; Wong, *et al.*, 2012) (Figure 12.1G & H). In most of these cases, juveniles are typically present within groups, and these are sexually undifferentiated (Cole, 1990). Within groups, there is typically a strict size-based hierarchy, with larger individuals dominating smaller individuals and subordinates getting progressively smaller down the hierarchy (Buston, 2003; Wong, *et al.*, 2007; Ang & Manica, 2010a) (Figure 12.1E-H).

Unlike in freshwater fish, groups of tropical marine fish form through the recruitment and settlement of larvae from the plankton after a pelagic larval phase of varying duration. In other words, offspring are not retained on their natal territory, which is the typical pattern in cooperative breeders. Therefore, relatedness is effectively zero within groups of most tropical marine fish (Buston, *et al.*, 2009), although there is recent evidence for occasional pairs of relatives within groups, possibly as a result of recruiting together (Buston, *et al.*, 2009). Since dispersal between groups may occur post-settlement, groups are dynamic and can form and dissolve (e.g. Wong, 2010). However, these species typically reside permanently within discrete patches of habitat (such as coral, anemone, sponges or on seawhips) (e.g. Allen, 1972) or within defended territories on the reef substrate (e.g. Robertson & Hoffman, 1977).

12.3 Why Fishes Form Social Groups

Groups form primarily for the purpose of protection, the search for and acquisition of resources such as food, the facilitation of locomotion, thermal homeostasis, and for reproduction. While all these functions, except for thermal homeostasis, can be observed in fish groups, like in other major taxa the most complex forms of social organization occur in the context of mating and brood care.

12.3.1 Resource Acquisition and Use

The highly social cichlids that arguably represent the pinnacle of intraspecific cooperative behavior in fish are characterized by joint reproduction and brood care. In some species, it seems impossible to raise offspring successfully without cooperative care involving more individuals than just a pair of breeders, which is partly due to the need of joint defense and manufacture of shelters (Taborsky, 2016). Reproductive skew in such groups varies among species (Taborsky, 2009), but typically most or all group members participate in the protection and care of eggs and young (Figure 12.2) (Martin & Taborsky, 1997; Heg & Bachar, 2006; Bruintjes & Taborsky, 2011). The most extensively studied species, *N. pulcher*, for instance, occurs in rocky habitats around the sublittoral zone of Lake Tanganyika (Duftner, *et al.*, 2007). Group territories consist of a breeding shelter and a variable number of additional shelters that are typically shared by a dominant breeding pair and on average 5 to 6 subordinate helpers (Taborsky, 1984, 2016; Balshine, *et al.*, 2001; Werner, *et al.*, 2003). A small space around these shelters is also part of the defended area, but territories do not contain significant amounts of food for adults. Virtually all subordinate group members help to defend, maintain, and improve the shelters and territory, and to care for the brood. Hence, the crucial resources for group members are the shelters jointly defended and maintained in their territories. This is different in another cooperative breeder living in similar habitat, *N. obscurus*, where group members consume most of their food within their territory (Tanaka, *et al.*, 2015).

Tropical marine fishes that form stable social groups reside on coral reefs which are spatially heterogeneous habitats. Corals and anemones represent fixed, defensible resources in the form of food, shelter and breeding sites, and hence promote territoriality and site fidelity in some species. For example, in group living coral gobies of the genus

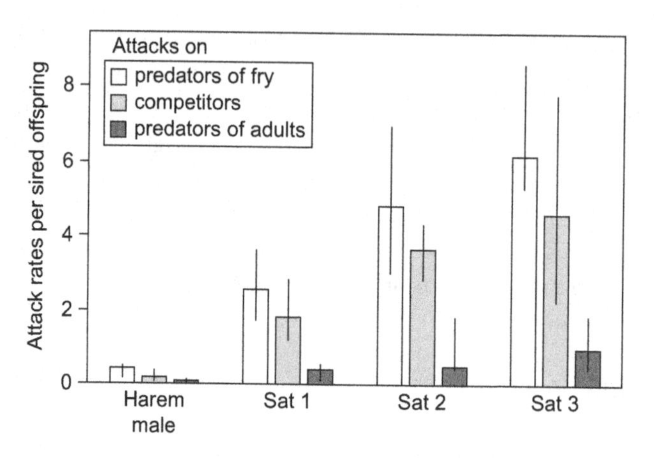

Figure 12.2 Defense effort per offspring produced by polygynous harem males (HM) and helper males ("satellites" of decreasing dominance rank: Sat 1, Sat 2, Sat 3) of the West-African cichlid *Pelvicachromis pulcher*. Attack rates (medians and quartiles) against intruding predators of fry and adults and towards foreign conspecifics were calculated per produced offspring (number of fertilized eggs) per day, for one entire breeding cycle (*sensu* Martin & Taborsky, 1997). This shows that the most subordinate helpers (Sat 3) had to work hardest for the number of offspring they could sire, whereas the harem owners (HM) showed very little defense effort when compared to the great number of young they produced. After Martin & Taborsky (1997).

Gobiodon and *Paragobiodon*, and coral crouchers of the genus *Caracanthus*, individuals reside within corals, maneuvering among the branches but rarely leaving the corals themselves (Kuwamura, *et al.*, 1994; Munday, 2001; Wong, *et al.*, 2007). Gobies from the genus *Paragobiodon* have been observed to be aggressive towards introduced conspecifics and thereby defend the coral (Lassig, 1977; Wong, *et al.*, 2008a), and group members within corals have been observed to defend sub-territories from other group members (Lassig, 1977; Wong, *personal observation*). Hence, corals and anemones provide critical resources without which the fish are unable to survive.

Subordinate group members can also benefit from residing in groups through inheriting the territory and breeding position in the future, as demonstrated in the cooperatively breeding cichlid *N. pulcher* (Balshine-Earn, *et al.*, 1998; Dierkes, *et al.*, 2005). Similarly, in all marine fish species for which territory inheritance has been studied, subordinates do have the opportunity to become dominant breeders within the group upon death or disappearance of the dominant group member (Wong, *et al.*, 2007). In species like *Amphiprion percula* (Pomacentridae), subordinates are heavily site-attached, therefore queuing is the only option for attaining breeding status.

12.3.2 Predator Avoidance

In *N. pulcher*, sexually mature group members feed primarily on zooplankton, mainly crustaceans, in the water column (Taborsky & Limberger, 1981; Gashagaza, 1988; Bruintjes, *et al.*, 2010). Hence, the main function of territories is not privileged access to food, except for small young, but protection from predators. Safety for group members is accomplished by the secured access to shelters and by the defense performed by other group members, especially by the dominant breeders (Taborsky, 1984; Heg, *et al.*, 2004a). Therefore, a major reason for group formation rests on the fact that all group members participate heavily in territory defense against conspecifics (Bruintjes & Taborsky, 2008) and predators (Taborsky & Limberger, 1981; Heg & Taborsky, 2010; Zöttl, *et al.*, 2013a). A strategic sharing of risk among group members is suggested by a significant positive correlation between group size and the per capita attack frequencies of breeders and large helpers against experimentally presented large predators (Heg & Taborsky, 2010). Such behavior apparently reduces predation risk, since members of large groups survived more likely despite their higher defense effort (Heg, *et al.*, 2004a). Additionally, large groups persist over longer time periods than smaller ones (Heg, *et al.*, 2005a). Predation risk is clearly the ecological constraint responsible for delayed dispersal in these cichlids (Heg, *et al.*, 2004a; Groenewoud, *et al.*, 2016).

Apart from breeding in groups, *N. pulcher* groups assemble in colonies (Taborsky & Limberger, 1981; Heg, *et al.*, 2008), which again indicates a protective role of social cohesion (Jungwirth & Taborsky, 2015). Experimental presentation of vacant territories inside and outside of colonies revealed a strong preference for vacancies in *N. pulcher* arising within colonies; territories provided at the edge of a colony were not taken (Heg, *et al.*, 2008). Benefits of coloniality probably resemble the benefits of group living. Aggregations of territories afford enhanced protection due to dilution effects, and group members save defense effort due to the defense activities of neighbors (Jungwirth, *et al.*, 2015a). Experimental exposure of predators to groups located at different densities in a colony revealed a strategic reduction of defense effort at high densities, which resulted in similar defense frequencies against predators regardless of colony density, and a reduction in defense effort for group members living in high density areas (Jungwirth, *et al.*, 2015a).

Predation also shapes many aspects of the behavior and life history of coral reef fishes (Stewart & Jones, 2001). As many social species are small bodied, this renders them particularly vulnerable to predation, especially when outside the protection of their microhabitat or territory (Herler, *et al.*, 2011). Predation risk could thereby act as a constraint on the movement of individuals between habitat patches or neighboring territories, resulting in the formation and maintenance of social groups. Indeed, an experimental manipulation of dispersal distances in the social coral-goby *Paragobiodon xanthosomus* demonstrated that subordinates were more likely to disperse to another coral that was only 10 centimeters as opposed to 100 centimeters away (Wong, 2010), suggesting that dispersal distance and predation risk associated with proximity could play a role in the maintenance of social groups in these fishes.

12.3.3 Homeostasis

Thermal homeostasis plays little role in why fishes form social groups because they are largely poikilothermic and live their entire lives underwater, where body contact would hardly affect temperature exchange between group members. Some fishes such as tuna produce considerable heat during fast locomotion (Carey & Lawson 1973), but there is no connection known between body temperature and the social hunting of these animals. Group formation may help to save energy, however, when members of schools make use of the wake created by the swimming movement of others (Videler, 1993; Johansen, *et al.*, 2010; Burgerhout, *et al.*, 2013).

12.3.4 Mating

In addition to the potential benefits derived from resource acquisition and protection, group members may also gain from mating opportunities. Specifically, sexually mature subordinates in cooperatively breeding freshwater fishes may benefit from participating in reproduction (Martin & Taborsky, 1997; Awata, *et al.*, 2005; Hellmann, *et al.*, 2015; Taborksy, 2016), though reproductive skew is generally high (Taborsky, 2001, 2009). In *N. pulcher*, between 10 and 20 percent of broods on average contain offspring produced by subordinate group members. Female helpers produce roughly 15 percent of young in a territory, whereas the share of young produced by male helpers is about 5 percent (Taborsky, 2016). A reproductive function of group formation is obvious in species where adult males join a breeding pair for producing and rearing offspring (satellite males, cooperative polyandry, e.g. Martin & Taborsky, 1997; Kohda, *et al.*, 2009).

For some social coral reef fishes, groups clearly do not form for the purposes of mating because reproduction is monopolized by just two group members regardless of group size (e.g. Kuwamura, *et al.*, 1994; Wong, *et al.*, 2007). However, for marine species that form groups in which multiple group members breed (Figure 12.1G & H), the ability to acquire current direct reproduction could be a primary driver of sociality, at least from the subordinates' perspective. For the social coral-dwelling damselfish *Dascyllus aruanus*, for instance, microsatellite analysis of parentage revealed that smaller and more subordinate individuals received a significant, albeit lower proportional share of group reproduction (Wong, *et al.*, 2012). However, there were even some instances where subordinate group members obtained a larger reproductive share than their dominants (Wong, *et al.*, 2012), demonstrating that direct reproduction in groups could be an extremely attractive reason to live in groups.

Dominant breeders in social marine fish will usually aggressively chase subordinates away from nest sites, suggesting that they do not benefit from subordinate helping behavior (Mizushima, *et al.*, 2000). Instead, dominants may benefit from joint territory defense against intruding conspecifics, as exemplified by the joint nest defense of parental nest builders and satellite males in Mediterranean wrasses (Taborsky, *et al.*, 1987; Stiver & Alonzo, 2013). There has been some suggestion of this possibility also in the social coral-dwelling gobies, *Paragobiodon* spp, whereby as Lassig (1977) reported, "several residents combined in attack against large intruders and in so doing disregarded their own territorial boundaries."

12.3.5 Offspring Care

Cooperative behavior in fish reproduction is not limited to conspecific associations that have the purpose of obtaining mates or fertilizations. Cooperation also occurs between parents tending eggs or young (reviewed in Taborsky, 1994). In freshwater fish, groups may form for the purpose of caring for offspring that were produced independently (i.e. where different parents jointly raise and protect their respective offspring). This has been observed in Asian, American, and African cichlids (Ward & Wyman, 1977; McKaye & McKaye, 1977; Ribbink, *et al.*, 1981). Occasionally, such temporary groups forming for the purpose of joint brood care may consist also of members of different species, such as in cyprinids (Hankinson, 1920), or between cichlids and catfish (McKaye, *et al.*, 1992).

Alternatively, groups consisting of dominant breeders and subordinate individuals that may or may not participate in reproduction themselves may share the duties of tending and protecting offspring. This form of cooperative brood care is primarily shown by Lake Tanganyikan cichlids such as *N. pulcher*, where the dominant breeders benefit from the shared effort in territory defense and maintenance, and from the subordinates' care and protection of young (Brouwer, *et al.*, 2005). Subordinate group members, by contrast, benefit mainly from the essential protection gained in the territory, and from some chance to participate in reproduction. Kin selection benefits derived from the protection of non-descendent relatives accrue only to a minor extent, because mature helpers are rarely related to the beneficiaries of their help (Dierkes, *et al.*, 2005; Stiver, *et al.*, 2005), and experiments revealed that relatedness even decreases the propensity of subordinates to help caring for eggs (Zöttl, *et al.*, 2013b). Costs to subordinate group members in this species mainly result from the steady aggression of dominants demanding their help and from the energetic costs of helping (Grantner & Taborsky, 1998; Taborsky & Grantner, 1998), which induces significant growth costs (Taborsky, 1984; Heg, *et al.*, 2004b). This system is characterized by continual negotiations of dominants and subordinates about the amount of help in brood care the latter need to pay in order to be tolerated in the territory (Taborsky, 1985; Bergmüller & Taborsky, 2005; Bergmüller, *et al.*, 2005; Heg & Taborsky, 2010; Zöttl, *et al.*, 2013b; Fischer, *et al.*, 2014). Similar reciprocal relationships may be responsible for the cooperation observed among group members in other social cichlids, as the degrees of relatedness between helpers and beneficiaries are generally low.

Only one species of marine fish is known to exhibit biparental care (*Acanthochromis polyacanthus*, Whiteman & Côté, 2004), but on the whole, and where any care is given, it is usually paternal (Perrone & Zaret, 1979). Hence, the lack of cooperative breeding in marine fishes is consistent with the idea that biparental care is a precondition for the evolution of this form of social organization. For other forms of group living, it appears that sociality has evolved multiple times owing to the diverse taxonomic spread of sociality in marine fishes, suggesting that marine environments, especially coral reefs, may set the stage for the evolution of social group living and complexity beyond the sharing of brood care duties.

12.4 The Role of Ecology in Shaping Sociality in Fishes

In fishes, as in other major taxa, habitat characteristics and the type of food consumed and its distribution may have important consequences on social structure and behavior.

In addition, in highly social fish species such as cooperatively breeding cichlids, predation pressure is a crucial factor selecting for certain forms of sociality (Heg, *et al.*, 2004a).

12.4.1 Habitat and Environment

Cooperatively breeding cichlids use a wide range of habitats for shelter and breeding, ranging from holes and crevices in rocks, through sandy substrates interspersed with stones, to accumulations of empty gastropod shells. Most species seem to be specialized in using one particular habitat type among this diversity, but interestingly, *N. pulcher* uses the entire range (Groenewoud, *et al.*, 2016). The presence and extent of sand is of particular importance in this regard because the necessity to dig out burrows from underneath stones in order to create shelters for hiding and breeding can strongly affect group composition. While rock habitat can provide shelters that are easily accessible, their number and size cannot be modified. Hence, the number of group members may be limited by the number and size of available holes and clefts. In contrast, fish using habitats that allow modifying the number of available shelters by digging may not suffer from such limitation.

Another important habitat characteristic concerns the distribution of food. Some cooperatively breeding cichlids, including *N. multifasciatus* and *obscurus,* find most of their food within their territories (Kohler, 1998; Tanaka, *et al.*, 2015), which might affect group size and composition. In contrast, *N. pulcher* feeds upon zooplankton, so their food is mainly obtained outside the territory. Global food availability has been experimentally shown to be important for the social interactions within groups of this species (Bruintjes, *et al.*, 2010), but food distribution and abundance are of minor importance for the utility of groups.

Predation risk is clearly the most important ecological factor determining social structure and behavior in at least some of the cooperatively breeding cichlids. Small specimens of *N. pulcher*, for instance, have little chance to survive outside of groups (Heg, *et al.*, 2004a), where they benefit from access to safe shelters and the antipredator defense of larger group members (Taborsky, 1984). A comparison between eight different populations of this species revealed that predation risk explains substantial variation in body size, social structure and behavior of this species, even on a small spatial scale (Groenewoud, *et al.*, 2016). A comparison between two populations of *N. obscurus* also suggested a strong role of predation risk for the social structure of this species. The intensity of predation explained the timing and distance of dispersal, and the body size at which helpers left their natal territory (Tanaka, 2014).

Habitat factors also have an important influence on many aspects of reef fish behavior, including social and reproductive behaviors. The observation that group size is positively correlated with the size of suitable habitat, which has been reported in a wide range of both socially despotic and egalitarian coral reef fishes, supports the hypothesis that the availability of suitable habitat plays an important role in the formation of groups (Holbrook, *et al.*, 2000; Wong, *et al.*, 2012). Experimental manipulations of habitat patch size in coral dwelling species have indicated the causality of this relationship, with habitat patch size influencing group size (Fricke, 1980).

Furthermore, the degree of variation in habitat patch size appears to have profound effects on the occurrence of group living in benthic marine fishes. For example, with increasing size differences between two corals, individuals of the social coral goby *Paragobiodon xanthosomus* increasingly chose group living as a non-breeding subordinate on the larger coral over immediate reproduction as a breeder on the smaller coral (Wong, 2010). Various lines of evidence support a role of limited habitat availability (i.e. through saturation and occupation) in determining the social organization of coral reef fishes, particularly for habitat-specialists. High habitat saturation at the group level can apparently also promote the maintenance of group living in marine fish species, because in *Paragobiodon xanthosomus*, experimental removals of subordinates from neighboring corals resulted in group dissolution through the dispersal of individuals to these experimentally unsaturated corals. (Wong, 2010). These patterns therefore suggest that low habitat availability at the population level promotes the evolution of sociality because a lack of available suitable habitat may prevent individuals from breeding independently elsewhere, just like in cooperative breeders outside of fishes (Koenig, *et al.*, 1992).

12.4.2 Biogeography

With nearly 2,000 species, the cichlids are one of the most speciose families of bony fishes. However, cooperative breeding has been hitherto recorded only from roughly twenty-five species of this family, almost exclusively confined to the Lamprologini of Lake Tanganyika (Taborsky, 1994, 2016; Heg & Bachar, 2006). Some form of cooperative brood care has been observed in Asian, American, and West African cichlids, where either satellite males join dominant breeders to participate in reproduction and brood care (Martin & Taborsky, 1997), or where pairs that have produced offspring independently join forces to defend their young collectively (McKaye & McKaye, 1977; Ward & Wyman, 1977; Ribbink, *et al.*, 1981). However, only Lake Tanganyika cichlids show the form of cooperative breeding that is commonly found in birds and occasionally in mammals, where groups contain breeders and non-breeders that all collaborate in the duties of brood care, territory defense and maintenance. It is important to note that in the majority of cichlids, broods are raised in the mouths of their parents (Breder & Rosen, 1966). This is also the prevailing pattern of brood care in the huge species flock of cichlids that have radiated in the Great Lakes of East Africa (Fryer & Iles, 1972). Mouthbrooding is an exceptionally efficient form of brood care, by which usually one, occasionally two parents can successfully protect their young until independence. There is virtually no scope for alloparental care. Only in the oldest of the three Great Lakes of East Africa, Lake Tanganyika, there is a tribe of an estimated 70 to 100 so-called "substrate-brooding" species that raise their offspring biparentally. Between a quarter and a third of these Lamprologini exhibit cooperative breeding (Taborsky, 1994, 2016; Heg & Bachar, 2006).

12.4.3 Niches

Group composition in cooperatively breeding cichlids may partly depend upon the environmental potential for niche construction. As we have described earlier, some *N. pulcher* colonies live on sand, where they dig out shelters under rocks that are dispersed on the sandy bottom. A comparison between different populations of *N. pulcher* showed that the number of small-sized helpers in a group increased markedly with the amount of sand cover in the environment (Groenewoud, *et al.*, 2016). This is due to the potential to dig out shelters in habitats with sandy bottom, which in turn enables many more individuals to safely hide within a territory. In addition, sand increases the mortality risk of small offspring (Taborsky & Limberger, 1981), which means that the demand for help increases when the bottom is sandy (i.e. in order to dig out breeding shelters, Bruintjes & Taborsky, 2011). As sandy habitat is used also by other cooperatively breeding cichlids (Kohler, 1998; Heg & Bachar, 2006; Heg, *et al.*, 2008; Tanaka, 2014), the environmental potential for niche construction might be an important evolutionary driver of social organization in cooperatively breeding fishes in general.

12.5 The Role of Evolutionary History in Shaping Sociality in Fishes

We argue that two intrinsic features serve as a precondition for the evolution of cooperative breeding in fishes: (1) a sufficient behavioral repertoire and complexity in social interactions; and (2) the existence of biparental brood care (see Komdeur & Ekman, 2010). Cichlids are exemplary for their enormous behavioral repertoire and flexibility, so if any group of fish meets the first of these conditions, it is to be members of this family. The second condition is met mainly by substrate-brooding species, as most of the mouthbrooders show uniparental care. Our hypothesis regarding the two above mentioned preconditions for the evolution of cooperative breeding is corroborated by the fact that in the lamprologine substrate-brooders of Lake Tanganyika, cooperative breeding seems to have independently evolved at least four or five times (Heg & Bachar, 2006; H. Tanaka, personal communication). The ecological trigger for the evolution of cooperative breeding in this tribe is most likely the high predation pressure through piscivorous fishes, as discussed above.

In tropical marine fishes, social species (as defined by those that form stable groups in which individuals develop dominance relationships) have so far been identified in at least seven families. Solitary or pair-forming species also occur in these families, suggesting that phylogenetic constraints do not play a substantial role in the development of social group living in coral-reef fishes. While it is impossible to accurately estimate the proportion of social species within this group owing to a lack of data on the social organization of many species, it appears that sociality has likely evolved multiple times in this group, suggesting that tropical marine environments, especially coral reefs, set the stage for the evolution of sociality in marine fishes.

II SOCIAL TRAITS

12.6 Traits of Social Species

A particular feature of fish in comparison with many other taxa exemplifying high levels of sociality (e.g. mammals, birds, and social insects) is the fact that the vast majority of fishes does not stop growing after reaching maturity. Therefore, in social fishes, groups usually consist of both small and large members which strongly affects behavior and social structure. In addition, sex determination is variable in teleosts, which adds to the diversity of reproductive and social patterns found in this taxon. The most significant feature differentiating freshwater from marine fishes is their diverging dispersal pattern. Most teleosts living in freshwater have limited dispersal due to the lack of planktonic stages, whereas marine fish are subject to much greater mixing among and between populations. This has important effects on social structure, as we will discuss later.

12.6.1 Cognition and Communication

Group members of cooperative cichlids specialize in various duties in dependence of their body size, dominance status, and relatedness (Bruintjes & Taborsky, 2011). This requires elaborate recognition and communication. In *N. pulcher*, helpers recognize kin through chemical cues (Le Vin, *et al.*, 2010), and relatedness asymmetries between helpers and male and female breeders, which are caused by sex-specific replacement rates of breeders, result in respective adjustments of helping effort (Stiver, *et al.*, 2005). Helpers related to female breeders were found to invest more in cooperative defense than helpers unrelated to female breeders, whereas helpers related to the male breeder showed lower defense effort than unrelated ones, as the latter need to pay more rent for being allowed to stay in the territory (Stiver, *et al.*, 2005). This was corroborated by studies in which the relatedness between helpers and breeders was experimentally manipulated, revealing that the cooperative effort of unrelated helpers exceeds that of related helpers considerably (Stiver, *et al.*, 2005; Zöttl, *et al.*, 2013b), which can be explained by reciprocal trading of different commodities. Such mutual exchange of services involves individual recognition, individual attribution of received behaviors and specific social memory.

Specialization in divergent duties within cooperative groups may depend upon context, and upon individual aptitude, condition, and personality. In *N. pulcher*, group members differ consistently in their propensity to perform certain tasks such as territory maintenance, defense, and exploration (Bergmüller & Taborsky, 2007; Schürch & Heg, 2010), implying strategic specialization at different life history stages (Bergmüller & Taborsky, 2007). Thereby, tasks are distributed efficiently among group members with different aptitudes (Bruintjes & Taborsky, 2011). In addition, subordinates vary consistently in their propensity to cooperate (Schürch & Heg, 2010). Personality type in this species explains the propensity to invest in helping behavior even better than does relatedness (Le Vin, *et al.*, 2011). Importantly, the existence of both temporal and persistent individual specializations of different group members in a variety of collaborative tasks demands permanent surveillance of other group members, and appropriate adjustments when individuals decide about their response to environmental and social challenges (Bruintjes & Taborsky, 2011; Riebli, *et al.*, 2012), which requires sufficient cognitive abilities. The consistent and heritable behavioral types of individuals (Chervet, *et al.*, 2011) partly explain communication and interactions among group members (Hamilton & Ligocki, 2012), as well as group composition.

In social reef fish, visual and tactile cues are used to detect and communicate with other species residing within corals, particularly crustaceans (Lassig, 1977), and chemical cues are used by corals to lure fish to attack and eat invading algae (Dixson & Hay, 2012). However, communication between fish within corals has yet to be quantified.

12.6.2 Lifespan and Longevity

Longer lifespans resulting in the overlapping of generations within groups and populations can ultimately drive habitat saturation and the evolution of sociality (known as the "life history" hypothesis, Arnold & Owens, 1998). A key prediction of this hypothesis is that social species are longer lived than non-social species. In general, cooperative breeders are usually long-lived, which coincides with late maturation and rather low fecundity (Arnold & Owens, 1998). This hypothesis appears to hold also for cooperatively breeding fishes, but the available data are scarce and there is no comparative data available from cichlids differing in social complexity. In *N. pulcher*, individually marked subjects exceeding five years of age were found in the field (A. Jungwirth, M. Zöttl & M. Taborsky, *unpublished data*) while in the aquarium individuals easily exceed 10 years of age, which seems remarkable for a fish of just 6 centimeters in length.

Amongst coral reef fishes, the diversity of lifespans seems enormous. In general, larger-bodied species are the most long-lived, with some reaching maximum ages of up to 45 years (Choat & Axe, 1996). In contrast, maximum lifespans of small-bodied species range from a few years to as little as 8 weeks (Depczynski & Bellwood, 2006; Hernamen & Munday, 2007; Herler, *et al.*, 2011). As yet, there have been no formal analyses of longevity between non-social and social coral reef fishes, making it difficult to assess how longevity might influence the evolution of sociality. Importantly, such a comparison must control for body size and phylogeny, since both could profoundly affect longevity. Available reports do suggest that the lifespan of pair-forming (i.e. non-social) species might be shorter than that of group living (i.e. social) species, at least within genera (e.g. in *Gobiodon*, *Amphiprion* and *Premnas*, Moyer, 1986; Munday, 2001; Buechler, 2005; Buston & Garcia, 2007). The non-social goby *Gobiodon histrio* has been reported to live for at least 4 years (Munday, 2001), whereas the social goby *G. okinawae* has a longer reported maximum lifespan of approximately 13 years (the latter was measured under aquarium conditions; Randall & Delbeek, 2009). Similarly in anemonefish, the loosely social *Amphiprion clarkia* has been reported to live up to 12 years (Moyer, 1986) and the strictly asocial anemonefish *Premnas biaculeatus* has been reported to live for 17 years (Buechler, 2005). In contrast, the strictly social species *A. percula* and *A. melanopus* have been reported to live for up to 30 and 38 years, respectively (Buechler, 2005; Buston & Garcia, 2007). These limited and largely anecdotal reports hint at the possibility that variation in social organization could be related to differential longevity in coral reef fishes.

12.6.3 Fecundity

In cooperatively breeding cichlids, clutch sizes vary between a few dozen and approximately 200 eggs, which does not seem to differ strongly from closely related non-social species of similar body size. However, the number of helpers in the group significantly influences clutch size in *N. pulcher*. Breeders with helpers produce larger clutches than those without helpers (Taborsky, 1984), which is apparently partly due to a strategic reduction of egg size when dominants take advantage of the help of many subordinate group members (Taborsky, *et al.*, 2007). Breeding females were shown to adjust their brood care effort to the amount of brood care help they received from subordinate group members, suggesting load-lightening effects and negotiation over investment in brood care (Zöttl, *et al.*, 2013c).

In social marine fishes, clutch sizes have been so far reported in the hundreds (Wong, *et al.*, 2008a; Buston & Elith, 2011) to up to several thousand eggs (Wong, *et al.*, 2012). Clutch size appears most commonly related to the amount of resources i.e. food availability (Wong, *et al.*, 2008a) and life history traits, such as growth rates (Buston & Elith, 2011), body size (Buston & Elith, 2011, but see Wong, *et al.*, 2008a), and breeding experience (Buston & Elith, 2011). In contrast, clutch size does not appear related to the presence, number, or phenotype of subordinates in a group (Buston & Elith, 2011), supporting the notion that breeders do not greatly benefit from the presence of non-breeders.

12.6.4 Age at First Reproduction

In *N. pulcher*, sexual maturity occurs at a body size of approximately 3.5 centimeters (standard length; Taborsky, 1985), which corresponds to an age of a bit less than one year under natural conditions (Skubic, *et al.*, 2004). This size is much too small, however, to obtain dominant breeder status in a group. Instead, both males and females of this size and age may start to take part in reproduction of the dominant breeders of their own and neighboring groups, as outlined above.

In marine fishes, age at first reproduction is typically governed by social conditions rather than occurring at a set age (Munday, *et al.*, 2006). The role of such social factors has been most rigorously demonstrated via breeder removal experiments, in which dominant breeders are removed resulting in the rapid maturation and subsequent breeding of previously immature, non-breeding subordinates (Wong, *et al.*, 2007). Many coral reef fishes have the capacity to change sex, commencing breeding as one sex rather than the other, in response to changes in social conditions. For example, removal of breeder males in protogynous species, such as cleaner wrasse (Labriidae), results in the sex change of the largest breeding female into the new male (Nakashima, *et al.*, 2000).

12.6.5 Dispersal

In freshwater, fish dispersal is often an active process involving the voluntary movement of young. In contrast to non-social freshwater fishes, where dispersal occurs at an early stage, the young of cooperative breeders delay dispersal enormously, which is comparable to cooperatively breeding birds and mammals. Typically, the fish are already sexually mature when dispersing. Sometimes individuals disperse into another group while retaining subordinate status, but such group switches are usually well prepared by repeated visits of potential target groups (Jungwirth, *et al.*, 2015b). Prospecting the environment before dispersal increases the survival chances of individual *N. pulcher* (Jungwirth, *et al.*, 2015b). Once an individual has decided to leave the natal group, it reduces its helping effort at home (Bergmüller, *et al.*, 2005; Zöttl, *et al.*, 2013d). It can afford to save effort in this way because if dominants decide that rent payment is not sufficient, eventual eviction from the territory is not so costly any more (Bergmüller, *et al.*, 2005).

Dispersal in marine fishes occurs via a planktonic larval phase in most species, whereby larvae disperse upon hatching and enter the pelagic zone for extended periods (often up to several months; Thorson, 1950; Leis, 1991). During this time, the larvae are widely dispersed by prevailing currents before settling back to the reef (Caley, *et al.*, 1996). As such, social groups of marine fishes are highly unlikely to exhibit significant patterns of above average relatedness (Buston, *et al.*, 2009). Further, groups are not formed via "delayed dispersal" of offspring from natal territories, a central tenet of cooperative breeding theory (Koenig, *et al.*, 1992). Rather, sociality arises via the settlement of larval recruits into groups or via the dispersal of individuals to groups post-settlement (Sweatman, 1983). These settlers are unrelated to most if not all other group members, although molecular verification of this is lacking for most species (but see Buston, *et al.*, 2009; Bernardi, *et al.* 2012). Therefore, there is no overlap of generations within a group and group members are typically unrelated (Buston, *et al.*, 2009). Post-settlement, many species are capable of secondary dispersal and movement to different locations (Sweatman, 1983; Hattori, 1991; Kuwamura, *et al.*, 2011). Secondary dispersal however is typically limited, and depends on many factors, including prevailing ecological conditions, such as the availability of suitable alternative territories, distance to suitable dispersal areas and the quality of surrounding habitats (Wong, 2010).

Even if ecological conditions favor dispersal, acceptance of dispersers into a new group or habitat is not guaranteed in marine fishes. Groups are usually not "free-entry"; rather, the identity and number of individuals that are allowed to enter a social group is strictly controlled. Social conflict between group members has been shown to constrain maximum group size in a range of social species (Buston & Cant, 2006; Ang & Manica, 2010b) and hence presumably influences the benefits of dispersal. Dispersers (or settlers) have been shown to prefer coral colonies containing conspecifics (*Dascyllus* spp., Sweatman, 1983; Booth, 1992). Apart from potential anti-predator benefits, such preference may arise because the presence of adults indicates the appropriateness of the site for growth, survival and reproduction

(Sweatman, 1983). Joiners can also refine their dispersal decisions based upon more subtle aspects of group structure, including relative body sizes, familiarity of conspecifics within the group (Jordan, *et al.*, 2010), and the avoidance of aggression from residents (cf. Taborsky, 1985). In *D. aruanus*, juveniles typically settle in coral colonies that are separate but less than 150 centimeters away from an adult colony, thereby avoiding the stress incurred from adult aggression yet still reaping some of the benefits of joining an existing adult group obtained via close proximity (Ben-Tsvi, *et al.*, 2009). From the resident's perspective, some adults do not aggressively exclude juveniles (Sweatman, 1983), suggesting that they may benefit from the additional group members. Conversely, residents of other species do aggressively exclude juveniles (Ben-Tsvi, *et al.*, 2009), although once again, settler exclusion may depend on more subtle aspects of group demographics (Jordan, *et al.*, 2010). All in all, even amongst social species there appear to be variations in the costs and benefits of dispersal that are driven by complex social interactions between settlers and residents.

12.6.6 Other Traits: Body Size

The most important trait affecting social interactions and intraspecific relationships in fish is body size. In contrast to other major taxa such as mammals, birds, and insects, in most fish species individuals do not stop growing when sexually mature, so typically social partners exhibit a large range of body sizes (Taborsky, 1999, 2008). This causes asymmetries among them, which usually results in strict, size-dependent dominance hierarchies and sometimes specialized social roles (Taborsky, *et al.*, 1986; Hamilton, *et al.*, 2005; Bruintjes & Taborsky, 2011; Reddon, *et al.*, 2011). This is of particular significance in group living species like *N. pulcher*, because if tasks are shared by several group members, there is scope for a size-dependent polyethism (Taborsky & Limberger, 1981; Taborsky, *et al.*, 1986; Bruintjes & Taborsky, 2011). Growth in this species depends on social rank, group composition, and cooperative effort, which is very energy demanding (Taborsky, 1984; Grantner & Taborsky, 1998; Riebli, *et al.*, 2012). The flexibility of growth rates and hence body sizes in fishes also enables subordinates to remain smaller than their dominants for extended periods of time. Such size-structuring is actually a form of conflict resolution, whereby subordinates regulate their growth to avoid growing too large and hence becoming a threat to their immediate dominants in the queue for breeding (Heg, *et al.*, 2004b; Buston & Cant, 2006; Wong, *et al.*, 2007).

12.7 Traits of Social Groups

12.7.1 Genetic Structure

In cooperative cichlids, relatedness among group members varies greatly both between and within species (Kohler, 1998; Awata, *et al.*, 2005; Dierkes, *et al.*, 2005; Taborsky, 2009), which affects cooperation among group members (Stiver, *et al.*, 2005; Zöttl, *et al.*, 2013b). Reproduction may be highly skewed or rather balanced among group members (Taborsky, 2009), but this variation can neither be easily explained by within-group relatedness and potential inbreeding avoidance (Stiver, *et al.*, 2008), nor by the variation in the availability of alternative breeding options (Heg, *et al.*, 2006).

In *N. pulcher*, groups consist of a mixture of related and unrelated individuals, which results from a combination of delayed dispersal of young with frequent breeder replacement, after which the helpers usually stay with the new territory owners (Balshine-Earn, *et al.*, 1998; Dierkes, *et al.*, 2005). In addition, relatedness between breeders and helpers is reduced by the production of extra-pair young by helpers of both sexes, and by unrelated neighboring males (Taborsky, 1985, 2016; Stiver, *et al.*, 2009; Hellmann, *et al.*, 2015). Experiments have revealed that relatedness among group members does affect the propensity of subordinates to expend cooperative effort, but opposite to predictions derived from kin selection theory. In the first of these studies, the overall cooperative effort of unrelated helpers exceeded that of related helpers about ten-fold (Stiver, *et al.*, 2005). In the second study, groups were experimentally assembled to contain either only related or only unrelated female helpers. Unrelated helpers showed significantly more direct egg care than related helpers. Cleaning and fanning eggs are purely altruistic behaviors that do not provide any direct fitness benefits to alloparents. In addition, unrelated helpers also invested more in digging out the breeding shelter than related ones, which is the energetically most demanding behaviour in these fish (Grantner & Taborsky, 1998). Hence, relatedness among group members in *N. pulcher* reduces cooperation rather than enhancing it (Zöttl, *et al.*, 2013b).

Social groups of coral reef fishes are generally thought to completely lack kin structure, thereby contrasting with the relatedness characteristics of social cichlids and many terrestrial social taxa. This lack of genetic relatedness arises because, as we discussed above, many benthic or sedentary marine organisms have a pelagic larval stage. Despite this generality, some instances of kinship could arise between group members in social coral reef fishes. The traditional assumption that reef fish larvae are simply swept around at the mercy of prevailing currents has been debunked by a series of studies demonstrating that larvae often settle back onto their natal reefs, either due to larval retention or recolonization (Swearer, *et al.*, 2002; Jones, *et al.*, 2005). Further, larvae of various social reef fishes use both visual and chemical cues on which to base settlement preferences (e.g. Booth, 1992; Arvelund, *et al.*, 1999). Indeed, a microsatellite analysis of kinship within social groups of the coral-dwelling damselfish, *Dascyllus aruanus* (Pomacentridae), revealed a subtle opportunity for kin selection within groups. Although mean within-group relatedness was not significantly greater than zero, a comparison of pair-wise relatedness values between individuals within groups revealed that some were closely related (Buston, *et al.*, 2009). These individuals were generally small and similar-sized, suggesting they had recruited together to the same group, generating the potential for kin selection to influence the evolution of social behavior (Figure 12.3). This remarkable observation highlights the need to verify the assumption of a lack of relatedness within social groups of marine fishes in general and to

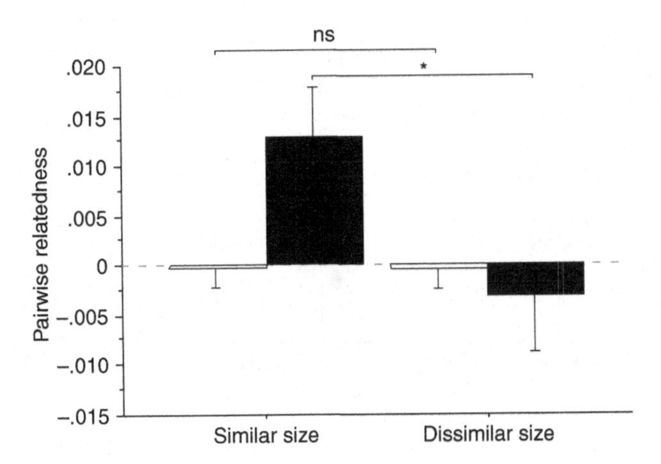

Figure 12.3 Relationship between relatedness and body size between members of the same group (black bars) and different groups (white bars) in the damselfish *Dascyllus aruanus*. Small group members show a significantly greater relatedness among each other than large group members, which might indicate joint recruitment. Nevertheless, the absolute relatedness levels are low. From Buston, *et al.*, (2009).

study the potential existence of kin recognition, particularly in light of testing the applicability of kin selection theory towards understanding the evolution of sociality in the marine environment.

12.7.2 Group Structure, Breeding Structure and Sex Ratio

The most important social unit in cooperatively breeding cichlids is the breeding group, which consists of dominant breeders, subordinate helpers and the young produced in the territory (Taborsky & Limberger, 1981; Martin & Taborsky, 1997; Heg, *et al.*, 2005a, b). Group sizes differ between species and the numbers of immature helpers is particularly variable, but the number of reproductively mature helpers in a group is generally low (i.e. on average between 1.3 and 2.3 mature helpers in four species where the respective information is available, Taborsky, 2009) (Figure 12.4). All group members in these species may participate in territory defense and maintenance, and in direct brood care such as cleaning and fanning eggs and larvae. The groups are size structured (Dey, *et al.*, 2013), and the division of labor depends on the size of subordinates (Bruintjes & Taborsky, 2011) and on their participation in reproduction (Kohda, *et al.*, 2009).

Cooperatively breeding groups of *Neolamprologus* species (*pulcher, savory, multifasciatus*) are often assembled in aggregations or colonies (Limberger, 1983; Kohler, 1998; Heg, *et al.*, 2005a,b, 2008), which seems to improve survival chances (Heg, *et al.*, 2008) and the economy of anti-predator behavior (Jungwirth, *et al.*, 2015a). In contrast, groups of the cooperatively breeding cichlids *Julidochromis marlieri* and *J. ornatus* are rather distributed with territory borders that are typically not contiguous

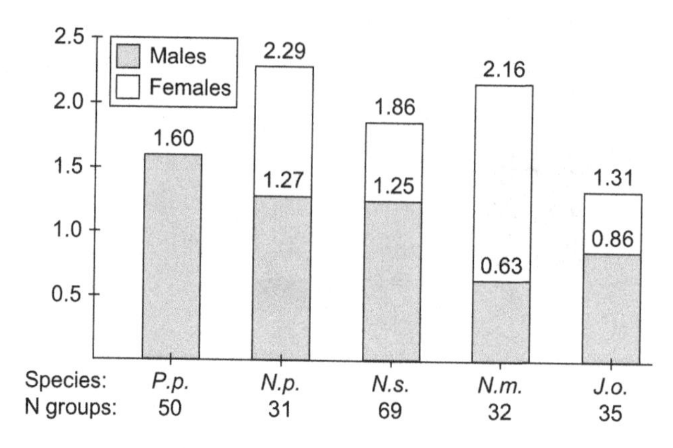

Figure 12.4 Average number and sex ratio of mature brood care helpers within groups of five species of cichlids, including *Pelvicachromis pulcher* (*P.p.*; Martin & Taborsky, 1997), *Neolamprologus pulcher* (*N.p.*; Dierkes, *et al.*, 2005), *N. savoryi* (*N.s.*; Heg, *et al.*, 2005b), *N. multifasciatus* (*N.m.*; Kohler, 1998; Taborsky, 2001) and *Julidochromis ornatus* (*J.o.*; Awata, *et al.*, 2005; Heg & Bachar, 2006). From Taborsky (2009).

(Taborsky, 1994; Yamagishi & Kohda, 1996; Heg & Bachar, 2006; M. Taborsky, personal observations). Due to this different spatial structure of populations the exchange of group members and reproductive participation of neighbors are hence likely to differ between cooperatively breeding *Neolamprologus* and *Julidochromis* species (Awata, *et al.*, 2005; Stiver, *et al.*, 2009).

Social groups of marine fishes are also not random aggregations of individuals; rather, individuals within groups interact regularly and are organized into a dominance hierarchy. In all the studied cases so far, these hierarchies are size-based. Many of these subordinates are of a size where they are reproductively capable, although in socially despotic species they are reproductively suppressed by dominant breeders (Buston, 2003; Wong, *et al.*, 2007) (Figure 12.1E & F). In contrast to social cichlids and most terrestrial taxa, hierarchies of social coral reef fishes are single sex hierarchies owing to the occurrence of sequential hermaphroditism (sex change) exhibited in these species. As in social cichlids, individuals can inherit the breeding position within the group, or disperse to locate better opportunities for rank ascension or breeding (Hattori, 1991), although their extreme site attachment coupled with a lack of habitat vacancies and difficulties in predicting group structure in alternative groups typically means that peaceful queuing is the only viable option (Wong, *et al.*, 2007).

Conflict over rank within queues plays a major role in determining social structure at the group level in social coral reef fishes. For each species studied to date, there exist remarkably strict size ratios (body size of subordinate/dominant) between subordinates and their immediate dominants in the queue (Buston & Cant, 2006; Wong, *et al.*, 2007; Ang & Manica, 2010a,b). These size ratios are maintained via the regulation of subordinate growth rates through the modulation of feeding rates

(Wong, *et al.*, 2008b; Ang & Manica, 2010b), and this form of subordinate cooperation is enforced by the threat of eviction (Buston & Cant, 2006; Wong, *et al.*, 2007) and/or direct aggression by dominants (Ang & Manica, 2010b). This is similar in cooperatively breeding cichlids. Experiments revealed that in *N. pulcher*, large male helpers keep a certain size margin to the dominant male breeder, whereas female helpers are much less affected by the dominant female's size (Heg, *et al.*, 2004b; Heg, 2010).

In contrast to many social birds and mammals, subordinate group members in cooperatively breeding cichlids may be immature or mature, and individuals of either one or both sexes help in brood care and other duties (only mature males in *Pelvicachromis pulcher*, Martin & Taborsky, 1997; immatures, and mature males and females in *N. pulcher*, *savory* and *multifasciatus*, and in *J. marlieri, ornatus* and *transcriptus*, Taborsky 1985, 1994; Yamagishi & Kohda 1996; Kuwamura, 1997; Kohler, 1998; Sunobe, 2000; Awata, *et al.*, 2005; Heg, *et al.*, 2005b; Heg & Bachar, 2006; Dierkes, *et al.*, 2008). In the case of mature male and female helpers, the sex ratios of subordinate group members diverge between species, being male biased in *J. ornatus* and *N. savoryi*, and female biased in *N. multifasciatus* and partly in *N. pulcher* (Taborsky, 1985, 2001; Kohler, 1998; Awata, *et al.*, 2005; Heg, *et al.*, 2005b; Heg & Bachar, 2006). If the sex ratio of subordinates within groups is biased in one direction, it may be biased in the opposite direction in similar-sized, non-breeding individuals living outside of groups (Taborsky, 1985; Heg & Bachar, 2006).

Sex ratios of helpers within groups may also vary between different populations of a species; in a northern population of *N. pulcher*, for instance, the sex ratio of reproductively mature non-breeders was 2:1 (females:males) within breeding groups, whereas it was exactly opposite (1:2) in non-breeding aggregations in the vicinity of breeding groups (Taborsky, 1985). In contrast, in a southern population of the same species, the sex ratio of subordinates within breeding groups was slightly biased towards males (0.65:1, not significantly different from 1:1, Balshine, *et al.*, 2001; Dierkes, *et al.*, 2008). This population divergence relates to a significant difference in the social organization of non-reproductives between these populations; in the north of Lake Tanganyika, many non-reproductives stay in localized, stable aggregations in the vicinity of breeding groups (Taborsky & Limberger, 1981; Limberger, 1983), whereas in the south, non-group members are rare and do not aggregate (Heg, *et al.*, 2005a).

The sexes of mature helpers are important for the potential costs from reproductive competition with dominant breeders. Male helpers pose a greater cost to male breeders than female helpers do for female breeders, because fertilizations by male helpers reduce the reproductive success of male breeders, which is not the case if female helpers lay eggs within the breeders' territory (Taborsky, 1985, 2016; Heg & Hamilton, 2008). Accordingly, experiments revealed that the dominant breeders adjust their behavior in response to the sexes of their helpers in *N. pulcher* (Mitchell, *et al.*, 2009).

All social coral reef fishes studied to date exhibit some form of hermaphroditism, and this cardinal life history trait determines the sex ratio observed within groups. Socially despotic species that exhibit monogamy within groups show either *protogynous* (i.e. female to male), protandrous (i.e. male to female), or bidirectional (both ways) sex change (reviewed in Warner, 1984; Munday, *et al.*, 2006). On the other hand, socially egalitarian species are predominantly protogynous (Schwarz & Smith, 1990; Nakashima, *et al.*, 2000; Hamaguchi, *et al.*, 2002; Cole, 2003; Asoh, 2003), with some capacity for bidirectional sex change (Sunobe & Nakazono, 1999; Manabe, *et al.*, 2007; Kuwamura, *et al.*, 2011). In protogynous species, juveniles differentiate into females (this is usually true for bi-directional sex changers too) before assuming the male role; conversely for protandrous species, juveniles differentiate into males (Sadovy & Shapiro, 1987; Cole & Hoese, 2001). Adults can then change sex when the social conditions are appropriate; typically, upon death or disappearance of the most dominant group member, the next largest group member will change sex to replace the dominant (Munday, *et al.*, 2006). Thus, the sex ratio of groups is heavily female biased in protogynous species, male biased in protandrous species (Sadovy & Shapiro, 1987), and typically 1:1 in bi-directional species (Nakashima, *et al.*, 2000; Cole & Hoese, 2001).

III SOCIAL SYNTHESIS

12.8 A Summary of Fish Sociality

The range of social patterns in fish encompasses virtually the entire variety of sociality found in animals, probably with the only exception that there are no eusocial fishes known (but see Ruxton, *et al.*, 2014). Social organization in fish ranges from loose aggregations with highly dynamic membership, which usually serve a limited set of purposes, to durable and complex groups consisting of individuals varying in age, sex, reproductive and social roles. Individuals of different sizes face divergent opportunities. For example, the reproductive differentiation of males is often greatly determined by body size, with large, "bourgeois" males investing in the defense of resources, attraction of mates and in brood care, whereas small "parasitic" males specialize in exploiting the effort of resource holders (Taborsky, 1994, 2008). If group membership is essential for survival, small individuals may pay to be allowed to stay within a protected territory by investing in different duties that serve primarily its large and dominant owners (Taborsky, 1985; Bergmüller & Taborsky, 2005; Heg & Taborsky, 2010; Zöttl, *et al.*, 2013a; Fischer, *et al.*, 2014). If small group members do not pose net costs to dominants, they may be allowed to stay and queue for dominance even when not paying rent for residence (Mitchell, 2003; Wong, *et al.*, 2007). Similarly, satellite males participating in reproduction at bourgeois males' nests can compensate for the costs they cause to the latter by behavioral effort such as territory defense, which may either neutralize or overcompensate the costs they occasion (Taborsky, *et al.*, 1987; Oliveira, *et al.*, 2002; Hamilton & Taborsky, 2005; Stiver & Alonzo, 2013).

Despite the fact that both freshwater and marine fishes exhibit sociality, there are striking differences in the characteristics of sociality between these groups. Most notably, subordinates of tropical coral reef fishes do not increase breeder fitness by providing alloparental care within groups (Ang & Manica, 2011), as is performed by subordinates of freshwater cichlids via territory defense, maintenance, and broodcare (Taborsky, 1984, 2016). The fact that the eggs of marine fish hatch and larvae disperse into the water column also likely precludes the formation of kin groups and the occurrence of natal philopatry (Buston, *et al.*, 2009). This has consequences for the role, or lack thereof, of indirect fitness benefits in promoting sociality (Wong & Buston, 2013). Further, differences in social structure arise owing to the occurrence of hermaphroditism in marine species, which introduces a uniquely biased sex ratio within groups that is not apparent in freshwater species (Munday, *et al.*, 2006).

12.9 Comparative Perspectives on Fish Sociality

Two peculiarities of fish make a difference in the forms of sociality that we find in this taxon in comparison to other highly social animals, such as many insects, birds, and mammals: (1) indeterminate growth; and (2) the flexibility of sex determination. As most fish species do not stop growing when sexually mature, there is usually a great range of sizes in the adult population, which substantially influences reproductive and social opportunities and roles of individuals (Taborsky, 1994, 1999; Bruintjes & Taborsky, 2011; Wong & Buston, 2013). Similarly, the fact that the sexes of fish may be determined by genetic, environmental, or social mechanisms in different species (Devlin & Nagahama, 2002) creates an enormous variation in reproductive and social patterns (Warner, 1984; Ross, 1990; Munday, *et al.*, 2006).

Another characteristic of fish social systems apparently differentiating them from other animal taxa is the high variance and low average level of relatedness within groups (Dierkes, *et al.*, 2005; Buston, *et al.*, 2009; Bernardi, *et al.*, 2012; Kamel & Grosberg, 2013). This raises the question of what evolutionary mechanisms can generate cooperation among social partners, as the prevailing explanation for cooperative breeding and other forms of seemingly altruistic behaviour is based on the preferential interaction among close kin (Hamilton, 1964). Kin selection often does not appear to explain cooperation among conspecifics in fish, partly because interacting partners are unrelated among each other. In groups consisting of related and unrelated individuals, the degree of cooperation may even be higher among unrelated social partners that trade different commodities and services against each other, than among close relatives (Stiver, *et al.*, 2005; Zöttl, *et al.*, 2013b). This represents one of the most obvious examples of reciprocal commodity trading among animals (Taborsky, 2013; Taborsky, *et al.*, 2016), illustrating that negotiation and trading can be more effective drivers of cooperation and altruism than kin selection (Quinones, *et al.*, 2016).

Finally, the most important ecological factors causing group formation and cooperation in fish apparently diverge from other cooperative vertebrates. In most birds and mammals, the major cause for delayed dispersal and cooperative brood care seems to be the limitation of suitable breeding territories (Koenig, *et al.*, 1992).

In these taxa, groups usually defend long-term, all-purpose territories containing all the resources required to survive and reproduce (Koenig & Dickinson, 2016). In contrast, in cooperatively breeding fish the major ecological factor selecting for group formation seems to be predation risk (Taborsky, 2016). This can be exemplified by the ecology of cooperative breeding in the hitherto most extensively studied social cichlid, *N. pulcher* (Taborsky, 1984, 2016; Balshine, *et al.*, 2001; Heg, *et al.*, 2004a; Groenewoud, *et al.*, 2016). In this species, experimental exposure of groups to different types of predators revealed that both dominants and subordinates benefit from each other's presence and collaboration (Taborsky, 1984; Heg, *et al.*, 2004a; Bruintjes & Taborsky, 2011). The fact that in nature almost all reproducing pairs have brood care helpers suggests that in addition, successful raising of young requires alloparental care and protection from predators, which was confirmed by a field experiment (Brouwer, *et al.*, 2005). Furthermore, this species breeds in colonies, and long-term surveys and experiments in nature revealed that on top of group membership, the affiliation to a breeding colony is important for survival and group persistence (Heg, *et al.*, 2005a, 2008; Jungwirth & Taborsky, 2015; Jungwirth, *et al.*, 2015a). Hence, in *N. pulcher* predation risk is apparently the primary ecological factor responsible for group living, colonial breeding, and cooperative brood care (Taborsky, 2016), which may apply also to other cooperative cichlids (Tanaka, 2014).

12.10 Concluding Remarks

Given that about half of the estimated 66,000 species of vertebrates on earth are fishes, the knowledge we have about the social systems of just a few dozen species conveys a clear message: sociality of in fish is appallingly understudied. The few well studied systems already demonstrate that the social systems of fish are exceptionally diverse, encompassing virtually the entire range of social systems found in vertebrates at large. Given that fish constitute the basis of vertebrate evolution, this group would deserve much greater attention if we wish to unravel general principles of sociality. In addition, social behavior of fish can be studied experimentally, both under highly controlled laboratory conditions and in the field, which is rare among other vertebrates. Hence, the most important message this review may entail is a plea for directing strong future research effort on the ultimate and proximate mechanisms of sociality in fish.

The overview we attempted to give here on social systems of fish both in freshwater and marine environments cannot be comprehensive for the simple reason that our knowledge of fish sociality is so limited. We have tried to give a glimpse of the enormous diversity of fish social systems and of the principles of social evolution that can be scrutinized exceptionally well in this taxon. However, we stress that this can provide only a very cursory insight, indeed. We look forward to learning much more about the diversity of sociality and the underlying mechanisms based on studies of many more fish model systems in the future. This will also enable the application of proper comparative analysis of social patterns within and between fish taxa, which is currently hampered by a lack of basic life history information.

References

Allen, G. R. (1972) *The Anemonefishes: Their Classification and Biology. 2nd ed.* Neptune City: T.F.H. Publications

Ang, T. Z. & Manica, A. (2010a) Aggression, segregation and stability in a domiance hierarchy. *Proceedings of the Royal Society of London B*, **277**, 1337–1343.

(2010b) Unavoidable limits on group size in a body size-based linear hierarchy. *Behavioral Ecology*, **21**, 819–825.

(2011) Effect of the presence of subordinates on dominant female behaviour and fitness in hierarchies of the dwarf angelfish *Centropyge bicolor*. *Ethology*, **117**, 1111–1119.

Arnold, K. E. & Owens, I. P. F. (1998) Cooperative breeding in birds: A comparative test of the life history hypothesis. *Proceedings of the Royal Society of London B*, **265**, 739–745.

Arvelund, M., McCormick, M. I., Fautin, D. G., & Bildsoe, M. (1999) Host recognition and possible imprinting in the anemonefish *Amphprion melanopuls* (Pisces: Pomacentridae). *Marine Ecology Progress Series*, **188**, 207–218.

Asoh, K. (2003) Gonadal development and infrequent sex change in a population of the humbug damselfish, *Dascyllus aruanus*, in continuous coral-cover habitat. *Marine Biology*, **142**, 1207–1218.

Awata, S., Munehara, H., & Kohda, M. (2005) Social system and reproduction of helpers in a cooperatively breeding cichlid fish (*Julidochromis ornatus*) in Lake Tanganyika: Field observations and parentage analyses. *Behavioral Ecology and Sociobiology*, **58**, 506–516.

Balshine, S. & Buston, P. M. (2008) Cooperative behaviour in fishes. *In*: Magnhagen, C., Braithwaite, V.A., Forsgren, E., & Kapoor, B.G. (eds.) *Fish Behaviour Boca Raton*, USA: CRC Press, pp. 437–484.

Balshine, S., Leach, B., Neat, F., Reid, H., Taborsky, M., *et al.* (2001) Correlates of group size in a cooperatively breeding cichlid fish. *Behavioral Ecology and Sociobiology*, **50**, 134–140.

Balshine-Earn, S., Neat, F. C., Reid, H., & Taborsky, M. (1998) Paying to stay or paying to breed? Field evidence for direct benefits of helping behavior in a cooperatively breeding fish. *Behavioral Ecology*, **9**, 432–438.

Baylis, J. R. (1974) The behavior and ecology of *Herotilapia multispinosa* (Teleostei, Cichlidae). *Zeitschrift für Tierpsychologie*, **34**, 115–146.

Ben-Tzvi, O., Kiflawi, M., Polak, O., & Abelson, A. (2009) The effect of adult aggression on habitat selection by settlers of two coral-dwelling damselfishes. *PLoS ONE*, **4**, e5511.

Bergmüller, R., Heg, D., & Taborsky, M. (2005) Helpers in a cooperatively breeding cichlid stay and pay or disperse and breed, depending on ecological constraints. *Proceedings of the Royal Society of London B*, **272**, 325–331.

Bergmüller, R. & Taborsky, M. (2007) Adaptive behavioural syndromes due to strategic niche specialization. *BMC Ecology*, **7**, 12.

Bernardi, G., Beldade, R., Holbrook, S. J., & Schmitt, R. J. (2012) Full-sibs in cohorts of newly settled coral reef fishes. *PLoS ONE*, **7**, e44953.

Biedermann, P. H. (2014) Evolution of cooperation in ambrosia beetles. *Mitteilungen der Deutschen Gesellschaft fur Allgemeine und Angewandte Entomologie*, **19**, 191–201.

Bisazza, A., Marconato, A., & Marin, G. (1989) Male competition and female choice in *Padogobius martensi* (Pisces: Gobiidae). *Animal Behaviour*, **38**, 406–413.

Booth, D. J. (1992) Larval settlement patterns and preferences by domino damselfish *Dascyllus albisella* Gill. *Journal of Experimental Marine Biology and Ecology*, **155**, 85–104.

Bourke, A. F. G. (2011) *Principles of Social Evolution*. Oxford, UK: Oxford University Press.

Breder, C. M. & Rosen, D. E. (1966) *Modes of Reproduction in Fishes*. Garden City, NY: Natural History Press.

Brouwer, L., Heg, D., & Taborsky, M. (2005) Experimental evidence for helper effects in a cooperatively breeding cichlid. *Behavioral Ecology*, **16**, 667–673.

Bruintjes, R. & Taborsky, M. (2008) Helpers in a cooperative breeder pay a high price to stay: Effects of demand, helper size and sex. *Animal Behaviour*, **75**, 1843–1850.

(2011) Size-dependent task specialization in a cooperative cichlid in response to experimental variation of demand. *Animal Behaviour*, **81**, 387–394.

Bruintjes, R., Hekman, R., & Taborsky, M. (2010) Experimental global food reduction raises resource acquisition costs of brood care helpers and reduces their helping effort. *Functional Ecology*, **24**, 1054–1063.

Bshary, R. & Grutter, A. S. (2006) Image scoring and cooperation in a cleaner fish mutualism. *Nature*, **441**, 975–978.

Bshary, R., Hohner, A., Ait-el-Djoudi, K., & Fricke, H. (2006) Interspecific communicative and coordinated hunting between groupers and giant moray eels in the Red Sea. *PLoS Biology*, **4**, 2393–2398.

Buechler, K. (2005) An evaluation of the geographic variation in the life history and behaviour of anemonefishes: A common garden approach. Ph.D. Thesis, James Cook University, Australia.

Burchard, J. E. Jr. (1965) Family structure in the dwarf cichlid *Apistogramma trifasciatum* Eigenmann and Kennedy. *Zeitschrift für Tierpsychologie*, **22**, 150–162.

Burgerhout, E., Tudorache, C., Brittijn, S. A., Palstra, A. P., Dirks, R. P., *et al.* (2013) Schooling reduces energy consumption in swimming male European eels, *Anguilla anguilla* L. *Journal of Experimental Marine Biology and Ecology*, **448**, 66–71.

Buston, P. M. & Cant, M. A. (2006) A new perspective on size hierarchies in nature: Patterns causes and consequences. *Oecologia*, **149**, 362–372.

Buston, P. M. & Elith, J. (2011) Determinants of reproductive success in dominant pairs of clownfish: A boosted regression tree analysis. *Journal of Animal Ecology*, **80**, 528–538.

Buston, P. M. & García, M. B. (2007) An extraordinary life span estimate for the clown anemonefish. *Journal of Fish Biology*, **70**, 1710–1719.

Buston, P. M., Fauvelot, C., Wong, M. Y. L., & Planes, S. (2009) Genetic relatedness in groups of humbug damselfish *Dascyllus aruanus*: Small, similarly-sized individuals may be close kin. *Molecular Ecology*, **18**, 4707–4715.

Caley, M. J., Carr, M. H., Hixon, M. A., Hughes, T. P., Jones, G. P., *et al.* (1996) Recruitment and the local dynamics of open marine populations. *Annual Review of Ecology, Evolution, and Systematics*, **27**, 477–500.

Carey, F. G. & Lawson, K. D. (1973) Temperature regulation in free-swimming bluefin tuna. *Comparative Biochemistry and Physiology*, **44**, 375–392.

Chervet, N., Zöttl, M., Schurch, R., Taborsky, M., & Heg, D. (2011) Repeatability and heritability of behavioural types in a social cichlid. *International Journal of Evolutionary Biology*, **2011**, 321729.

Choat, J. H. & Axe, L. M. (1996) Growth and longevity in acanthurid fishes; An analysis of otolith increments. *Marine Ecology Progress Series*, **134**, 15–26.

Clifton, K. E. (1990). The costs and benefits of territory sharing for the Caribbean coral reef fish, *Scarus iserti. Behavioral Ecology and Sociobiology*, **26**, 139–147.

Cole, K. S. (1990) Patterns of gonad structure in hermaphroditic gobies (Teleostei: Gobiidae). *Environmental Biology of Fishes*, **28**, 125–142.

(2003) Hermaphroditic characteristics of gonad morphology and inferences regarding reproductive biology in Caracanthus (Teleostei, Scorpaeniformes). *Copeia*, **2003**, 68–80.

Cole, K. S. & Hoese, D. F. (2001) Gonad morphology, colony demography and evidence for hermaphroditism in *Gobiodon okinawae* (Gobiidae). *Environmental Biology of Fishes*, **61**, 161–173.

Constantz, G. D. (1979) Social dynamics and parental care in the tessellated darter (Pisces: Percidae). *Proceedings of the Academy of Natural Science of Philadelphia*, **131**, 131–138.

Coolen, I., van Bergen, Y., Day, R. L., & Laland, K. N. (2003) Species difference in adaptive use of public information in sticklebacks. *Proceedings of the Royal Society of London B*, **270**, 2413–2419.

Cote, I. M. (2000) Evolution and ecology of cleaning symbioses in the sea. *Oceanography and Marine Biology*, **38**, 311–355.

Croft, D. P., James, R., Thomas, P. O. R., Hathaway, C., Mawdsley, D., *et al.* (2006) Social structure and co-operative interactions in a wild population of guppies (*Poecilia reticulata*). *Behavioral Ecology and Sociobiology*, **59**, 644–650.

Danchin, E., Giraldeau, L. A., Valone, T. J., & Wagner, R. H. (2004) Public information: From nosy neighbors to cultural evolution. *Science*, **305**, 487–491.

Daniels, R. A. (1979) Nest guard replacement in the Antarctic fish *Harpagifer bispinis*: Possible altruistic behavior. *Science*, **205**, 831–833.

Delcourt, J. & Poncin, P. (2012) Shoals and schools: Back to the heuristic definitions and quantitative references. *Reviews in Fish Biology and Fisheries*, **22**, 595–619.

Depczynski, M. & Belwood, D. R. (2006) Ectremes, plasticity and invariance in vertebrate life history traits: Insights from coral reef fishes. *Ecology*, **87**, 3119–3127.

Devlin, R. H. & Nagahama, Y. (2002) Sex determination and sex differentiation in fish: An overview of genetic, physiological, and environmental influences. *Aquaculture*, **208**, 191–364.

Dey, C. J., Reddon, A. R., O'Connor, C. M., & Balshine, S. (2013) Network structure is related to social conflict in a cooperatively breeding fish. *Animal Behaviour*, **85**, 395–402.

Diaz-Munoz, S. L., DuVal, E. H., Krakauer, A. H., & Lacey, E. A. (2014) Cooperating to compete: Altruism, sexual selection and causes of male reproductive cooperation. *Animal Behaviour*, **88**, 67–78.

Dierkes, P., Heg, D., Taborsky, M., Skubic, E., & Achmann, R. (2005) Genetic relatedness in groups is sex-specific and declines with age of helpers in a cooperatively breeding cichlid. *Ecology Letters*, **8**, 968–975.

Dierkes, P., Taborsky, M., & Achmann, R. (2008) Multiple paternity in the cooperatively breeding fish *Neolamprologus pulcher. Behavioral Ecology and Sociobiology*, **62**, 1581–1589.

Dixson, D. L. & Hay, M. E. (2012) Corals chemically cue mutualistic fishes to remove competing seaweeds. *Science*, **338**, 804–807.

Duftner, N., Sefc, K. M., Koblmueller, S., Salzburger, W., Taborsky, M., *et al.* (2007) Parallel evolution of facial stripe patterns in the *Neolamprologus brichardilpulcher* species complex endemic to Lake Tanganyika. *Molecular Phylogenetics and Evolution*, **45**, 706–715.

Dugatkin, L. A. & Godin, J. G. J. (1992) Predator inspection, shoaling and foraging under predation hazard in the Trinidadian guppy, *Poecilia reticulata. Environmental Biology of Fishes*, **34**, 265–276.

Emlen, S. T. (1984) Cooperative breeding in birds and mammals. *In*: Krebs, J. R. & Davies, N.B. (eds.) *Behavioural Ecology*. Oxford: Blackwell Science, pp. 305–339

Fischer, S., Zöttl, M., Groenewoud, F., & Taborsky, B. (2014) Group-size-dependent punishment of idle subordinates in a cooperative breeder where helpers pay to stay. *Proceedings of the Royal Society of London B*, **281**, 20140184.

Fricke, H.W. (1980) Control of different mating systems in a coral reef fish by one environmental factor. *Animal Behaviour*, **28**, 561–569

Fryer, G. & Iles, T. D. (1972) *The Cichlid Fishes of the Great Lakes of Africa: Their Biology and Evolution*. Edinburgh: Oliver & Boyd.

Gashagaza, M. M. (1988) Feeding activity of a tanganyikan cichlid fish *Lamprologus brichardi*. *African Study Monographs*, **9**, 1–9.

Grantner, A. & Taborsky, M. (1998) The metabolic rates associated with resting, and with the performance of agonistic, submissive and digging behaviours in the cichlid fish *Neolamprologus pulcher* (Pisces: Cichlidae). *Journal of Comparative Physiology B*, **168**, 427–433.

Groenewoud, F., Frommen, J. G., Josi, D., Tanaka, H., Jungwirth, A., & Taborsky, M. (2016) Predation risk drives social complexity in cooperative breederes. *Proceedings of the National Academy of Sciences USA*, **113**, 4104–4109.

Hamaguchi, Y., Sakai, Y., Takasu, F., & Shigesada M. (2002) Modeling spawning strategy for sex change under social control in haremic angelfishes. *Behavioral Ecology*, **13**, 75–82.

Hamilton, I. M., Heg, D., & Bender, N. (2005) Size differences within a dominance hierarchy influence conflict and help in a cooperatively breeding cichlid. *Behaviour*, **142**, 1591–1613.

Hamilton, I. M. & Ligocki, I. Y. (2012) The extended personality: Indirect effects of behavioural syndromes on the behaviour of others in a group living cichlid. *Animal Behaviour*, **84**, 659–664.

Hamilton, I. M. & Taborsky, M. (2005) Unrelated helpers will not fully compensate for costs imposed on breeders when they pay to stay. *Proceedings of the Royal Society of London B*, **272**, 445–454.

Hamilton, W. D. (1964) The genetical evolution of social behaviour I & II. *Journal of Theoretical Biology*, **7**, 1–52.

Handegard, N. O., Boswell, K. M., Ioannou, C. C., Leblanc, S. P., Tjostheim, D. B., *et al.* (2012) The dynamics of coordinated group hunting and collective information transfer among schooling prey. *Current Biology*, **22**, 1213–1217.

Hankinson, T. L. (1920) Report on the investigations of fish of the Galien River, Berrien County, Michigan. *Occ.Pap.Mus.Zool.Univ.Mich.*, 89, 1014.

Hattori, A. (1991) Socially controlled growth and size-dependent sex change in the anemonefish *Amphiprion frenatus* in Okinawa, Japan. *Japanese Journal of Ichthyology*, **38**, 165–177.

Heg, D. (2010) Status-dependent and strategic growth adjustments in female cooperative cichlids. *Behavioral Ecology and Sociobiology*, **64**, 1309–1316.

Heg, D. & Bachar, Z. (2006) Cooperative breeding in the Lake Tanganyika cichlid *Julidochromis ornatus*. *Environmental Biology of Fishes*, **76**, 265–281.

Heg, D. & Hamilton, I. M. (2008) Tug-of-war over reproduction in a cooperatively breeding cichlid. *Behavioral Ecology and Sociobiology*, **62**, 1249–1257.

Heg, D. & Taborsky, M. (2010) Helper response to experimentally manipulated predation risk in the cooperatively breeding cichlid *Neolamprologus pulcher*. *PLoS ONE*, **5**, e10784.

Heg, D., Bachar, Z., Brouwer, L., & Taborsky, M. (2004a) Predation risk is an ecological constraint for helper dispersal in a cooperatively breeding cichlid. *Proceedings of the Royal Society of London B*, **271**, 2367–2374.

Heg, D., Bender, N., & Hamilton, I. (2004b) Strategic growth decisions in helper cichlids. *Proceedings of the Royal Society of London B*, **271**, S505-S508.

Heg, D., Brouwer, L., Bachar, Z., & Taborsky, M. (2005a) Large group size yields group stability in the cooperatively breeding cichlid *Neolamprologus pulcher*. *Behaviour*, **142**, 1615–1641.

Heg, D., Bachar, Z., & Taborsky, M. (2005b) Cooperative breeding and group structure in the Lake Tanganyika cichlid *Neolamprologus savoryi*. *Ethology*, **111**, 1017–1043.

Heg, D., Bergmüller, R., Bonfils, D., Otti, O., Bachar, Z., *et al.* (2006) Cichlids do not adjust reproductive skew to the availability of independent breeding options. *Behavioral Ecology*, 17, 419–429.

Heg, D., Heg-Bachar, Z., Brouwer, L., & Taborsky, M. (2008) Experimentally induced helper dispersal in colonially breeding cooperative cichlids. *Environmental Biology of Fishes*, **83**, 191–206.

Hellmann, J. K., Ligocki, I. Y., O'Connor, C. M., Reddon, A. R., Garvy, K. A., *et al.* (2015) Reproductive sharing in relation to group and colony-level attributes in a cooperative breeding fish. *Proceedings of the Royal Society of London B*, **282**, 20150954.

Henson, S. A. & Warner, R. R. (1997) Male and female alternative reproductive behaviours in fishes: A new approach using intersexual dynamics. *Annual Review of Ecology, Evolution, and Systematics*, **28**, 571–592.

Herler, J., Munday, P. M., & Hernaman, V. (2011) Gobies on coral reefs. *In*: Patzner, R. A., Van Tassell, J. L., Kovačić, M., & Kapoor, B. G. (eds.). *The Biology of Gobies*. Channel Islands, U.K: Science Publishers, pp. 493–529.

Hernamen, V. & Munday, P. L. (2007) Evolution of mating systems in coral reef gobies and constraints on mating system plasticity. *Coral Reefs*, **26**, 585–95.

Holbrook, S. J., Forrester, G. E., Schmitt, R. J. (2000) Spatial patterns in abundance of a damselfish reflect availability of suitable habitat. *Oecologia*, **122**, 109–120

Ioannou, C., Guttal, V., & Couzin, I. (2012) Predatory fish select for coordinated collective motion in virtual prey. *Science*, **337**, 1212–1215.

Johansen, J., Vaknin, R., Steffensen, J., & Domenici, P. (2010) Kinematics and energetic benefits of schooling in the labriform fish, striped surfperch *Embiotoca lateralis*. *Marine Ecology Progress Series*, **420**, 221–229.

Jones, A. G., Östlund-Nilsson, S., & Avise, J. C. (1998) A microsatellite assessment of sneaked fertilizations and egg thievery in the fifteenspine stickleback. *Evolution*, **52**, 848–858.

Jones, G. P., Planes, S., & Thorrold, S. R. (2005) Coral reef fish larvae settled close to home. *Current Biology*, **15**, 1314–1318.

Jordan, L. A., Avolio, C., Herbert-Read, J. E., Krause, J., Rubenstein, D. I., *et al.* (2010) Group structure in a restricted entry system is mediated by both resident and joiner preferences. *Behavioral Ecology and Sociobiology*, **64**, 1099–1106.

Jungwirth, A. & Taborsky, M. (2015) First- and second-order sociality determine survival and reproduction in cooperative cichlids. *Proceedings of the Royal Society of London B*, **282**: 20151971.

Jungwirth, A., Josi, D., Walker, J., & Taborsky, M. (2015a) Benefits of coloniality: Communal defense saves anti-predator effort in cooperative breeders. *Functional Ecology*, **29**, 1218–1224.

Jungwirth, A., Walker, J., & Taborsky, M. (2015b) Prospecting precedes dispersal and increases survival chances in cooperatively breeding cichlids. *Animal Behaviour*, **106**, 107–114.

Kamel, S. J. & Grosberg, R. K. (2013) Kinship and the evolution of social behaviours in the sea. *Biology Letters*, **9**, 20130454.

Kellogg, K. A., Markert, J. A., Stauffer, J. R., & Kocher, T. D. (1998) Intraspecific brood mixing and reduced polyandry in a maternal mouth-brooding cichlid. *Behavioral Ecology*, **9**, 309–312.

Koenig, W. D. & Dickinson, J. L. (eds.) (2016) *Cooperative Breeding in Vertebrates: Studies in Ecology, Evolution, and Behavior*. Cambridge: Cambridge University Press.

Koenig, W. D., Pitelka, F. A., Carmen, W. J., Mumme, R. L., & Stanback, M. T. (1992) The evolution of delayed dispersal in cooperative breeders. *Quarterly Review of Biology*, **67**, 111–150.

Kohda, M., Heg, D., Makino, Y., Takeyama, T., Shibata, J. Y., *et al.* (2009) Living on the wedge: Female control of paternity in a cooperatively polyandrous cichlid. *Proceedings of the Royal Society of London B*, **276**, 4207–4214.

Kohler, U. (1998) *Zur Struktur und Evolution des Sozialsystems von Neolamprologus multifasciatus (Cichlidae, Pisces), dem kleinsten Schneckenbuntbarsch des Tanganjikasees*. Aachen: Shaker Verlag.

Komdeur, J. & Ekman, J. (2010) Adaptations and constraints in the evolution of delayed dispersal: Implications for cooperation. *In*: Szekely, T., Moore, A. J., & Komdeur, J. (eds.) *Social Behaviour. Genes, Ecology and Evolution*. Cambridge, UK: Cambridge University Press, pp. 306–327.

Kraak, S. B. M. & Groothuis, T. G. G. (1994) Female preference for nests with eggs is based on the presence of the eggs themselves. *Behaviour*, **131**, 189–206.

Kuwamura, T. (1997) The evolution of parental care and mating systems among Tanganyikan cichlids. *In*: Kawanabe, H., Hori, M., & Nagoshi, M. (eds.) *Fish Communities in Lake Tanganyika*. Kyoto: Kyoto Univ. Press, pp. 57–86.

Kuwamura, T., Nakashima, Y., & Yogo, Y. (1994). Population dynamics of goby *Paragobiodon echinocephalus* and host coral *Stylophora pistillata*. *Marine Ecology Progress Series*, **103**, 17–23.

Kuwamura, T., Suzuki S., & Kadota, T. (2011) Reversed sex change by widowed males in polygynous and protogynous fishes: Female removal experiments in the field. *Naturwissenschaften*, **98**, 1041–1048.

Lassig, B. R. (1977) Field Observations on the reproductive behaviour of *Paragobiodon* spp. (Gobiidae) at Heron Island Great Barrier Reef. *Marine Behaviour and Physiology*, **3**, 283–293.

Le Vin, A., Mable, B., & Arnold, K. (2010) Kin recognition via phenotype matching in a cooperatively breeding cichlid, *Neolamprologus pulcher*. *Animal Behaviour*, **79**, 1109–1114.

Le Vin, A., Mable, B., Taborsky, M., Heg, D., & Arnold, K. (2011) Individual variation in helping in a cooperative breeder: relatedness versus behavioural type. *Animal Behaviour*, **82**, 467–477.

Leis, J. M. (1991) The pelagic phase of coral reef fishes: Larval biology of coral reef fishes. *In*: Sale, P. F. (ed). *The Ecology of Fishes on Coral Reefs*. San Diego: Academic Press, pp 183–230.

Liberman, T., Genin, A., & Loya, Y. (1995) Effects on growth and reproduction of the coral *Stylophora pistillata* by the mutualistic damselfish *Dascyllus marginatus*. *Marine Biology*, **121**, 741–746.

Limberger, D. (1983) Pairs and harems in a cichlifd fish, *Lamprologus brichardi*. *Zeitschrift für Tierpsychologie*, **62**, 115–144.

Maan, M. E. & Taborsky, M. (2008) Sexual conflict over breeding substrate causes female expulsion and offspring loss in a cichlid fish. *Behavioural Ecology*, **19**, 302–208.

Manabe, H., Ishimura, M., Shinomiya, A., & Sunobe, T. (2007) Field evidence for bi-directional sex change in the polygynous gobiid fish *Trimma okinawae*. *Journal of Fish Biology*, **70**, 600–609.

Martin, E. & Taborsky, M. (1997) Alternative male mating tactics in a cichlid, *Pelvicachromis pulcher*: A comparison of reproductive effort and success. *Behavioral Ecology and Sociobiology*, **41**, 311–319.

McKaye, K. R. & McKaye, N. M. (1977) Communal care and kidnapping of young by parental cichlids. *Evolution*, **31**, 674–681.

McKaye, K. R., Mughogho, D. E., & Lovullo, T. J. (1992) Formation of the selfish school. *Environmental Biology of Fishes*, **35**, 213–218.

Milinski, M., Kulling, D., & Kettler, R. (1990) Tit for tat: Sticklebacks (*Gasterosteus aculeatus*) "trusting" a cooperating partner. *Behavioral Ecology*, **1**, 7–11.

Mitchell, J. S., Jutzeler, E., Heg, D., & Taborsky, M. (2009) Dominant members of cooperatively-breeding groups adjust their behaviour in response to the sexes of their subordinates. *Behaviour*, **146**, 1665–1686.

Mizushima, N., Nakashima, Y., & Kuwamura, T. (2000) Semilunar spawning cycle of the humbug damselfish *Dascyllus aruanus*. *Journal of Ethology*, **18**, 105–108.

Moyer, J. T. (1986) Longevity of the anemonefish *Amphiprion clarkia* at Miyake-jima, Japan with notes on four other species. *Copeia*, **1986**, 135–139.

Munday, P. L. (2001) Fitness consequences of habitat use and competition among coral-dwelling fishes. *Oecologia*, **128**, 585–593.

Munday, P. L., Buston, P. M., & Warner, R. R. (2006) Diversity and flexibility of sex-change strategies in animals. *Trends in Ecology and Evolution*, **21**, 89–95.

Nakashima, Y., Sakai, Y., Karino, K., & Kuwamura, T. (2000) Female–female spawning and sex change in a haremic coral-reef fish, *Labroides dimidiatus*. *Zoological Science*, **17**, 967–970.

Oliveira, R. F., Carvalho, N., Miranda, J., Goncalves, E. J., Grober, M., *et al.* (2002) The relationship between the presence of satellite males and nest-holders' mating success in the Azorean rock-pool blenny *Parablennius sanguinolentus parvicornis*. *Ethology*, **108**, 223–235.

Partridge, B. L., Johansson, J., & Kalish, J. (1983) The structure of schools of giant bluefin tuna in Cape Cod Bay. *Environmental Biology of Fishes*, **9**, 253–262.

Perrone, M. & Zaret, T. M. (1979) Parental care patterns of fishes. *The American Naturalist*, **113**, 351–361.

Pitcher, T. J. & Parrish, J. K. (1993) Functions of shoaling behaviour. *In*: Pitcher, T. J. (ed.) *Behaviour of Teleost Fishes*. London: Champman & Hall, pp. 363–439.

Quiñones, A. E., van Doorn, S., Pen, I., Weissing, F. J., Taborsky, M. (2016) Negotiation and appeasement can be more effective drivers of sociality than kin selection. *Proceedings of the Royal Society of London B*, **371**, 20150089.

Randall, J. E. & Deelbeek, J. C. (2009) Comments on the extremes of longevity in fishes, with special reference to the Gobiidae. *Proceedings of the California Academy of Sciences*, **60**, 447–44.

Reddon, A. R., Voisin, M. R., Menon, N., Marsh-Rollo, S. E., Wong, M. Y., *et al.* (2011) Rules of engagement for resource contests in a social fish. *Animal Behaviour*, **82**, 93–99.

Ribbink, A. J., Marsh, A. C., & Marsh, B. A. (1981) Nest-building and communal care of young by *Tilapia rendalli* Dumeril (*Pisces, Cichlidae*) in Lake Malawi. *Environmental Biology of Fishes*, **6**, 219–222.

Riebli, T., Taborsky, M., Chervet, N., Apolloni, N., Zuercher, Y., *et al.* (2012). Behavioural type, status and social context affect behaviour and resource allocation in cooperatively breeding cichlids. *Animal Behaviour*, **84**, 925–936.

Robertson, D. R. & Hoffman, S. G. (1977) The roles of female mate choice and predation in the mating systems of some tropical Labroid fishes. *Zeitschrift fur Tierpsychologie*, **45**, 298–320.

Roopin, M., Thornhill, D. J., Santos, S. R., & Chadwick, N. E. (2011) Ammonia flux, physiological parameters, and Symbiodinium diversity in the anemonefish symbiosis on Red Sea coral reefs. *Symbiosis*, **53**, 63–74.

Ross, R. M. (1990) The evolution of sex-change mechanisms in fishes. *Environmental Biology of Fishes*, **29**, 81–93.

Ruxton, G. D., Humphries, S., Morrell, L. J., & Wilkinson, D. M. (2014) Why is eusociality an almost exclusively terrestrial phenomenon? *Journal of Animal Ecology*, **83**, 1248–1255.

Sadovy, Y. & Shapiro, D. (1987) Criteria for the diagnosis of hermaphroditism in fishes. *Copeia*, **1987**, 136–156.

Schaedelin, F. C., van Dongen, W. F., & Wagner, R. H. (2013) Nonrandom brood mixing suggests adoption in a colonial cichlid. *Behavioral Ecology*, **24**, 540–546.

Schürch, R. & Heg, D. (2010) Life history and behavioural type in the highly social cichlid *Neolamprologus pulcher*. *Behavioral Ecology*, **21**, 588–598.

Schwarz, A. L. & Smith, C. L. (1990) Sex change in the dameselfish *Dascyllus reticulatus* (Richardson) (Perciformes: Pomacentridae). *Bulletin of Marine Science*, **46**, 790–798.

Sefc, K. M., Hermann, C. M., Taborsky, B., & Koblmueller, S. (2012) Brood mixing and reduced polyandry in a maternally mouthbrooding cichlid with elevated among-breeder relatedness. *Molecular Ecology*, **21**, 2805–2815.

Skubic, E., Taborsky, M., McNamara, J. M., & Houston, A. I. (2004) When to parasitize? A dynamic optimization model of reproductive strategies in a cooperative breeder. *Journal of Theoretical Biology*, **227**, 487–501.

Solomon, N. G. & French, J. A. (1997) *Cooperative Breeding in Mammals*. Cambridge: Cambridge University Press.

Stewart, B. D. & Jones, G. P. (2001) Associations between the abundance of piscivorous fishes and their prey on coral reefs: Implications for pre-fish mortality. *Marine Biology*, **138**, 383–397.

Stiver, K. A. & Alonzo, S. H. (2013) Does the risk of sperm competition help explain cooperation between reproductive competitors? A study in the ocellated wrasse (*Symphodus ocellatus*). *The American Naturalist*, **181**, 357–368.

Stiver, K. A., Dierkes, P., Taborsky, M., Gibbs, H. L., & Balshine, S. (2005) Relatedness and helping in fish: Examining the theoretical predictions. *Proceedings of the Royal Society of London B*, **272**, 1593–1599.

Stiver, K., Fitzpatrick, J., Desjardins, J., Neff, B., Quinn, J., *et al.* (2008) The role of genetic relatedness among social mates in a cooperative breeder. *Behavioral Ecology*, **19**, 816–823.

Stiver, K., Fitzpatrick, J., Desjardins, J., & Balshine, S. (2009) Mixed parentage in *Neolamprologus pulcher* groups. *Journal of Fish Biology*, **74**, 1129–1135.

Sunobe, T. (2000) Social structure, nest guarding and interspecific relationships of the cichlid fish *Julidochromis marlieri* in Lake Tanganyika. *African Study Monographs*, **21**, 83–89.

Sunobe, T. & Nakazono, A. (1999) Mating system and hermaphroditism in the gobiid fish, *Priolepis cincta*, at Kagoshima, Japan. *Ichthyological Research*, **46**, 103–105.

Swearer, S. E., Thorrold, S. R., Shima, J. S., Hellberg, M. E., Jones, G. P., *et al.* (2002) Evidence of self-recruitment in demersal marine populations. *Bulletin of Marine Science*, **70**, 251–272.

Sweatman, H. P. A. (1983) Influence of conspecifics on choice of settlement sites by larvae of two pomacentrid fishes (*Dascyilus aruanus* and *D. reticulatus*) on coral reefs. *Marine Biology*, **75**, 225–229.

Taborsky, B., Skubic, E., & Bruintjes, R. (2007) Mothers adjust egg size to helper number in a cooperatively breeding cichlid. *Behavioral Ecology*, **18**, 652–657.

Taborsky, M. (1984) Broodcare helpers in the cichlid fish *Lamprologus brichardi* – their costs and benefits. *Animal Behaviour*, **32**, 1236–1252.

(1985) Breeder-helper conflict in a cichlid fish with broodcare helpers - an experimental analysis. *Behaviour*, **95**, 45–75.

(1987) Cooperative behaviour in fish: Coalitions, kin groups and reciprocity. *In*: Ito, Y., Brown, J. L. & Kikkawa, J. (eds.) *Animal Societies: Theories and Facts.* Tokyo: Japanese Science Society Press, pp. 229–237.

(1994) Sneakers, satellites, and helpers: Parasitic and cooperative behaviour in fish reproduction. *Advances in the Study of Behavior*, **23**, 1–100.

(1997) Bourgeois and parasitic tactics: Do we need collective, functional terms for alternative reproductive behaviours? *Behavioral Ecology and Sociobiology*, **41**, 361–362.

(1999) Conflict or cooperation: What determines optimal solutions to competition in fish reproduction? *In:* Oliveira, R., Almada, V. & Goncalves, E. (eds.) *Behaviour and Conservation of Littoral Fishes.* Lisboa: Instituto Superior de Psicologia Aplicada, pp. 301–349.

(2001) The evolution of parasitic and cooperative reproductive behaviours in fishes. *Journal of Heredity*, **92**, 100–110.

(2008) Alternative reproductive tactics in fish. *In*: Oliveira, R. F., Taborsky, M. & Brockmann, H. J. (eds.) *Alternative Reproductive Tactics: An Integrative Approach.* Cambridge, UK: Cambridge University Press, pp. 251–299.

(2009) Reproductive skew in cooperative fish groups: Virtue and limitations of alternative modeling approaches. *In*: Hager, R. & Jones, C. *Reproductive Skew in Vertebrates: Proximate and Ultimate Causes.* Cambridge, UK: Cambridge University Press, pp. 265–304.

(2013) Social evolution: Reciprocity there is. *Current Biology*, **23**, R486-R488.

(2016) Cichlid fishes: A model for the integrative study of social behavior. *In*: Koenig, W. D. & Dickinson, J. L. (eds.) *Cooperative Breeding in Vertebrates.* Cambridge, UK: Cambridge University Press, pp. 272–293.

Taborsky, M. & Grantner, A. (1998) Behavioural time-energy budgets of cooperatively breeding *Neolamprologus pulcher* (Pisces: Cichlidae). *Animal Behaviour*, **56**, 1375–1382.

Taborsky, M. & Limberger, D. (1981) Helpers in fish. *Behavioral Ecology and Sociobiology*, **8**, 143–145.

Taborsky, M. & Neat, F. (2010) Fertilization mode, sperm competition and cryptic female choice shape primary and secondary sexual characters in fish. *In*: Leonard, J. & Córdoba-Aguilar, A. (eds.) *The Evolution of Primary Sexual Characters in Animals.* Oxford, UK: Oxford University Press, pp. 379–408.

Taborsky, M., Hert, E., Siemens, M., & Stoerig, P. (1986) Social behaviour of *Lamprologus* species: Functions and mechanisms. *Annals of the Museum, Royal African Centre for Science, Zoology*, **251**, 7–11.

Taborsky, M., Hudde, B., & Wirtz, P. (1987) Reproductive behaviour and ecology of *Symphodus (Crenilabrus) ocellatus*, a European wrasse with four types of male behaviour. *Behaviour*, **102**, 82–118.

Taborsky, M., Frommen, J. G., & Riehl, C. (2016) Correlated pay-offs are key to cooperation. *Philosophical Transactions of the Royal Society B*, **371**: 20150084.

Tanaka, H. (2014) Social structure, dispersal and helping behaviour in the cooperatively breeding cichlid Neolamprologus obscurus. Ph.D. Thesis, Osaka City University, Japan.

Tanaka, H., Heg, D., Takeshima, H., Takeyama, T., Awata, S., *et al.* (2015) Group composition, relatedness, and dispersal in the cooperatively breeding cichlid *Neolamprologus obscurus*. *Behavioral Ecology and Sociobiology*, **69**, 169–181.

Thorson, G. (1950) Reproductive and larval ecology of marine bottom invertebrates. *Biological Reviews*, **25**, 1–45.

Thresher, R. E. (1985) Brood-directed parental aggression and early brood loss in the coral reef fish, *Acanthochromis polyacanthus* (Pomacentridae). *Animal Behaviour*, **33**, 897–907.

Trivers, R. L. (1985) *Social Evolution*. Menlo Park, CA. (USA): Benjamin/Cummings Publishing Comp. Inc.

Tunstrom, K., Katz, Y., Ioannou, C. C., Huepe, C., Lutz, M. J., *et al.* (2013) Collective states, multistability and transitional behaviour in schooling fish. *PLoS Computational Biology*, **9**, e1002915.

van Rooij, J. M., de Jong, E., Vaandrager, F., & Videler, J. J. (1996) Resource and habitat sharing by the stoplight parrotfish, *Sparisoma viride*, a Caribbean reef herbivore. *Environmental Biology of Fishes*, **47**, 81–91.

Videler, J. J. (1993) *Fish Swimming*. London: Chapman & Hall.

Ward, J. A. & Wyman, R. A. (1977) Ethology and ecology of cichlid fishes of the genus *Etroplus* in Srilanka: Preliminary findings. *Enviromnetal Biology of Fishes*, **2**, 137–145.

Warner, R. R. (1984) Mating behavior and hermaphroditism in coral reef fishes: The diverse forms of sexuality found among tropical marine fishes can be viewed as adaptations to their equally diverse mating systems. *American Scientist*, **72**, 128–136.

Weihs, D. (1973) Hydromechanics of fish schooling. *Nature*, **241**, 290–291.

Werner, N. Y., Balshine, S., Leach, B., & Lotem, A. (2003) Helping opportunities and space segregation in cooperatively breeding cichlids. *Behavioral Ecology*, **14**, 749–756.

Whiteman, E. A. & Côté, I. M. (2004) Monogamy in marine fishes. *Biological Reviews*, **79**, 351–375.

Wilson, E. O. (1975) *Sociobiology*. Cambridge, MA: Belknap Press.

Wisenden, B. D. (1999) Alloparental care in fishes. *Reviews in Fish Biology and Fisheries*, **9**, 45–70.

Wisenden, B. D. & Keenleyside, M. H. A. (1992) Intraspecific brood adoption in convict cichlids: Mutual benefit. *Behavioral Ecology and Sociobiology*, **31/4**, 263–269.

Wisenden, B. D., Mammenga, E. A., Storseth, C. N., & Berglund, N. J. (2014) Odour tracking by young convict cichlids and a mechanism for alloparental brood amalgamation. *Animal Behaviour*, **93**, 201–206.

Wong, M. Y. L. (2010). Ecological constraints and benefits of philopatry promote group living in a social but non-cooperatively breeding fish. *Proceedings of the Royal Society of London Series B*, 277, 353–358.

Wong, M. Y. L., Buston, P. M., Munday, P. L., & Jones, G. P. (2007) The threat of punishment enforces peaceful cooperation and stabilizes queues in a coral-reef fish. *Proceedings of the Royal Society of London B*, **274**, 1093–1099.

Wong, M. Y. L., Buston, P. M., Munday, P. L., & Jones, G. P. (2008a) Monogamy when there is potential for polygyny: Tests of multiple hypotheses in a group living fish. *Behavioral Ecology*, **19**, 353–361.

Wong, M. Y. L., Munday, P. L., Buston, P. M., & Jones, G. P. (2008b) Fasting or feasting in a fish social hierarchy. *Current Biology*, **18**, R372–373.

Wong, M. Y. L., Fauvelot, C., Planes, S., & Buston, P. M. (2012) Discrete and continuous reproductive tactics in a hermaphroditic society. *Animal Behaviour*, **84**, 897–906.

Yamagishi, S. & Kohda, M. (1996) Is the cichlid fish *Julidochromis marlieri* polyandrous? *Ichthyological Research*, **43**, 469–471.

Yanagisawa, Y. (1985) Parental strategy of the cichlid fish *Perissodus microlepis*, with particular reference to intraspecific brood "farming out". *Environmental Biology of Fishes*, **12**, 241–249.

Yanagisawa, Y. & Ochi, H. (1986) Step-fathering in the anemonefish *Amphiprion clarkii*: A removal study. *Animal Behaviour*, **34**, 1769–1780.

Zöttl, M., Frommen, J. G., & Taborsky, M. (2013a) Group size adjustment to ecological demand in a cooperative breeder. *Proceedings of the Royal Society of London B*, **280**, 20122772.

Zöttl, M., Heg, D., Chervet, N., & Taborsky, M. (2013b) Kinship reduces alloparental care in cooperative cichlids where helpers pay-to-stay. *Nature Communications*, **4**, 1341.

Zöttl, M., Fischer, S., & Taborsky, M. (2013c) Partial brood care compensation by female breeders in response to experimental manipulation of alloparental care. *Animal Behaviour*, **85**, 1471–1478.

Zöttl, M., Chapuis, L., Freiburghaus, M., & Taborsky, M. (2013d) Strategic reduction of help before dispersal in a cooperative breeder. *Biology Letters*, **9**, 20120878.

13 Sociality in Lizards

Martin J. Whiting and Geoffrey M. While

Overview

Lizards, snakes and amphisbaenians (worm lizards) form a monophyletic group (the squamate reptiles), which contains 9,712 species (Uetz & Hošek, 2015) in 61 families (Wiens, *et al.*, 2012). New species are constantly being described, particularly with the advent of modern molecular systematics and improved access to remote regions. Consequently, this group is likely to be considerably larger in the future (Pyron, *et al.*, 2013). Not only is this a taxonomically diverse group of terrestrial vertebrates, but species occupy a wide range of habitats and ecosystems, and occur on all continents except Antarctica. Furthermore, they span a wide range of body sizes and forms from miniature chameleons and geckos that perch comfortably on a matchstick, to reticulated pythons in excess of 6 m in length. While snakes have traditionally been viewed as a group separate from lizards (e.g. different suborders in traditional taxonomic terms), they are in fact embedded within lizards such that some lizards are more closely related to snakes than they are to other lizards (Wiens, *et al.*, 2012; Pyron, *et al.*, 2013).

Squamate reptiles display a wide array of life history strategies, reproductive tactics, and social behaviors that frequently bring males and females into conflict and which are invariably further influenced by high levels of male contest competition (Baird, 2013). As a consequence, most species are polygynous, polyandrous, or both (polygynandry), and many are territorial (Stamps, 1977). However, some species also occur in stable social aggregations (Gardner, *et al.*, 2016). Of these, a small proportion live in family groups with a socially and mostly genetically monogamous parental unit. Interestingly, a single radiation of lizards in Australia (*Egernia* group of scincid lizards) (Figure 13.1), commonly forms long-term pair bonds and have a particularly high incidence of family

We dedicate this chapter to the memory of Mike Bull, in recognition of his immense contribution to our understanding of lizard social systems and natural history in general.

We thank the myriad of researchers that have toiled long and hard in the field and lab in gathering the data presented in this chapter. For comments on earlier drafts we are grateful to the editors, Richard Byrne, Daniel Noble, Adam Stow, Céline Frère, and Julia Riley. We also thank the reviewers (Alison Davis Rabosky and Dave Chapple) and the editors for their very helpful feedback that greatly improved this chapter. Stewart MacDonald, David O'Connor and Alison Davis Rabosky are thanked for contributing photos. M. Whiting was supported by Macquarie University and the Australian Research Council. G. While was supported by the University of Tasmania and the Australian Research Council.

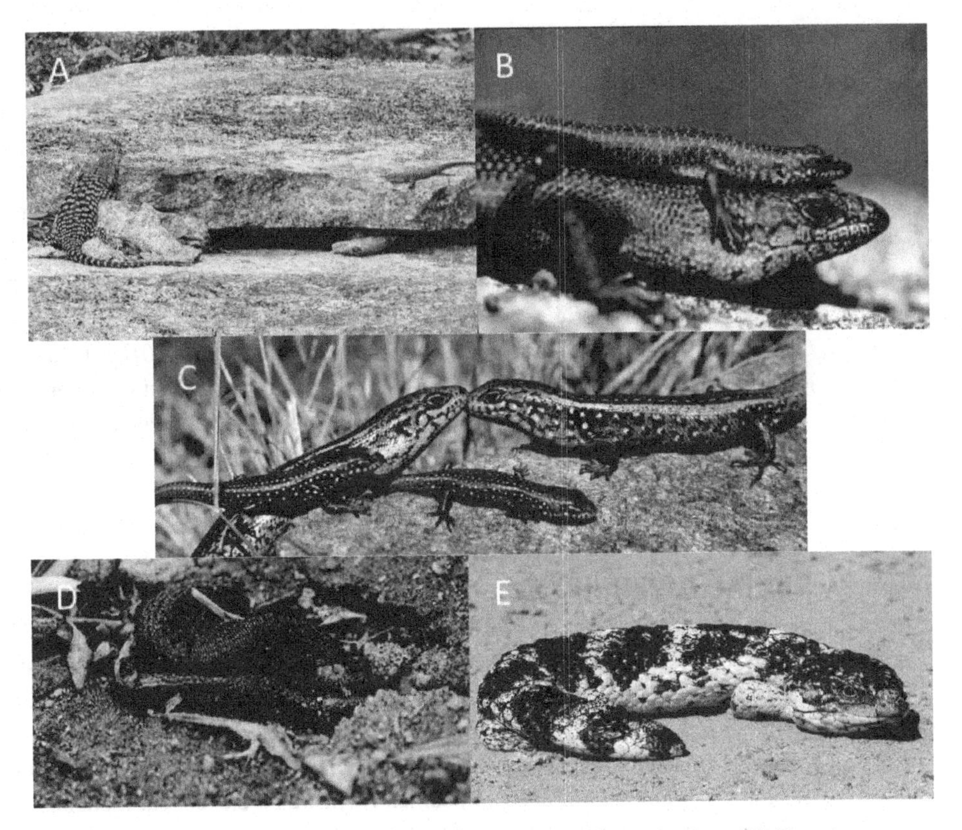

Figure 13.1 Representative lizard species with different social organization. (A) *Egernia cunninghami* adult and juvenile on rock outcrop close to refuge, photo © Stewart MacDonald; (B) *Egernia saxatilis*: adult and offspring, photo © David O'Connor; (C) *Egernia whitii* family: adult pair and offspring, photo © Geoff While; (D) *Xantusia vigilis*: adult female and juvenile, photo © Alison Davis Rabosky (see Davis, *et al.*, 2011); and (E) *Tiliqua rugosa* (sleepy lizard), photo © Stewart MacDonald. All these species live in family groups except the sleepy lizard, which pairs up before and during the breeding season and which has lifelong pair bonds.

living (Chapple, 2003; While, *et al.*, 2015; Gardner, *et al.*, 2016). The *Egernia* group is a Melanesian-Australian radiation (Gardner, *et al.*, 2008) consisting of seven genera (*Egernia*, *Liopholis*, *Lissolepis*, *Bellatorias*, *Cyclodomorphus*, *Tiliqua*, and *Corucia*), and is so called because the former *Egernia* genus was paraphyletic and split into four monophyletic genera. This review will focus primarily upon the evolution of kin-based sociality in lizards with a heavy emphasis on the *Egernia* group. We draw upon data from snakes only when it helps inform the evolution of kin-based sociality in "typical" lizards. Our review is therefore admittedly biased towards lizards because our overall goal is to advance our understanding of the evolution of sociality in general, but kin-based sociality in particular.

Lizards provide a unique opportunity to understand the early evolution of vertebrate sociality because (1) social behavior is relatively simple and easily quantifiable, (2) social behavior is not obligate (e.g. most species exhibit facultative or temporary forms of group living), and (3) there is enough variation in social strategies, both within and between species, to allow for meaningful tests. In lizards, mating systems vary from monogamy to polygynandry, parental care if present is typically through parent-offspring association (i.e. presence of parent deters potential predators, particularly conspecific adults), with social bonds and interactions between parents and offspring typically less complex than in many avian and mammalian systems (While, *et al.*, 2014a). As such, they offer great potential as models for understanding transitions to more complex forms of social structure and for uncovering the mechanisms that triggered the initial origins as well as the maintenance of family living and sociality in animals (Chapple, 2003; Doody, *et al.*, 2013; While, *et al.*, 2015). To understand the evolution of kin-based sociality in lizards, it is important to first consider the nature of social interactions in lizards more generally, since social selection (*sensu* Lyon & Montgomerie, 2012) acts on all aspects of social interactions that might lead to pair bonding and group formation (Kavaliers & Choleris, 2013).

I SOCIAL DIVERSITY

Social systems have previously been organized into three broad categories: (1) social organization; (2) social structure; and (3) mating system (Kappeler & van Schaik, 2002; Kappeler, *et al.*, 2013). Much of what we know about the social and reproductive behavior of lizards is through studies of social structure and mating systems (Brattstrom, 1974; Fox, *et al.*, 2003). Specifically, lizards are excellent models for studies of territorial behavior, alternate reproductive tactics, and sexual selection more generally because they are frequently diurnal, easy to catch and follow, and exhibit strong site fidelity (Fox, *et al.*, 2003). In fact, the first significant publication of lizard behavior was a monograph of lizard mating behavior and sexual selection by Noble & Bradley (1933). Studies of lizard social structure have focused mainly on: (1) contest competition and rival recognition (Stamps, 1977; Fox & Baird, 1992; Whiting, 1999; Whiting, *et al.*, 2003; Whiting, *et al.*, 2006; Carazo, *et al.*, 2008; Sacchi, *et al.*, 2009; Umbers, *et al.*, 2012); (2) male alternate reproductive tactics (hereafter ARTs) (Wikelski, *et al.*, 1996; Sinervo & Lively, 1996; Whiting, *et al.*, 2009; Noble, *et al.*, 2013); (3) mate preference and mate choice (Olsson, *et al.*, 1994; Olsson & Madsen, 1995; Tokarz, 1995); and (4) communication in the form of static and dynamic visual signals, chemical signals/cues, and sometimes vocal signals (Martins, 1993; Ord, *et al.*, 2002; Ord & Martins, 2006; Hibbitts, *et al.*, 2007).

In addition to this large body of literature on social structure, lizards are increasingly becoming the focus of sexual selection studies using genetic parentage testing (Uller & Olsson, 2008; Wapstra & Olsson, 2014). From this work, we have detailed knowledge of mating systems, and in numerous cases, spatial and social organization. Lizards are therefore particularly useful organisms for studying the extent to which sexual selection influences patterns of social organization and social behavior. However, our understanding of social complexity in lizards – the extent to which they are capable of true individual recognition, kin recognition, forming associations and alliances, and tracking of third party relationships – is in its infancy compared to studies in other taxonomic groups.

13.1 How Common is Sociality in Lizards?

Some form of aggregation has been documented in 94 species from 22 families, while stable social aggregations have been documented in 18 species from 7 families (Graves & Duvall, 1995; Mouton, *et al.*, 1999; Mouton, 2011; Gardner, *et al.*, 2016). However, our knowledge of the relatedness of individuals in these groups is generally scant. In some species, aggregations may be daily/nightly sleeping refuges or occur seasonally, and social bonds within these species are presumably weak, absent, or potentially strong (e.g. female rattlesnakes at rookeries with related females and offspring, Schuett, *et al.*, 2016). The vast majority of lizards, particularly those that experience any form of sexual conflict or sexual selection (e.g. contest competition), are expected to form social bonds with other individuals in their neighborhood independently of relatedness (e.g. Strickland, *et al.*, 2014). In this context, population social structure occurs when individuals are linked through non-random associations with one another (Whitehead, *et al.*, 2005; Croft, *et al.*, 2008; Strickland, *et al.*, 2014).

In contrast, kin-based (i.e. family) groups have only been identified in a single higher-level group of lizards (Scincoidea, *sensu* Pyron, *et al.*, 2013). Within this group or lineage are an additional four lineages that constitute major families: Xantusiidae (three subfamilies), Gerrhosauridae (two subfamilies), Cordylidae (two subfamilies), and Scincidae (three subfamilies). Kin-based sociality has thus far only been documented in the Xantusiidae and Scincidae, but it could very well be present in the remaining two families because species are known to live in stable conspecific aggregations but relatedness has not yet been tested. Only a single species of xantusiid, *Xantusia vigilis*, has been shown to have kin-based sociality (Figure 13.2) while nine species of the *Egernia* group in three genera (*Egernia, Liopholis, Bellatorias*) have thus far been confirmed to live in stable family groups (Figure 13.1, Table 13.1). However, field observations of lizard groups found in burrows, rock crevices, tree cavities, and underneath cover items which specifically consist of a single adult pair and associated juveniles have been reported from at least eight additional species representing four more squamate families (reviewed in Chapple, 2003; Davis, *et al.*, 2011; Gardner, *et al.*, 2016). Thus, future relatedness studies may uncover a more widespread distribution of kin-based sociality across lizards.

13.2 Forms of Sociality in Lizards

Lizard societies take a number of forms that include groups without kin structure, monogamous pairs with kin-based family living, and extended parent-offspring associations.

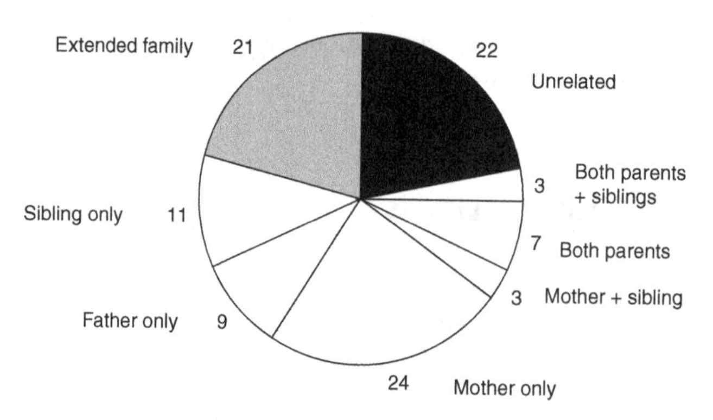

Figure 13.2 Frequency of group composition according to relatedness in the desert night lizard (*Xantusia vigilis*). Reproduced from Davis, *et al.* (2011).

13.2.1 Social Organization and Structure Independent of Kin

Examining the link between social organization and mating system is a good first step towards understanding the different forms of sociality in lizards. In terms of mating systems, multiple mating is common in lizards and appears to be the case for almost every lizard species examined (Uller & Olsson, 2008; Wapstra & Olsson, 2014). In this respect, most territorial species have traditionally been classified as polygynous (Bull, 2000) because a subset of males appear to secure the majority of paternity. However, this type of categorization ignores the fact that in almost every species for which there has been genetic paternity testing of offspring, females have been shown to mate multiply (Uller & Olsson, 2008). Most species are therefore polygynandrous rather than polygynous. Polyandry in lizards has been suggested to be driven by indirect (genetic) benefits, despite little actual evidence for this (Uller & Olsson, 2008). More recently, it has been proposed that polyandry may have evolved in lizards through the combined effects of mate encounter frequency and sexual conflict over mating rates. Females are less likely to resist mating with multiple males if the costs are relatively low (Uller & Olsson, 2008) and/or if females are able to exclude the sperm of less desirable males through postcopulatory means (Olsson, *et al.*, 1994). In these systems, there is no pair bond or association, and females may mate with close kin but avoid the costs of inbreeding through sperm selection (Olsson, *et al.*, 1994).

In classical territorial systems, more than one female may occupy a male's territory. However, while residents may keep rival males at bay and exclude them from their territory, they cannot prevent females from visiting other males or copulating with sneaker males out of view (Wikelski, *et al.*, 1996). In these systems, the social organization is not kin-based, but there are social bonds and therefore there is social structure. However, it is unclear whether territoriality is driving sociality (a common assumption in lizard studies), or even the degree to which it influences sociality in lizards. For example, in the non-family living Australian eastern water dragon, *Intellagama lesueurii*, there are social bonds within and between the sexes. The strongest bonds

Table 13.1 A summary of family living in lizards defined by long-term, stable groups consisting of a parental unit and one or more generations of offspring. Parental units are largely monogamous for at least the season and often across seasons. We have not included species that have parent-offspring associations <6 months in duration and/or studies in which there is no genetic confirmation of group membership such as in the Cordylidae. This list is therefore less inclusive than Davis, *et al.* (2011). We have not indicated levels of extra-pair paternity because of differences in the way these data have been collected (see text). The sleepy lizard, *Tiliqua rugosa*, does not associate with its offspring but is included here because it forms life-long stable pairs and is part of the clade containing the *Egernia* group. We also include the few members of the *Egernia* group that we know are solitary. *Cyclodomorphus* is the only member of the clade containing the *Egernia* group not listed here because family living has not been reported (or properly tested for).

Family	Species	Solitary	Nuclear family	Extended family	Group size	Reference
Scincidae	*Liopholis inornata*	X			N/A	Chapple, 2003
	Liopholis kintorei		X			McAlpin, *et al.*, 2011
	Liopholis whitii		X		2–6	Chapple & Keogh 2005, 2006; While, *et al.*, 2009b, 2011
	Tiliqua rugosus	X			2	
	Tiliqua adelaidensis	X			N/A	
	*Bellatorias major**		?	?	3–8	Osterwalder, *et al.*, 2004
	Bellatorias frerei		X (genetic evidence)			Fuller, *et al.*, 2005
	*Egernia kingii***		X			Masters & Shine, 2003
	Egernia cunninghami		X		2–27	Stow & Sunnock, 2004; Stow, *et al.*, 2001
	Egernia saxatilis		X		2–14	O'Connor & Shine, 2003
	Egernia striolata		X		2–4	Duckett, *et al.*, 2012; Bull & Bonnett, 2004
	Egernia stokesii			X	2–17	Duffield & Bull, 2002; Gardner, *et al.*, 2001, 2002, 2007
	Lissolepis coventryi	X			N/A	Chapple, 2003
	Gnypetoscincus queenslandiae		X (loose)		2–9	Sumner, 2006
	*Corucia zebrata****	X			2–5	Hagen, *et al.*, 2013
Xantusiidae	*Xantusia vigilis*****		X	X	2–20	Davis, *et al.*, 2011; Zweifel & Lowe, 1966

* Male-female bonds, some association with juveniles, telemetry study.

** Based on one family observed for 10 years.

*** Mostly solitary, some mixed groups.

**** 22% (26/117 groups) occur in unrelated groups, 57% (67/117 groups) in nuclear families, and 21% (25/117 groups) in extended families.

are between females, although males and females form relatively strong bonds (Strickland, *et al.*, 2014). While females may show social affiliation with a particular male, they still mate multiply and fertilization favors males with higher levels of heterozygosity (i.e. less inbred males, Frère, *et al.*, 2015). Sociality in lizard systems that are not kin-based is likely more complex than we think, and not simply an artefact of territoriality.

13.2.2 Monogamous Pair Bonds and Kin-based Family Living

A much less common form of social organization in lizards is a single, long-term monogamous pair. There are two forms of long-term monogamy in lizards: (1) a single social unit consisting of a male and female but with no association to their offspring; and (2) a parental unit and one or more generations that live together over the long-term. In the former option, there is one well documented example of a male and female unit with no offspring association: the sleepy lizard, *Tiliqua rugosa* (Figure 13.1), which has been studied by Bull and colleagues for more than thirty years (Bull, 1994; Leu, *et al.*, 2015). This species forms long-term pair bonds that frequently exceed 10 years and one particular pair currently stands at more than 27 years as of 2013 (Leu, *et al.*, 2015). Males and females pair-up six to eight weeks before they copulate and then separate shortly thereafter. However, there is no parental care or parent-offspring association (Leu, *et al.*, 2011). Pairs that have been together longer tend to come together earlier in the breeding season, breed sooner, and are thought to co-ordinate their reproductive behaviors more effectively (Leu, *et al.*, 2015). In the lizard group *Egernia*, sociality consisting of a parental unit and kin is common. The *Egernia* group consists of three genera in which stable family groups with sometimes multiple cohorts of offspring have been documented (Table 13.1, Figure 13.1). Additionally, kin-based family groups have also been documented in the North American desert night lizard (Xantusiidae: *Xantusia vigilis*). Groups in this species typically consist of kin and at least one parent that refuge together in decaying logs or under rocks (Davis, *et al.*, 2011; Davis Rabosky, *et al.*, 2012).

13.2.3 Nest Attendance, Brooding, and Parent-Offspring Associations

Solitary species that do not readily fall into a sociality category may nevertheless exhibit parental care and form short-term bonds with their offspring during a brief postpartum phase. Among squamate reptiles, these species may be oviparous or viviparous. Egg-laying species will either actively brood their eggs or simply remain at the nest and many species will actively defend their nests against potential predators (Somma, 2003; Huang, 2006; Stahlschmidt & DeNardo, 2011; While, *et al.*, 2014a). In the lizard *Eutropis longicauda*, females cannot recognize their own eggs but they defend their nests as an extension of territorial behavior (Huang & Pike, 2011). Parental care then ceases immediately, or shortly after the emergence of young. In viviparous species (and some oviparous species), there can be a more prolonged association between offspring and parents that may persist from days (in some snakes) to a few years (in some members of the *Egernia* group, Greene, *et al.*, 2002, Chapple, 2003; Somma, 2003; While, *et al.*, 2014a). For example, gravid female live-bearing North American pitvipers (more than 33 species in 8 genera) are particularly well-known for firstly aggregating at rookeries to give birth and secondly for forming mother-offspring associations. These associations typically last until they first shed their skin, sometime in the second week following birth (Greene, *et al.*, 2002). Female rattlesnakes are also more aggressive during the time that they are in association with their offspring (Greene, *et al.*, 2002). In timber rattlesnakes, gravid females are more likely to associate with female kin when they aggregate at rookeries, while neonates stay close to their mother (Clark, 2004). Timber rattlesnakes are also capable of kin recognition (Clark, 2004). Collectively, these and other snake species show the hallmarks of sociality but do not form stable long-term groups. Nonetheless, they may prove valuable for understanding the early evolution of parental care and sociality in reptiles (Schuett, *et al.*, 2016).

13.3 Why Lizards Form Social Groups

A wide range of lizard and snake species aggregate for ecological reasons (reviewed in Graves & Duvall, 1995; Gardner, *et al.*, 2016). These aggregations range in size from just a few individuals to several thousand and occur either because optimal refuges are limited and/or because of the physiological benefits of grouping (Graves & Duvall, 1995; Lancaster, *et al.*, 2006; Mouton, 2011). The types of shelters that promote aggregations in squamates are those that offer thermal benefits (Shah, *et al.*, 2003; Shah, *et al.*, 2004), refuge from predators (Mouton, 2011), hibernacula in temperate regions with cold winters (Gregory, 1984; Graves & Duvall, 1995), or oviposition/rookery sites (Graves & Duvall, 1995; Doody, *et al.*, 2009). While the primary reason for an aggregation may be ecological (e.g. refuge) or physiological (e.g. temperature/water loss), there could still be social benefits from grouping. Furthermore, it has been hypothesized that grouping behavior as a result of abiotic factors could act as an initial trigger for the evolution of sociality (Graves & Duvall, 1995; Shah, *et al.*, 2003, Lancaster, *et al.*, 2006; Davis Rabosky, *et al.*, 2012). Several studies have tested whether lizards will still group when offered an abundance of shelters under uniform environmental conditions in the lab. The results from these studies are mixed. For example, the cordylid lizard *Platysaurus broadleyi* occupies communal refuges consisting of more than 100 individuals in the wild (M. Whiting, *unpublished data*) but does not group in captivity and males avoid sharing refuges. Conversely, both the gecko *Nephrurus milii* and the cordylid *Ouroborus cataphractus* will continue to aggregate under uniform conditions in the lab (Shah, *et al.*, 2003; Visagie, *et al.*, 2005). At present, the potential role of environmental factors in enhancing social affiliation, and thereby potentially favoring sociality, is poorly understood.

In contrast, explanations for the emergence, maintenance and diversification of more stable social organization (i.e. those found in the *Egernia* group) have focused mostly on explanations analogous to those put forward for family-living birds and mammals (e.g. Komdeur, 1992; Koenig, *et al.*, 1992; Hatchwell & Komdeur, 2000). Simply put, social organization emerges as a result of constraints imposed by the availability of key habitat features such as permanent retreat sites. Thus, while there may be additional benefits of grouping together similar to those identified for other less stable aggregations of lizards (and other organisms), it is unlikely that these benefits are the source of selection on lizards' social organization itself. However, *X. vigilis* that aggregated had both higher reproductive success and survival compared to solitary individuals (Davis Rabosky, *et al.*, 2012), suggesting that a key feature can also be the simple presence of the conspecifics themselves. Below, we address the potential benefits of aggregating first, before detailing the important role that ecology plays in mediating the evolution of social organization within the *Egernia* group and *X. vigilis*.

13.3.1 Resource Acquisition and Use

A key component of the ecology of the *Egernia* group is its reliance on permanent shelter sites (Table 13.2). These shelter sites are either established structures such as rock outcrops or tree hollows (Chapple, 2003; Michael, *et al.*, 2010), or they have been constructed via the excavation of deep and complex burrow systems, sometimes by multiple generations of the same family (McAlpin, *et al.*, 2011). These structures tend to be patchily distributed across the landscape and separated from other such outcrops by unsuitable habitat (Duffield & Bull, 2002; O'Connor & Shine, 2003). However, the extent of heterogeneity in habitat availability differs markedly between species. In burrowing species, suitable habitat can be relatively homogenous, separated by a matter of meters. In contrast, for species that live on rocky outcrops (e.g. *E. striolata* and *E. stokesii*; Ttable 13.2), patches of suitable habitat can be separated by distances of 50 meters and often much more (Gardner, *et al.*, 2001), providing considerable barriers to dispersal. As a consequence of this patchy distribution, acquisition of shelter sites has been suggested to be the key factor that influenced the emergence and diversification of social organization in the *Egernia* group. Indeed, most of the family-living species live on either rock outcrops or trees (Table 13.2) with only one exception (*L. kintorei*). In contrast, all the *Tiliqua* are solitary except *T. rugosa*, which has life-time pair bonds during the breeding season, and do not live in discrete habitats.

The majority of studies on the role of permanent shelter sites on the emergence of social organization have focused on the extent to which habitat availability constrains dispersal. However, some authors have suggested that group organization may influence the ability of individuals to acquire high quality shelter sites. For example, Michael, *et al.* (2010) found group size in *Egernia striolata* to be positively correlated with habitat quality (independently of crevice size), with larger social groups found in the highest quality areas of the rock outcrops (e.g. the top, with greatest access to basking sites and reduced predation risk). They suggested that larger groups of lizards might be better able to defend high-quality home ranges than solitary individuals. However, an equally likely scenario is that more extensive habitat supports more individuals and therefore has larger groups. Interestingly, there is the possibility that family groups can usurp other groups and gain access to higher quality resources such as basking areas. *Egernia saxatilis* families will control basking areas at the expense

Table 13.2 Ecological characteristics of family living lizards. All species are viviparous. See Greer (1989) and Chapple (2003) for additional ecological information. Age at maturity is the year in which individuals become sexually mature. We include the same solitary species from Table 13.1 here. All species are in the family Scincidae except *X. vigilis*, which is in the Xantusiidae.

Species	Rock-dwelling	Semi-arboreal	Terrestrial	Longevity (yrs)	Litter size	Age at maturity (yrs)	Refuge type	Habitat	Reference
*Liopholis inornata**			X		1–4	2	Burrow	Sandy, xeric	Chapple, 2003
Liopholis kintorei			X		1–7	2	Burrow	Sandy, xeric	McAlpin, *et al.*, 2011; Chapple, 2003
Liopholis whitii			X	13	1–4	2	Rock crevices, logs	Coastal heath, grasslands, dry sclerophyll forest, open woodland	Chapple & Keogh, 2005, 2006 While, *et al.*, 2009b, 2011
*Tiliqua rugosus***			X	>50	1–3	5	Burrows, vegetation	Chenopod shrubland, xeric	Bull, *et al.*, 1993; Leu, *et al.*, 2011; M. Bull *unpublished data*
*Tiliqua adelaidensis**			X	9	1–4	2	Burrow	Grassland, spider burrows	Milne, *et al.*, 2002; Schoefield, *et al.*, 2014; Hutchinson, *et al.*, 1994
Bellatorias major			X	11–23	3–7		Hollow logs	Rainforest, wet sclerophyll forest	Osterwalder, *et al.*, 2004; Shea, 1999
Bellatorias frerei			X				Burrows, hollow logs	Well-watered forest, seasonally dry forest	Fuller, *et al.*, 2005
Egernia kingii			X	>10	2–11	3	Rock crevice	Heath	Masters & Shine, 2003; Chapple, 2003
Egernia cunninghami	X			>50***	4–8	5	Rock crevice	Open *Eucalyptus* woodland, exposed rock outcrops	Stow & Sunnocks, 2004; Stow, *et al.*, 2001
Egernia saxatilis	X	X			1–4		Rock crevice	Open *Eucalyptus* woodland, exposed rock outcrops	O'Connor & Shine, 2003

Table 13.2 (*cont.*)

Species	Rock-dwelling	Semi-arboreal	Terrestrial	Longevity (yrs)	Litter size	Age at maturity (yrs)	Refuge type	Habitat	Reference
Egernia striolata	X	X			1–7	2–3	Crevices in rock, logs, tree hollows	Open *Eucalyptus* woodland, temperate semi-arid woodland	Duckett, *et al.*, 2010; Bull & Bonnett, 2004
Egernia stokesii	X	X		>12	1–8	5–6	Rock crevice	Rocky outcrops in grassland	Duffield & Bull, 2002; Gardner, *et al.*, 2001, 2002, 2007
*Lissolepis coventryi**			X	>8	1–4	2–3	Vegetation, burrows	Wetlands, swampy heath	Clemenn, *et al.*, 2004; Chapple, 2003
Gnypetoscincus queenslandiae			X	ca. 10	1–5	6.5	Rotting logs	Rainforest	Sumner, 2006
Corucia zebrata		arboreal			1		Tree hollows	Rainforest	Hagen, *et al.*, 2013
Xantusia vigilis			X	8–10	1–2	2–3	Logs, under rocks	Joshua tree woodland, desert	Davis, *et al.*, 2010; Zweifel & Lowe, 1966

* Solitary species.
** Pairs up before and during breeding season.
*** Based on one captive animal; see text.

of a rival family during staged trials in the lab (O'Connor & Shine, 2004), but the degree to which this might occur in the wild is unknown.

In addition to asking why and how sociality, and in particular, family living evolved, we should also ask why family living is so rare among squamate reptiles. The answer to the latter question may have something to do with diet and foraging mode. Almost all species are primarily insectivorous (Vitt, *et al.*, 2003) and adopt either a sit-and-wait foraging mode or an active/wide-foraging mode although these differences are not always clear-cut (McBrayer, *et al.*, 2007). Actively foraging lizards move through the landscape and use vomerolfaction (i.e. the ability to detect prey chemicals through tongue-flicks and processing of chemical cues in the Jacobsen's organ in the roof of the mouth) to find hidden prey, which they then retrieve (Cooper, 1994). For these species, there is likely to be less competition for a food source that is mostly scattered if they forage alone. Furthermore, there may be additional costs to coordinating with other individuals such as attracting the attention of predators and any cognitive and behavioral constraints associated with coordinating movements. Furthermore, active foragers have significantly higher metabolic rates than ambush foragers (Brown & Nagy, 2007). This foraging-centric lifestyle may preclude group or family living, although this idea has received no significant attention in the literature and would be difficult to test beyond a correlation or comparative analysis.

Ambush foraging lizards that are group living have the added challenge of localized competition for food. To some degree, this competition may be ameliorated though excursions to prey patches away from the group. For example, the armadillo lizard, *Ouroborus* (formerly *Cordylus*) *cataphractus*, lives in stable, large aggregations (commonly 2 to 6, but up to 55 individuals, Effenberger & Mouton, 2007) and is thought to reduce food competition by binge feeding at termite foraging ports away from the group, particularly during the dry season (Shuttleworth, *et al.*, 2008, 2013). The family-living members of the *Egernia* group are typically omnivorous or even, herbivorous, as adults (Chapple, 2003). Although we know very little about their foraging behavior, they are not active foragers although they likely make short excursions to feed on plant matter. Furthermore, they are frequently confined to rock outcrops or trees, which are discrete units of habitat normally associated with sit-and-wait foraging lizards. While their foraging mode and diet may be an outcome of non-social factors and may have preceded family living, being bound to a discrete resource in the case of ambush foragers, could facilitate kin-based sociality.

13.3.2 Predator Avoidance

The stable social aggregations observed in the *Egernia* group have also been suggested to provide benefits in terms of reduced predation risk. Enhanced vigilance and reduced predation risk were first proposed to explain the prolonged associations between male and female *T. rugosa* (Bull & Pamula, 1998). Positive benefits of group organization on anti-predatory behavior have also been documented in large *Egernia* social groups. For example, groups of both *E. stokesii* and *E. cunninghami* have been shown to detect predators earlier than solitary individuals (Eifler, 2001;

Lanham & Bull, 2004). The complex burrow and crevice networks used by many species are likely to reduce predation risk further.

13.3.3 Homeostasis

In addition to general habitat constraints, much of the behavior and life history of lizards and snakes more broadly is driven by thermal requirements dictated by their environment (Pianka & Vitt, 2003). For example, temperature has a profound effect on growth rates, as well as reproductive mode, frequency, and activity (Vitt & Seigel, 1985; James & Shine, 1988; Warner & Shine, 2007; Radder, *et al.*, 2008). Furthermore, in more temperate regions there are a limited number of suitable over-wintering sites and this results in aggregations of individuals in hibernation or during temperatures unsuitable for activity. The close proximity of these individuals as a consequence of physiological constraints may have a bearing on sociality that is little appreciated. Thermoregulatory benefits have been associated with social aggregations in *Egernia* and *X. vigilis*. In winter, *X. vigilis* that aggregated, and juveniles in particular, experienced significant thermal benefits that translated into higher fitness (Davis Rabosky, *et al.*, 2012). In *E. stokesii*, group size has a positive effect on heat retention, with larger groups maintaining higher nighttime body temperatures than smaller groups (Lanham, 2001). Grouping behavior in *Egernia* group species may also have indirect effects on homeostasis by allowing increased investment in thermoregulatory behavior due to decreased investment in other behaviors like vigilance. For example, *Egernia stokesii* exhibit reduced vigilance behavior and flight initiation distances in groups, which allowed individuals more time to bask (Lanham & Bull, 2004). Finally, the reliance on permanent shelter and crevice sites also provides individuals homeostatic benefits as burrows act to both reduce temperature oscillations as well as enhance humidity (Henzell, 1972; Webber, 1979).

13.3.4 Mating

Mating benefits are not only one of the key drivers of social behavior observed in many lizard species (Graves & Duvall 1995), but they are also thought to be responsible for the relatively simple polygynous mating systems based on territoriality or dominance hierarchies, characteristic of the majority of reptile social systems (Stamps, 1983). As detailed earlier, social organization in the *Egernia* group emerges as a result of restricted dispersal rather than direct benefits associated with increased access to mates. Nevertheless, variation in social organization within populations largely emerges as a result of a male's ability to monopolize areas that can attract multiple females. For example, *Liopholis whitii* (Figure 13.1) males that occupy areas with high rock crevice availability are more likely to acquire multiple female partners (e.g. polygynous social groups) than those in areas with fewer rock crevices (Chapple & Keogh, 2006; G. While, *personal observations*). In *X. vigilis*, social behavior greatly impacts male reproductive success. Of 230 neonates molecularly assigned to 123 sires, none were the offspring of males that were known to be solitary the preceding winter (approximately a third of all males, Davis Rabosky, *et al.*, 2012).

Seasonal monogamy can be driven by a wide variety of factors including obligate parental care or a shortage of receptive females, which in turn may drive mate-guarding behavior (Lukas & Clutton-Brock, 2013). However, different factors could be driving long-term monogamy. The mate familiarity hypothesis suggests that long-term monogamy is adaptive because it results in better coordination of reproductive behavior, particularly in organisms exhibiting biparental care (e.g. Black, 1996; Black, 2001; Mariette & Griffith, 2012). Nevertheless, long-term monogamy in non-avian taxa that do not provision their young could still be adaptive if it facilitates the priming and/or coordination of male and female reproductive cycles and behavior. For example, in the sleepy lizard, *Tiliqua rugosa*, long-term partners pair up earlier in the season and have their young earlier in the year (Leu, *et al.*, 2015). Although this could give juveniles a head start, and therefore, a potential fitness advantage, this hypothesis remains untested. Similarly, there may be benefits to stable long-term monogamy in coordinating reproduction in the *Egernia* group. To our knowledge, the benefits of mate familiarity and coordinated reproduction have never previously been considered for lizard systems other than the sleepy lizard. This element could be important in explaining some of the variance in fitness benefits of monogamy and should be considered in future studies.

13.3.5 Offspring Care

Although the level of parent–offspring interaction in lizards that exhibit kin-based social organization is substantially lower than in other vertebrate species (e.g. full parental provisioning, Clutton-Brock, 1991), it nevertheless shows a greater level of parent–offspring interaction than previously reported in the majority of squamate species. For example, delayed juvenile dispersal is a key feature of both *X. vigilis* (Davis, *et al.*, 2011) and *Egernia* group social organization (Gardner, *et al.*, 2001; O'Connor & Shine, 2003; Stow & Sunnucks, 2004; Davis, *et al.*, 2011). Thus, the emergence of sociality within these groups is closely linked to the benefits of delayed dispersal of offspring. Moreover, lizards are capable of social learning and in water skinks, *Eulamprus quoyii*, young lizards are more likely to pay attention to social information than older individuals (Noble, *et al.*, 2014). For species that live in family groups, there could be significant benefits to social learning, particularly in juveniles and subadults.

Offspring may benefit from associations with parents in a number of ways. First, offspring may gain increased access to basking locations, foraging opportunities, and retreat sites (Bull & Baghurst, 1998; O'Connor & Shine, 2004; but see Langkilde, *et al.*, 2007). Furthermore, the presence of high levels of genetic relatedness within social groups in species such as *E. stokesii* suggests that offspring may gain from prolonged parental care through the inheritance of territories (Gardner, *et al.*, 2001). Alternatively, offspring may benefit from extended parent-offspring interactions via a reduction in the risk of infanticide and conspecific aggression (O'Connor & Shine, 2004; Sinn, *et al.*, 2008). *Egernia* group species frequently live in highly saturated environments whereby aggression towards conspecifics is common. Indeed, infanticide has been identified as a key cause of offspring mortality in a number of these species (Post, 2000; Lanham & Bull, 2000; O'Connor & Shine, 2004). Parents of most species vigorously and aggressively defend their home range from conspecifics (Chapple, 2003; O'Connor & Shine, 2004), providing a significant benefit to the offspring who reside within it. Experimental evidence from *E. saxatilis* has shown that the presence of a parent nearly eliminates all of the aggression displayed towards its offspring by unrelated adults (O'Connor & Shine, 2004). Work on both *E. saxatilis* and *L. whitii* has shown that female aggression is heightened during periods of post-partum parent offspring association, when offspring are presumably most at risk (O'Connor & Shine, 2004; Sinn, *et al.*, 2008). As a consequence, the extent of aggression a female displays towards a conspecific is the key predictor of offspring survival in *L. whitii* (Sinn, *et al.*, 2008).

13.3.6 Other Reasons: Parasites and Infection Risk

One additional potential factor suggested to explain the emergence of social living in the *Egernia* group is parasite prevalence and risk of infection. Parasites have been suggested to select both for and against social living. For example, group living can increase the risk of infection by providing parasites with easy transmission between individuals (Altizer, *et al.*, 2003). Alternatively, group living can decrease parasite transmission through dilution effects (Mooring & Hart, 1992), cooperation in ectoparasite removal (Wikelski, 1999), or through minimizing contact with individuals outside the social group whom may carry novel parasites (Bull & Burzacott, 2006; Godfrey, *et al.*, 2009). In the *Egernia* group, early work suggested that long-term monogamy played a crucial role in reducing the risk of parasite transmission by minimizing interactions with other individuals. Studies of "divorce" (i.e. planned separation) in *T. rugosa* provide some evidence for this hypothesis: males who separated from partners one season to the next had significantly higher parasite prevalence than males who retained partners (Bull & Burzacott, 2006). In contrast, parasite prevalence in *E. stokesii* had no effect on group size (Godfrey, *et al.*, 2006). However, both an individual's position within a transmission network (based upon shared shelter sites) as well as the level of within-group relatedness have been shown to be strongly related to the risk of infection from parasites (Godfrey, *et al.*, 2006, 2009). Work on other social and less social species is required to fully appreciate the role that parasite prevalence, along with some of the other factors detailed above, play in mediating social organization.

13.4 The Role of Ecology in Shaping Sociality in Lizards

As mentioned above, ecology is likely to have played a central role in mediating the evolution of social organization in lizards, particularly within the *Egernia* group.

13.4.1 Habitat and Environment

Habitat constraints are fundamental to explaining the emergence of kin based social organization within the *Egernia* group and potentially lizards more generally. While life history traits may act as a powerful precursor to social living, variation in habitat characteristics are likely to ultimately influence between and within species variation in sociality (Duffield & Bull, 2002). As detailed above, species in the *Egernia* group are typically found in close association with permanent shelter and crevice sites from which they undertake the majority of their basking, feeding, and social activities (Chapple, 2003, Table 13.2). Because of a dependence on such sites, populations are typically highly saturated with intense competition over a limited number of patchily distributed retreat sites that results in high levels of conspecific aggression and high juvenile mortality (Chapple, 2003; O'Connor & Shine, 2004; Langkilde, *et al.*, 2005). Habitat heterogeneity is therefore likely to influence the two key components of social organization in the *Egernia* group: the composition of long-term stable pair bonds and delayed juvenile dispersal.

Many species of *Egernia* (e.g. *E. stokesii*, *E. cunninghami*, *E. saxatilis* and *E. striolata*) live in rocky outcrops, isolated from other species and populations by unsuitable habitat (Duffield & Bull, 2002). Given that large scale dispersal is generally low within the genus (Stow, *et al.*, 2001; Duffield & Bull, 2002; Chapple & Keogh, 2005), a lack of available crevice sites within an outcrop means that the only option open for individuals is to share crevices (Duffield & Bull, 2002; Chapple, 2003). The resulting social heterogeneity will in turn dictate the mating system by influencing the extent to which males can monopolize females (monogamy versus polygyny), the opportunity for extra-pair copulations (Uller & Olsson, 2008; Wapstra & Olsson, 2014), and the costs of undertaking extra-pair copulations. Importantly, social density and resource availability related to habitat suitability are also likely to influence patterns of offspring dispersal by mediating the costs and benefits of delayed dispersal for offspring and prolonged care for parents. Kin selection could then influence tolerance of offspring within the natal home range, as opposed to the crevices of other individuals, ultimately resulting in the formation of closely related family groups (While, *et al.*, 2009a).

An obvious prediction is that higher levels of social organization should evolve where dispersal between available habitat patches is limited. This prediction can be tested both empirically (e.g. by carrying out large-scale field experiments in enclosures where habitat availability is manipulated) and by comparing social organization between populations of the same species that use different habitats (e.g. *E. striolata*, which lives on trees or rocks). A study across 44 populations of *E. striolata* confirmed a strong environmental component to variation in social organization, with lizard group size correlating with various attributes of the rock outcrops (Michael, *et al.*, 2010). In contrast, a study across seven populations of *E. stokesii* found strong conservatism in social group composition (Gardner, *et al.*, 2007). Furthermore, populations of *E. cunninghami* did not differ in social organization between a fragmented and non-fragmented habitat although there were differences in dispersal between the populations (Stow & Sunnucks, 2004). These results suggest that social organization in the *Egernia* group may be influenced by phylogenetic constraint (Gardner, *et al.*, 2007). Testing the relative roles of phylogenetic history and current ecology remains a major challenge for future research on lizard social organization (Gardner, *et al.*, 2016).

13.4.2 Biogeography

Species in the *Egernia* group are widespread across the entire range of the Australian continent and occupy a range of biomes (Chapple, 2003; Gardner, *et al.*, 2008). Chapple & Keogh (2004) showed that the *Liopholis* genus consists of two major clades that have undergone parallel adaptive radiations: those that are obligate burrowers in sandy deserts in the interior and those that are temperate-adapted rock-dwelling species. The arid zone *Liopholis* are the product of a single origin and subsequent radiation. More detailed biogeographic study of the remaining *Liopholis* will help clarify the role of history and the environment in generating the current distribution and diversity within the group. Interestingly, throughout these parallel adaptive radiations sociality has remained relatively conserved. In lizards more generally, species living in stable aggregations, which may include family groups, are distributed across all continents and a range of latitudes, suggesting that biogeography may not be as important as other factors.

13.4.3 Niches

Family-living lizards appear to occupy a relatively narrow and specialized habitat niche centered on a structural refuge consisting of either a rock crevice/outcrop, a burrow, or a refuge in a tree. Tree skinks (*E. striolata*) are slightly more general and will use both rock and upright or fallen trees. Likewise, many lizard species lacking kin-based sociality may occupy a similar niche, particularly on other continents. To properly understand the potential role of the niche, including all its dimensions, in influencing social organization and structure, we need to test for, and survey, a broader spectrum of species for sociality.

13.5 The Role of Evolutionary History in Shaping Sociality in Lizards

Given the enormous species richness of lizards, family living is extraordinarily rare overall (Table 13.1). Recent advances in our understanding of higher-level snake and lizard relationships (Wiens, *et al.*, 2012; Pyron, *et al.*, 2013) has helped depict an evolutionary picture of squamate reptile sociality. Based on extensive molecular sequencing, eight major groupings of squamates have been identified. Within these groups are a much larger number of lineages, each of which may include multiple families and subfamilies (Pyron, *et al.*, 2013). Interestingly, family living has been identified in only one of these major clades, the Scincoidea. Having said that, it is likely that family living will be documented in other clades in the future (Doody, *et al.*, 2013), such as the Iguania clade (e.g. Tropiduridae: *Liolaemus*, E. Santoyo-Brito & S. Fox, *unpublished data*) or even Gekkota (Diplodactylidae: *Hoplodactylus*, Barry, *et al.*, 2014), due to field observations of group age/sex composition suggesting family structure but for which relatedness has not yet been examined. There are four families within Scincoidea (Xantusiidae, Gerrhosauridae, Cordylidae and Scincidae), and each is further divided into multiple subfamilies. The Cordylidae has numerous instances of species living in stable aggregations (Mouton, 2011). There is a distinct possibility that family living occurs in some of these species, but this remains to be confirmed by detailed genetic and population study. Similarly, the Gerrhosauridae are also known to aggregate, but family living has not been examined in this group. The observation that juveniles are frequently seen in association with adults is suggestive of family living (M. Whiting, *personal observations*). Finally, there is a single case of family living in the Xantusiidae, with the possibility of others (Davis, *et al.*, 2011).

The vast majority of instances of family living are all in the Scincidae and more specifically, in the *Egernia* group. The likely ancestor of the *Egernia* group is the prehensile-tailed skink *Corucia zebrata* from the Solomon Islands, from which it diverged about 25 ma (Skinner, *et al.*, 2011). *Corucia zebrata* does not live in family groups, although individuals nearest to one another are more closely related than expected by chance (Hagen, *et al.*, 2013). In contrast, family living is widespread across the *Egernia* group (Table 13.1). Evidence for family living has been produced in almost every study targeting sociality in this complex thus far, although this likely represents a bias towards species suspected of kin-based sociality based on observations of group living. Although the relationships among the *Egernia* group are mostly resolved (Gardner, *et al.*, 2008), a comprehensive phylogeny with greater population coverage is currently in preparation and will give a clearer perspective on the origins of kin-based sociality in the group. Furthermore, many of the gaps will likely be filled in the coming years now that sociality is receiving increased focus in this complex. Overall, family living has evolved only a few times independently in lizards.

II SOCIAL TRAITS

13.6 Traits of Social Species

Lizards with kin-based sociality tend to have numerous reproductive, life history, and social traits in common. Below we discuss the most important traits for understanding social evolution in lizards.

13.6.1 Cognition and Communication

Lizards have been the subjects of considerable recent study with respect to learning ability. Not only are lizards capable of behavioral flexibility (Leal & Powell, 2012; Clark, *et al.*, 2014) but also social learning (Noble, *et al.*, 2014; Kis, *et al.*, 2014). The *Egernia* group may be an excellent model system with which to test social intelligence theory – the idea that social behavior sets the stage for increasingly complex cognition (Byrne & Bates, 2007) – because sociality is variable and they offer a unique phylogenetic control with closely related species varying in their degree of sociality.

Territorial lizards (particularly agamids and iguanids) are often conspicuously colored (males) and use dynamic visual signals to communicate at a distance (Ord &

Martins, 2006). This is consistent with a polygynous mating system. Conversely, the *Egernia* group and *Xantusia vigilis* lack obvious sex-based color differences and are not known for using conspicuous dynamic visual signals (Zweifel & Lowe, 1966; Chapple, 2003). This may be more in line with a monogamous mating system, although males do defend core areas. *Egernia*, and indeed many species of lizards, tongue-flick to acquire social information, which may be important for recognizing kin (Bull, *et al.*, 1999; Bull, *et al.*, 2001). This information can be obtained directly from the substrate or the animal itself. With the exception of color-based and dynamic visual signals, the mode of communication used by family-living lizards is therefore not unique compared to many less social and non-family living species.

13.6.2 Lifespan and Longevity

Kin-based sociality and/or monogamy in lizards are accompanied by increased longevity. The *Egernia* group are known to be long-lived, although data are limited to a relatively few long-term studies (Table 13.2). In the case of the monogamous sleepy lizard, individuals are able to live to be more than 50 years of age (C. Bull, *unpublished data*). Likewise, almost all the *Egernia* for which there are data are thought to live for more than 10 years. For example, *Egernia stokesii* live more than 25 years (Gardner, *et al.*, 2016), and a captive *Egernia cunninghami* is still alive after more than 50 years (P. Harlow, *unpublished data*). Finally, free-living desert night lizards, *X. vigilis*, can live at least 8 to 10 years and likely many more years than that based on undetectable growth in older individuals (Zweifel & Lowe, 1966; Davis, *et al.*, 2011). In contrast, a large proportion of lizards in less social species may only live 5 years or less, some are annual, and others only live for a few months (Pianka & Vitt, 2003; Karsten, *et al.*, 2008; Wilson, 2012;). Longevity is also accompanied by a suite of correlated life history traits including delayed maturity and high reproductive investment in relatively few offspring (Pianka & Vitt, 2003). As life history characteristics are less conserved in lizards than in birds, the *Egernia* group (and lizards in general) may provide an excellent model system for teasing apart the relative influence of habitat and life history traits – or their interaction – in promoting the evolution of family living (Chapple, 2003).

13.6.3 Fecundity

Fecundity (measured as relative clutch mass) can be correlated with and constrained by a range of factors including lifespan, body size, habitat use, and foraging mode (Huey & Pianka, 1981; Vitt, 1981; Pianka & Vitt, 2003; Uller & While, 2014). For example, females of many lizard species may have large clutch sizes and relative clutch mass, which in turn is linked to their ambush foraging strategy and even their diet (Vitt & Price, 1982). Conversely, the *Egernia* group all has relatively small litters of large offspring (mode: 1 to 4; range: 1 to 11) and in this sense, are more akin to traditional *K*-selected organisms. Furthermore, compared to many other lizard species, they are typically large-bodied, slow to mature (typically 2 to 3 years, but up to 5 years), skip opportunities to reproduce (i.e. not reproduce every year), and invest more in individual offspring (Chapple, 2003, Table 13.2). *Xantusia vigilis* shows the same patterns of small litter size (1 to 2, very rarely 3), of proportionately large offspring, late maturity (2–3 years), and missed reproduction especially in years of low rainfall (Miller, 1951; Zweifel & Lowe, 1966). In general, the prediction that post-hatching parental care is more likely to be associated with high quality offspring is supported in family-living species although once again, it has a phylogenetic bias.

13.6.4 Age at First Reproduction

In general, lizard reproduction has two extremes: fast maturing (often oviparous) species that mature rapidly and breed (and die) at a young age, and late maturing (often viviparous) species that mature slowly, breed later in life, and have better prospects for living much longer (Pianka & Vitt, 2003). Members of the *Egernia* group tend to be late maturating with variation in time to reach maturity between species related to body size such that small-medium sized species mature in 2 to 3 years, while larger species such as *E. cunninghami* may take at least five years (reviewed in Chapple, 2003).

13.6.5 Dispersal

Delayed dispersal is a feature of all family-living lizards. Some species will disperse after their first year, while others may stay in a family group for 3 or more years (Gardner, *et al.*, 2001; O'Connor & Shine, 2003; Stow & Sunnucks, 2004; Davis, *et al.*, 2011). Ecological drivers of dispersal are largely unstudied, although *E. cunninghami* in deforested habitat had reduced dispersal ability (Stow, *et al.*, 2001). In *X. vigilis*, cross-fostering experiments show that juveniles placed with unrelated individuals were more likely to disperse and move greater distances than juveniles released with genetic relatives (i.e. mothers and siblings), suggesting kin presence itself can actively promote delayed dispersal (Davis, 2012). Clear dispersal patterns with regards to sex are yet to be documented, with some species exhibiting no sex bias in dispersal (Bull & Cooper, 1999; Gardner, *et al.*, 2012), others showing a male bias (Gardner, *et al.*, 2001, Stow, et al., 2001), and still others showing evidence of both (Chapple & Keogh, 2005). It has been hypothesized that the predominant mating system of a population will dictate which sex disperses, with male-biased dispersal favored in polygynous mating systems (e.g. many mammal species) and female-biased or no sex-biased dispersal favored in monogamous mating systems (e.g. many bird species) (Greenwood, 1980; Perrin & Mazalov, 2000). Recent studies suggest that this relationship is more complex, and that social systems play a key role in dictating the intensity and direction of dispersal (Galliard, *et al.*, 2003; Lawson Handley & Perrin, 2007). Therefore, the within-population variation in both mating and social systems exhibited by *Egernia* group species may be responsible for the lack of a clear sex-biased dispersal pattern (e.g. Chapple & Keogh, 2005).

13.6.6 Other Traits: Reproductive Mode, Parental Care and Family-Living

All species that live in family groups are viviparous, a mode of reproduction present in only about 17 percent of squamates (Pyron & Burbrink, 2014). While there are numerous instances of parental care among oviparous species through brooding or nest protection (Somma, 2003), post-partum care is largely restricted to viviparous species (While, *et al.*, 2014a). The prolonged connection between parent and offspring, and their immediate contact following birth, may serve to reinforce a connection between mother and offspring (Davis, *et al.*, 2011), and set the stage for the emergence of family living. It is also possible that maternal manipulation of hormone delivery to offspring may be crucial to fostering parent-offspring social bonds, although this has not been investigated in squamate reptiles. This connection is likely fostered through chemical recognition because squamates are known for their acute chemosensory ability (Halpern, 1992). Furthermore, chemical cues play an important role in social behaviors including kin– and mother–offspring recognition (Main & Bull, 1996; O'Connor & Shine, 2006). Indeed, neonate rattlesnakes have been recorded to contact their mother using face-wiping and tongue-flicking at much higher frequencies following birth (Graves, *et al.*, 1987), and this might help imprint her chemical signature and facilitate maternal recognition. Ultimately, experimental testing of the role of viviparity in forging mother-offspring associations might be possible, particularly in populations of lizards that vary in reproductive mode.

13.7 Traits of Social Groups

Lizards that live in stable social groups tend to have high levels of genetic structure, delayed juvenile dispersal, and pair bonds that are stable across seasons. Nevertheless, there is considerable variation across species in the composition of groups and the form of their pair bond and mating system.

13.7.1 Genetic Relatedness

Egernia group species are characterized by social groups with long-term stability, high levels of genetic monogamy, and delayed juvenile dispersal. As such, they are characterized by strong genetic structure (Gardner, *et al.*, 2001; Chapple & Keogh, 2005; McAlpin, *et al.*, 2011; While, *et al.*, 2014b). Indeed, in all species studied to date, within-group relatedness is significantly greater than between-group relatedness (e.g. Stow & Sunnucks, 2004; Fuller, *et al.*, 2005). This results primarily from delayed juvenile dispersal (often multiple cohorts), which leads to high levels of relatedness between adults and offspring within a social group (see also Davis, *et al.*, 2011 for similar patterns in *Xantusia*). However, in species that live in extended families, there is also evidence of either retention of offspring in the social group into adulthood or preferential association with kin. For example, in large social groups of *E. stokesii*, within group relatedness between adult females was extremely high (0.25 less than *r* less than 0.55), strongly suggesting the possibility that these groups comprise mothers and their adult daughters (Gardner, *et al.*, 2001).

One of the major challenges associated with increased genetic structure in social groups is the risk of inbreeding depression. Species in the *Egernia* group exhibit a number of mechanisms to deal with this. First, recognition of kin, by both parents and offspring, has been documented in several species (Bull, *et al.*, 1994; Main & Bull, 1996; Bull, *et al.*, 1999, 2001; O'Connor & Shine, 2006). However, the mechanism of kin recognition may differ among species (O'Connor & Shine, 2006). Second, there is strong evidence of non-random mate choice with respect to relatedness (Bull & Cooper, 1999; Gardner, *et al.*, 2001; Chapple & Keogh, 2005; While, *et al.*, 2014b). In some species, this occurs at the social pair level, with pairs less related to one another than expected by chance (Bull & Cooper, 1999; Gardner, *et al.*, 2001; Chapple & Keogh, 2005). In other species, pairs are more related to one another than expected by chance, but females choose extra-pair partners who are significantly less related (While, *et al.*, 2014b). As a consequence, extra-pair offspring have significantly higher genetic heterozygosity compared to their half-siblings (While, *et al.*, 2014b). Taken together, these results suggest considerable plasticity within the genus with respect to mate choice, which has implications for the genetic structure within social groups.

13.7.2 Group Structure, Breeding Structure and Sex Ratio

Species in the *Egernia* group are characterized primarily by long-term stable adult pair-bonds and delayed juvenile dispersal resulting in the formation of small family groups consisting of adults and their offspring (Chapple, 2003). However, despite the fact that these broad traits are relatively consistent across the *Egernia* group, there is considerable variation in group size and composition both within and among species. There are four main categories of group organization that have been identified within the *Egernia* group: (1) species that are largely solitary (e.g. *Tiliqua adelaidenssis*, Schofield, *et al.*, 2014; *Liopholis inornata*, Daniel, 1998; *Lissolepis coventryi*, Taylor, 1995); (2) species in which adults pair-bond during the breeding season (e.g. *T. rugosa*, Bull, 2000); (3) species which live in small family groups (e.g. *L. whitii*, *E. saxatilis*, O'Connor & Shine, 2003); and (4) species which live in large stable extended family groups (e.g. *E. cunninghami*, Stow & Sunnucks, 2004; *E. stokesii*, Gardner, *et al.*, 2001). Each of these categories of group organization differ in both the nature and formation of adult pair bonds as well as extent of delayed juvenile dispersal.

As mentioned previously, the best example of consistent, potentially life-long pair bonds is the sleepy lizard, *T. rugosa*. Such extended long-term monogamy within an entire activity season has since been shown to form the basis for social organization within most species in the *Egernia* group. Specifically, species that live in small family groups tend to be characterized by long-term socially monogamous pair bonds (e.g. *L. whitii, E. saxatilis, L. kintorei*). However, in all of these species, polygynous social groups where some males form pair bonds with multiple females (typically 2 to 5 females) are also common, albeit often at a lower frequency than monogamous pair bonds. For example, 30 percent of social groups in *Egernia whitii* are characterized by a single male sharing his crevice site with up to three females (While, *et al.*, 2009b).

In other species, variation in social organization includes not only monogamous and polygynous social groups, but also aggregations containing multiple adults of both sexes. These social groups typically contain one unrelated breeding pair as well as several adult offspring of the breeding pair. Perhaps the best example of this is *E. stokesii*, where up to 11 adults (both males and females; up to 17 individuals total, including offspring) can co-occur in a stable social aggregation sharing a single crevice site (Gardner, *et al.*, 2002; Gardner, *et al.*, 2001; Duffield & Bull, 2002). Other species, such as *E. cunninghami*, *E. mcpheei* and *E. striolata* also live in large communal groups containing multiple adults (Stow, *et al.*, 2001; Chapple, 2003; Stow & Sunnucks, 2004).

Although long-term data are only available for a small number of species, these monogamous, polygynous, and polygynandrous pair bonds exhibit surprising stability across seasons. For example, sleepy lizard pair bonds appear to be life-long (Leu, *et al.*, 2015). Similarly, *L. whitii* pairs exhibit considerable stability across years with some pairs together for 10 years (G. While, *unpublished data*; see also Chapple & Keogh, 2006; While, *et al.*, 2009b), constituting the majority of those adults' reproductive lifespan (up to 13 years; G. While, *unpublished data*). Studies over several breeding seasons have confirmed strong social stability between years in *E. cunninghami* (Stow & Sunnucks, 2004) and *E. saxatilis* (O'Connor & Shine, 2003). Unsurprisingly, separation by choice (i.e. not mortality) is extremely rare in these systems. In *L. whitii*, while there are moderate levels of unplanned separation (i.e. mortality of approximately 40 percent) only 15 percent of pairs across 8 years ended in separation by choice (G. While, *unpublished data*), which suggests a remarkable level of pair fidelity.

Delayed offspring dispersal and prolonged parent-offspring association are also key features of the *Egernia* group, but they vary considerably both within and among species. At a broad level, parent-offspring association in the *Egernia* group ranges from species that do not exhibit any parental tolerance of offspring (e.g. *L. inornata*, Daniel, 1998; *T. rugosa*, Bull & Baghurst, 1998) to those that predominantly tolerate a single offspring or cohort of offspring (e.g. *L. whitii*, Chapple & Keogh, 2006; While, *et al.*, 2009a; *Liopholis slateri*, Fenner, *et al.*, 2012), to species that tolerate multiple cohorts of offspring resulting in the formation of multigenerational family groups (e.g. *E. cunninghami*, Stow, *et al.*, 2001; *E. saxatilis*, O'Connor & Shine, 2003; *E. stokesii*, Gardner, *et al.*, 2012) This variation in tolerance of offspring tends to be closely related to variation in social organization such that larger adult social groups appear to have greater tolerance of offspring (e.g. multigenerational family groups). However, this is not always the case, as predominantly solitary species such as *T. adelaidensis*, exhibit some, albeit low levels, of maternal care despite the complete lack of adult group structure (Schofield, *et al.*, 2014).

Similar to species in the *Egernia* group, social organization in *Xantusia vigilis* is also characterized by long-term male–female pair bonds and delayed juvenile dispersal. However, these groups are a little different because they are seasonal, reforming each winter after more solitary behavior in the summer. The specific composition of social groups also varies among individuals within a population. For example, group size varies from between 2 to 18 individuals, with social organization including both nuclear families and extended family groups (Davis, *et al.*, 2011). Social groups also show moderate stability with 29 percent of groups stable across consecutive years, with stability being higher for nuclear family groups than for extended family groups (Davis, *et al.*, 2011). However, at least four breeding pairs and their multiple cohorts of offspring formed "dynasties", re-aggregating in the same groups underneath the same logs every winter for up to 4 years (Davis, *et al.*, 2011).

Unlike obligate cooperatively breeding and eusocial species, the social groups in the *Egernia* group do not display high reproductive skew. In general, sex ratios are relatively even within groups and all females and most males in a population have the opportunity to breed (While, *et al.*, 2011), although reproduction opportunities in *X. vigilis* seem to favor social over solitary males (Davis, *et al.*, 2012). However, there is still considerable variation in breeding structure within and among populations, which has the potential to place strong selective pressure on the stability of social systems within the group.

Social groups are characterized by relatively high levels of genetic monogamy. As with many other socially monogamous systems, while genetic monogamy is the rule, species in the *Egernia* group also exhibit some level of extra-pair mating. Levels of extra-pair paternity differ considerably both within and among populations. For example, in *E. cunninghami*, only 2.6 percent of litters have extra-pair offspring (Stow & Sunnucks, 2004). Levels of extra-pair paternity are low for other species including 12 to 26 percent in *L. whitii* (Chapple & Keogh, 2005; While, *et al.*, 2009b), 20 percent in *E. saxatilis* (O'Connor & Shine, 2003), 25 percent in *E. stokesii* (Gardner, *et al.*, 2001), and 19 percent in *T. rugosa* (Bull, *et al.*, 1998). In contrast, less social and non-family living species appear to have higher levels of genetic polyandry. In *T. adelaidensis*, 75 percent of offspring within litters are the result of multiple mating (Schofield, *et al.*, 2014). These patterns of predominant social and genetic monogamy with moderate levels of extra-pair paternity closely mirror those observed for many other family-living species (Griffith, *et al.*, 2002; Cornwallis, *et al.*, 2010; Lukas & Clutton-Brock, 2012).

III SOCIAL SYNTHESIS

Group living brings individuals into close contact and invariably, reproductive or resource-based conflict. Consequently, groups may vary in their stability depending on the environmental and social factors that influence fitness. Lizards have been identified as a useful model for understanding the transition to "complex" sociality because some species delay dispersal and have elementary parental care coupled with monogamy and family living. Surprisingly, sociality has only evolved within one major squamate lineage (Scincoidea, *sensu* Pyron, *et al.*, 2013) although future work could reveal additional instances.

13.8 A Summary of Lizard Sociality

Within lizards, various forms of territoriality have evolved many times over among all the major lizard radiations (Stamps, 1977) and in snakes (Webb, *et al.*, 2015). Polygyny is a general feature of these systems, although males may also defend territories in monogamous species (Chapple, 2003; While, *et al.*, 2009a). In territorial systems, males and females are likely to form social bonds, as are females that share space (Strickland, *et al.*, 2014). These non-kin based social bonds are therefore likely to be the most common form of social structure in lizards, although most studies do not quantify these bonds directly (but see Strickland, *et al.*, 2014). Instead, social structure is inferred by examining patterns of spatial overlap. Typically, we learn that males maintain exclusive space overlapping multiple females while females commonly overlap each other in space. These social bonds are likely to be relatively stable over time because males of many different species have high site fidelity and typically return to the same territories every breeding season. Unfortunately, most studies of lizard sexual selection do not focus on social relationships *per se*, but instead, on reproductive and sometimes morph-specific tactics. Likewise, parent–offspring associations, a hallmark of sociality, could be far more common than we realize because so little attention has been directed towards this possibility. Finally, many species over-winter communally in hibernacula for which they have strong fidelity. We know little about the potential social bonds that might form between potential mates in these systems, particularly since hibernation precedes mating (Graves & Duvall, 1995; Schuett, *et al.*, 2016).

Parental care through association between mother and offspring is also likely to be far more common than thus far documented because these links can be cryptic. Unlike in mammals and birds, there is little or only subtle direct interaction between parents and offspring in lizards. For example, many researchers may not know the relatedness of juveniles and adults in the field unless mothers are brought into captivity to give birth before being released back into the wild. We believe that this factor alone has likely resulted in a disconnect between researchers of social behavior in lizards and the possibility of parental care. Thus far, almost all known instances of mother-offspring association are in viviparous species, including a substantial number of rattlesnake species (Greene, *et al.*, 2002; Schuett, *et al.*, 2016). This parent–offspring association may result in direct protection against predators or infanticide, particularly from unrelated adults (O'Connor & Shine, 2004). This obvious fitness advantage has likely resulted in its independent evolution many times across the entire phylogeny of squamate reptiles.

There is only a single well-documented case of long-term monogamy in a lizard in which there is no parental care. Unlike lizards that live in family groups, the sleepy lizard, *Tiliqua rugosa*, lives in relatively homogenous habitat and is wide-ranging. Their pair bonds are remarkably stable (current record: more than 27 years) and one advantage of long-term monogamy appears to be effective coordination of reproduction between paired males and females (Leu, *et al.*, 2015).

The pinnacle of sociality in lizards is kin-based family living in the *Egernia* group of lizards and the North American xantusiid: *Xantusia vigilis*. These few instances of family living are all from a single major radiation of lizards (Scincoidea) in which long-term stable group living has been documented in multiple genera. It is highly likely that future genetic and field studies will uncover more instances of family living, such as in the African Cordylidae or the South American tropidurid genus *Liolaemus*. In the former systems, parental units are mostly monogamous although multiple mating does occur (Uller & Olsson, 2008; Wapstra & Olsson, 2014). One or more generations of kin delay dispersal, and males appear more likely to disperse than females, although this is not always clear and data are needed from additional species. Post-hatching parental care is primarily in the form of parent–offspring associations, although cases of direct parental care have been reported (While, *et al.*, 2014a).

An hypothesis for the evolution of kin-based sociality in lizards that has received empirical support is that family living is best explained as an interaction between indirect male enforcement of female fidelity and kin-based benefits through protection against predation or infanticide (While, *et al.*, 2009a). Kin benefits and enforcement of female fidelity are intimately linked because any offspring sired by the non-resident male will be forced to a restricted area of the parental home range and potentially pay a fitness cost through aggressive interactions from the resident male, reduced food acquisition, growth rates, and/or future survival (While, *et al.*, 2009a). Furthermore, while we know very little about parental care in the majority of species, juveniles are more likely to be attacked by adults other than their parents, particularly when their parents are not around, in at least the two species in which this has been studied (O'Connor & Shine, 2004; While, *et al.*, 2009a). The family-living species of the *Egernia* group frequently occur in patchy habitat where high density can limit dispersal opportunity and where venturing away from protection can be very risky. This helps set the stage for group stability and reinforcement of parent–offspring and sibling–sibling bonds and parental care, particularly in a group that is long-lived and in some species, slow to mature, relative to most other species of lizards (Figure 13.3). Increased pair stability and low female polyandry across years will increase the average relatedness of offspring within and among cohorts, which reduces competition and could lead to the evolution of larger social colonies as found in some species of *Egernia* (e.g. *Egernia stokesii*, Gardner, *et al.*, 2001).

13.9 Comparative Perspectives on Lizard Sociality

Lizards, and to a lesser degree snakes, exhibit a wide range of life history strategies, social organization, social structure, and mating systems (Stamps, 1977; Pianka & Vitt, 2003; Vitt, *et al.*, 2003). Sociality in lizards overlaps most with fishes, birds, and mammals, but bares little resemblance to the eusocial insects. Lizards fundamentally differ from these groups in that parental care is uncommon and when it occurs, is largely cryptic (except egg brooding) and typically involves tolerance of offspring (and therefore access to key resources) and protection against infanticide although direct protection of offspring is possible (While, *et al.*, 2014a). Furthermore, unlike birds and mammals, lizards do not provision their young and therefore, do not breed

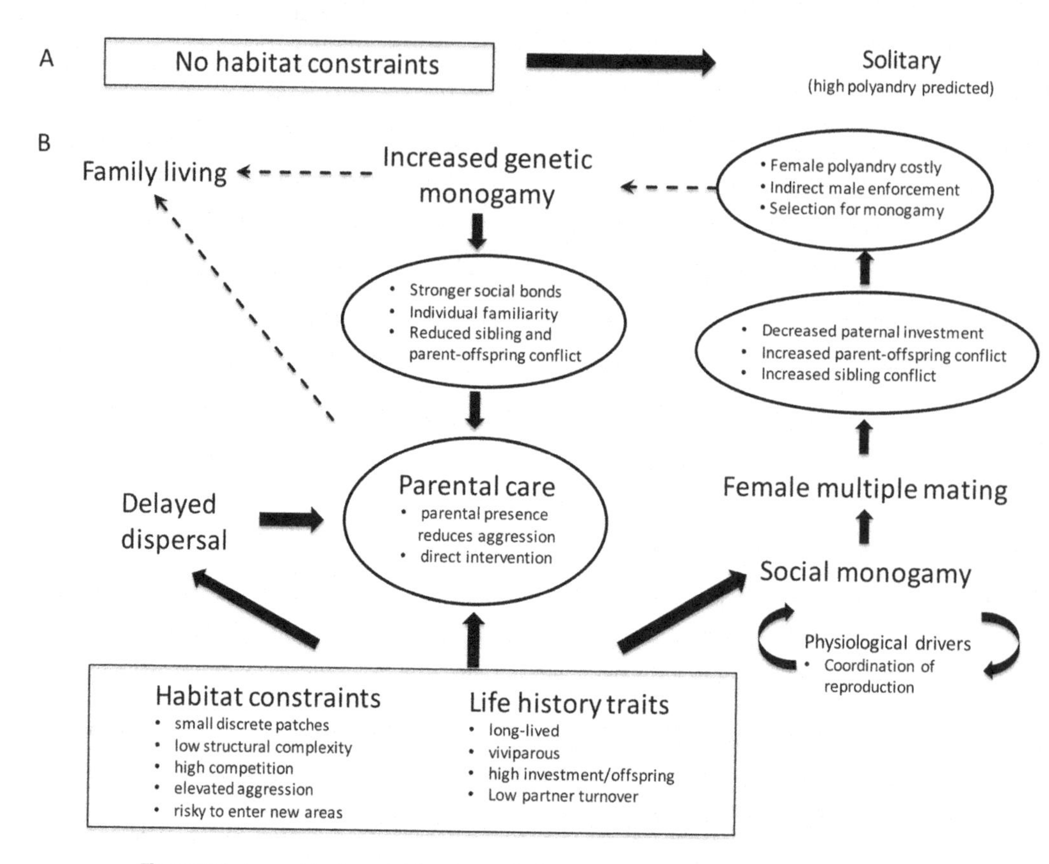

Figure 13.3 A graphical model for the evolution and maintenance of sociality in lizards based upon our knowledge of multiple species from the *Egernia* group and adapted and expanded from While, *et al.* (2009a). (A) When habitat availability is not constrained or discrete at a fine scale, species are predicted to be more solitary and to mate multiply. We have not depicted this graphically, but these species would have roughly opposite life history traits compared to more social species (i.e. oviparous, low investment/offspring, comparatively reduced longevity). (B) When habitat is discrete or patchy, species are more likely to form aggregations. Many species occur on discrete habitat patches such as trees and rocks and are dependent on refuges. Habitat availability is variable in the wild, but in cases where higher quality patches are limited, there may be greater competition for resources and more frequent contact with conspecifics. Dispersal under these conditions can be costly because unfamiliar juveniles (i.e. non-kin) are frequently attacked by adults and therefore, there is a risk of infanticide. Evidence for heightened aggression against non-kin has been found in *Egernia saxatilis* (O'Connor & Shine, 2004) and *Liopholis whitii* (While, *et al.*, 2009a). Delayed dispersal also sets the stage for the formation and strengthening of parent–offspring bonds, regardless of whether kin recognition is genetic or based on familiarity/ phenotype matching. Furthermore, the stage is set for parental care either by simple association with a parent (thereby deterring attacks from unrelated adults) or through direct protection. Finally, male parental tolerance of related offspring helps promote female fidelity because unrelated offspring are excluded from the male's territory thereby potentially imposing a fitness cost on the mother. Life history traits also set the stage for sociality. Lizard species with kin-based sociality are all viviparous, often slow to mature, invest relatively heavily in offspring production, and are long-lived. Monogamy likely initially evolved as a consequence of limited habitat availability (i.e. saturation) interacting with life history and resulting in low mate availability and/ or breeder turnover. Under these circumstances, it would pay to remain with the same mate, and monogamy may have been preceded by simple mate guarding of females by males. This scenario could result in a feedback loop where monogamy is reinforced through better coordination of breeding (Leu, *et al.*, 2015) and/or reduced promiscuity by females through indirect costs to her offspring.

cooperatively. Lizard family groups typically contain a long-term monogamous parental unit that is also territorial, similar to that found in certain mammals (e.g. numerous species of small African antelope, Skinner & Chimimba, 2006), and may contain multiple generations of offspring. In mammals, social monogamy is derived from an ancestral state in which females were solitary and males attempted to overlap the ranges of multiple females. This ancestral state is similar to many polygynous systems in lizards where territorial males control resources that encompass the home ranges of multiple females. However, the transition to kin-based sociality in lizards has likely followed a different path to that of mammals. In mammals, transitions to social monogamy are a consequence of intolerance among breeding females coupled with intense competition for resources and low population density. Mating systems and monogamy in particular, are frequently linked to parental care in birds (Cornwallis, *et al.*, 2010) and mammals (Lukas & Clutton-Brock, 2012) and this also appears to be the case in lizards.

13.10 Concluding Remarks

A suite of traits and abiotic factors correlate with kin-based sociality in lizards. These include use of a discrete habitat patch or refuge (e.g. rocks or trees) or in the case of some sand-living lizard species, a burrow complex. Males defend a core area such as a crevice, tree hollow, or burrow. All lizards that live in stable family groups have a viviparous reproductive mode, many are omnivorous or herbivorous, are generally long-lived, and they may have delayed maturity. These traits in combination appear to be an important precursor for kin-based sociality (Figure 13.3), although any inferences we make must be tempered by the phylogenetic bias represented by this clade. If future genetic testing reveals more instances of kin-based sociality across the lizard phylogeny then this bias will be reduced and the importance of habitat and life history traits will be amplified.

Many of the arguments that we make for the evolution of kin-based sociality in lizards are analogous to those proposed to explain the evolution of advanced forms of social behavior (e.g. cooperative breeding, eusociality) in birds, mammals, and insects (Hughes, *et al.*, 2008; Cornwallis, *et al.*, 2010; Lukas & Clutton-Brock, 2012). Recent research has suggested that to truly understand the conditions under which group formation and social life initially emerged and evolved, we need to move away from systems in which sociality is derived, and in which individuals exhibit obligate or permanent forms of group living (Smiseth, *et al.*, 2003; Falk, *et al.*, 2014). Instead, we should focus on identifying the nature of social behavior and organization in species that exhibit facultative and/or temporary forms of social grouping. To this end, lizards, and the *Egernia* group in particular, offer an outstanding model system. In only a few other systems is there the combination of relatively simply ancestral forms of parental care and social behavior coupled with both within- *and* among-species diversity in social complexity (see also Chapters 8 and 9). Taken collectively, this is a system that could inform on the basic triggers required to transition to kin-based family living and relatively complex sociality. There is a rich potential for new discovery and we anticipate that lizards will feature more prominently in the social behavior literature in the future.

References

Altizer, S., Nunn, C. L., Thrall, P. H., *et al.* (2003) Social organization and parasite risk in mammals: Integrating theory and empirical studies. *Annual Review of Ecology and Systematics*, **34**, 517–547.

Baird, T. A. (2013) Lizards and other reptiles as model systems for the study of contest behaviour. In: Hardy, I. C. W. & Briffa, M. (eds.) *Animal Contests*. Cambridge, UK: Cambridge University Press, pp. 258–286

Barry, M., Shanas, U., & Brunton, D. H. (2014) Year-round mixed-age shelter aggregations in Duvaucel's geckos (*Hoplodactylus duvaucelii*). *Herpetologica*, **70**, 395–406.

Black, J. M. (1996) Introduction: Pair bonds and partnerships. In: Black, J. M. (ed.) *Partnerships in Birds: The Study of Monogamy*. Oxford, UK: Oxford University Press, pp. 3–20.

(2001) Fitness consequences of long-term pair bonds in barnacle geese: Monogamy in the extreme. *Behavioral Ecology*, **12**, 640–645.

Brattstrom, B. H. (1974) The evolution of reptilian social behavior. *American Zoologist*, **14**, 35–49.

Brown, T. K. & Nagy, K. A. (2007) Lizard energetics and the sit-and-wait vs. wide-foraging paradigm. *In:* Reilly, S. M., McBrayer, L. D., & Miles, D. B. (eds.) *Lizard Ecology: The Evolutionary Consequences of Foraging Mode*. Cambridge, UK: Cambridge University Press, pp. 120–140.

Bull, C. M. (1994) Population dynamics and pair fidelity in sleepy lizards. In: Vitt, L. J. & Pianka, E. R. (eds.) *Lizard Ecology: Historical and Experimental Perspectives*. Princeton, NJ: Princeton University Press, pp. 159–174.

(2000) Monogamy in lizards. *Behavioural Processes*, **51**, 7–20.

Bull, C. M. & Baghurst, B. C. (1998) Home range overlap of mothers and their offspring in the sleepy lizard, *Tiliqua rugosa*. *Behavioral Ecology & Sociobiology*, **42**, 357–362.

Bull, C. M. & Bonnett, M. (2004) *Egernia striolata* (tree skink) reproduction. *Herpetological Review*, **35**, 389.

Bull, C. M. & Burzacott, D. A. (2006) The influence of parasites on the retention of long-term partnerships in the Australian sleepy lizard, *Tiliqua rugosa*. *Oecologia*, **146**, 675–680.

Bull, C. M. & Cooper, S. J. B. (1999) Relatedness and avoidance of inbreeding in the lizard, *Tiliqua rugosa*. *Behavioral Ecology and Sociobiology*, **46**, 367–372.

Bull, C. M. & Pamula, Y. (1998) Enhanced vigilance in monogamous pairs of the lizard, Tiliqua rugosa. *Behavioral Ecology*, **9**, 452–455.

Bull, C. M., Pamula, Y., & Schulze, L. (1993) Parturition in the sleepy lizard *Tiliqua rugosa*. *Journal of Herpetology*, **27**, 489–492.

Bull, C. M., Doherty, M., Schulze, L. R., & Pamula, Y. (1994) Recognition of offspring by females of the Australian skink, *Tiliqua rugosa*. *Journal of Herpetology*, **28**, 117–120.

Bull, C. M., Cooper, S. J. B., & Baghurst, B. C. (1998) Social monogamy and extra-pair fertilization in an Australian lizard, *Tiliqua rugosa*. *Behavioral Ecology and Sociobiology*, **44**, 63–72.

Bull, C. M., Griffin, C. L., & Johnston, G. R. (1999) Olfactory discrimination in scat-piling lizards. *Behavioral Ecology*, **10**, 136–140.

Bull, M. C., Griffin, C. L., Bonnett, M., Gardner, M. G., & Cooper, S. J. (2001) Discrimination between related and unrelated individuals in the Australian lizard *Egernia striolata. Behavioral Ecology and Sociobiology*, **50**, 173–179.

Byrne, R. W. & Bates, L. A. (2007) Sociality, evolution and cognition. *Current Biology*, **17**, R714-R723.

Carazo, P., Font, E., & Desfilis, E. (2008) Beyond "nasty neighbours" and "dear enemies?" Individual recognition by scent marks in a lizard (Podarcis hispanica). *Animal Behaviour*, **76**, 1953–1963.

Chapple, D. G. (2003) Ecology, life-history, and behavior in the Australian Scincid genus *Egernia*, with comments on the evolution of complex sociality in lizards. *Herpetological Monographs*, **17**, 145–180.

Chapple, D. G. & Keogh, J. S. (2004) Parallel adaptive radiations in and temperate Australia: Molecular phylogeography and systematics of the *Egernia whitii* (Lacertilia: Scincidae) species group. *Biological Journal of the Linnean Society*, **83**, 157–173.

(2005) Complex mating system and dispersal patterns in a social lizard, *Egernia whitii. Molecular Ecology*, **14**, 1215–1227.

(2006) Group structure and stability in social aggregations of White's skink, *Egernia whitii. Ethology*, **112**, 247–257.

Clark, B. F., Amiel, J. J., Shine, R., Noble, D. W. A., & Whiting, M. J. (2014) Colour discrimination and associative learning in hatchling lizards incubated at "hot" and "cold" temperatures. *Behavioral Ecology and Sociobiology*, **68**, 239–247.

Clark, R. W. (2004) Kin recognition in rattlesnakes. *Proceedings of the Royal Society of London. Series B*, **271**, S243-S245.

Clemann, N., Chapple, D. G., & Wainer, J. (2004) Sexual dimorphism, diet, and reproduction in the swamp skink, *Egernia coventryi. Journal of Herpetology*, **38**, 461–467.

Clutton-Brock, T. H. (1991) *The Evolution of Parental Care*. Princeton, NJ: Princeton University Press.

Cooper, W. E. J. (1994) Prey chemical discrimination, foraging mode, and phylogeny. In: Vitt, L. J. & Pianka, E. R. (eds.) *Lizard Ecology: Historical and Experimental Perspectives*. Princeton, NJ: Princeton University Press, pp. 95–116.

Cornwallis, C. K., West, S. A., & Davis, K. E. (2010) Promiscuity and the evolutionary transition to complex societies. *Nature*, **466**, 969–972.

Croft, D. P., James, R., & Krause, J. (2008). *Exploring Animal Social Networks*. Princeton, NJ: Princeton University Press.

Daniel, M. C. (1998) *Aspects of the ecology of Rosen's Desert Skink, Egernia inornata, in the Middleback Ranges, Eyre Peninsula*. Honours thesis, Adelaide, Australia: University of Adelaide.

Davis, A. R. (2012) Kin presence drives philopatry and social aggregation in juvenile desert night lizards (*Xantusia vigilis*). *Behavioral Ecology*, **23**, 18–24.

Davis, A. R., Corl, A., Surget-Groba, Y., & Sinervo, B. (2011) Convergent evolution of kin-based sociality in a lizard. *Proceedings of the Royal Society B-Biological Sciences*, **278**, 1507–1514.

Davis Rabosky, A. R., Corl, A., Liwanag, H. E. M., Surget-Groba, Y., & Sinervo, B. (2012) Direct fitness correlates and thermal consequences of facultative aggregation in a desert lizard. *Plos One*, **7**, e40866.

Doody, J. S., Freedberg, S., & Keogh, J. S. (2009) Communal egg-laying in reptiles and amphibians: Evolutionary patterns and hypotheses. *Quarterly Review of Biology*, **84**, 229–252.

Doody, J. S., Burghardt, G. M., & Dinets, V. (2013) Breaking the social–non-social dichotomy: A role for reptiles in vertebrate social behavior research? *Ethology*, **119**, 95–103.

Duckett, P. E., Morgan, M. H., & Stow, A. J. (2012) Tree-dwelling populations of the skink *Egemia striolata* aggregate in groups of close kin. *Copeia*, **2012**, 130–134.

Duffield, G. A. & Bull, M. C. (2002) Stable social aggregations in an Australian lizard, *Egernia stokesii*. *Naturwissenschaften*, **89**, 424–427.

Effenberger, E. & Mouton, P. L. N. (2007) Space use in a multi-male group of the group living lizard. *Journal of Zoology*, **272**, 202–208.

Eifler, D. (2001) *Egernia cunninghami* (Cunningham's Skink), escape behavior. *Herpetological Review*, **32**, 40.

Falk, J., Wong, J. M. Y., Kolliker, M., & Meunier, J. (2014) Sibling competition in earwig families provides insights into the early evolution of social life. *The American Naturalist*, **183**, 547–557.

Fenner, A. L., Pavey, C. R., & Bull, C. M. (2012) Behavioural observations and use of burrow systems by an endangered Australian arid-zone lizard, Slater's skink (*Liopholis slateri*). *Australian Journal of Zoology*, **60**, 127–132.

Fox, S. F. & Baird, T. A. (1992) The dear enemy phenomenon in the collared lizard, *Crotaphytus collaris,* with a cautionary note on experimental methodology *Animal Behaviour*, **44**, 780–782.

Fox, S. F., Mccoy, J. K., & Baird, T. A. (2003) *Lizard Social Behavior*, Baltimore, MD: Johns Hopkins University Press.

Frère, C. H., Chandrasoma, D., & Whiting, M. J. (2015) Polyandry in dragon lizards: Inbred paternal genotypes sire fewer offspring. *Ecology and Evolution*, **5**, 1686–1692.

Fuller, S. J., Bull, C. M., Murray, K., & Spencer, R. J. (2005) Clustering of related individuals in a population of the Australian lizard, *Egernia frerei*. *Molecular Ecology*, **14**, 1207–1213.

Galliard, J. L., Ferrière, & Clobert, J. (2003) Mother–offspring interactions affect natal dispersal in a lizard. *Proceedings of the Royal Society B*, **270**, 1163–1169.

Gardner, M. G., Bull, C. M., Cooper, S. J. B., & Duffield, G. A. (2001) Genetic evidence for a family structure in stable social aggregations of the Australian lizard *Egernia stokesii*. *Molecular Ecology*, **10**, 175–183.

Gardner, M. G., Bull, C. M., & Cooper, S. J. B. (2002) High levels of genetic monogamy in the group living Australian lizard *Egernia stokesii*. *Molecular Ecology*, **11**, 1787–1794.

Gardner, M. G., Bull, C. M., Fenner, A., Murray, K., & Donnellan, S. C. (2007) Consistent social structure within aggregations of the Australian lizard, *Egernia stokesii* across seven disconnected rocky outcrops. *Journal of Ethology*, **25**, 263–270.

Gardner, M. G., Hugall, A. F., Donnellan, S. C., Hutchinson, M. N., & Foster, R. (2008) Molecular systematics of social skinks: Phylogeny and taxonomy of the *Egernia* group (Reptilia: Scincidae). *Zoological Journal of the Linnean Society*, **154**, 781–794.

Gardner, M. G., Godfrey, S. S., Fenner, A. L., Donnellan, S. C., & Bull, C. M. (2012) Fine-scale spatial structuring as an inbreeding avoidance mechanism in the social skink *Egernia stokesii*. *Australian Journal of Zoology*, **60**, 272–277.

Gardner, M. G., Pearson, S. K., Johnston, G. R., & Schwarz, M. P. (2016) Group living in squamate reptiles: A review of evidence for stable aggregations. *Biological Reviews*, **91**, 925–936.

Godfrey, S. S., Bull, C. M., Murray, K., & Gardner, M. G. (2006) Transmission mode and distribution of parasites among groups of the social lizard *Egernia stokesii*. *Parasitology Research*, **99**, 223–230.

Godfrey, S., Bull, C., James, R., & Murray, K. (2009) Network structure and parasite transmission in a group living lizard, the gidgee skink, *Egernia stokesii*. *Behavioral Ecology and Sociobiology*, **63**, 1045–1056.

Graves, B. M. & Duvall, D. (1995) Aggregation of squamate reptiles associated with gestation, oviposition, and parturition. *Herpetological Monographs*, **9**, 102–119.

Graves, B. M., Carpenter, G. C., & Duvall, D (1987). Chemosensory behaviors of neonate prairie rattlesnakes, *Crotalus vitridis*. *The Southwestern Naturalist*, **32**, 515–517.

Greene, H. W., May, P. G., Hardy, D. L., *et al.* (2002) Parental behavior of vipers. In: Schuett, G. W., Hoggren, M., Douglas, M. E., & Greene, H. W. (eds.) *Biology of the Vipers*. Eagle Mountain, UT: Eagle Mountain Publishing, pp. 179–206.

Greenwood, P. J. (1980) Mating systems, philopatry, and dispersal in birds: A review of interspecific variation and adaptive function. *Molecular Ecology*, **11**, 2195–2212.

Greer, A. E. (1989) *The Biology and Evolution of Australian Lizards*, Chipping Norton, NSW: Surrey Beattie & Sons.

Gregory, P. T. (1984) Communal denning in snakes. In: Seigel, R. A., Hunt, L. E., Knight, J. L., Malaret, L., & Zuschlag, N. L. (eds.) *Vertebrate Ecology and Systematics: A Tribute to Henry S. Fitch*. Lawrence, KS: *The University of Kansas Museum of Natural History Special Publication NO. 10*, pp. 57–75.

Griffith, S. C., Owens, I. P. F., & Thuman, K. A. (2002) Extra pair paternity in birds: A review of interspecific variation and adaptive function. *Molecular Ecology*, **11**, 2195–2212.

Hagen, I. J., Herfindal, I., Donnellan, S. C., & Bull, C. M. (2013) Fine scale genetic structure in a population of the prehensile tailed skink, *Corucia zebrata*. *Journal of Herpetology*, **47**, 308–313.

Halpern, M. (1992) Nasal chemical senses in reptiles: Structure and function. In: Gans, C. & Crews, D. (eds.) *Biology of the Reptilia: Brain, Hormones, and Behavior*. Chicago, IL: University of Chicago Press.

Hatchwell, B. J. & Komdeur, J. (2000) Ecological constraints, life history traits and the evolution of cooperative breeding. *Animal Behavior*, **59**, 1079–1086.

Henzell, R. P. (1972) *Adaptation to aridity in lizards of the Egernia whiteii species-group*. PhD thesis. Adelaide, Australia: University of Adelaide.

Hibbitts, T. J., Whiting, M. J., & Stuart-Fox, D. M. (2007) Shouting the odds: Vocalization signals status in a lizard. *Behavioral Ecology and Sociobiology*, **61**, 1169–1176.

Huang, W. S. (2006) Parental care in the long-tailed skink, *Mabuya longicaudata*, on a tropical Asian island. *Animal Behaviour*, **72**, 791–795.

Huang, W. S. & Pike, D. A. (2011) Does maternal care evolve through egg recognition or directed territoriality? *Journal of Evolutionary Biology*, **24**, 1984–1991.

Huey, R. B. & Pianka, E. R. (1981) Ecological consequences of foraging mode. *Ecology*, **62**, 991–999.

Hughes, W. O. H., Oldroyd, B. P., Beekman, M., & Ratnieks, F. L. W. (2008) Ancestral monogamy shows kin selection is key to the evolution of eusociality. *Science*, **320**, 1213–1216.

Hutchinson, M. N., Milne, T., & Croft, T. (1994) Redescription and ecological notes on the pygmy bluetongue, *Tiliqua adelaidensis* (Squamata: Scincidae). *Transactions of the Royal Society of South Australia*, **118**, 217–226.

James, C. & Shine, R. (1988) Life history strategies of Australian lizards: A comparison between the tropics and the temperate zone. *Oecologia*, **75**, 307–316.

Kappeler, P. M. & van Schaik, C. P. (2002) Evolution of primate social systems. *International Journal of Primatology*, **23**, 707–740.

Kappeler, P. M., Barrett, L., Blumstein, D. T., & Clutton-Brock, T. H. (2013) Constraints and flexibility in mammalian social behaviour: Introduction and synthesis. *Philosophical Transactions of the Royal Society of London B*, **386**, 20120337.

Karsten, K. B., Andriamandimbiarisoa, L. N., Fox, S. F., & Raxworthy, C. J. (2008) A unique life history among tetrapods: An annual chameleon living mostly as an egg. *Proceedings of the National Academy of Sciences of the United States of America*, **105**, 8980–8984.

Kavaliers, M. & Choleris, E. (2013) Neurobiological correlates of sociality, mate choice, and learning. *Trends in Ecology & Evolution*, **28**, 4–5.

Kis, A., Huber, L., & Wilkinson, A. (2014). Social learning by imitation in a reptile (*Pogona vitticeps*). *Animal Cognition*, **18**, 325–331.

Koenig, W. D., Pitelka, F. A., Carmen, W. J., Mumme, R. L., & Stanback, M. T. (1992) The evolution of delayed dispersal in cooperative breeders. *Quarterly Review of Biology*, **67**, 111–150.

Komdeur, J. (1992) Importance of habitat saturation and territory quality for evolution of cooperative breeding in the Seychelles warbler. *Nature*, **358**, 493–495.

Lancaster, J. R., Wilson, P., & Espinoza, R. E. (2006) Physiological benefits as precursors of sociality: Why banded geckos band. *Animal Behaviour*, **72**, 199–207.

Langkilde, T., Lance, V. A., & Shine, R. (2005) Ecological consequences of agonistic interactions in lizards. *Ecology*, **86**, 1650–1659.

Langkilde, T., O'Connor, D., & Shine, R. (2007) Benefits of parental care: Do juvenile lizards obtain better-quality habitat by remaining with their parents? *Austral Ecology*, **32**, 950–954.

Lanham, E. J. (2001) *Group living in the Australian skink, Egernia stokesii*. Honours dissertation. Flinders University.

Lanham, E. J. & Bull, C. M. (2000) Maternal care and infanticide in the Australian skink, *Egernia stokesii*. *Herpetological Review*, **31**, 151–152.

(2004) Enhanced vigilance in groups in *Egernia stokesii*, a lizard with stable social aggregations. *Journal of Zoology*, **263**, 95–99.

Lawson Handley, L. J., & Perrin, N. (2007) Advances in our understanding of mammalian sex-biased dispersal. *Molecular Ecology*, **16**, 1559–1578.

Leal, M. & Powell, B. J. (2012) Behavioural flexibility and problem-solving in a tropical lizard. *Biology Letters*, **8**, 28–30.

Leu, S. T., Kappeler, P. M., & Bull, C. M. (2011) Pair-living in the absence of obligate biparental care in a lizard: Trading-off sex and food? *Ethology*, **117**, 758–768.

Leu, S. T., Burzacott, D., Whiting, M. J., & Bull, C. M. (2015) Mate familiarity drives long-term monogamy in a lizard: Evidence from a thirty-year field study. *Ethology* **121**, 720–728.

Lukas, D. & Clutton-Brock, T. H. (2012) Cooperative breeding and monogamy in mammalian societies. *Proceedings of the Royal Society of London B*, **279**, 2151–2156.

(2013) The evolution of social monogamy in mammals. *Science*, **341**, 526–530.

Lyon, B. E. & Montgomerie, R. (2012) Sexual selection is a form of social selection. *Philosophical Transactions of the Royal Society B*, **367**, 2266–2273.

Main, A. R. & Bull, C. M. (1996). Mother–offspring recognition in two Australian lizards, *Tiliqua rugosa* and *Egernia stokesii*. *Animal Behaviour*, **52**, 193–200.

Mariette, M. M. & Griffith, S. C. (2012) Nest visit synchrony is high and correlates with reproductive success in the wild zebra finch *Taeniopygia guttata*. *Journal of Avian Biology*, **43**, 131–140.

Martins, E. P. (1993) A comparative study of the evolution of *Sceloporus* push-up displays. *The American Naturalist*, **142**, 994–1018.

Masters, C. & Shine, R. (2003) Sociality in lizards: Family structure in free-living King's Skinks *Egernia kingii* from southwestern Australia. *Australian Zoologist*, **32**, 377–380.

McAlpin, S., Duckett, P., & Stow, A. (2011) Lizards cooperatively tunnel to construct a long-term home for family members. *PLOS ONE*, **6**, e19041.

McBrayer, L. D., Miles, D. B., & Reilly, S. M. (2007) The evolution of the foraging mode paradigm in lizard ecology. In: Reilly, S. M., McBrayer, L. B., & Miles, D. B. (eds.) *Lizard Ecology: The Evolutionary Consequences of Foraging Mode*. Cambridge, UK: Cambridge University Press, pp. 508–521.

Michael, D. R., Cunningham, R. B., & Lindenmayer, D. B. (2010) The social elite: Habitat heterogeneity, complexity and quality in granite inselbergs influence patterns of aggregation in *Egernia striolata* (Lygosominae: Scincidae). *Austral Ecology*, **35**, 862–870.

Miller, M. R. (1951) Some aspects of the life history of the Yucca Night Lizard, *Xantusia vigilis*. *Copeia*, **1951**, 114–120.

Milne, T., Bull, C. M., & Hutchinson, M. N. (2002) Characteristics of litters and juvenile dispersal in the endangered Australian skink *Tiliqua adelaidensis*. *Journal of Herpetology*, **36**, 110–112.

Mooring, M. S. & Hart, B. L. (1992) Animal grouping for protection from parasites: Selfish herd and encounter–dilution effects. *Behaviour*, **123**, 173–193.

Mouton, P. L. F. N., Flemming, A. F., & Kanga, E. M. (1999) Grouping behaviour, tail-biting behaviour and sexual dimorphism in the armadillo lizard (*Cordylus cataphractus*) from South Africa. *Journal of Zoology*, **249**, 1–10.

Mouton, P. L. N. (2011) Aggregation behaviour of lizards in the arid western regions of South Africa. *African Journal of Herpetology*, **60**, 155–170.

Noble, G. K. & Bradley, H. T. (1933) The mating behavior of lizards; Its bearing on the theory of sexual selection. *Annals of the New York Academy of Sciences*. 35, 25–100.

Noble, D. W. A., Wechmann, K., Keogh, J. S., & Whiting, M. J. (2013) Behavioral and morphological traits interact to promote the evolution of alternative reproductive tactics in a lizard. *The American Naturalist*, **182**, 726–742.

Noble, D. W. A., Byrne, R. W., & Whiting, M. J. (2014) Age-dependent social learning in a lizard. *Biology Letters*, **10**, 20140430.

O'Connor, D. E. & Shine, R. (2003) Lizards in "nuclear families": A novel reptilian social system in *Egernia saxatilis* (Scincidae). *Molecular Ecology*, **12**, 743–752.

(2004) Parental care protects against infanticide in the lizard *Egernia saxatilis* (Scincidae). *Animal Behaviour*, **68**, 1361–1369.

(2006) Kin discrimination in the social lizard *Egernia saxatilis* (Scincidae). *Behavioral Ecology*, **17**, 206–211.

Olsson, M. & Madsen, T. (1995) Female choice on male quantitative traits in lizards: Why is it so rare? *Behavioural Ecology and Sociobiology*, **36**, 179–184.

Olsson, M., Madsen, T., Shine, R., Gullberg, A., & Tegelstrom, H. (1994) Rewards of promiscuity. *Nature*, **372**, 230–230.

Ord, T. J. & Martins, E. P. (2006) Tracing the origins of signal diversity in anole lizards: Phylogenetic approaches to inferring the evolution of complex behaviour. *Animal Behaviour*, **71**, 1411–1429.

Ord, T. J., Blumstein, D. T., & Evans, C. S. (2002) Ecology and signal evolution in lizards. *Biological Journal of the Linnean Society*, **77**, 127–148.

Osterwalder, K., Klingenböck, A., & Shine, R. (2004) Field studies on a social lizard: Home range and social organization in an Australian skink, *Egernia major*. *Austral Ecology*, **29**, 241–249.

Perrin, N. & Mazalov, V. (2000) Local competition, inbreeding, and the evolution of sex-biased dispersal. *The American Naturalist*, **155**, 116–127.

Pianka, E. R. & Vitt, L. J. (2003) *Lizards: Windows to the Evolution of Diversity Berkeley*, CA: University of California Press.

Post, M. J. (2000). The captive husbandry and reproduction of the Hosmer's skink *Egernia hosmeri*. *Herpetofauna*, **30**, 2–6.

Pyron, R. A. & Burbrink, F. T. (2014) Early origin of viviparity and multiple reversions to oviparity in squamate reptiles. *Ecology Letters*, **17**, 13–21.

Pyron, R. A., Burbrink, F. T., & Wiens, J. J. (2013) A phylogeny and revised classification of Squamata, including 4161 species of lizards and snakes. *BMC Evolutionary Biology*, **13**, 93.

Radder, R. S., Elphick, M. J., Warner, D. A., Pike, D. A., & Shine, R. (2008) Reproductive modes in lizards: Measuring fitness consequences of the duration of uterine retention of eggs. *Functional Ecology*, **22**, 332–339.

Sacchi, R., Pupin, F., Gentilli, A., Rubolini, D., Scali, S., Fasola, M., & Galeotti, P. (2009) Male–male combats in a polymorphic lizard: Residency and size, but not color, affect fighting rules and contest outcome. *Aggressive Behavior*, **35**, 274–83.

Schofield, J. A., Gardner, M. G., Fenner, A. L., & Bull, C. M. (2014) Promiscuous mating in the endangered Australian lizard *Tiliqua adelaidensis*: A potential windfall for its conservation. *Conservation Genetics*, **15**, 177–185.

Schuett, G. W., Clark, R. W., Repp, R. A., Amarello, M., & Greene, H. W. (2016) Social behavior of rattlesnakes: A shifting paradigm. In: Schuett, G. W., Feldner, M. J., Reiserer, R. S., & Smith, C. S. (eds.) *Rattlesnakes of Arizona*. Rodeo, NM: Eco Publishing, pp 155–236.

Shah, B., Shine, R., Hudson, S., & Kearney, M. (2003) Sociality in lizards: Why do thick-tailed geckos (*Nephrurus milii*) aggregate? *Behaviour*, **140**, 1039–1052.

Shah, B., Shine, R., Hudson, S., & Kearney, M. (2004) Experimental analysis of retreat-site selection by thick-tailed geckos *Nephrurus milii*. *Austral Ecology*, **29**, 547–552.

Shea, G. M. (1999) Morphology and natural history of the Land Mullet *Egernia major* (Squamata: Scincidae). *Australian Zoologist*, **31**, 351–364.

Shuttleworth, C., Mouton, P. L. N., & Van Niekerk, A. (2013) Climate and the evolution of group living behaviour in the armadillo lizard (*Ouroborus cataphractus*). *African Zoology*, **48**, 367–373.

Shuttleworth, C., Mouton, P. L. N., & Van Wyk, J. H. (2008) Group size and termite consumption in the armadillo lizard, *Cordylus cataphractus*. *Amphibia-Reptilia*, **29**, 171–176.

Sinervo, B. & Lively, C. (1996) The rock-paper-scissors game and the evolution of alternative male strategies. *Nature*, **380**, 240–243.

Sinn, D. L., While, G. M., & Wapstra, E. (2008) Maternal care in a social lizard: Links between female aggression and offspring fitness. *Animal Behaviour*, **76**, 1249–1257.

Skinner, A., Hugall, A. F., & Hutchinson, M. N. (2011) Lygosomine phylogeny and the origins of Australian scincid lizards. *Journal of Biogeography*, **38**, 1044–1058.

Skinner, J. D. & Chimimba, C. T. (2006) *Mammals of the Southern African Subregion*. Cambridge, UK: Cambridge University Press.

Smiseth, P. T., Darwell, C. T., & Moore, A. J. (2003) Partial begging: An empirical model for the early evolution of offspring signalling. *Proceedings of the Royal Society of London B*, **270**, 1773–1777.

Somma, L. A. (2003) *Parental Behavior in Lepidosaurian and Testudinian Reptiles: A Literature Survey*. Malabar, FL: Krieger Publishing Company.

Stahlschmidt, Z. R. & Denardo, D. F. (2011) Parental care in snakes. In: Aldridge, R. D. & Sever, D. M. (eds.) *Reproductive Biology and Phylogeny of Snakes*. Boca Raton, FL: CRC Press, pp. 673–702.

Stamps, J. A. (1977) Social behavior and spacing patterns in lizards. In: Gans, C. & Tinkle, D. W. (eds.) *Biology of the Reptilia. Volume 7. Ecology and Behavior A*. New York, NY: Academic Press, pp. 265–334.

(1983) Sexual selection, sexual dimorphism, and territoriality. In: Huey, R. B., Pianka, E. R., & Schoener, T. W. (eds.) *Lizard Ecology: Studies of a Model Organism*. Cambridge, MA: Harvard University Press. 169–204.

Stow, A. J. & Sunnucks, P. (2004) Inbreeding avoidance in Cunningham's skinks (*Egernia cunninghami*) in natural and fragmented habitat. *Molecular Ecology*, **13**, 443–447.

Stow, A. J., Sunnucks, P., Briscoe, D. A., & Gardner, M. G. (2001) The impact of habitat fragmentation on dispersal of Cunningham's skink (*Egernia cunninghami*): Evidence from allelic and genotypic analyses of microsatellites. *Molecular Ecology*, **10**, 867–878.

Strickland, K., Gardiner, R., Schult, A. J., & Frère, C. H. (2014) The social life of eastern water dragons: Sex differences, spatial overlap and genetic relatedness. *Animal Behaviour*, **97**, 53–61.

Sumner, J. (2006) Higher relatedness within groups due to variable subadult dispersal in a rainforest skink, *Gnypetoscincus queenslandiae*. *Austral Ecology*, **31**, 441–448.

Taylor, M. (1995) Back to the swamp: Completion of the swamp skink project. *Thylacinus*, **20**, 15–17.

Tokarz, R. R. (1995) Mate choice in lizards: A review. *Herpetological Monographs*, **9**, 17–40.

Uller, T. & Olsson, M. (2008). Multiple paternity in reptiles: Patterns and processes. *Molecular Ecology*, **17**, 2566–2580.

Uller, T. & While, G. (2014) The evolutionary ecology of reproductive investment in lizards. In: Rheubert, J. L., Siegel, D. S., & Trauth, S. E. (eds.) *Reproductive Biology and Phylogeny of Lizards and Tuatara*. Boca Raton, FL: CRC Press, pp. 425–447.

Umbers, K. D. L., Osborne, L., & Keogh, J. S. (2012) The effects of residency and body size on contest initiation and outcome in the territorial dragon, *Ctenophorus decresii*. *PLOS ONE*, **7**, e47143.

Uetz, P. & Hosek, J. (eds). The Reptile Database, www.reptile-database.org, accessed September 27, 2015.

Visagie, L., Mouton, P. L. N., & Bauwens, D. (2005) Experimental analysis of grouping behaviour in cordylid lizards. *Herpetological Journal*, **15**, 91–96.

Vitt, L. J. (1981) Lizard reproduction: Habitat specificity and constraints on relative clutch mass. *American Naturalist*, **117**, 506–514.

Vitt, L. J. & Price, H. J. (1982) Ecological and evolutionary determinants of relative clutch mass in lizards. *Herpetologica*, **38**, 237–255.

Vitt, L. J. & Seigel, R. A. (1985) Life-history traits of lizards and snakes. *American Naturalist*, **125**, 480–484.

Vitt, L. J., Pianka, E. R., Cooper, W. E., & Schwenk, K. (2003) History and the global ecology of squamate reptiles. *American Naturalist*, **162**, 44–60.

Wapstra, E. & Olsson, M. (2014) The evolution of polyandry and patterns of multiple paternity in lizards. In: Rheubert, J. L., Siegel, D. S., & Trauth, S. E. (eds.) *Reproductive Biology and Phylogeny of Lizards and Tuatara*. Boca Raton, FL: CRC Press, pp. 564–589.

Warner, D. A. & Shine, R. (2007) Fitness of juvenile lizards depends on seasonal timing of hatching, not offspring body size. *Oecologia*, **154**, 65–73.

Webb, J. K., Scott, M. L., Whiting, M. J., & Shine, R. (2015) Territoriality in a snake. *In review*.

Webber, P. (1979) Burrow density, position and relationship of burrows to vegetation coverage shown Rosen's desert skink *Egernia inornata* (Lacertilia: Scincidae). *Herpetofauna*, **10**, 16–20.

While, G., Uller, T., & Wapstra, E. (2009a) Family conflict and the evolution of sociality in reptiles. *Behavioural Ecology*, **20**, 245–250.

(2009b) Within-population variation in social strategies characterize the social and mating system of an Australian lizard, *Egernia whitii*. *Austral Ecology*, **34**, 938–949.

(2011) Variation in social organization influences the opportunity for sexual selection in a social lizard. *Molecular Ecology*, **20**, 844–852.

While, G. M., Halliwell, B., & Uller, T. (2014a) The evolutionary ecology of parental care in lizards. In: Rheubert, J. L., Siegel, D. S., & Trauth, S. E. (eds.) *Reproductive Biology and Phylogeny of Lizards and Tuatara*. Boca Raton, FL: CRC Press, pp. 590–619.

While, G. M., Uller, T., Bordogna, G., & Wapstra, E. (2014b) Promiscuity resolves constraints on social mate choice imposed by population viscosity. *Molecular Ecology*, **23**, 721–732.

While, G. M., Chapple, D. G., Gardner, M. G., Uller, T., & Whiting, M. J. (2015) *Egernia* lizards. *Current Biology*, **25**, R593-R595.

Whitehead, H., Bejder, L., & Ottensmeyer, C. (2005) Testing association patterns: Issues arising and extensions. *Animal Behaviour*, **69**, e1-e6.

Whiting, M. J. (1999). When to be neighbourly: Differential agonistic responses in the lizard *Platysaurus broadleyi*. *Behavioral Ecology and Sociobiology*, **46**, 210–214.

Whiting, M. J., Nagy, K. A., & Bateman, P. W. (2003) Evolution and maintenance of social status signalling badges: Experimental manipulations in lizards. In: Fox, S. F., McCoy, J. K. & Baird, T. A. (eds.) *Lizard Social Behavior*. Baltimore, MD: Johns Hopkins University Press.

Whiting, M. J., Stuart-Fox, D. M., O'Connor, D., *et al.* (2006) Ultraviolet signals ultra-aggression in a lizard. *Animal Behaviour*, **72**, 353–363.

Whiting, M. J., Webb, J. K., & Keogh, J. S. (2009) Flat lizard female mimics use sexual deception in visual but not chemical signals. *Proceedings of the Royal Society B-Biological Sciences*, **276**, 1585–1591.

Wiens, J. J., Hutter, C. R., Mulcahy, D. G., Noonan, B. P., Townsend, T. M., *et al.* (2012) Resolving the phylogeny of lizards and snakes (Squamata) with extensive sampling of genes and species. *Biology Letters*, **8**, 1043–1046.

Wikelski, M. (1999) Influences of parasites and thermoregulation on grouping tendencies in marine iguanas. *Behavioral Ecology*, **10**, 22–29.

Wikelski, M., Carbone, C., & Trillmich, F. (1996). Lekking in marine iguanas: Female grouping and male reproductive strategies. *Animal Behaviour*, **52**, 581–596.

Wilson, S. K. (2012) *Australian Lizards: A Natural History*. Collingwood, Vic: CSIRO Publishing.

Zweifel, R. G. & Lowe, C. H. (1966) The ecology of a population of *Xantusia vigilis*, the desert night lizard. *American Museum Novitates*, **2247**, 1–57.

14 Social Synthesis

Opportunities for Comparative Social Evolution

Dustin R. Rubenstein and Patrick Abbot

Overview

As we have learned throughout the chapters of this book, sociality – defined broadly as cooperative group living (Chapter 1) – occurs in diverse animal species. Through consideration of the traits of these social species, as well as those of the groups that they form, some broad-scale similarities and differences start to emerge. Here we begin to explore some of the ways that life history shapes – or is shaped by – sociality, by summarizing the traits of social species and social groups. We use Hamilton's rule to guide our analysis, and note as others have, that one challenge to synthesis has been the degree to which invertebrate and vertebrate biologists have emphasized different parameters of this equation (Elgar, 2015). We argue that traits of the social group should be used to describe social organization, and that traits of social species be used to describe social syndromes. We then introduce a simple categorization scheme that uses just four key social traits of the group, and emphasize two emergent social syndromes. We highlight a number of areas ripe for trait-based comparative work, particularly in an age of genomics. Just as this book is meant to be a starting point for dialogue about comparative perspectives on the evolution of sociality, so too is this chapter meant to be a first attempt at using life history data from the diversity of animal taxa containing social species to generate new social synthesis, ideas, hypotheses, and research agendas. A trait-based approach is particularly important as we enter the genomic era because it will help guide a true comparative evolutionary approach for studying sociality, especially if we apply a systems approach.

We acknowledge all of the authors who wrote chapters for this book for their hard work and dedicated study of social animals. Their thorough and insightful descriptions of social animals set the stage for not only this chapter, but for future attempts at social synthesis. We also thank the many taxonomic specialists who helped review the chapters. Dustin Rubenstein (IOS-1121435, IOS-1257530, IOS-1439985) and Patrick Abbot (IOS-1147033) were both supported by the US National Science Foundation, and thank Michelle Elekonich for her assistance. We also acknowledge the National Evolutionary Synthesis Center (NESCent), which supported many of the authors. We thank Judith Korb, Eileen Lacey, and Dan Rubenstein for constructive comments on previous versions of this chapter. We thank Brandi Cannon and Ian Hewitt for helping to copy-edit not only this chapter, but all of the chapters in the book.

14.1 Patterns of Social Diversity

The honeybee is one of the most recognizable of social organisms. Within a colony, a single queen can produce up to tens of thousands of worker bees that divide tasks and cooperatively coexist. Yet, as we learned in Chapter 3, most bees are not social; the majority of bees are solitary creatures. In contrast to the bees are the ants, a group closely related to bees in which, as we learned in Chapter 2, all of the more than 15,000 species are eusocial. Similarly, the nearly 3000 species of termites (Chapter 5) are also all eusocial. In most taxonomic groups, however, social species occur only in a subset of lineages. For example, in shrimps (Chapter 8) and freshwater fishes (Chapter 12), all of the known social species occur within a single genus. In wasps (Chapter 4), another close relative of the bees, sociality occurs in only in 3 of 37 families, and in spiders (Chapter 7), only a few dozen of the nearly 50,000 described species are social. Indeed, sociality occurs in less than 2 percent of all insects, and in only about 5 percent of mammals and 9 percent of birds (Wilson, 1971; Cockburn, 2006; Lukas & Clutton-Brock, 2012).

If sociality is so rare, why is studying its evolution so important? As we highlighted in Chapter 1, an obvious reason is the fact that we are social, as are many of the most charismatic megafauna on earth. But social species are often ecologically and evolutionarily important. Ants and termites dominate the terrestrial habitats in which they occur, and account for about half of all of the biomass of the planet's biological diversity (Wilson, 1990; Hölldobler & Wilson, 1990; Wilson, 2012). Many of the eusocial insects also exemplify one of the major evolutionary transitions in life, as individuals cooperate and coordinate their behavior to from a collective, "superorganism" in some of the most extreme cases (Maynard Smith & Szathmáry, 1995; Hölldobler & Wilson, 2008; Queller & Strassman, 2009; Bourke, 2011). As has been pointed out by others, there are few traits that do not affect social interactions one way or another (Szèkely, *et al.*, 2010). The study of sociality is therefore not only integral to the study of biology, but to all life on earth.

To any casual observer, it would seem that sociality has been well studied over the past century. However, one point that many of the chapters in this book have made clear is that not all social species have been given equal attention. There has been a greater focus on the Hymenoptera (ants, bees, and wasps) than on any other group of social organisms (Elgar, 2015). Many of the authors in this book make an explicit call for greater work on some of the least studied organisms. For example, Hultgren, *et al.* (Chapter 8) and Taborsky & Wong (Chapter 12) argue that sociality in marine species is much less studied than in terrestrial or freshwater species. In the vertebrates, social lizards (Chapter 13) are much more poorly studied than cooperatively breeding mammals (Chapter 10) or birds (Chapter 11). Even within the Hymenoptera, it is remarkable how much diversity remains unexplored. Clearly, there is a need for comparative perspectives on animal social evolution that embrace the full diversity of social animals (see previous treatments in Rubenstein & Wrangham, 1986; Choe & Crespi, 1997; Korb & Heinze, 2008a). The "bottom-up" approach that we advocated for in Chapter 1 (i.e. using a trait-based approach to identify similarities and differences in the traits of social species and the groups that they form, rather than a "top-down" approach that prescribes a theoretical framework based upon the hypothesized reason that groups form) can help inform us about what to study in each of these species. The data compiled here should illuminate which areas and types of life history traits need greater attention, even in the best-studied social species.

A look back over the chapters in this book shows that they clearly encompass a daunting amount of information. The goal of this concluding chapter is to begin synthesizing this information by comparing and contrasting social phenotypes within and across lineages of animals, by examining some of the social traits covered in detail in each of the other chapters of this book. We begin by exploring past efforts at social synthesis, arguing that traits of the social group should be used to describe social organization and that traits of social species be used to describe social syndromes. Building upon these past efforts, we introduce a simple categorization scheme that uses just four key social traits of the group: (1) group structure; (2) reproductive structure; (3) alloparental care; and (4) genetic structure. We next discuss two social syndromes that have been identified in recent years that link vertebrates and invertebrates: (1) central place foraging; and (2) fortress defense. We then summarize what we have learned in our readings of these chapters, emphasizing three life history traits that we believe offer the greatest potential for future comparative work: (1) longevity; (2) fecundity; and (3) developmental mode. We highlight these traits of social species because they have received a great deal of previous attention (in a variety of species), yet there is no consensus on how they broadly relate to sociality, because different measures are often used in different taxa or studies (i.e. for longevity and fecundity), or they have not been compared formally across lineages (e.g. developmental mode). We then discuss one area of social organization that appears ripe for further study: communal breeding. Communal breeders, which are common in both invertebrates and vertebrates, are ideal for tests of the origins of group living and the evolutionary transitions among different forms of social organization because of the way in which direct fitness effects of group living can be determined without the complication of derived traits such as a division of labor. Paradoxically, communal breeders are poorly studied, perhaps because they have been seen as "way stops" on the road to cooperative breeding or eusociality. We argue that a greater effort to study communal breeders and social transitions is needed to develop a comprehensive theory for complex animal sociality. Finally, we discuss how two social syndromes (i.e. central place foragers and fortress defenders) offer an opportunity to explore the monogamy hypothesis (Boomsma, 2007) and further test the idea that high genetic relatedness among offspring is essential for the evolution of eusociality, castes, and perhaps the evolution of sociality more broadly. We use Hamilton's rule to guide much of our synthesis, and note as others have that one challenge to synthesis has been the degree to which invertebrate and vertebrate biologists have emphasized different parameters of this equation (Herbers, 2009; Elgar, 2015). In doing so, we highlight areas to be tested empirically and comparatively with further field study and new molecular tools and techniques, including a systems approach to studying social evolution.

14.2 Social Synthesis: A Trait-Based Approach

Consistent themes have emerged in our understanding of the ecological factors that shape the most derived forms of sociality in various vertebrate and invertebrate taxa (Wilson, 1975; Evans, 1977; Abe, 1991; Clutton-Brock, *et al.*, 2009; Sherman, 2013). Principle among these is that while social species may be taxonomically diverse, many share a common set of ecological and life history traits. The implication is that social species can be categorized by the social traits that they share, with the result being that there may be only two or three broad social types, or "syndromes" (Alexander, *et al.*, 1991; Crespi, 1994; Queller & Strassmann, 1998; Korb & Heinze, 2008b). But as appealing as the idea may be that social animals fit into distinct bins, these categories are hypotheses, not givens. As we describe later, the study of sociality in invertebrates and vertebrates has developed along separate traditions. This means that efforts at synthesis – including this one – inevitably paint with broad brush strokes. As all of the authors in this book emphasize, there are exceptions in every group, and still many unknowns. An important consequence of these separate traditions is that, as we discussed in Chapter 1, finding a common terminology remains a work in progress, and without one, true social synthesis remains challenging.

Finding a common terminology is no simple task. Scientists studying animal sociality love to use jargon. We use specialized terms like "pleometrosis" (i.e. colony foundation by several queens) or "supernumerary" (i.e. a helper in birds). We also use very different terms to describe the variation in social organization within disparate taxonomic groups (e.g. quasisocial, semisocial, communal, singular breeding, plural breeding, monogynous, polygynous), often for social organizations that may have at least superficially similar structures (Rubenstein, *et al.*, 2016). At the most basic level, even determining what constitutes a group varies among taxa. For example, in primates (Chapter 9), a pair is considered to be a type of group, but in birds (Chapter 11), where the majority of species form monogamous pair bonds, a social group is defined as having three or more individuals. Indeed, perhaps one of the most striking observations to be made from this book is that the attributes used to describe social organization in each taxonomic group vary greatly among chapters. Figure 14.1 illustrates the terms that each group of authors chose to use to describe the forms of sociality within their taxon. There is surprisingly little if any overlap in the terminology used to describe social organization in each chapter, even for chapters that emphasize closely related taxa (e.g. Hymenoptera or mammals). This illustrates a longstanding problem in this field: taxon-specific terminology used to define social organization often clouds attempts at social synthesis.

Part of this confusion likely stems from the type of traits that researchers use to categorize different social organizations and species. That is, some attempts at social synthesis are based upon similarities in the traits of social *species* (i.e. traits that can be measured in a single individual), whereas others are based upon similarities in the traits of social *groups* (i.e. traits that can only be quantified by looking at a group of interacting individuals). It would be useful to have a classification scheme that accounts for both types of traits. For example, "social organization" could be used to describe the

Ants	polygyny	monogyny	social parasitism	
Bees	communal and quasisocial	eusocial		
Wasps	communal	facultative	obligate	swarm-founding
Termites	one piece life type	separate piece life type		
Aphids & Thrips	non-social group	communal	eusocial	
Spiders	periodic social	permanent social		
Shrimps	communal	eusocial		
Primates	pair-living	group living		
Mammals	plural w/ care	plural w/o care	singular	
Birds	joint-nesting	plural	singular	
Fishes	group living	cooperative breeding		
Lizards	non-kin-based	kin-based	parent-offspring	

Figure 14.1 The terms that each group of authors in this book used to describe the forms of social organization within their taxon.

structure of animal societies from the traits of the groups that they form, whereas "social syndrome" could be used to describe the type or flavor of animal societies from the traits of social species. Below, after first summarizing how previous researchers have attempted to categorize different forms of animal societies with species- and group-level traits, we outline what such an organizational scheme might look like.

14.2.1 Social Organization: Social Classification Using the Traits of Social Groups

One of the first attempts to categorize different social organization was Michener's (1969) comparison of social behavior in bees. He developed a hierarchical categorization based largely upon the traits of social groups that was later modified by Wilson (1971) to separate social insect species based upon the presence or absence of castes, cooperative care of offspring, and overlapping generations. Recognizing that these same criteria described sociality in cooperatively breeding vertebrates, Sherman, *et al.* (1995) used an alternative approach to categorize social species based upon the division of reproduction within social groups. Their "eusociality continuum" idea arrayed social species along a spectrum of reproductive skew (i.e. reproductive sharing), though critics emphasized the differences in social organization more than the similarities (Crespi & Yanega, 1995). Both of these approaches categorized social species based largely upon the traits of the groups that they form. Yet, employing a completely different approach, Helms Cahan, *et al.* (2002) advocated viewing a species' social organization as the result of a series of decisions that individuals make about whether or not to disperse from their natal territory, whether to co-breed or refrain from breeding, and whether or not to provide alloparental care. This social trajectory approach was based more upon individual decision rules and traits of species, rather than upon patterns of social behaviors and traits of their groups (*sensu* Michener, 1969; Wilson, 1971; Sherman, *et al.*, 1995). It expanded upon the decision rules often used to study helping behavior in cooperatively breeding bird societies: (1) to stay or disperse; and (2) to help or not (Emlen, 1982; Dickinson & Hatchwell, 2004).

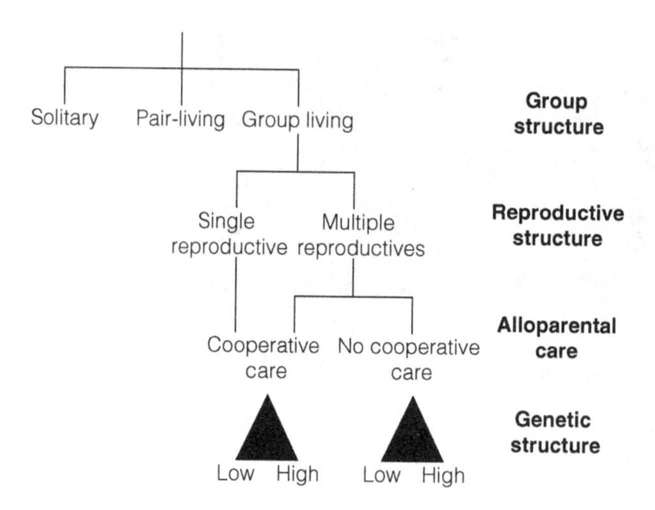

Figure 14.2 Categorization of social species using four key social traits of the group: (1) group structure; (2) reproductive structure; (3) alloparental care; and (4) genetic structure. Group structure describes whether individuals live solitarily, form a pair, or form a group (of more than two individuals). Reproductive structure describes whether one versus more than one female breeds in a group. Alloparental care describes whether individuals cooperate to care for young. Genetic structure describes the relatedness among group members, and in this context in particular, the relatedness among breeding females. The black triangles describe the range of genetics structures (from low to high) within a group.

Although these and many other authors have proposed ways to categorize variation in social organization within and among taxonomic groups, one of the clear challenges in developing a unifying framework to account for variation in social organization is that vertebrate and invertebrate biologists cannot seem to agree. As we (Chapter 1) and others (e.g. Wcislo, 1997) have noted, one reason for this may be the overabundance of terms used to describe social structure (e.g. Figure 14.1), which makes it difficult to compare and contrast across animal lineages. Figure 14.2 represents our attempt to categorize social species using four key social traits of the group: (1) group structure; (2) reproductive structure; (3) alloparental care; and (4) genetic structure. Our approach is more akin to the social trajectory approach used by Michener (1969) and Wilson (1971) to make a hierarchy of social organizations, but it also employs some of the decision rule logic adopted by Helms Cahan, *et al.* (2002) and others.

Group structure describes whether individuals live solitarily, form a pair, or form a group (of more than two individuals). Although for most species this distinction is obvious, for some it may be less clear. For example, a species of bird with bi-parental care at a nest would be considered to form a pair, but a species with uniparental care would be considered to live solitarily. Similarly, a non-eusocial species of aphid in which a single breeding female lives with her offspring in a gall prior to their dispersal would be considered to live solitarily. *Reproductive structure* describes whether one versus more than one female in a group breeds and produces offspring. Social species are typically divided into those characterized by a single breeding female per group (i.e. monogyny in insects, singular breeding in vertebrates) versus those with multiple breeding females per group (i.e. polygyny in insects, plural breeding in vertebrates) (West-Eberhard, 1978; Brown, 1987; Keller & Vargo, 1993; Keller & Reeve 1994; Solomon & French, 1997). This distinction in social organization between societies with one versus more than one breeding female may represent an important evolutionary divide in both insects and vertebrates (Rubenstein, *et al.* 2016). *Alloparental care* describes whether individuals cooperate to care for young that are not their own. Implicit in this concept of alloparental care is the idea that groups of individuals who provide cooperative care of young consist of overlapping generations. *Genetic structure* describes the relatedness among group members and, in particular, the relatedness among breeding females (i.e. whether and how related they are to each other). Genetic relatedness among group mates is influenced both by the number of female breeders in the group (i.e. reproductive structure), but also by the number of mates that each female has (Rubenstein, 2012). So while genetic structure is related to the group structure, they describe slightly different things.

Can these four traits capture the majority of the social variation across taxonomic groups? If we approach this question only thinking about the traits of the group, then we believe that they can. After all, the basic demographic and reproductive structures (i.e. numbers and patterns of breeding versus non-breeding group members) of animal societies are similar in both vertebrates and invertebrates (Rubenstein, *et al.*, 2016). For example, although all ants are eusocial, the way that colonies form and the number of queens in a colony helps distinguish among the forms of sociality in ant societies (Chapter 1). Moreover, birds and mammals are categorized by social groups with one versus more than one breeding female, but one factor that further distinguishes one form from another is whether the groups with multiple breeding pairs cooperatively care for offspring or not (Chapters 8–10). Additionally, the factors that influence the genetic structure within groups are the same for invertebrates and vertebrates: the number of breeders and the number of mates that each breeder has (Rubenstein, 2012).

It is clear that whether the breeders or other members of a group are related has important implications for levels of cooperation and conflict within the group, as well as the potential for kin-directed benefits of cooperation (Rubenstein, 2012; Boomsma, *et al.*, 2014; Rubenstein, *et al.*, 2016). So while these four social traits of the group – group structure, reproductive structure, alloparental care, and genetic structure – do not capture all of the social diversity of life, we believe that they represent a relatively simple way to classify social organization in most invertebrates and vertebrates in a unified way. Within specific taxonomic groups, species can be further sub-categorized by other group traits as taxonomic specialists see fit. More generally, however, using traits of the group to describe similarities among social species gives us a straightforward way to classify social organization in diverse taxonomic groups and across lineages. Looking forward, as more data are accumulated about a diversity of social creatures, such a scheme is a natural fit for comparative databases that compile the critical features that characterize animal societies (Starr, 2006).

14.2.2 Social Syndromes: Social Classification Using the Traits of Social Species

Paralleling efforts by Michener and others to classify similar forms of social organization in animals based upon similarities in the traits of their groups were attempts to compare and contrast social species based upon shared ecological patterns and their general life histories (Clutton-Brock, *et al.*, 2009). This approach was based more upon the idea of using traits of social species than using traits of the groups that they form. For example, all eusocial Hymenoptera, separate type termites, and cooperatively-breeding vertebrates (except the eusocial mole-rats) are "central place foragers" (i.e. some group members forage outside of the nest or domicile, and may delay or forego reproduction altogether). Offspring within these groups may develop relatively slowly and are dependent upon parents for food and protection. And, as Evans (1977) described, a foraging solitary wasp requires basic adaptations for protecting larval brood while she is foraging (Chapter 8). At the group level, cooperative nest defense is one such mechanism for protecting offspring not only in insects (Wilson, 1971; Strassmann, *et al.*, 1988), but also in communal breeding birds (Vehrencamp, 1978). Thus, the idea is that social species share, and can be categorized by, features such as communal defense or brood care, which are themselves a consequence of a series of species-level traits that we describe further below.

Gadagkar (1990) brought together these basic life history traits of social species and proposed that a primary benefit of sociality in insects was insurance against complete reproductive failure (e.g. as would occur upon death of a solitary female that provisions and protects her young, or other such emergencies, West-Eberhard, 1975; Queller, 1989). Ecological and life history traits such as helpless offspring and long development could act as important predispositions to highly derived sociality, and that assured fitness returns provided by helping relatives could "tilt the scale" towards helping behaviors. The subsequent discovery of eusocial snapping shrimp that spend nearly their entire lives inside marine sponges only reinforced the generality of this model (Duffy, 1996). While predation may influence sociality in some vertebrate groups like birds (Poiani & Pagel, 1997) and fishes (Groenewoud, *et al.*, 2016), the association between sociality and predation risk is strongest for the vertebrate and insect species that live inside their food resources for most, if not all, of their lives.

Subsequent discoveries of sociality in insect species such as ambrosia beetles (Kent & Simpson, 1992), as well as work on the differences in social features across termites (reviewed in Korb, 2007; Korb, *et al.*, 2012; Chapter 5) and other taxonomic groups more broadly (summarized in Korb & Heinze, 2008a,b), formed the basis of Crespi (1994) and Queller & Strassmann's (1998) proposition of a fundamental division in the organization of sociality in insects. Because all social Hymenoptera and higher termites are central place foragers with helpless offspring that are provisioned by workers and that delay or forego reproduction, the basic function of sociality in these species is to provide "life insurance" against costs and risks associated with foraging and providing care (West-Eberhard, 1975). In contrast, other social insects are "fortress defenders" that live in or near their food and have mobile offspring that require little or no provisioning, but are vulnerable to predation and parasitism because they live in aggregations from which easy escape is not possible (Crespi, 1994). This basic dichotomy in social syndromes – fortress defense versus central place foraging, or life insurance – helps to link many of the primitively eusocial invertebrate and vertebrate taxa. With the descriptions of remarkably consistent patterns of fortress defense in species such as snapping shrimps (Chapter 8), aphids and thrips (Chapter 6), polyembryonic wasps (Cruz, 1981; Chapter 4), one-piece type or wood-dwelling termites (Chapter 5), and the naked mole-rat (Kent & Simpson, 1992; Chapter 10), it is clear that these two syndromes are ecologically and taxonomically dissimilar, with decisive consequences for how social groups are organized and function (Korb & Heinze, 2008b). Moreover, these syndromes may also help link social vertebrates and invertebrates. That is, even the naked mole-rat, which is often considered eusocial, is essentially a fortress defender that lives almost entirely underground in tunnels connected to its tuber food resources (Jarvis, 1981; Alexander, *et al.*, 1991; O'Rainin, *et al.*, 2000).

To reiterate, however, these alternative social syndromes are hypotheses. Many authors have recognized similarities between social invertebrates and vertebrates that suggest additional syndromes could exist. For example, Andersson (1984) and later Brockmann (1997) recognized fundamental similarities in how ecological constraints shape patterns of cooperative breeding in birds and wasps. These species do not fit neatly into the central place foraging/fortress defense dichotomy: they lack true workers and tend to form small cooperative groups in which all individuals are totipotent with opportunities for both direct and indirect fitness returns from helping (Sherman, 2013). Korb & Heinze (2008b) suggested a third social syndrome that captures these features of cooperatively breeding vertebrates and wasps that Andersson (1984) and Brockmann (1997) initially pointed out. Additionally, open questions remain about how a focus upon conflict within these societies might also generate new ways to categorize sociality and introduce other ways to identify convergent social features (Hart & Ratnieks, 2004; West, *et al.*, 2015). Nonetheless, the division of social species into social syndromes, or categories based upon the species-level traits that they share, has the potential to lead to breakthroughs in social synthesis, particularly in the age of genomics, as we describe below.

14.3 Opportunities for a Trait-Based Approach to Comparative Social Evolution

We have argued that taking a trait-based approach to social synthesis will allow us to make important insights into comparative social evolution. But what are those insights, and what are the opportunities for social synthesis? Below, we focus upon just three of the traits that we believe offer the greatest potential for future comparative work: (1) longevity; (2) fecundity; and (3) developmental mode. Although these traits have received previous attention, there is no consensus on how they broadly relate to sociality because different measures are often used in different taxa or studies (i.e. for longevity and fecundity), or they have not yet been compared across lineages (e.g. developmental mode). Additionally, one area of social organization that appears ripe for further study is a focus on communal breeding species. This intermediate form of social organization lends itself to comparative study across taxonomic groups and lineages, and opens up a broader discussion on social transitions. Finally, the two social syndromes we discussed earlier (i.e. central place foraging and fortress defense) offer opportunities to explore the monogamy hypothesis (Boomsma, 2007) and the idea that high genetic relatedness among offspring is essential for the evolution of eusociality, castes, and perhaps sociality more broadly.

14.3.1 Social Traits: Life History and the Potential for Social Synthesis

A striking pattern that emerges from the chapters in this book is the degree to which the authors agree on the importance and promise of life history approaches to understanding sociality (Starr, 2006). Although it has long been noted that social groups must solve the problems of allocation and scheduling that face individual organisms (e.g. Richards, 1953; Oster & Wilson, 1978), life history perspectives on social evolution have remained consistently underdeveloped (Starr, 2006; Heinze, 2006; Chapter 3). Wilson (1971) observed that many eusocial insects occupying stable environments have a suite of traits such as longer-lived or perennial colonies, slower reproduction, and low juvenile mortality that are characteristic of species with "slow" life histories (i.e. k-selected, in contrast to the "fast" life histories of r-selected species, with shorter lives and faster reproduction, Promislow & Harvey, 1990; Starr, 2006; Dobson, 2007). Social animals may thus be those predisposed towards characteristic life histories, and exhibit similar responses to the mechanisms of density regulation (Pen & Weissing, 2000; Tsuji, 2006), parental care in the context of multiple bouts of reproduction within nests (Trumbo, 2013), or spatial structures that favor kin interactions (Lion, *et al.*, 2011). As tantalizing as these ideas may be, broad theoretical perspectives on insect life cycles remain notoriously challenging (Tsuji & Tsuji, 1996), and empirical patterns in life history may not always prove to have much explanatory power (Schwarz, *et al.*, 2006). In vertebrates, the trend of linking studies of sociality with those of life history has a longer history, and consequently many of the ideas are more developed than in insects (Horn & Rubenstein, 1978). We briefly focus upon three life history traits that stand out in our synthesis among taxonomic groups (summarized in Table 14.1) and offer potential for future comparative work: (1) longevity; (2) fecundity; and (3) developmental mode.

Researchers studying cooperatively breeding vertebrates noted that social vertebrates exhibit life history and demographic traits that differ from their non-social counterparts (Brown, 1987; Hatchwell & Komdeur, 2000). In cooperatively breeding birds, the life history hypothesis is one of several related ideas that link life history traits such as longevity and fecundity to sociality. These hypotheses were derived from empirical observations that cooperative breeding in birds is associated with factors that cause habitat saturation (e.g. reduced adult mortality) and reduce the cost of helping relative to dispersing and breeding independently (Horn & Rubenstein, 1978). For example, sociality is associated with longer life spans and reduced clutch sizes in a number of avian cooperative breeders (Arnold & Owens, 1998; Cockburn, 1998). However, recent comparative analyses have indicated that the patterns derived from earlier studies are

Table 14.1 Summary of key life history traits in the taxonomic groups discussed in this book. In some cases, little or no data are available, and for all traits where data are available on diversity within taxonomic groups, exceptions can be common.

	Life span/longevity	Fecundity	Developmental mode
Ants	queens longer-lived relative to workers or non-social insects; queens longer lived in monogynous species than polygynous species	high fecundity in queens; higher individual fecundity in monogynous than polygynous colonies;	holometabolous
Bees	queens longer lived than workers or non-social insects	higher in social than non-social species	holometabolous
Wasps	queens longer lived than workers or non-social insects	higher in social than non-social species	holometabolous
Termites	breeders longer-lived than non-breeders, particularly in separate type species	physograstric queens common; fertility of both primary male and female reproductive extremely high; lower in one piece life species (small colonies) and higher in separate type species (large colonies)	hemimetabolous; relatively slow juvenile development in presence of reproductives
Aphids & Thrips	galling phase tends to be longer-lived in social than non-social species, implying breeders longer-lived than those in non-social species; no data on reproductive life spans	lower in eusocial species relative to non-social species	hemimetabolous; social aphids tend to express slow juvenile development; longer gall duration in thrips
Spiders	social species may take longer to develop	lower in social species than non-social species	hemimetabolous-like; social species with slow juvenile development
Shrimps	unknown	lower in eusocial than non-eusocial species	eusocial with crawling larvae that delay dispersal; non-social with planktonic larvae
Primates	positive correlation between longevity, body size and brain size; sex expressing parental care longer lived than non-caring sex	fertility inversely related to body size	slow, extended juvenile development
Non-primate mammals	no relationship between cooperative breeding and maximum longevity	general trends of higher fecundity with group size and body size	cooperative breeders tend to have slow developing offspring requiring extended parental care
Birds	annual survival higher in cooperative breeders; cooperative breeders are longer lived than non-cooperative breeders	possible trend of smaller clutch sizes in cooperative breeders	cooperative breeding more common in species with altricial offspring

Table 14.1 (*cont.*)

	Life span/longevity	Fecundity	Developmental mode
Fish	anecdotal evidence that social species are longer lived	anecdotal evidence of lower fecundity in social species	late maturation in cooperatively breeding species
Lizards	kin-based sociality and/or monogamy associated with increased longevity	lower fecundity in social species	*Egernia* tend to have late maturing offspring

either more ambiguous than previously thought (Blumstein & Møller, 2008; Beauchamp, 2014), or in need of more detailed study (Downing, *et al.*, 2015). Moreover, in mammals, it is unclear if there is any broad association between longevity and sociality (Lukas & Clutton-Brock, 2012). Indeed, sociality is just one of many ways of reducing adult relative to juvenile mortality that might favor repeated breeding and long life, so it is in some ways not surprising that these relationships vary among taxa.

There remains a surprisingly slow integration of life history perspectives on longevity and social evolution in insects (Bourke & Franks, 1995; Kipyatkov, 2006). The life cycles of eusocial Hymenoptera and termites exhibit clear associations between patterns of colony growth (i.e. single versus multiple bouts of reproduction) and means of colony founding (i.e. independent founding by single breeders and monogyny versus dependent founding by swarms of breeders and polygyny) (Keller, 1991; Starr, 2006). Insects also clearly exhibit trade-offs among these and other life history traits. Colony founding is a vulnerable period in any insect lifecycle. The survival advantages of large body size may underlie a trade-off in ant queen size at colony foundation and queen number (Wiernasz & Cole, 2003). However, some highly derived eusocial insects defy life history rules by not exhibiting any apparent trade-off between life history traits such as longevity and fecundity (Keller & Genoud, 1997; Remolina & Hughes, 2008; Parker, 2010). Theoretical studies have also produced contrasting predictions regarding the association between sociality and life history evolution (Kokko & Ekman, 2002; Koykka & Wild, 2015). Thus, life history traits may predispose some taxonomic groups towards sociality, but those same traits undoubtedly have evolved as a consequence of sociality in others, or simply may evolve capriciously in different insect groups (Schwarz, *et al.*, 2006).

Both kin selection and demographic models of senescence can predict either positive or negative associations between sociality and various measures of survival or longevity (Gadagkar, 1991; Bourke, 2007). Positive associations derive from either direct benefits of social groups (i.e. benefits of intergenerational transfer of resources or decreased extrinsic mortality due to helping, Lee, 2003) or inclusive fitness benefits (i.e. kin-selected extension of post-reproductive life spans, Cohen, 2004; Coxworth, *et al.*, 2015). Bourke (2007) cited a model by Carey & Judge (2001) that showed how selection for parental care can lead to reduced fecundity and lifespan extension (i.e. a slow life history) that accounts for the observed patterns in many cooperatively breeding vertebrates, including how reduced adult mortality favors ecological conditions for delayed juvenile dispersal and helping (Arnold & Owens, 1998; Hatchwell & Komdeur,

2000). However, the situation is more complicated in social insects with castes because of how breeders are protected from extrinsic sources of mortality, while workers are not only exposed to increased risk while foraging (Hartmann & Heinze, 2003; Lopez-Vaamonde, *et al.*, 2009), but in cases such as the fortress defenders, are practically designed to die.

Despite arguments for kin-selected benefits to shorter life spans (Tallamy & Brown, 1999; Bourke, 2007), the overall picture is of a positive association between sociality and annual adult survival or lifespan. In primates (Chapter 9), birds (Chapter 11), fishes (Chapter 12), and lizards (Chapter 13), there is either clear or anecdotal evidence that monogamy, helping, and sociality are associated with measures of longevity. Remarkably, this pattern is mirrored in each of the invertebrate taxa discussed in this book. Indeed, even in the caste-based eusocial insects, workers tend to be longer-lived than solitary insects of similar body sizes (Table 14.1). Despite the theoretical plausibility of an association between sociality and shortened life spans (i.e. if intergenerational transfer of resources is facilitated by short adult life spans, Bourke, 2007), there is little evidence of the generality of such an association (Trumbo, 2013). Clearly, this is an area ripe for broad-scale comparative analysis across lineages and disparate taxonomic groups, though as we learned in each of the chapters, the way lifespan or longevity is quantified in different taxa is often very different. Even within just the birds, for example, researchers cannot agree how to quantify longevity, and using different measures of mean versus maximum lifespan can lead to very different conclusions (Blumstein & Møller, 2008; Beauchamp, 2014; Downing, *et al.*, 2015). Thus, before we can conduct true, broad-scale comparative analysis of the relationship between social behavior and lifespan across different taxonomic groups or lineages, researchers first need to be clear about how to quantify and measure longevity in different species.

The relationship between sociality and fecundity is less clear even than the relationship between sociality and longevity. Although there is evidence that cooperative breeding is associated with reduced fecundity in primates (Chapter 9), birds (Chapter 11), lizards (Chapter 13), spiders (Chapter 7), aphids and thrips (Chapter 6), shrimps (Chapter 8), and anecdotally, fishes (Chapter 12), the opposite is true in the advanced eusocial Hymenoptera (Chapters 2–4) and higher termites (Chapter 5). The reasons for these taxon-specific differences are complex, but they are likely related to the disruptive effects of the division of labor and extreme reproductive skew on fecundity/longevity trade-offs between breeders and non-breeders. Moreover, not only does it remain unclear, and even somewhat controversial, how fecundity evolves in social animals, whether fecundity is even a cause or consequence of being social has proven difficult to disentangle (Härdling & Kokko, 2003; Koykka & Wild, 2015). On the one hand, low fecundity may predispose species towards helping, but on the other, helping should have a positive effect on fitness. Additionally, conflicts of interest between breeders and non-breeders may also contribute to variation in the effects of sociality on fecundity (Bourke, 2007). Many of the species discussed in this book offer precisely the kind of variation required for phylogenetically-informed tests of the relationships between life history traits such as body size, fecundity, and longevity across multiple origins of sociality.

Another theme that emerges from the book is that developmental mode is a defining social trait in many taxonomic groups. For example, in some groups like the shrimps (Chapter 8), the mode of development differs between eusocial and non-eusocial species, where the former develop as crawling larvae that remain in the host sponge and the latter as planktonic larvae that disperse in the water column. Similarly, slow development in termites may be a driver of sociality in this group (Chapter 5). More generally, differences in the mode of development among insect taxa may at least partially explain differences in the form of social organization that a species adopts. For example, in holometabolic species with complete metamorphosis, offspring are altricial, and in holometabolic social insects, they require care, either parental or alloparental. In contrast, in hemimetabolic species with incomplete metamorphosis where offspring resemble adults, offspring are precocial and generally do not require active provisioning (Korb, 2008). This distinction in the mode of development may be particularly important in the evolution of eusociality, and possibly in the evolution of cooperative breeding as well.

Although vertebrates show different patterns of development from insects, developmental mode may also be related to sociality in birds. Vertebrate offspring show gradual development like the hemimetabolic insects, but many also need extended parental or alloparental care like the holometabolic insects. In birds, cooperative breeding occurs more frequently in altricial species (i.e. young are undeveloped and require care and feeding) than precocial species (i.e. young are mature and capable of movement after birth) (Cockburn, 2006). Altriciality in birds has been argued to play a key role in the evolution of cooperative breeding behavior because transitions to cooperation occur more frequently in altricial lineages (Ligon & Burt, 2004), though complex alloparental care still occurs in some precocial species (Hatchwell, 2009). Part of the relationship between mode of development and sociality may relate to ecology and the costs of rearing young. For example, early hypotheses for the evolution of cooperative breeding in vertebrates suggested that costs of rearing young in harsh environments promoted sociality (Emlen, 1982; Koenig, *et al.* 2011, 2016). Indeed, this hypothesis may at least partially explain the broad-scale patterns of sociality in birds where cooperatively breeding species occur in more temporally variable and unpredictable environments than non-cooperatively breeding species (Rubenstein & Lovette, 2007; Jetz & Rubenstein, 2011). Ultimately, further comparative studies in both insects and vertebrates are needed to understand if differences in the form of social organization within and among taxonomic groups (and/or different social syndromes) are related to how young develop. Comparing and contrasting holometabolic and hemimetabolic species, as well as altricial and precocial species, offers one promising area for synthesis across lineages. In particular, exploring how these and other traits relate to the social syndromes described above, as well as to ecology, will be informative.

14.3.2 Social Organization: The Importance of Communal Societies

We often tend to study the most derived forms of sociality in the taxonomic groups in which we work. Yet, the “less” social species may offer great potential for understanding not only how sociality evolves, but also if it evolves in similar ways across taxa. A term that occurs repeatedly – though is used inconsistently – in the literature and even more so in the chapters of this book is the word "communal" (e.g. Chapters 4, 6, 8, 10 and 11). In this book, "communal" is used in both vertebrates and invertebrates to describe very different types of animal societies. In general, however, the word communal is used to describe societies in which multiple females (often unrelated to each other) breed in the same domicile, but – at least in invertebrates – without cooperative care (e.g. various bees, wasps, shrimps, and thrips). We note, however, that in vertebrates – both birds (Brown, 1987) and mammals (Solomon & French, 1997) – communal breeding often involves some form of communal care of young, much like quasisociality in insects (Michener, 1969). Nonetheless, this distinction between cooperative offspring care, or the lack-there-of, in eusocial versus communal species appears to be an important one, particularly in the insects.

Studying communal breeding species offers the opportunity to examine intermediate steps in the evolution towards eusociality and other derived forms of sociality. As Korb & Heinze (2008b) pointed out, thinking about evolutionary transitions towards eusociality may be as important as studying the social endpoints. Although they argued for thinking about transitions between social syndromes, understanding transitions between different forms of social organization is also important. This approach harkens back to the ideas of Michener (1969, 1974) and the parasocial route to eusociality in insects, and of Emlen (1995) and Emlen, *et al.* (1995) and the extended family model of sociality in vertebrates. Both of these ideas examined transitions – or the lack thereof – between different forms of social organization. Empirical tests of these ideas have been limited, but evidence from bees suggests that eusociality and communal breeding represent different evolutionary endpoints (Wcislo & Tierney, 2009; Chapter 3). These alternative social trajectories may be common in social vertebrates and invertebrates alike (Rubenstein, *et al.* 2016). Ultimately, thinking about the role of communal species in the evolution of higher forms of social organization may be informative if we are to understand how advanced social species evolve.

14.3.3 Social Syndromes: From Hamilton's Rule to Fortress Defense and Central Place Foraging

More than any other theoretical framework of social evolution, inclusive fitness theory generally, and Hamilton's Rule specifically (*r*b more than c; Hamilton, 1964), has guided empirical studies of social vertebrates and invertebrates alike, though with an emphasis on different parameters of the equation in the different lineages (Elgar, 2015). Comparative analyses of the costs and benefits of sociality emerged early from the study of vertebrates (Crook, 1964; Lack, 1947). Decades of empirical work on cooperatively breeding birds and mammals tended to emphasize these constraints on independent breeding (e.g. Emlen, 1982; Stacey & Koenig, 1990; Koenig, *et al.*, 1992; Koenig & Dickinson, 2016) as well as the benefits of grouping (e.g. Alexander, 1974; Stacey & Ligon, 1991). In contrast, early work on invertebrates focused more upon issues of sex allocation and conflict in the haplodiploid societies of the Hymenoptera, and how relatedness resolved problems of freeloaders and sterile workers (Trivers & Hare, 1976).

Over the past few years, the taxonomic divide over which parts of Hamilton's Rule are most often tested empirically in different lineages has begun to blur, as studies of vertebrate sociality continue to emphasize relatedness more, and those of invertebrate sociality tend to emphasize ecology more. For example, a renewed focus on the role that lifetime monogamy has played in the evolution of eusociality (Boomsma, 2007, 2009, 2013) has led to testing of the relationship between sociality and polyandry/promiscuity comparatively in both insects (Hughes, *et al.*, 2008) and vertebrates alike (Cornwallis, *et al.*, 2010; Lukas & Clutton-Brock, 2012). At the same time, comparative studies relating environmental factors to the evolution of sociality in birds (Rubenstein & Lovette, 2007; Jetz & Rubenstein, 2011) have generated parallel studies in Hymenoptera (Kocher, *et al.*, 2014; Purcell, *et al.*, 2015; Sheehan, *et al.*, 2015).

Despite great interest in the monogamy hypothesis in different taxonomic groups (e.g. Hughes, *et al.*, 2008; Cornwallis, *et al.*, 2010; Lukas & Clutton-Brock, 2012), various difficulties and shortcomings have been pointed out in insects (Nonacs, 2014), mammals (Kramer & Russell, 2014), and birds (Dillard & Westneat, 2016). Dillard & Westneat (2016) argued for taking a more holistic approach, and rather than focusing just upon genetic relatedness, instead suggested considering the potential covariance between the variables in Hamilton's Rule. For example, ecologically-driven covariance (i.e. the interaction between mating system and environmental variation) could also explain the relationship between monogamy and cooperation, at least in vertebrates (Dillard & Westneat, 2016). Moreover, factors other than (lifetime) monogamy influence relatedness among offspring and are undoubtedly important in the evolution of both eusociality and cooperative breeding (Rubenstein, 2012). In many eusocial species that lack a worker class, relatedness among group members is often high (sometimes even higher than 0.5), likely because of inbreeding (reviewed in Aviles & Purcell, 2012). These patterns in eusocial species with high relatedness but no true sterile worker class suggest that high genetic relatedness by itself is not sufficient for the evolution of sterile castes. In other words, the central place foragers have evolved "classically eusocial societies" defined by a true worker class, whereas the fortress defenders have evolved "primitively eusocial societies" lacking a worker class, despite often having relatedness among offspring as high as in the classically eusocial central place foragers. Thus, studying primitively eusocial, and perhaps communal, species offers an opportunity to refine our theoretical understanding of how monogamy leads to the evolution of sociality.

Ultimately, the devil is in the details when it comes to testing these ideas, and rather than glossing over lesser-studied species or those that do not fit cleanly into the framework, we suggest that they deserve particular attention. Two of the taxa that we work on (aphids and shrimps) exemplify this – neither of these primitively eusocial species fits neatly into the monogamy hypothesis framework. Eusocial and clonal aphids evolved from ancestors with low within-colony relatedness, and many non-eusocial aphids live in groups with high relatedness (Abbot, 2009; Chapter 6). Similarly, many eusocial snapping shrimps have extremely high within-colony relatedness, but workers in all species appear to be totipotent and monogamous (Chak, *et al.*, 2015;

Rubenstein, *et al., unpublished data*). We argue that focusing upon the life history differences within and between the classically eusocial, central place foragers and the primitively eusocial, fortress defenders may help to codify the monogamy hypothesis further (Starr, 2006). For example, considering the developmental differences between holometabolic and hemimetabolic species with their different forms of parental care will be informative (i.e. central place foragers tend to produce workers with larval development before breeders, whereas fortress defenders produce offspring with direct development that can grow into breeders; Table 14.1). This may represent a fundamental life history difference (i.e. life insurance against the costs of foraging and parental care) that could help drive the evolution of castes independently of genetic relatedness among offspring. Termites may make an ideal system within which to address this issue because the group contains some species that are central place foragers and others that are fortress defenders (Korb, 2007; Chapter 5). Only by studying societies other than the Hymenoptera (Costa, 2006), including communal breeders and other species with multiple breeding females (Rubenstein, 2012), can we hope to assess the generality of the monogamy hypothesis in insects, let alone in other lineages (e.g. Cornwallis, *et al.*, 2010; Lukas & Clutton-Brock, 2012). Moreover, considering the roles of ecology and other life history traits or pre-adaptations (Dillard & Westneat, 2016), as well as the roles of cooperation and conflict in influencing worker sterility (Nonacs, 2014), within the monogamy framework will also be informative. Finally, addressing this issue within the context of social syndromes may offer an ideal opportunity for true social synthesis across very different types of organisms.

14.4 Life History Traits and the Genomics Era: The Future of Comparative Social Evolution

Like much of science in the twenty-first century, the study of sociality has entered an age of genomics. As it becomes increasingly cheaper and easier to sequence and assemble complete transcriptomes (i.e. all expressed mRNA) – and even genomes – of non-model organisms (Calisi & MacManes, 2015), we are poised for a massive effort in comparative evolutionary understanding of sociality (Blumstein, *et al.*, 2010; Hofmann, *et al.*, 2014; Rubenstein & Hofmann, 2015). Yet, although it is clearly an exciting time to be working in this area, we lack a general framework for how best to do this in the genomics era. Considering both the frameworks of social organization (based upon traits of the social group) and of social syndromes (based upon traits of the social species) offers a complimentary approach to studying the molecular mechanisms underlying sociality and the social transitions towards advanced forms of sociality. For example, studies of social structure and genetic architecture in ants have shown that polygynous and monogynous colonies of the same species have distinct haplotypes whose loci occur together on a "social chromosome," a non-combining region of the genome (Wang, et al., 2013; Purcell, *et al.*, 2014). Additionally, a forward genomic approach can be used to identify the molecular bases of social phenotypes. For example, functional genomic studies in eusocial Hymenoptera (e.g. Toth, *et al.*, 2007; Smith, et al., 2008) and termites (e.g. Terrapon, *et al.*, 2014) have identified genes and gene modules associated with different social and reproductive behaviors. Moreover, we are beginning to understand how gene regulatory networks (e.g. Bloch & Grozinger, 2011) and epigenetic mechanisms (e.g. Yan, *et al.*, 2014) influence the expression and transitions between castes and other social phenotypes. These approaches are even being applied comparatively across lineages, as conserved genetic toolkits involved in independent evolutions of social behaviors have been identified in vertebrates and insects alike (Rittshof, *et al.*, 2014).

The possibilities for comparative genomic work are seemingly endless, particularly as we expand outside of the Hymenoptera and other model organisms (Taborsky, *et al.*, 2015). One of the most important developments in the study of social insects has been the emergence of the field of "sociogenomics" (Robinson, *et al.*, 2005), which has subsequently expanded to include vertebrates (Hofmann, *et al.*, 2014; Rubenstein & Hofmann, 2015). In the coming years, sociogenomic research will bear on everything from cognition and aging to reproductive biology. The rapid emergence of new genomes of primitively social "transitional" species offers the immediate opportunity to apply completely unique perspectives on social evolution and social evolutionary transitions. For example, by sequencing ten bee genomes, Kapheim, *et al.* (2015) showed that there is no single genomic route to eusociality, and that evolutionary transitions in sociality have independent – though similar – genetic bases. Yet, as we begin to learn more about the molecular mechanisms that underlie sociality and social behavior more generally, we must first gain a better understanding of the similarities and differences in the traits that define social species. Taking a social syndrome approach based upon the traits of social species (Hofmann, *et al.*, 2014) to do this may be more informative that simply choosing model organisms or well-studied systems (Blumstein, *et al.*, 2010; Taborsky, *et al.*, 2015), as is often done in comparative studies (e.g. Rittshof, *et al.*, 2014). For example, if the classically eusocial central place foragers and the primitively eusocial fortress defenders truly represent distinct types of social syndromes, or if monogynous and polygynous species represent distinct social evolutionary outcomes, then we need to design genomic and transcriptomic studies that compare disparately related species with similar syndromes or social organizations, and these need to be coupled to studies using experimental approaches in natural populations.

From genomics to ecology, social diversity offers new opportunities for experiments and synthesis, but describing and accounting for variation in behavior and life history in natural populations remain essential. In the 1990's, easy access to molecular markers produced a technology-fueled rush of estimates of genetic relatedness within social taxa. It is worthwhile to reflect upon what we should make of those datasets today, and what collectively we learned from them. How has our understanding of the biology of social taxa changed over the past two decades, and do we interpret those genetic data in the same way now? How have the theories and concepts that motivated those studies changed? We are in the middle of another technology-fueled rush, thanks to ready access to –omics data. These data will add to the complexity of what we know about social complexity itself, and we risk being overwhelmed. To better manage the flood of data, and to be more intentional about the comparative data that we collect, we echo

Moore, *et al.* (2010) in calling for something akin to a systems approach to the study of social evolution. Such an approach would involve the integration of multiple types of data (Blumstein, *et al.*, 2010). While inevitably reducing social behavior to various data-friendly parts (e.g. a tissue-specific transcriptome), a critical element of a systems approach is that it is egalitarian in its prioritization of data: genomics does not trump ecology. Rather, as we learned from the chapters of this book, ecology not only tells us what our functional data mean for the organisms from which it derives, but knowing something about our organisms in their natural environments – how they behave and how they vary from species to species and from place to place – allows us to sew together the various parts of disparate datasets into a composite whole. This integration is the the ultimate source of new questions and new directions of study.

In summary, many of the exciting opportunities outlined here will be lost without the insights that only natural history can provide. What the authors in this volume have shown is just how promising the prospects are for the next generation of biologists who are fascinated by social behavior. It is not hard to imagine the enthusiasm that William Morton Wheeler, Alexander Skutch, Niko Tinbergen, or William D. Hamilton would have had at these prospects. We have outlined just a few agendas for comparative research in this final chapter. Within the pages of this book lie many more ideas for potential projects, Ph.D. dissertations, and research careers. Indeed, the prospects for synthetic and comparative analyses of the evolution of sociality in animals are brighter than ever.

References

Abbot P. (2009) On the evolution of dispersal and altruism in aphids. *Evolution*, **63**, 2687–2696.

Abe, T. (1991) Ecological factors associated with the evolution of worker and soldier castes in termites. *Annual Review of Entomology*, **9**, 101–107.

Alexander, R. D. (1974) The evolution of social behavior. *Annual Review of Ecology, Evolution, and Systematics*, **5**, 325–383.

Alexander, R. D., Noonan, K. M., & Crespi, B. J. (1991) The evolution of eusociality. *In*: Sherman, R. W. Jarvis, J. U. M., & Alexander, R. D. (eds.) *The Biology of the Naked Mole Rat*. Princeton, NJ: Princeton University Press, pp. 3–44.

Allman, J., Rosin, A., Kumar, R., & Hasenstaub, A. (1998) Parenting and survival in anthropoid primates: Caretakers live longer. *Proceedings of the National Academies of Sciences USA*, **95**, 6866–6869.

Andersson, M. (1984) The evolution of eusociality. *Annual Review of Ecology and Systematics*, **15**, 165–189.

Arnold, K. E. & Owens, I. P. F. (1998) Cooperative breeding in birds: A comparative test of the life history hypothesis. *Proceedings of the Royal Society of London B*, **265**, 739–745.

Aviles, L. & Purcell, J. (2012) The evolution of inbred social systems in spiders and other organisms: From short-term gains to long-term evolutionary dead ends? *In:* Brockmann, J., Roper, S. T., Naguib, M., Mitani, J. C., & Simmons, L. W. (eds). *Advances in the Study of Behavior, Vol. 44*. Burlington, VT: Academic Press, pp. 99–133.

Beauchamp, G. (2014) Do avian cooperative breeders live longer? *Proceedings of the Royal Society of London B*, **281**, 2014.0844.

Bloch, G. & Grozinger, C. M. (2011) Social molecular pathways and the evolution of bee societies. *Philosophical Transactions of the Royal Society B*, **366**, 2155–2170.

Blumstein, D. T. & Møller, A. P. (2008). Is sociality associated with high longevity in North American birds? *Biology Letters*, **4**, 146–148.

Blumstein D. T., Ebensperger, L. A., Hayes, L. D., Vásquez, R.A., Ahern, T. H., *et al.* (2010) Towards an integrative understanding of social behavior: New models and new opportunities. *Frontiers in Neuroscience*, **4**, 1–9.

Boomsma, J. J. (2007) Kin selection vs. sexual selection: Why the ends do not meet. *Current Biology*, **17**, R673–R683.

(2009) Lifetime monogamy and the evolution of eusociality. *Philosophical Transactions of the Royal Society B*, **364**, 3191–3207.

(2013) Beyond promiscuity: Mate-choice commitments in social breeding. *Philosophical Transactions of the Royal Society B*, **368**, 20120050.

Boomsma, J. J., Huszar, D. B., & Pederson, J. S. (2014) The evolution of multiqueen breeding in eusocial lineages with permanent physically differentiated castes. *Animal Behaviour*, **92**, 231–252.

Bourke, A. F. G. (2007) Kin selection and the evolutionary theory of aging. *Annual Review of Ecology, Evolution, and Systematics*, **38**, 103–128.

(2011) *Principles of Social Evolution*. Oxford: Oxford University Press.

Bourke, A. F. G. & Franks, N. R. (1995) *Social Evolution in Ants*. Princeton: Princeton University Press.

Brockmann, H. J. (1997) Cooperative breeding in wasps and vertebrates: The role of ecological constraints. *In*: Choe, J. C., & Crespi, B. J. (eds.) *Evolution of Social Behavior in Insects and Arachnids*. Cambridge: Cambridge University Press, pp. 347–371.

Brown, J. L. (1987) *Helping and Communal Breeding in Birds: Ecology and Evolution*. Princeton: Princeton University Press.

Calisi, R. M. & MacManes, M. D. (2015) RNAseq-ing a more integrative understanding of animal behavior. *Current Opinion in Behavioral Sciences*, **6**, 65–68.

Carey J. R. & Judge D. S. (2001) Life span extension in humans is self-reinforcing: A general theory of longevity. *Population and Development Review*, **27**, 411–436.

Chak, T. C. S, Duffy, J. E., & Rubenstein, D. R. (2015) Reproductive skew drives patterns of sexual dimorphism in sponge-dwelling snapping shrimps. *Proceedings of the Royal Society of London B*, **282**, 20150342.

Choe, J. C. & Crespi, B. (1997) *The Evolution of Social Behavior in Insects and Arachnids*. Cambridge: Cambridge University Press.

Clutton-Brock, T., West, S., Ratnieks, F., & Foley, R. (2009) The evolution of society. *Philosophical Transactions of the Royal Society B*, **364**, 3127–3133.

Cockburn, A. (1998) Evolution of helping behaviour in cooperatively breeding birds. *Annual Review of Ecology and Systematics*, **29**, 141–177.

(2006) Prevalence of different modes of parental care in birds. *Proceedings of the Royal Society of London B*, **273**, 1375–1383.

Cohen, A. A. (2004) Female postreproductive lifespan: A general mammalian trait. *Biological Reviews*, **79**, 733–750.

Cornwallis, C. K., West, S. A., Davis, K. E., & Griffin, A. S. (2010) Promiscuity and the evolutionary transition to complex societies. *Nature*, **466**, 969–972.

Costa, J. T. (2006) *The Other Insect Societies*. Cambridge, MA: The Belknap Press of Harvard University Press.

Coxworth, J. E., Kim, P. S., McQueen, J. S., & Hawkes, K. (2015) Grandmothering life histories and human pair bonding. *Proceedings of the National Academy of Sciences USA*, **112**, 11806–11811.

Crespi, B. J. (1994) Three conditions for the evolution of eusociality: Are they sufficient? *Insectes Sociaux*, **41**, 395–400.

Crespi, B. J. & Yanega, D. (1995) The definition of eusociality. *Behavioral Ecology*, **6**, 109–115.

Crook, J. H. (1964) *The Evolution Of Social Organisation And Visual Communication in the Weaver Birds (Ploceinae)*. Brill, Leiden: Behaviour Supplements, no. 10.

Cruz, Y. R. (1981) A sterile defender morph in a polyembryonic hymenopterous parasite. *Nature*, **294**, 446–447.

Dickinson, J. L. & Hatchwell, B. J. (2004) Fitness consequences of helping. *In*: Koenig, W. & Dickinson, J. L. (eds.) *Ecology and Evolution of Cooperative Breeding in Birds*. Cambridge: Cambridge University Press, pp. 48–66.

Dillard, J. E. & Westneat, D. F. (2016) Disentangling the correlated evolution of monogamy and cooperation. *Trends in Ecology and Evolution*, **7**, 503–513.

Dobson, F. S. (2007) A lifestyle view of life-history evolution. *Proceedings of the National Academy of Sciences USA*, **104**, 17565–17566.

Downing, P. A., Cornwallis, C. K., & Griffin, A. S. (2015) Sex, long life and the evolutionary transition to cooperative breeding in birds. *Proceedings of the Royal Society of London B*, **282**, 20151663.

Duffy, J. E. (1996) Eusociality in a coral-reef shrimp. *Nature*, **381**, 512–514.

Elgar, M. A. (2015) Integrating insights across diverse taxa: Challenges for understanding social evolution. *Frontiers in Ecology and Evolution*, **3**, 124.

Emlen, S. T. (1982) The evolution of helping. 1. An ecological constraints model. *The American Naturalist*, **119**, 29–39.

Emlen, S. T. (1995) An evolutionary theory of the family. *Proceedings of the National Academy of Sciences USA*, 92, 8092–8099

Emlen, S. T., Wrege, P. H., & Demong N. J. (1995) Making decisions in the family: An evolutionary perspective. *American Scientist*, **83**, 148–157

Evans, H. E. (1977) Commentary: Extrinsic versus intrinsic factors in the evolution of insect sociality. *BioScience*, **27**, 613–617.

Gadagkar, R. (1990) Evolution of eusociality: The advantage of assured fitness returns. *Philosophical Transactions of the Royal Society B*, **329**, 17–25.

(1991) Demographic predisposition to the evolution of eusociality: A hierarchy of models. *Proceedings of the National Academy of Sciences USA*, **88**, 10993–10997.

Groenewoud, F., Frommen, J. G., Josi, D., Tanaka, H., Jungwirth, A., & Taborsky, M. (2016) Predation risk drives social complexity in cooperative breederes. *Proceedings of the National Academy of Sciences USA*, **113**, doi:10.1073/pnas.1524178113.

Hamilton, W. D. (1964) The genetical evolution of social behaviour. *I and II. Journal of Theoretical Biology*, **7**, 1–52.

Härdling R. & Kokko H. (2003) Life-history traits as causes or consequences of social behaviour: Why do cooperative breeders lay small clutches? *Evolutionary Ecology Research*, **76**, 1373–1380.

Hart, A. G. & Ratnieks, F. L. W. (2004) Crossing the taxonomic divide: Conflict and its resolution in societies of reproductively totipotent individuals. *Journal of Evolutionary Biology*, **18**, 383–395.

Hartmann, A. and Heinze, J. (2003) Lay eggs, live longer: Division of labor and life span in a clonal ant species. *Evolution*, **57**, 2424–2427.

Hatchwell, B. J. (2009) The evolution of cooperative breeding in birds: Kinship, dispersal and life history. *Philosophical Transactions of the Royal Society B*, **364**, 3217–3227.

Hatchwell, B. J. & Komdeur, J. (2000) Ecological constraints, life history traits and the evolution of cooperative breeding. *Animal Behaviour*, **59**, 1079–1086.

Heinze, J. (2006) Life in a nutshell: Social evolution in formicoxenine ants. *In:* Kipyatkov, V. (ed.) *Life Cycles in Social Insects: Behaviour, Ecology and Evolution*. St. Petersburg, Russia: St. Petersburg University Press, pp. 49–61.

Helms Cahan, S., Blumstein, D. T., Sundstrom, L., Liebig, J., & Griffin, A. (2002) Social trajectories and the evolution of social behavior. *Oikos*, **96**, 206–216.

Herbers, J. M. (2009) Darwin's "one special difficulty": Celebrating Darwin 200. *Biology Letters*, **5**, 214–217.

Hofmann H. A., Beery, A. K., Blumstein, D. T., Couzin, I. D., Earley, R. L., *et al.* (2014) An evolutionary framework for studying mechanisms of social behavior. *Trends in Ecology and Evolution*, **29**, 581–589.

Hölldobler, B. & Wilson, E. O. (1990) *The Ants*. Cambridge, MA: Harvard University Press.

(2008) *The Superorganism: The Beauty, Elegance, and Strangeness of Insect Societies*. New York: W. W. Norton & Company.

Horn, H. S. & Rubenstein, D. I. (1978) Behavioural adaptations and life history. *In*: Krebs, J. R. & Davies, N. B. (eds.) *Behavioural Ecology: An Evolutionary Approach*. Oxford, UK: Blackwell Scientific Publications, pp. 279–298.

Hughes, W. O. H., Oldroyd, B. P., Beekman, M., & Ratnieks, F. L. W. (2008) Ancestral monogamy shows kin selection is key to the evolution of eusociality. *Science*, **320**, 1213–1216.

Jarvis, J. U. M. (1981). Eusociality in a mammal: Cooperative breeding in naked mole-rat colonies. *Science*, **212**, 571–573.

Jetz, W. & Rubenstein, D. R. (2011) Environmental uncertainty and the global biogeography of cooperative breeding in birds. *Current Biology*, **21**, 72–78.

Kapheim, K. M., Pan, H., Li, C., Salzberg, S. L., Puiu, D., *et al.* (2015) Social evolution. *Genomic signatures of evolutionary transitions from solitary to group living. Science*, **348**, 1139–1143.

Keller, L. (1991) Queen number, mode of colony founding, and queen reproductive success in ants (Hymenoptera Formicidae). *Ethology Ecology & Evolution*, **3**, 307–316.

Keller, L. & Genoud, M. (1997) Extraordinary lifespans in ants: A test of evolutionary theories of ageing. *Nature*, **389**, 958–960.

Keller, L. & Reeve, H. K. (1994) Partitioning of reproduction in animal societies. *Trends in Ecology and Evolution*, **9**, 98–102.

Keller, L. & Vargo, E. L. (1993) Reproductive structure and reproductive roles in colonies of eusocial insects. *In:* Keller, L. (eds.) *Queen Number and Sociality in Insects*. Oxford: Oxford University Press, pp. 16–44.

Kent, D. S. & Simpson, J. A. (1992) Eusociality in the beetle *Austroplatypus incompertus* (Coleoptera:Curculionidae). *Naturwissenschaften*, **79**, 86–87.

Kipyatkov, V. (2006) *Life Cycles in Social Insects: Behaviour, Ecology and Evolution*. St. Petersburg, Russia: St. Petersburg University Press.

Kocher, S. D., Pellissier, L., Veller, C. Purcell, J., Nowak, M. A., *et al.* (2014) Transitions in social complexity along elevational gradients reveal a combined impact of season length and development time on social evolution. *Proceedings of the Royal Society of London B*, **281**, 20140627.

Koenig, W. D., Pitelka, F. A., Carmen, W. J., Mumme, R. L., & Stanback, M. T. (1992) The evolution of delayed dispersal in cooperative breeders. *Quarterly Review of Biology*, **67**, 111–150.

Koenig, W. D., Walters, E. L., & Haydock, J. (2011) Variable helper effects, ecological conditions, and the evolution of cooperative breeding in the acorn woodpecker. *The American Naturalist*, **178**, 145–158.

Koenig, W. D., Dickinson, J. L., & Emlen, S. T. (2016). Synthesis: Cooperative breeding in the twenty-first century. In: Koenig, W. D., & Dickinson, J. L. (eds.) *Cooperative Breeding in Vertebrates: Studies of Ecology, Evolution, and Behavior*. Cambridge: Cambridge University Press, pp. 353–374.

Koenig, W. D. & Dickinson, J. L. (2016) *Cooperative Breeding in Vertebrates: Studies of Ecology, Evolution, and Behavior*. Cambridge: Cambridge University Press.

Kokko, H. & Ekman, J. (2002) Delayed dispersal as a route to breeding: Territorial inheritance, safe havens, and ecological constraints. *The American Naturalist*, **160**, 468–484.

Korb, J. (2007). Termites. *Current Biology*, **17**, R995–R999.

(2008) The ecology of social evolution in termites. *In*: Korb, J. & Heinze, J. (eds.) *Ecology of Social Evolution*. Berlin: Springer-Verlag, pp. 151–174.

Korb, J. & Heinze, J. (2008a) *Ecology of Social Evolution*. Berlin: Springer-Verlag.

(2008b) The ecology of social life: A synthesis. *In*: Korb, J. & Heinze, J. (eds.) *Ecology of Social Evolution*. Berlin: Springer-Verlag, pp. 245–259.

Korb, J., Buschmann, M., Schafberg, S., Liebig, J., & Bagnères, A. G. (2012) Brood care and social evolution in termites. *Proceedings of the Royal Society of London B*, **279**, 2662–2671.

Koykka, C. & Wild, G. (2015) The association between the emergence of cooperative breeding and clutch size. *Journal of Evolutionary Biology*, **29**, 58–76.

Kramer, K. L. & Russell, A. F. (2014) Kin-selected cooperation without lifetime monogamy: Human insights and animal implications. *Trends in Ecology and Evolution*, **29**, 600–606.

Lack, D. (1947) *Darwin's Finches*. Cambridge: Cambridge University Press.

Lee, R. D. (2003) Rethinking the evolutionary theory of aging: Transfers, not births, shape senescence in social species. *Proceedings of the National Academy of Sciences USA*, **100**, 9637–9642.

Lehmann, L. & Rousset, F. (2010) How life history and demography promote or inhibit the evolution of helping behaviours. *Philosophical Transactions of the Royal Society B*, **365**, 2599–2617.

Ligon, J. D. & Burt, D. B. (2004) Evolutionary origins. *In*: Koenig, W. D., & Dickinson, J. L. (eds.) *Ecology and Evolution of Cooperative Breeding in Birds*. Cambridge, UK: Cambridge University Press, pp. 5–34.

Lin, N. (1964) Increased parasite pressure as a major factor in the evolution of social behavior in halictine bees. *Insectes Sociaux*, **11**, 187–192.

Lion, S. B., Jansen, V. A. A., & Day, T. (2011) Evolution in structured populations: Beyond the kin versus group debate. *Trends in Ecology and Evolution*, **26**, 193–201.

Lopez-Vaamonde C., Raine N. E., Koning J. W., Brown R. M., Pereboom J. J. M., *et al.* (2009) Lifetime reproductive success and longevity of queens in an annual social insect. *Journal of Evolutionary Biology*, **22**, 983–996.

Lukas, D. & Clutton-Brock, T. H. (2012) Cooperative breeding and monogamy in mammalian societies. *Proceedings of the Royal Society of London B*, **279**, 2151–2156.

Maynard Smith, J. & Szathmary, E. (1995) *The Major Transitions in Evolution*. Oxford: Oxford University Press.

Michener, C. D. (1969) The evolution of social behavior of bees. *Annual Review of Entomology*, **14**, 299–342.

(1974) *The Social Behavior of the Bees: A Comparative Study*. Cambridge, MA: Harvard University Press.

Moore, A. J., Szèkely, T. & Komdeur, J. (2010) Prospects for research in social behaviour: Systems biology meets behaviour. *In:* Szèkely T., Moore A. J., & Komdeur, J. (eds.) *Social Behaviour: Genes, Ecology and Evolution*. Cambridge, UK: Cambridge University Press, pp. 539–550.

Nonacs, P. (2014) Resolving the evolution of sterile worker castes: A window on the advantages and disadvantages of monogamy. *Biology Letters*, **10**, 20140089.

O'Riain, M. J., Jarvis, J. U. M., Alexander, R. D., Buffenstein, R., & Peeters, C. (2000) Morphological castes in a vertebrate. *Proceedings of the National Academy of Sciences USA*, **97**, 13194–13197.

Oster G. F. & Wilson, E. O. (1978) *Caste and Ecology in the Social Insects*. Princeton: Princeton University Press.

Parker, J.D. (2010) What are social insects telling us about aging? *Myrmecological News*, **13**, 103–110.

Pen, I. & Weissing, F. J. (2000) Towards a unified theory of cooperative breeding: The role of ecology and life history re-examined. *Proceedings of the Royal Society of London B*, **267**, 2411–2418.

Poiani, A & Pagel, M. (1997) Evolution of avian cooperative breeding: Comparative tests of the nest predation hypothesis. *Evolution*, **51**, 226–240.

Promislow, D. E. & Harvey, P. H. (1990) Living fast and dying young: A comparative analysis of life-history variation among mammals. *Journal of Zoology*, **220**, 417–437.

Purcell, J., Brelsford, A., Wurm, Y., Perrin, N., & Chapuisat, M. (2014) Convergent genetic architecture underlies social organization in ants. *Current Biology*, **24**, 2728–2732.

Purcell, J., Pellissier, L., & Chapuisat, M. (2015) Social structure varies with elevation in an Alpine ant. *Molecular Ecology*, **24**, 498–507.

Queller, D. C. (1989) The evolution of eusociality: Re-productive head starts of workers. *Proceedings of the National Academy of Sciences USA*, **86**, 3224–3226.

Queller, D. C. & Strassmann, J. E. (1998) Kin selection and social insects. *BioScience*, **48**, 165–175.

(2009) Beyond society: The evolution of organismality. *Philosophical Transactions of the Royal Society B*, **364**, 3143–3155.

Remolina, S. C. & Hughes, K. A. (2008) Evolution and mechanisms of long life and high fertility in queen honey bees. *Age*, **30**, 177–185.

Richards, O. W. (1953) The care of the young and the development of social life in the Hymenoptera. *Transactions of the Ninth International Congress of Entomology, Amsterdam*, **12**, 135–138.

Rittschof, C. C., Bukhari, S. A., Sloofman, L. G., Troy, J. M., Caetano-Anolles, D., *et al.* (2014) Neuromolecular responses to social challenge: Common mechanisms across mouse, stickleback fish, and honey bee. *Proceedings of the National Academy of Sciences USA*, **111**, 17929–1934.

Robinson, G. E., Grozinger, C. M., & Whitfield, C. W. (2005) Sociogenomics: Social life in molecular terms. *Nature Reviews Genetics*, **6**, 257–270.

Rubenstein, D. R. (2012). Family feuds: Social competition and sexual conflict in complex societies. *Philosophical Transactions of the Royal Society B*, **367**, 2304–2313.

Rubenstein, D. R. & Hofmann, H. A. (2015) Proximate pathways underlying social behavior. *Current Opinion in Behavioral Sciences*, **6**, 154–159.

Rubenstein, D. R. & Lovette, I. J. (2007) Temporal environmental variability drives the evolution of cooperative breeding in birds. *Current Biology*, **17**, 1414–1419.

Rubenstein, D. R. & Wrangham, R. W. (1986) *Ecological Aspects of Social Evolution*. Princeton: Princeton University Press.

Rubenstein, D. R., Botero, C. A., & Lacey, E. A. (2016) Discrete but variable structure of animal societies leads to the false perception of a social continuum. *Royal Society Open Science*, **3**, 160147.

Schwarz, M. P., Tierney, S. M., & Chapman, T. W. (2006) Phylogenetic analyses of life history traits in allodapine bees and social evolution. *In:* Kipyatkov, V. (ed.) *Life Cycles in Social Insects: Behaviour, Ecology and Evolution*. St. Petersburg, Russia: St. Petersburg University Press, pp. 147–155.

Sheehan, M. J., Botero, C. A., Hendry, T. A., Sedio, B. E., Jandt, J. M. *et al.* (2015) Different axes of environmental variation explain the presence vs. extent of cooperative nest founding associations in *Polistes* paper wasps. *Ecology Letters*, **18**, 1057–1067.

Sherman, P. W. (2013) Richard Alexander, the naked mole rat, and the evolution of eusociality. *In:* Summers, K. & Crespi, B. J. (eds.) *Human Social Evolution: The Foundational Works of Richard D. Alexander*. Oxford, UK: Oxford University Press, pp. 55–62.

Sherman, P. W., Lacey, E. A., Reeve, H. K., & Keller, L. (1995) The eusociality continuum. *Behavioral Ecology*, **6**, 102–108.

Smith, C. R., Toth, A. L., Suarez, A. V., & Robinson, G. E. (2008) Genetic and genomic analyses of the division of labour in insect societies. *Nature Reviews Genetics*, **9**, 735–748.

Solomon, N. G. & French, J. A. (1997) *Cooperative Breeding in Mammals*. Cambridge, MA: Cambridge University Press.

Stacey, P. B. & Koenig, W. D. (1990) *Cooperative Breeding in Birds: Long-term Studies of Ecology and Behavior*. Cambridge: Cambridge University Press.

Stacey, P. B. & Ligon, J. D. (1991) The benefits-of-philopatry hypothesis for the evolution of cooperative breeding: Variation in territory quality and group size effects. *The American Naturalist*, **137**, 831–846.

Starr, C. (2006) Steps toward a general theory of the colony cycle in social insects. *In:* Kipyatkov, V. (ed.) *Life Cycles in Social Insects: Behaviour, Ecology and Evolution*. St. Petersburg, Russia: St. Petersburg University Press, pp. 1–20.

Strassmann, J. E., Queller, D. C., & Hughes, C. R. (1988) Predation and the evolution of sociality in the paper wasp *Polistes bellicosus*. *Ecology*, **69**, 1497–1505.

Szèkely T., Moore A. J., & Komdeur, J. (2010) The uphill climb of sociobiology: Towards a new synthesis. *In:* Szèkely T., Moore A. J., & Komdeur, J. (eds.) *Social Behaviour: Genes, Ecology and Evolution*. Cambridge: Cambridge University Press, pp. 1–5.

Taborsky, M., Hofmann, H. A., Beery, A. K., Blumstein, D. T., Hayes, L. D. *et al.* (2015) Taxon matters: Promoting integrative studies of social behavior. *Trends in Neuroscience*, **38**, 189–191.

Tallamy, D. & Brown, W. (1999) Semelparity and the evolution of maternal care in insects. *Animal Behaviour*, **57**, 727–730.

Terrapon, N., Li, C., Robertson, H. M., Ji, L., Meng, X., Booth, W. *et al.* (2014) Molecular traces of alternative social organization in a termite genome. *Nature Communications*, **5**, 1–12.

Toth, A. L., Varala, K., Newman, T. C., Miguez, F. E., Hutchison, S. K., *et al.* (2007) Wasp gene expression supports an evolutionary link between maternal behavior and eusociality. *Science*, **318**, 441–444.

Trivers R. L. & Hare H. (1976) Haplodiploidy and the evolution of the social insects. *Science*, **191**, 249–263.

Trumbo, S. T. (2013) Maternal care, iteroparity, and the evolution of social behavior: A critique of the semelparity hypothesis. *Evolutionary Biology*, **40**, 613–626.

Tsuji, K. (2006) Life history strategy and evolution of insect societies: Age structure, spatial distribution and density dependence. *In:* Kipyatkov, V. (ed.) *Life Cycles in Social Insects: Behaviour, Ecology and Evolution*. St. Petersburg, Russia: St. Petersburg University Press, pp. 21–36.

Tsuji, K. & Tsuji, N. (1996) Evolution of life history strategies in ants: Variation in queen number and mode of colony founding. *Oikos*, **76**, 8–92.

Vehrencamp, S. L. (1978) The adaptive significance of communal nesting in groove-billed anis (*Crotophaga sulciostris*). *Behavioral Ecology and Sociobiology*, **4**, 1–33.

Wang, J. Y., Wurm, Y., Nipitwattanaphon, M., Riba-Grognuz, O., Huang, Y.-C., *et al.* (2013) A Y-like social chromosome causes alternative colony organization in fire ants. *Nature*, **493**, 664–668.

Wcislo, W. T. (1997). Are behavioral classifications blinders to natural variation? *In:* Choe, J. C. & Crespi, B. (eds.) *The Evolution of Social Behavior in Insects and Arachnids*. Cambridge: Cambridge University Press, pp. 8–13.

Wcislo, W.T. & Tierney, S. M. (2009) The evolution of communal behavior in bees and wasps: An alternative to eusociality. *In:* J Gadau, J., & Fewell, J. (eds.) *Organization of Insect Societies: From Genome to Sociocomplexity*, Cambridge MA: Harvard University Press, pp. 148–169.

West, S. A., Murray, M. G., Machado, C. A., Griffin, A. S., & Herre, E. A. (2001) Testing Hamilton's rule with competition between relatives. *Nature*, **409**, 510–513.

West, S. A., Fisher, R. M., Gardner, A., & Kiers, E. T. (2015) Major evolutionary transitions in individuality. *Proceedings of the National Academy of Sciences USA*, **112**, 10112–10119.

West-Eberhard, M. J. (1975) The evolution of social behavior by kin selection. *The Quarterly Review of Biology*, **50**, 1–33.

West-Eberhard, M. J. (1978) Polygyny and the evolution of social behavior in wasps. *Journal of the Kansas Entomological Society*, **51**, 832–856.

Wiernasz D. C. & Cole B. J. (2003) Queen size mediates queen survival and colony fitness in harvester ants. *Evolution*, **57**, 2179–2183.

Wilson, E. O. (1971) *The Insect Societies*. Cambridge, MA: The Belknap Press of Harvard University Press.

(1975) *Sociobiology: The New Synthesis*. Cambridge, MA: Harvard University Press.

(1990) *Success and Dominance in Ecosystems: The Case of the Social Insects*. Oldendorf/Luhe, Germany: Ecology Institute.

(2012) *The Social Conquest of Earth*. New York: Liverlight Publishing Corporation.

Yan, H., Simola, D. F., Bonasio, R., Liebig, J., Berger, S. L., *et al.* (2014) Eusocial insects as emerging models for behavioural epigenetics. *Nature Reviews Genetics*, **15**, 677–688.

Glossary

Allee effect: Positive association between average individual fitness and population size, or positive density dependence, implying lower fitness in smaller populations.
allomaternal: Offspring care by individuals other than the mother.
alloparental care: Offspring care by individuals other than the parents.
bivoltinism: Two broods or generations of an organism in a year.
caste: Set of individuals in a given social group of a species with consistent differences related to their functional roles.
central-place foraging: Foraging that requires leaving a domicile to gather food. Life insurers are central-place foragers.
claustral founding: Female reproductives seal themselves into a cavity after mating and produce offspring without foraging, instead using nutrients stored within their own body.
communal: Multiple reproductives breeding in the same domicile and primarily caring for their own young rather than those of neighboring females.
cleptoparasitism: A form of feeding in which one animal takes prey or other food from another. In social insects, this can include laying eggs in the nests of conspecifics (e.g. brood parasitism in vertebrates).
cooperative breeding: When more than two individuals care for offspring, typically with the aid of non-breeding helpers.
dependent founding: Colony foundation by a one or more reproductive females with help from a subset of individuals from a parental colony.
diapause: Delayed development in response to regularly and recurring periods of adverse environmental conditions.
diplodiploidy: Sexual reproduction with diploid males, and used to parallel the term "haplodiploidy" with haploid males.
direct development: Development without metamorphosis, when the newborn offspring is similar in most regards to the adult.
ergatoid neotenic: Reproductive termite derived from a worker.
eusociality: Society defined by overlapping generations, cooperative care of young, and a reproductive division of labor.
exaptation: A term used to describe a trait whose function differs from the function originally favored by natural selection. Sometimes synonymous with "pre-adaptation."

fortress defense: Describes a form of sociality associated with a suite of life history traits involving development near or within defensible domiciles or food resources, and the production of specialized offspring that express aggressive defense of the group against predators, parasites, and competitors.
group: Any set of individuals of the same species that remain together for a period of time while interacting with one another more than with other conspecifics.
haplodiploidy: Sexual reproduction with haploid males. In Hymenoptera, males develop from unfertilized eggs and are haploid, whereas females develop from fertilized eggs and are diploid.
helper: Non-breeding individual that aides in the care of offspring. Synonymous with "supernumerary."
hemimetabolism: Gradual development of the same body plan via molts through juveniles stages to adult.
hermaphroditism: The coexistence of both female and male sex organs in the same individual.
holometabolism: Complete process of metamorphosis in which an insect develops through all four stages as embryo, larva, pupa, and imago (i.e. the final and fully developed adult stage of an insect, typically winged).
homeostasis: Property of a system in which variables are regulated so that internal conditions remain stable and relatively constant.
inclusive fitness theory: The total fitness of an organism that sums two components: (1) the direct fitness derived from reproduction; and (2) the indirect fitness that depends upon social interactions with relatives.
independent founding: Colony foundation by one or more reproductive females without help from individuals from a parental colony.
joint-nesting: Multiple reproductive females breeding together in the same domicile.
kin neighborhood: Kin structured populations that extend beyond the group.
kin selection: Selection on traits that include both direct and indirect components of fitness. Often used synonymously with "inclusive fitness."
kin structure: Genetic structure within a group.
lek: Site where females visit to select a mate among males displaying on resource-free territories.
mass-provisioning: A form of parental behavior in which an adult female stores all of the food for each of her offspring in a small chamber before egg-laying.
mating system: Refers to who mates with whom.
matriline: Offspring of a single female.
monandry: Single mating by the reproductive female of a group.
monogamy: One male mates with one female (and vice versa) in a breeding season or bout.
monogyny: When there is only a single reproductive female per group.
mutualism: Symbiotic interaction between different species that is mutually beneficial.
neotenic: Reproductive derived from a worker or nymph, primarily associated with termites.

nepotism: Altruism towards close kin, including offspring.
niche: The range of each environmental variable within which a species can exist and reproduce.
nymph: The immature stage in insects with gradual metamorphosis (hemimetabolism).
nymphoid neotenic: Reproductive termite derived from a nymph.
oligogyny: When the number of reproductive females in a group varies over time, often characterized by tolerance by non-reproductives towards more than one reproductive female, and antagonism among reproductive females.
paedogenesis: Attaining sexual maturity at a younger developmental stage. Synonymous with "progenesis."
parasocial: A term used to encompass forms of social aggregation that do not include eusociality (e.g. communal, quasisocial, and semisocial groups).
parthenogenesis: Type of asexual reproduction in which female offspring develop from an unfertilized egg.
patriline: Offspring of a single male.
philopatry: Tendency of an organism to stay in, or habitually return to, a particular area.
plesiomorphic trait: A trait that represents a primitive state of evolution relative to another trait.
pleometrosis: Several young reproductives cooperatively start a new group.
plural breeding: When there is more than one reproductive female per group.
policing: When individuals prevent group mates from producing offspring in the presence of established reproductives.
polyandry: When a female has more than one mate.
polydomy: Groups that inhabit more than one domicile simultaneously.
polyethism: Occurrence of two or more alternative phenotypes whose differences are induced by key differences in the environment.
polygamy: An individual having more than one mate.
polygyny: In reference to social behavior: when there are more than one reproductive females per group; in reference to mating behavior: when a male has more than one mate.
polygynandry: When both sexes have more than one mate.
polyphenism: A single genotype can develop into multiple, discrete phenotypes.
primary polygyny: When the presence of multiple female reproductives per group (polygyny) originates in cooperative colony founding (pleometrosis).
progressive provisioning: A form of parental behavior in which an adult feeds its larvae directly after they have hatched until development is complete.
protandry: Change from a male to a female in a hermaphrodite.
protogyny: Change from a female to a male in a hermaphrodite.
pseudergates: False worker, most commonly used in reference to termites.
quasisocial: Members of the same generation use the same domicile and cooperate in offspring care.
reproductive division of labor: When one or a few individuals in a group specialize on reproduction, and others are temporarily or permanently sterile.

reproductive skew: Quantification of how reproduction is shared among group members, where high skew means a small fraction of adult females in a group reproduce and low skew means most females in a group reproduce.
secondary polygyny: The presence of multiple female reproductives originates from the adoption of young reproductive females into an established colony.
semi-claustral founding: During the group founding phase, female reproductives forage to provide nutrients to their first offspring.
semisocial: Members of the same generation cooperate in offspring care and there is a reproductive division of labor (i.e. some individuals are primarily breeders and some are primarily workers).
serial polygyny: When a male has more than one mate in succession.
sexual conflict: Conflict between the evolutionary interests of individuals of the two sexes.
sexuparae: A winged female aphid that gives rise to either sexual females or both sexual females and males.
singular breeding: When there is only a single reproductive female per group.
social group: A collection of individuals that share similar characteristics, and are typically related.
social organization: Refers to who lives with whom in a group.
social parasitism: When a female reproductive enters and takes over an unrelated, established group and utilizes its workforce to rear its own offspring.
social system: Refers to the social bonds among group members.
social structure: Refers to the social relationships among group members.
sociality: Cooperative group living.
sociobiology: The study of the biological basis of all social behavior.
subsociality: When adults care for their offspring (either nymphs or larvae) for some period of time. Synonymous with "parental care" occurring after oviposition or parturition.
supernumerary: See "helper."
superorganism: Any society possessing features of organization analogous to the physiological properties of a single organism.
swarm-founding: Group foundation by organized swarms in which females filling different roles (i.e. castes) are highly dependent upon each other.
thelytokous parthenogenesis: Females are able to produce daughters from unfertilized eggs. This is in contrast to "arrhenotokous parthenogenesis" in haplodiploid animals where unfertilized eggs can develop into males.
trophallaxis: Transfer of food or other fluids among members of a group through mouth-to-mouth (stomodeal) or anus-to-mouth (proctodeal) feeding.
worker: Member of non-reproductive, laboring caste.

Index

Printed in the United States
By Bookmasters

III-V Integrated Circuit Fabrication Technology